BACTERIA AS
MULTICELLULAR
ORGANISMS

BACTERIA AS MULTICELLULAR ORGANISMS

Edited by

JAMES A. SHAPIRO

& MARTIN DWORKIN

New York   Oxford
Oxford University Press
1997

Oxford University Press

Oxford   New York

Athens   Auckland   Bangkok   Bogota   Bombay   Buenos Aires
Calcutta   Cape Town   Dar es Salaam   Delhi   Florence   Hong Kong
Istanbul   Karachi   Kuala Lumpur   Madras   Madrid   Melbourne
Mexico City   Nairobi   Paris   Singapore   Taipei   Tokyo   Toronto

and associated companies in
Berlin   Ibadan

Library of Congress Cataloging-in-Publication Data
Bacteria as multicellular organisms / edited by James A. Shapiro,
Martin Dworkin.
p.   cm.   Includes bibliographical references and index.
ISBN 0-19-509159-0
1. Bacteria–Ecology.   2. Microbial aggregation.
I.   Shapiro, James Alan, 1943–   .   II. Dworkin, Martin.
QR100.B325   1997
589.9′05–dc20   96-4288

1 3 5 7 9 8 6 4 2

Printed in the United States of America
on acid-free paper

Preface

This is the first book to deal explicitly with bacteria as multicellular rather than unicellular organisms. There were several impulses behind the decision to prepare a volume dedicated to the topic of bacterial multicellularity. Most immediate was the need to report on and summarize the results of two international conferences sponsored by the American Society for Microbiology on "Multicellular and Interactive Behavior of Bacteria: In Nature, Industry, and the Laboratory." These meetings were held in 1991 and 1993 at the Marine Biological Laboratory in Woods Hole, historically an important center for environmental microbiology. About 100 scientists from diverse backgrounds attended each conference. This venue for the ASM Conferences was particularly appropriate because the new emerging directions in microbial ecology have provided much of the impetus for what is presented in *Bacteria as Multicellular Organisms*.

As Chapter 1 recounts, there has been long-standing tension between single-cell and multicell views of bacteria because the two major historical traditions in microbiology have focused on different aspects of bacterial life. One was personified by the ideas of the great Russian soil microbiologist Sergei Winogradsky, who emphasized the interactive quality of the microbial flora in Nature. On the other hand, medical bacteriology, centered on Koch's postulates, emphasized the growth of pure cultures, each originating from a single bacterium. Koch and his school also campaigned for the "monomorphist" conception of each bacterial species as composed of a single cell type. The intellectual force of this medically oriented tradition, combined with its practical success in combating infectious diseases, created conditions in which the pure culture, single cell, ideal cell-type view came to dominate our theoretical conceptions of bacteria. Molecular biology emerged from medical bacteriology through studies of the transforming principle, bacteriophages, and genetic exchange; and the revolutionary impact of this new discipline reinforced the tendency to forge detailed explanations of bacterial activity on the basis of the isolated cell.

In contrast, environmental microbiologists have long recognized that most of the chemical transformations that maintain the biosphere involve the activities of complex multicomponent bacterial populations, often involving several distinct species. In recent years new molecular technologies have made it possible to study these populations *in situ* with great precision independently of the need to prepare laboratory cultures. Direct examination of bacteria in nature has revealed them to exist virtually always in films, mats, colonies, aggregates, and chains—rarely as isolated cells. Thus multicellularity is the natural viewpoint of the environmental microbiologist, and it is likely that the major questions of microbial ecology will be resolved within this perspective. It is ironic, but not surprising, that studies at the cellular and molecular levels of pure laboratory cultures have begun to reveal multiple systems of intercellular communication that affect all aspects of bacterial function, ranging from patterns of gene expression to decisions about cellular differentiation, such as sporulation.

Because this book is the first to take prokaryotic multicellularity as its theme, the chapters reflect a diversity of perspectives. As with all young fields, the study of bacteria as multicellular organisms has not yet developed its own sophisticated vocabulary and consensus views on fundamental issues. Years of single-cell conceptualization have created deeply ingrained habits and modes of expression in our discourse about bacteria. Indeed, intuitions bred during 130 years of pure culture microbiology have made it difficult for many microbiologists to consider differentiated populations as experimentally (perhaps even philosophically) legitimate systems for study. When we make statements about basic multicellular aspects of prokaryotic activity, we often do so in opposition to more conventional formulations. Thus the following chapters range in tone from orthodox molecular interpretations of the facts presented to calls for new theoretical paradigms. Personally, we relish this pluralism and hope that our readers also find a refreshing variety of intellectual fare.

All the chapters in this book are animated by a fascination with new and little-studied aspects of bacterial life. We have organized the book into five sections. Section I, entitled Conceptual Developments, includes two introductory chapters that we wrote. One, "Multiculturism Versus the Single Microbe" (Dworkin), outlines the historical debate between single-cell and multicell views of bacteria. The other, "Multicellularity: The Rule, Not the Exception; Lessons from *Escherichia coli* Colonies" (Shapiro), documents how even the most prototypical bacterial species provides abundant examples of multicellularity. Section II, Intercellular Communication, contains two chapters on well characterized molecules that serve as chemical carriers of information between bacteria: "Pheromone-Inducible Conjugation in *Enterococcus fae-*

calis: Mating Interactions Mediated by Chemical Signals and Direct Contact'' (Ruhfel, Leonard, and Dunny) and ''*N*-Acyl-L-Homoserine Lactone Autoinducers in Bacteria: Unity and Diversity'' (Dunlap).

Section III, Multicellular Life Styles, is the longest part of the book and is devoted to chapters providing detailed descriptions of multicellular life styles among different bacterial groups, each of which occupies a distinct ecological niche and highlights particular aspects of bacterial interactions: ''Cyanobacteria'' (Adams), ''Mycelial Life Style of *Streptomyces coelicolor* A3(2) and Its Relatives'' (Chater and Losick), ''*Proteus mirabilis* and Other Swarming Bacteria'' (Belas), ''Myxobacterial Multicellularity'' (Shimkets and Dworkin), and ''Oral Microbiology and Coaggregation'' (Kolenbrander).

Section IV, Examining Multicellular Populations, discusses the application of new technologies to analyzing structured bacterial populations and the profound theoretical implications the data have for microbial ecology: ''Flow Cytometry: Useful Tool for Analyzing Bacterial Populations Cell by Cell'' (Hauer and Eipel) and ''In Situ Analyses of Microbial Populations with Molecular Probes: Phylogenetic Dimension'' (Fry, Raskin, Sharp, Alm, Mobarry, and Stahl).

The book concludes with Section V, Physical View of Bacterial Multicellularity, containing four chapters: ''Physical and Genetic Consequences of Multicellularity in *Bacillus subtilis*'' (Mendelson, Salhi, and Li), ''Formation of Colony Patterns by a Bacterial Cell Population'' (Matsushita), ''Cooperative Formation of Bacterial Patterns'' (Ben-Jacob and Cohen), and ''Collective Behavior and Dynamics of Swimming Bacteria'' (Kessler and Wojciechowski). This final section deserves special comment because it is unusual to have chapters written largely by physicists in a book devoted to microorganisms. Bacterial colonies display patterns that are often similar to those found in inorganic systems, and they are suitable material for analysis by methods employed in the physical sciences. We anticipate that bacterial colonies will become choice experimental systems for studying fundamental problems of self-organization and pattern formation by complex systems for two reasons: (1) the ease and speed with which bacterial colonies can be grown under a wide variety of conditions; and (2) the possibilities bacteria offer for genetic manipulation. One of the great research advantages of microbial systems is that indescribably complex entities, living cells, can be modified one molecule at a time through genetic engineering techniques without disrupting their overall integrity. Such exquisite experimental control is not available in physical systems.

If our expectations are correct, bacteria (in their multicellular roles) will be at the forefront of the intellectual revolution that is rearranging the boundaries of traditional disciplines and merging the interests of physical and biological scientists. For its own reasons, bacteriology is developing from a highly reductionist discipline, focused on the individual microscopic cell and its constituent molecules, to a more holistic science, examining the interactive structures of bacterial populations and communication between bacteria and other organisms. In this regard, bacteriology fits into the contemporary shift of thinking in all areas of science toward a more organic, interactive view of natural phenomena.

Most of the chapters in this book deal with bacteria grown in the laboratory. One question that comes up repeatedly during discussions of bacterial multicellularity is whether the observed interactions are peculiar to the laboratory environment and so represent some kind of special situation without deep adaptive relevance. This question can be answered in two ways. The most direct response is to point out that multicellularity provides demonstrable advantages to bacteria in the laboratory, such as increased resistance to antagonists, and to extrapolate these advantages to natural situations. The deeper answer is to say that laboratory situations reveal the capabilities bacteria have for communication, coordination, and concerted action. These capabilities evolved before any bacteria were cultured and so must have contributed to survival in the wild. As pointed out in the chapter by Fry et al., studies of microbial ecology are poised to utilize new molecular and visualization technologies to generate a vast amount of information about bacteria in nature. This information must be integrated into coherent theories of bacterial growth, survival, and action; and these theories can be realistic only if they incorporate the results of laboratory studies demonstrating multicellular behavior.

From a thematic point of view, it would have been possible to include chapters on many more topics in this book. Important aspects of prokaryotic biology (e.g., biofilms, metabolic consortia, pathogenesis, symbiosis, and sporulation) depend on multicellular interactions and would have fit comfortably in this volume. Unfortunately, there was neither time nor space to present a fully comprehensive view of bacterial multicellularity. We regret not being able to include discussions of so many important and fascinating areas of bacterial life, but the gaps simply mean that there is plenty of material for future volumes. Our hope is that a growing awareness of multicellularity will prove fruitful in both research and practice. Medicine, ecology, and biotechnology are areas of great economic significance where bacterial populations are major actors. If this book helps stimulate new approaches to problem-solving based on an understanding of how bacteria enhance their powers through multicellular interactions, we will have done our job successfully.

James A. Shapiro
University of Chicago

Martin Dworkin
University of Minnesota

Contents

Contributors

DAVID G. ADAMS
Department of Microbiology
University of Leeds
Leeds, England

ELIZABETH W. ALM
Environmental Engineering and Sciences
Department of Civil Engineering
University of Illinois
Urbana, Illinois

ROBERT BELAS
Center of Marine Biotechnology
University of Maryland Biotechnology
  Institute
Baltimore, Maryland

ESHEL BEN-JACOB
School of Physics and Astronomy
Raymond & Beverly Sackler Faculty of
  Exact Sciences
Tel-Aviv University
Tel-Aviv, Israel

KEITH F. CHATER
John Innes Institute
Norwich, Norfolk, England

INON COHEN
School of Physics and Astronomy
Raymond & Beverly Sackler Faculty of
  Exact Sciences
Tel-Aviv University
Tel-Aviv, Israel

PAUL V. DUNLAP
Biology Department
Woods Hole Oceanographic Institution
Woods Hole, Massachusetts

GARY M. DUNNY
Department of Microbiology
University of Minnesota School
  of Medicine
Minneapolis, Minnesota

MARTIN DWORKIN
Department of Microbiology
University of Minnesota School of
* Medicine*
Minneapolis, Minnesota

HEINZ EIPEL
BASF AG
Ludwigshafen, Germany

NORMAN K. FRY
Environmental Health Engineering
Department of Civil Engineering
Northwestern University
Evanston, Illinois

BERNHARD HAUER
BASF AG
Ludwigshafen, Germany

JOHN O. KESSLER
Department of Physics
University of Arizona
Tucson, Arizona

PAUL E. KOLENBRANDER
Laboratory of Microbial Ecology
National Institute of Dental Research,
* NIH*
Bethesda, Maryland

BETTINA A.B. LEONARD
Institute for Advanced Studies in
* Biological Process Technology*
University of Minnesota
St. Paul, Minnesota

CHEN LI
Department of Molecular and Cellular
* Biology*
University of Arizona
Tucson, Arizona

RICHARD LOSICK
Biological Laboratories
Harvard University
Cambridge, Massachusetts

MITSUGU MATSUSHITA
Department of Physics
Chuo University
Kasuga, Bunkyo-Ku
Tokyo, Japan

NEIL H. MENDELSON
Department of Molecular and Cellular
* Biology*
University of Arizona
Tucson, Arizona

BRUCE K. MOBARRY
Environmental Health Engineering
Department of Civil Engineering
Northwestern University
Evanston, Illinois

LUTGARDE RASKIN
Environmental Engineering and Sciences
Department of Civil Engineering
University of Illinois
Urbana, Illinois

ROBERT E. RUHFEL
Perkin Elmer/ABD
Foster City, California

BACHIRA SALHI
Department of Molecular and Cellular
* Biology*
University of Arizona
Tucson, Arizona

JAMES A. SHAPIRO
Department of Biochemistry and
* Molecular Biology*
University of Chicago
Chicago, Illinois

RICHARD SHARP
Environmental Health Engineering
Department of Civil Engineering
Northwestern University
Evanston, Illinois

LAWRENCE J. SHIMKETS
Department of Microbiology
University of Georgia
Athens, Georgia

DAVID A. STAHL
Department of Civil Engineering
Northwestern University
Evanston, Illinois

MARTIN F. WOJCIECHOWSKI
Department of Ecology and Evolutionary
Biology
University of Arizona
Tucson, Arizona

I.

Conceptual Developments

1

Multiculturism Versus the Single Microbe

MARTIN DWORKIN

Microbiology is on the verge of a paradigm shift. The strictly Cartesian, reductionist strategy that has characterized most of twentieth century science in general and microbiology in particular has been immensely successful in generating a detailed understanding of the workings of the cell. It has faltered, however, when confronted with the problem of producing integrated explanations of the workings of the multicellular organism or of the interactions within a multimember population. Apropos, we have been taught that the microbe is a unicellular organism, but we have not been sufficiently taught that the microbe is also, almost always a member of a community of organisms, both micro- and macro-.

The goal of this book and of the Woods Hole meetings that set the stage for it is to persuade microbiologists that the view of bacteria as exclusively unicellular organisms has some serious drawbacks. From a practical point of view, this notion has interfered with our attempts to understand and control most of the multimember microbial transformations that take place in nature. This includes those involved in pathological transformations processes as well as those carrying out the chemical changes that are absolutely necessary for maintaining the dynamic equilibrium of the natural world. It is also interesting to consider the possibility that a better understanding of microbial cell−cell interactions may provide clues that could help us understand the phylogeny of conventional multicellularity.

There have been attempts in the past to examine multimember populations as an alternative, parallel way of looking at the behavior of bacteria in nature.

These studies are discussed later. The time for such explorations has not been quite right until now. Two conferences at Woods Hole (1991 and 1993) sponsored by the American Society for Microbiology reported the results of some of the most recent attempts to devise new approaches and experimental strategies to break out of this conceptual logjam.

It has become increasingly clear that the intellectual power of pure-culture microbiology must now become part of a dialectic with a more integrated examination of complex populations. Accordingly, we have assembled a spectrum of examples that illustrate the various interactions of bacteria with each other and with other organisms. They range from the interactions in the heterotypic oral flora to the homotypic interactions within an *Escherichia coli* colony to the structured, organized interactions in a myxobacterial fruiting body. Although we intend for the term ''multicellularity'' to refer to all of these interactions, we recognize that in the most traditional sense its use is somewhat more limited.

No attempt to set the stage for this sort of subject can avoid beginning with the inventor of bacteriology, Robert Koch, and with the logic of the events that led from the pre-Kochian intellectual chaos of pleomorphism to its tidy antithesis of monomorphism and pure culture microbiology. Koch not only invented the tools of the trade that made bacteriology possible, he crystallized the paradigm of the pure culture. The clarification that emerged from Koch's efforts swept away the vague and dualistic theories of von Naegli and Virchow and established rigorous parameters for determining the causal role of bacteria in natural and pathological transformations.

At that time, controversy swirled around the issue of pleomorphism. On one hand, the pleomorphists, led by the Swiss botanist Carl von Naegli (1877), claimed that there were no appropriate species distinctions among bacteria. ''For over ten years I have examined thousands of different fission organisms and (with the exception of the Sarcinae) I have been completely unable to distinguish even two distinct species.'' Failure to appreciate the fact that there were innumerable species of bacteria made it impossible to understand the notion of succession of types in mixed cultures and thus to provide a rational explanation for the changing variety of cell types in a natural population. Koch's brilliance led him to understand that the notion of pleomorphism—that any microbe could become any other—was inconsistent with the ideas of metabolic specificity that were already beginning to flow from Louis Pasteur's work. It led him to believe that infectious pathology could also be understood only in terms of microbial specificity, and that this understanding could be achieved only by means of pure cultures. Representing the monomorphists, Robert Koch was as explicit and demanding as was von Naegli: ''[O]ne fact was so prominent that I must regard it as constant, and . . . I look upon it as the most important result of my work. I refer to the differences which exist between pathogenic bacteria *and to the constancy of their characters* [italics added]'' (Koch, 1880). Koch's conceptual oversimplifications were necessary to unravel the tangled threads of causative factors and resultant pathologies. Until the idea of distinct and different species of bacteria was firmly established it would not be possible to assign proximal causes to complex effects. It was equally necessary to dispel the notion that bacteria could casually change their shape. To accomplish this reversal of ideas it was imperative to establish— and insist upon—pure culture microbiology.

The need for pure cultures had been earlier recognized by the mycologist Oskar Brefeld, who had achieved pure cultures of fungi by micomanipulating single fungal spores into a sterile medium. Brefeld left no room for argument: "If one doesn't work with pure cultures, only nonsense and *Penicillium glaucum* can come of it" (Stanier et al., 1957). Koch put this dictum into effect, and his efforts to simplify the factors in the causal chain of events that led to disease resulted in a series of dramatic successes that saw clarification of the etiology of anthrax, cholera, typhoid, meningitis, bubonic plague, and to some extent tuberculosis. There was a down side: Koch's towering dogmatism backed microbiology into a conceptual corner from which it is only now beginning to emerge. "He ended his career as an imperious and authoritarian father figure whose influence on bacteriology and medicine was so strong as to be downright dangerous" (Brock, 1988). Theobald Smith (1932) pointed out that, "In training his guns continually against the bacteriological superstitions of his day, Koch naturally became to a certain degree a victim of the recoil in going too far in insisting on the stability of form and function among bacteria."

Koch did what he had to do. The need to eliminate as many variables as possible from the tangled web of events that led to infectious disease made pure culture microbiology an absolute necessity. It was the field of microbiology itself that subsequently overreacted to Koch's necessary oversimplifications and reified them into disciplinary dogma. This dogma has made it difficult to deal either conceptually or technically with the reality of the interactions of bacteria with themselves, other bacteria, other organisms, and their substrata. There was insistence on the notion that anything other than a single typical form was a departure from some sort of Platonic view of the ideal bacterium, which substantially delayed appreciation of the developmental variability of bacteria. Moreover, sanctification of the pure culture method made many of the microbial events occurring outside the laboratory inaccessible to microbiology, technically or conceptually. It seemed as if the use of mixed cultures, deliberately or inadvertently, was grounds for excommunication.

It is not as though the Koch paradigm went unchallenged for the past hundred years. There has existed alongside this paradigm an alternative view of microbiology that emerged from the work and tradition established by the great Russian microbiologist, Sergei Winogradsky. His point of view is clearly expressed in the following passage from one of his papers on soil microbiology.

> [T]he evidence required in soil microbiology cannot be founded on the classical pure culture method as trustfully as in the case of chemical, industrial, or medical microbiology. Pure cultures of organisms, isolated years ago from the soil, grown on artificial media for an unlimited time on nutrients essentially different from those, the original wild species might

find at their disposal in the soil, severed for countless generations from the biological conditions of the soil—these cultures cannot tell us much about microbial activity in nature. The utmost the classical method can give us, is a few conventional notions only vaguely outlining what really happens.

Winogradsky (1935) then proceeded to make a few suggestions, including the following:

Make a special point of studying the reactions of the soil population as a whole, *since the competition between its components is the principal determinant of their individual functions* [italics added].

Winogradsky effectively used enrichment cultures and subsequent pure cultures derived from them to study and understand many of the metabolic processes in the soil, but he also remained a strong advocate of the classical botanical method of direct observations of mixed cultures. He wondered why this approach had been totally abandoned. With a mixture of insight and personal pique he commented (Winogradsky, 1937):

It may be that the explanation for this disregard can be found in the opinion sometimes encountered, that bacteriology practically begins with R. Koch and the gelatin method, all work prior or extraneous to that being in a sense prehistoric or negligible.

This approach—directly observing natural populations—led to Winogradsky's towering formulation of the concept of autotrophy, based on his experiments and observations with *Beggiatoa*, which he studied in mixed culture. It is interesting to wonder if Winogradsky would have tempered his remarks had he witnessed the immense insights that pure culture microbiology has generated.

The names Arthur T. Henrici, Selman Waksman, and Rene Dubos come to mind when considering this alternative view of microbiology. Henrici, who is one of the unsung heroes of microbiology, discovered and described *Caulobacter* and its life cycle. As a result, he retained a lifelong interest in the problems of bacterial life cycles and morphologic variation. In addition, he carried out the first detailed work on the relation between growth rate and morphologic variation in bacteria. Henrici (1928) was one of the first to recognize the ironic dilemma generated by the Koch revolution:

[T]he fundamental biological problems of the bacteria have been neglected. There are many reasons for this, but perhaps one of the most important has been the blind acceptance by the majority of bacteriologists of the Cohn-Koch dogma of the constancy of cell forms and the immutability of bacterial species, which has discouraged all investigation of problems of morphology, inheritance and variation in bacteria for a good many years.

One of the most apparent examples of an interaction between two species of microbes is the inhibition of one by a substance produced by the other (i.e., antibiosis). The most celebrated example of antibiosis as a manifestation of microbial interactions is, of course, the discovery of penicillin by Fleming

(1929). The idea that microorganisms competed with each other by means of excreted inhibitors had been proposed many years earlier. The first recorded reference seems to have been by Pasteur and Joubert (1877), who noted that an unidentified airborne bacillus inhibited the growth of *Bacillus anthracis* in culture.

Although it is still not clear whether antibiotics function as inhibitors in nature, the notion of antibiosis as a manifestation of microbial interactions gained momentum as a result of the work of Rene Dubos. Most of his career was spent as a medical microbiologist, but his training in soil microbiology forever disposed him to look on the isolated microbe with suspicion. In 1939 he described the isolation of a series of antimicrobial agents produced by *Bacillus brevis*. He named them gramicidin and tyrothricin and recognized the therapeutic potential of this approach. Thus began the modern era of antibiotics. His work, by the way, predated the epochal paper from the Oxford group describing the therapeutic use of penicillin (Chain et al., 1940). Dubos' work grew out of his understanding of the interactive nature of the microbial community.

> But microorganisms hardly ever exist in pure cultures under natural conditions. Except in the artificial environment created for them in the laboratory, they always live in association with other kinds of microorganisms, and also with all sorts of other living forms. To gain a better understanding of microbial life it is necessary, therefore, that we try to broaden the scope of our vision by moving from the artificial closed world of the laboratory to the open world [Dubos, 1962].

Selman Waksman, who won the Nobel prize in 1952 for his discovery of streptomycin, was a disciple of Winogradsky and wrote an interesting brief biography of him (Waksman, 1953a). As a soil microbiologist, Waksman appreciated the complex interactions among bacteria in nature; and his lifelong attention to antibiotics and antibiosis reflected his awareness of the extent to which bacteria in nature were competing and collaborating with each other.

> One must keep in mind, however, the fact that these numerous microorganisms act in the soil not in pure culture but in associations. The activities taking place in the test tube or flask under artificial laboratory conditions, especially in the case of cultures long kept in cultivation, may or may not take place in the soil, in the presence of numerous antagonistic and associative influences from other organisms [Waksman, 1928].

In an essay on the microbe as a biological system Waksman (1949) stated ''the impression that one gets from a perusal of the scientific literature in the field of microbiology is that microbes can hardly be studied except in pure cultures.'' In an address commemorating the centenary of the birth of Shibasaburo Kitasato, a pupil of Koch and the discoverer of the tetanus bacillus, Waksman (1953b) was explicit in describing a ''second approach'' to microbiology.

> In a mixed population one organism may exert a marked destructive effect on another. . . . Thus the pure culture concept of microbiology . . . resulting

from the work of Pasteur, Koch and Kitasato were supplemented and even partly replaced . . . by a new approach resulting from a better understanding of complex interrelationships among microorganisms.

Interest in microbe–microbe interactions was awakened by the work of G. F. Gause (1934), which represented the first attempt to approach the problem of understanding multimember microbial populations by means of mathematical analysis. Gause reconstructed multimember populations in batch cultures and then sought to describe the population dynamics by means of mathematical models. His efforts fell short, as they preceded any real understanding of the growth kinetics of even single-member populations. Nevertheless, Gause's classic, important book *The Struggle for Existence* followed by Jacques Monod's work (1942) on the mathematical analysis of the growth of *Escherichia coli* and the subsequent development of the chemostat (Herbert et al., 1956) set the stage for work that began to appear during the 1960s.

Some of the earliest attempts to analyze competitive interactions between bacteria were by Pfennig and Jannasch (1962), who were interested in the nature of the competitive interactions among marine bacteria. They attempted to understand the basis for stable competitions among marine microorganisms by examining the maximum growth rate as a function of substrate concentrations for a mixture of two bacteria. They then calculated that for certain pairs of bacteria their maximum growth rate–substrate relations would allow them to stably coexist at a particular substrate concentration. This outcome had earlier been predicted mathematically by Powell (1958).

Gause's original interest in predator–prey relations resurfaced shortly thereafter in the pioneering experiments of Tsuchiya and Fredrickson. This intellectual union between a microbiologist and a chemical engineer resulted in a series of chemostat experiments which began by examining the interaction between *E. coli* as the prey bacterium and *Dictyostelium discoideum* as the predatory eukaryotic microbe (Drake et al., 1968). This model system was subsequently expanded to a three-member population consisting of two prey bacteria competing for a single substrate and a protozoan feeding on both bacteria (Jost et al., 1973). Attempts to generate mathematical models that described and predicted the behavior of the populations, however, were unsuccessful. The models were simply unable to predict the nature of the oscillations in the populations or the conditions necessary for stabilization; nor could they rationalize the enigmatic persistence of a portion of the prey populations. Fredrickson (1991) concluded that ''factors not accounted for in the model must be acting.''

Their work was some 25 years ahead of its time, but in the final analysis there were two obstacles to the success of this approach. First was the difficulty of mimicking a natural ecosystem in a simple chemostat. Second was the difficulty of unearthing the hidden variables that had to be included in an effective mathematical model. Repeated failed attempts to come up with an accurate model caused Fredrickson and others to pull back from these studies and to rethink the approach, both conceptually and experimentally (e.g., Sambanis et al., 1987). Fredrickson (1991) has more recently expressed a cautious optimism that the goal can be achieved and that the availability of powerful analytical approaches such as Coulter Counters (which can also analyze size distribution), flow cytometers, image analysis, two-stage chemostats, confocal microscopy, and immense computer power expand substantially the questions one can ask in the laboratory. Nevertheless, it is still a matter of faith that by examining the variables one at a time, as we have been taught to do, we can eventually reconstruct an accurate, explanatory model of even so simple and defined a system as the single predator–prey relationship.

Bull and Slater (1982a) have edited a book whose primary emphasis was on studies of mixed populations; they attempted to describe in a systematic way the various approaches to dealing with microbial communities and have formulated a taxonomy of microbial interactions (Bull and Slater, 1982b). The problem is stated clearly in their opening chapter in the book: ''[S]tudies of mixed populations have taken very much longer to develop and to gain acceptance. Thus, the effect of the all-important biological factor—species interactions—on microbial behavior suffered neglect from all but a few microbiologists and in consequence remains the poorest understood facet of microbiology'' (Bull and Slater, 1982c).

There have been some important improvements in our understanding of microbial interactions at the biochemical level. Much of it started with the attempts to solve the enigma of *Methanobacillus omelianskii*. This organism was originally isolated by Barker (1940) as a methanogen capable of converting ethanol and bicarbonate to acetate and methane. The biochemical pathway of this conversion was obscure until it was demonstrated by Bryant et al. (1967) that *M. omelianskii* was a symbiotic association of two organisms. One of these organisms, subsequently named *Methanobacterium bryantii* (Balch et al., 1979), was the actual methanogen, capable of reducing bicarbonate to methane with molecular hydrogen. The other member of the symbiotic pair (called the S organism) was capable of oxidizing ethanol to acetate and hydrogen, but it carried out only limited transformation of ethanol in pure culture as it was self-inhibited by the hydrogen it produced. Thus the function of the methanogen in the mixed culture was to keep the partial pressure of hydrogen low by utilizing the hydrogen to reduce bicarbonate to methane. Under these conditions, the Gibbs free energy of ethanol oxidation is changed from an endothermic process requiring +9.6 kJ at one atmosphere of hydrogen to an exothermic one yielding -35.9 kJ at 10^{-4} atmospheres of hydrogen—thereby allowing the reaction to proceed. This process, known as interspecies hydrogen transfer, first demonstrated among the methanogens, has now been shown to occur also among the sulfate- and iron-reducing bacteria, acetogens, and fatty acid oxidizing bacteria (Schink, 1992); and it is likely a process of rather general occurrence. It was one of the first of a series of model systems illustrating how a metabolic collaboration between organisms could link them to each other in a consortium.

There has been considerable anxiety about the potential risks associated with the release of genetically engineered microorganisms. This worry has led to renewed interest in the question of the extent of genetic transfer among microorganisms in nature. (The transfer of genetic material between organisms is, in a sense, an ultimate interaction.) This interest has surfaced alongside a long-standing concern among evolutionary microbiologists about the role of genetic exchange in the development and maintenance of bacterial species.

It has been known for some time that different species and even genera can, under laboratory conditions, exchange genetic material (Reanney et al., 1982). There has been less certainty about the extent of these transfers in the natural environment. Bale et al. (1987) have shown that genetic exchange among species of *Pseudomonas* occurs on the surfaces of stones submerged in a flowing stream, and there have been numerous demonstrations of R factor exchange in soil, on plants, and in the intestinal tracts of animals (see Coughter and Stewart, 1989, for a general review of the subject). Chromosomal exchange between *Bacillus subtilis* and *Bacillus licheniformis* has been demonstrated to occur in sterile soil (Duncan et al., 1989) and, on the basis of genetic analysis, has been suggested as playing a role in the establishment of species identities (Istock et al., 1992).

The surface has only been scratched in the area of microbial pathogenesis. Whereas dramatic advances have been made in understanding infectious diseases that clearly are caused by invasion of a single foreign organism, other diseases that are caused by an imbalance among an indigenous population of microbes are still poorly understood. Although the Koch doctrine of how one thinks about the pathogenesis and etiology of infectious disease has locked medical microbiology into the dogma of one microorganism–one disease, medical microbiologists have long been aware that on one hand, microbes may conspire to produce an infectious disease that individual species cannot; on the other hand, they may interfere with each other's ability to do so. For example, the notion that a consortium of microbes could be responsible for a well defined pathogenic syndrome has been recognized for years by oral microbiologists struggling to unravel the etiology of periodontal disease (Socransky and Haffajee, 1990; Holt and Bramanti, 1991). The ability of the normal vaginal flora (consisting mainly of lactobacilli) to inhibit the growth and activity of potentially pathogenic gram-negative and yeast members of the flora has been well documented. The inhibitory effect has been attributed variously to generation of a low pH, competition for adherence, production of antimicrobial metabolites, and a number of other factors (Redondo-Lopez et al., 1990). A considerable amount of research has centered on the nature of the interactions among the rumen microbiota (Hungate, 1985) and the gastrointestinal flora of other vertebrates. Germ-free animals have been a useful, important tool in the latter area, but the complexity of the problem is illustrated by Freter's comment (1986):

> The gnotobiotic animal associated with one or a few bacteria is therefore no more appropriate as a valid model of bacterial interactions in (CV) conventional animals than in in vitro cultures, even though the fact that such animal experiments can be described by the term in vivo seems to lend a semantic touch of legitimacy to these undertakings. . . . Strictly

speaking, only one type of gnotobiotic animal would fulfill all criteria for relevance in studies of microbial interactions among the indigenous microflora, namely, the animal with a ''flora minus one.''

The problem of interactions among members of a microbial flora and their collective interaction with a host is an area of immense interest and complexity, limited only by the ability to devise conceptual and methodologic approaches to the problem. The formulation of thermodynamics led to an understanding of the properties of macroscopic matter without resource to the properties of their atomic and molecular building blocks. It is a reasonable analogy to consider that microbiological Carnots, Joules, Clausius, Kelvins, and Gibbs may be necessary to formulate an understanding of complex microbial interactions. Perhaps in the same sense that entropy is a measure of our ignorance about the composite atoms and molecules that comprise macroscopic mass, we may also have to deal with a biological equivalent that stands between some future understanding of complex microbiological interactions and the properties of the individual organisms.

References

Balch, W.E., Fox, G.E., Magrum, L.J., Woese, C.R., and Wolfe, R.S. (1979) Methanogens: reevaluation of a unique biological group. *Bacteriol. Rev.* **43**:260–296.

Bale, M.J., Fry, J.C., and Day, M.J. (1987) Plasmid transfer between strains of *Pseudomonas aeruginosa* on membrane filters attached to river stones. *J. Gen. Microbiol.* **133**:3099–3107.

Barker, H.A. (1940) Studies upon the methane fermentation. IV. The isolation and culture of *Methanobacillus omelianskii. Antonie Van Leeuwenhoek* **6**:210–220.

Brock, T.D. (1988) *Robert Koch: A Life in Medicine and Bacteriology.* Science Technical Publications, Madison, WI.

Bryant, M.P., Wolin, E.A., Wolin, M.J., and Wolfe, R.S. (1967) *Methanobacillus omelianskii*, a symbiotic association of two species of bacteria. *Arch. Mikrobiol.* **59**: 20–31.

Bull, A.T., and Slater, J.H. (1982a) *Microbial Interactions and Communities*, Vol. 1. Academic Press, Orlando, FL.

Bull, A.T., and Slater, J.H. (1982b) Microbial interactions and community structure. In *Microbial Interactions and Communities*, Vol. 1, pp. 13–44. Academic Press, Orlando, FL.

Bull, A.T., and Slater, J.H. (1982c) Historical perspectives on mixed cultures and microbial communities. In *Microbial Interactions and Communities*, Vol. 1, pp. 1–12, Academic Press, Orlando, FL.

Chain, E.B., Florey, H.W., Gardner, A.D., et al. (1940) Penicillin as a chemotherapeutic agent. *Lancet* **239**:226–228.

Coughter, J.P., and Stewart, G.J. (1989) Genetic exchange in the environment. *Antonie Van Leeuwenhoek* **55**:15–22.

Drake, J.F., Jost, J.L., Fredrickson, A.G., and Tsuchiya, H.M. (1968) The food chain. In *Bioregenerative Systems*, pp. 87–94. NASA SP-165. Government Printing Office, Washington, DC.

Dubos, R.J. (1939) Studies on a bactericidal agent extracted from a soil bacillus. 1. Preparation of the agent: its activity in vitro. *J. Exp. Med.* **70**:1–10.

Dubos, R.J. (1962) *The Unseen World.* Rockefeller Institute Press, New York.

Duncan, K.E., Istock, C.A., Graham, J.B., and Ferguson, N. (1989) Genetic exchange between *Bacillus subtilis* and *Bacillus licheniformis*: variable hybrid stability and the nature of bacterial species. *Evolution* **43**:1585–1609.

Fleming, A. (1929) On the antibacterial action of cultures of a *Penicillium*, with special reference to their use in the isolation of *B. influenzae. Br. J. Exp. Pathol.* **10**: 226–236.

Fredrickson, A.G. (1991) Segregated, structured, distributed models and their role in microbial ecology: a case study based on work done on the filter-feeding ciliate *Tetrahymena pyriformis. Microb. Ecol.* **22**:139–159.

Freter, R. (1986) Gnotobiotic and germfree animal systems. In *Bacteria in Nature*, Vol. II, J.S. Poindexter and E.R. Leadbetter (eds.), pp. 205–227. Plenum Press, New York.

Gause, G.F. (1934) *The Struggle for Existence.* Williams & Wilkins, Baltimore.

Henrici, A.T. (1928) *Morphologic Variation and the Rate of Growth of Bacteria.* Charles C Thomas, Springfield, IL.

Herbert, D., Elsworth, R., and Telling, R.C. (1956) The continuous culture of bacteria; a theoretical and experimental study. *J. Gen. Microbiol.* **14**:601–622.

Holt, S.C., and Bramanti, T.E. (1991) Factors in virulence expression and their role in periodontal disease expression. *Crit. Rev. Oral Biol. Med.* **2**:177–281.

Hungate, R.E. (1985) Anaerobic biotransformations of organic matter. In *Bacteria in Nature*, Vol. I, E.R. Leadbetter and J.R. Poindexter (eds.), pp. 39–95. Plenum Press, New York.

Istock, C.A., Duncan, K.E., Ferguson, N., and Zhou, X. (1992) Sexuality in a natural population of bacteria: *Bacillus subtilis* challenges the clonal paradigm. *Mol. Ecol.* **1**:95–103.

Jost, J.L., Drake, J.F., Fredrickson, A.G., and Tsuchiya, H.M. (1973) Interactions of *Tetrahymena pyriformis, Escherichia coli, Azotobacter vinelandii*, and glucose in a minimal medium. *J. Bacteriol.* **113**:834–840.

Koch, R. (1880) *Investigations into the Etiology of Traumatic Infectious Diseases*, translated by W. Watson Cheyne. New Sydenham Society, London.

Monod, J. (1942) *Recherches sur la Croissance des Cultures Bacteriennes.* Hermann et Cie, Paris.

Pasteur, L., and Joubert, J.F. (1877) Charbon et septicémie. *C.R. Acad. Sci. Paris* **85**: 101–115.

Pfennig, N., and Jannasch, H.W. (1962) Biologische Grungfragen bei der homokontinuierlichen Kultur von Mikroorganismen. *Ergebn. Biol.* **25**:93–135.

Powell, E.O. (1958) Criteria for the growth of contaminants and mutants in continuous culture. *J. Gen. Microbiol.* **18**:259–268.

Reanney, D.C., Roberts, W.P., and Kelly, W.J. (1982) Genetic interactions among microbial communities. In *Microbial Interactions and Communities*, Vol. 1, A.T. Bull and J.H. Slater (eds.). Academic Press, Orlando, FL.

Redondo-Lopez, V., Cook, R.L., and Sobel, J.D. (1990) Emerging role of lactobacilli in the control and maintenance of the vaginal bacterial microflora. *Rev. Infect. Dis.* **12**:856–872.

Sambanis, A., Pavlou, S., and Fredrickson, A.G. (1987) Coexistence of bacteria and feeding ciliates: growth of bacteria on autochthonous substrates as a stabilizing factor for coexistence. *Biotechnol. Bioeng.* **29**:714–728.

Schink, B. (1992) Syntrophism among prokaryotes. In *The Prokaryotes*, 2nd ed., A. Balows, H.G. Trüper, M. Dworkin, W. Harder, and K.-H. Schleiffer (eds.), pp. 276–299. Springer-Verlag, New York.

Smith, T. (1932) Koch's views on the stability of species among bacteria. *Ann. Med. Hist.* **NS4**:524–530.

Socransky, S.S., and Haffajee, A.D. (1990) Microbiological risk factors for destructive periodontal disease. In *Risk Assessment in Dentistry*, J.D. Bader (ed.). University of North Carolina Department of Dental Ecology, Chapel Hill.

Stanier, R.Y., Doudoroff, M., and Adelberg, E.A. (1957) *The Microbial World*, 1st ed. Prentice-Hall, Englewood Cliffs, NJ.

Von Naegli, C. (1877) *Die niederen Pilze in ihren Beziehungen zu den Infectionskrankheiten und der Gesundsheitspflege.* Oldenbourg, Munich.

Waksman, S.A. (1928) Nature, distribution, and function of soil microorganisms. In *The Newer Knowledge of Bacteriology and Immunology*, E.O. Jordan and I.S. Falk (eds.), University of Chicago Press, Chicago.

Waksman, S.A. (1949) The microbe as a biological system. In *Scientific Contributions of Selman A. Waksman*, H.B. Woodruff (ed.). Rutgers University Press, New Brunswick, NJ.

Waksman, S.A. (1953a) *Sergei N. Winogradsky: His Life and Work.* Rutgers University Press, New Brunswick, NJ.

Waksman, S.A. (1953b) The changing concept in microbiology. *Scientific Monthly* **76**: 127–133.

Winogradsky, S.N. (1935) The method in soil microbiology as illustrated by studies on *Azotobacter* and the nitrifying organisms. *Soil Sci.* **40**: 59–76.

Winogradsky, S.N. (1937) The doctrine of pleomorphism in bacteriology. *Soil Sci.* **43**: 327–340.

2

Multicellularity: The Rule, Not the Exception

Lessons from Escherichia coli *Colonies*

JAMES A. SHAPIRO

Bacterial multicellularity has been recognized for some time. Distinctively multicellular phenomena covered in other chapters in this book have long been well known to microbiologists. For example, intercellular communication in the form of DNA transfer was one of the key areas that led to the development of molecular biology. Yet the idea of bacteria as interactive multicellular organisms has not found acceptance as a fundamental tenet of microbiology because each example of multicellularity has been thought of as a unique, specialized adaptation grafted to the basic properties of autonomous single-cell bacteria. The prototypes of these isolated unicellular organisms have been *Escherichia coli* and *Bacillus subtilis*. The serendipitous observation that *E. coli* colonies are highly organized, differentiated structures (Shapiro, 1984b,c) made it clear that assumptions about the single-cell nature of bacterial life needed to be reexamined. A growing body of subsequent work on pattern formation in *E. coli*, *Salmonella typhimurium*, and *B. subtilis* has only reinforced this conclusion (Fujikawa and Matsushita, 1989; Budrene and Berg, 1991, 1995; Ben-Jacob et al., 1992; Blat and Eisenbach, 1995). The rapidly expanding knowledge of ubiquitous intercellular chemical communication systems complements work on pattern formation by revealing some of the tools bacteria use for regulating physiology and gene expression in response to each other (Kaiser and Losick, 1993; Kell et al., 1995).

This chapter summarizes some of the results on multicellular interactions in *E. coli* that affect colony morphogenesis, chemotaxis, and the processes of genetic change. It has been written as a complement to other reviews of the same material, where figures illustrating and substantiating many of the points made below can be found (Shapiro 1992a, 1994, 1995b).

Large Scale Patterns Visualized by Vital Staining and Macrophotography

My experience with bacterial multicellularity began when I tried to use *lacZ* genetic fusions and XGal (5-bromo-4-chloro-3-indolyl-β-D-galactoside) indicator medium to study gene expression in *Pseudomonas putida* and *Escherichia coli* colonies on petri dishes (Shapiro, 1984b,c). This combination of genetic engineering and histochemical staining permits visualization of differential gene expression because *lacZ* encodes the enzyme β-galactosidase, which hydrolyzes XGal to produce a nondiffusing blue dye. To my surprise, the XGal-stained colonies were like flowers, displaying striking, often beautiful patterns of β-galactosidase activity (Fig. 2.1). The patterns revealed that the colonies were multicellular communities composed of spatially organized biochemically differentiated bacteria.

There were two basic pattern elements visible in the photographs of XGal-stained colonies (Fig. 2.2). One element comprised the wedge-shaped, radially oriented sectors long familiar to bacterial geneticists as representing distinct clones of bacteria descended from a common ancestor. A sector became visible when a genetic change in the progenitor cell conferred distinct properties on its descendants, such as a change in the level of *lacZ* expression or in the ability to spread over the agar substrate. Formation of sectors reflected the vegetative nature of bacterial multiplication by cell division, although it is still far from clear how the geometry of cell division translates into the observed geometry of a sector.

The second basic colony pattern element comprised the sharply defined concentric rings displaying different levels of XGal staining. These rings were less familiar to bacteriologists and could not be explained as the consequences of vegetative growth and genetic switches. The cells within a ring did not constitute a clone. They were connected to each other by position rather than ancestry. The cells in one ring descended from cells in the previous (phenotypically distinct) ring and were the progenitors of cells in the succeeding (phenotypically distinct) ring. Thus two problems arose: What gave phenotypic coherence to the bacterial population in each ring, and how did the well demarcated phenotypic transitions occur from one ring to the next? As discussed below, part of the answer to the first question lies in the generation of chemical fields in the agar substrate, which provide cues to coordinate patterns of gene expression during colony development. We remain fairly ignorant about the second question, except to say that the bacteria evidently responded periodically to self-generated changes in their growth conditions (e.g., nutrient consumption, waste product accumulation, changes in oxygen tension and pH). It is likely that these self-generated changes also included changes in dedicated signaling/regulatory molecules, such as homoserine lactone autoinducers (Fu-

FIGURE 2.1   Three *E. coli* colonies stained for β-galactosidase activity on XGal indicator medium. (Approximately ×2.5)

qua et al., 1994; Huisman and Kolter, 1994; Kell et al., 1995) (see Chapter 4). That the bacterial responses to such changes were frequently abrupt and nonlinear could be seen in the sharp boundaries of the concentric rings, although some *lacZ* fusions did change in a gradual way as gradients formed during colony development.

Concentric ring patterns are an inherent aspect of *E. coli* colony formation on standard laboratory medium, where the bacteria are typically given saturat-

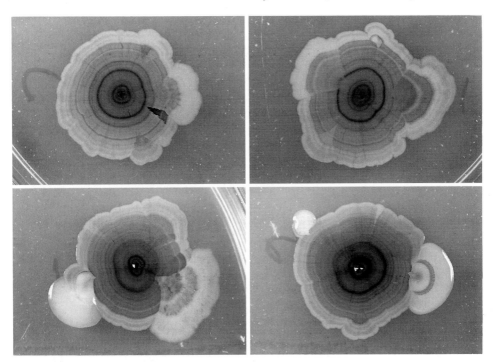

FIGURE 2.2 Organizational patterns in *E. coli* colonies revealed by XGal staining for β-galactosidase activity. These colonies carried a developmentally regulated *lacZ* fusion at 18.2 minutes on the chromosome and had been growing for 7 days at 37°C when these photographs were taken. All of these colonies were toothpick-inoculated with samples taken from different regions of a similar XGal-stained parental colony. Note the enzymatically distinct populations arranged in clonal sectors and in nonclonal concentric rings. (×4.7)

ing levels of nutrients. When XGal staining was used to visualize colony patterns, distinct underlying mechanisms could operate at the genetic level to control differential β-galactosidase expression in the various concentric rings. In some cases, stable fusion of *lacZ* to a differentially regulated genetic locus produced characteristic patterns. We have used the λplacMu system to generate hybrid sequences encoding β-galactosidase fusion proteins at different positions in the *E. coli* chromosome (Bremer et al., 1988), each displaying a distinct pattern of expression (Fig. 2.3). An interesting feature of Figure 2.3 is that colonies of different sizes produced by a particular strain all displayed the same β-galactosidase pattern. In other words, *lacZ* fusion expression did not shift simply in response to fixed threshold values of bacterial mass in the colony but altered proportionately to the final extent of colony growth. This scale-independence appeared to result from the dynamic nature of the pattern-forming system and may be interpreted in the following manner. Changes in β-galactosidase synthesis took place as bacterial growth created new chemical environments in the growth medium. These changes occurred most rapidly in

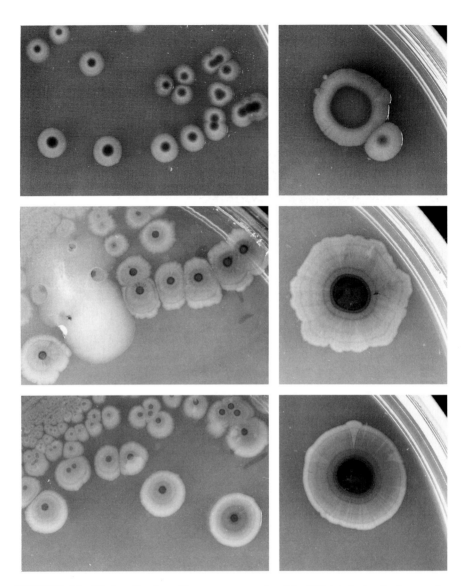

FIGURE 2.3 Distinct β-galactosidase patterns produced by XGal-stained colonies carrying independent *lacZ* fusions. Five stable *lacZ* fusion cultures were each inoculated twice onto adjacent areas of XGal indicator agar. One inoculum was streaked so a field of colonies developed, each from an individual colony-forming unit (left column); the other inoculum was left as a 1 μl spot containing about 10^5 cells so large colonies developed from a multicellular inoculum (right column). All pictures are at the same magnification. Except for the colony centers, which are different for the spot inoculations, all the colonies from a single strain in each row expressed the same nonclonal pattern, independently of size. Photographed after 6 days of incubation. (×4.5). Figure continues.

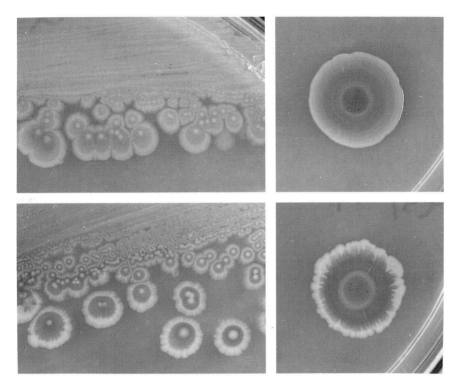

FIGURE 2.3   *Continued*

crowded zones, where the colonies were more limited in their expansion. Thus small colonies went through the same sequences of differential gene expression as large colonies but did so when there were fewer total bacteria.

A second molecular system for generating concentric ring patterns of β-galactosidase expression involved a somewhat different *lacZ* fusion system: the transposition and replication-competent Mu*dlac* elements constructed by Casadaban and his colleagues (Castilho et al., 1984). It was a surprise to discover that these elements generated ring patterns even when they were inserted into the chromosome in such a way that no active *lacZ* fusion was formed (Shapiro and Higgins, 1988, 1989). By examining the fate of Mu*dlac* DNA, it was discovered that there was a correlation between Mu*dlac* amplification and β-galactosidase synthesis. Because replication and transposition of Mu and its derivatives are inextricably connected (Shapiro, 1979), the observed correlation between replication and β-galactosidase synthesis suggested that the process of Mu*dlac* replication created active *lacZ* fusions. This hypothesis was confirmed by showing that genetic blocks to Mu*dlac* replication and transposition also prevented *lacZ* expression. An interesting consequence of the fact that β-galactosidase expression resulted from Mu*dlac* replication and transposition in these colonies was the connection of a specific biochemical differentiation to a process lethal to the differentiated cells. The process of Mu*dlac* replication grossly rearranged the genome (literally tying it in knots) and prevented normal cell division. Thus the cells expressing β-galactosidase could not produce progeny and were, effectively, terminally differentiated.

It is important to note that XGal staining was merely a convenient method for seeing colony patterns, not a requirement. The large-scale pattern features (sectors and rings) visualized with *lacZ* fusion systems could also be seen by reflected light photography without using genetic engineering or histochemical staining methods (Fig. 2.4) (Shapiro, 1984c, 1985b, 1992a, 1994). This finding meant that differentiation and spatial organization were integral features of the *E. coli* colony, not artifacts of the technologies that make them visible. In other words, even well fed colonies on normal laboratory medium have organization and structure. This point must be emphasized because there is a tendency to assume that the standard round bacterial colonies are ''simple'' accumulations of cells and that ordered cell populations form only in response to stressful situations. In fact, there may well be more biologically controlled order in round colonies than in the strikingly branched colonies formed by many bacteria under conditions of nutrient deprivation, where inorganic factors such as diffusion play a major role in pattern formation (see Chapters 13 and 14).

FIGURE 2.4   Concentric rings and sectors visualized by reflected light photography of *E. coli* colonies. These colonies were 6 days old when photographed. The top two colonies measured approximately 1 cm in diameter. The colonies at the bottom carried a mutation that resulted in more extensive colony expansion.

Monitoring Colony Development by Microscopy

By analogy with all other spatially organized biological systems, the *E. coli* colony must follow an intricate process of morphogenesis. This process could be appreciated only by following colony development over time. Observations with the light and scanning electron microscopes produced some intriguing results. One major surprise was the discovery that *E. coli* cells maximize cell-to-cell contact rather than individual cell access to substrate. This phenomenon was seen by examining the early phases of bacterial development on agar surfaces (Shapiro and Hsu, 1989; Shapiro, 1994).

Starting with a single cell, cell–cell interactions operated as soon as division had created a population of two siblings. The sister cells elongated by extending their inside poles along each other, with the outside poles remaining fixed in place. Division of the two cells during elongation produced a highly regular four-cell structure. [This structure was observed in bacterial growth on agar more than 85 years ago by Graham-Smith (1910).] Continued cell divisions and intimate side-by-side alignments created ordered monolayer microcolonies of several hundred highly aligned and tightly packed cells. At a certain point, cell divisions began to create a second and then subsequent cell layers next to the substrate, so the microcolonies acquired a well defined multilayered structure (Fig. 2.5). The tendency for maximizing cell–cell contacts was also observed when two microcolonies met; invariably they merged to create larger integrated microcolonies. At the point of initial contact between two microcolonies, highly ordered cell groups could frequently be observed (Fig. 2.6) (Shapiro, 1994).

It was possible to demonstrate the importance of cell–cell interactions during these initial phases by observing what happened to the progeny of densely inoculated bacteria. Bacteria neighboring a developing microcolony frequently attracted cells to deviate from their normal behavior pattern. For example, after the first cell division, one daughter could elongate by extending its outside pole to make contact with a neighbor rather than grow alongside its sibling (Shapiro and Hsu, 1989). It is important to note that *E. coli* cells on agar divided just as rapidly as their cousins in liquid suspension, despite their intimate side-by-side associations (Shapiro, 1992a). This result meant that *E. coli* has evolved cooperative systems needed to grow efficiently in crowded multicellular conditions.

The importance of cell–cell interactions creating ordered multicellular arrays was also clearly appreciated by observing colonies growing from multicellular inocula. It was convenient to inoculate colonies with small spots of liquid culture containing many thousands of cells (Shapiro, 1987, 1994). The cells tended to concentrate in a ring at the edge of the drop, where they remained as the liquid dried. Initially, the tightly packed cells lacked order, but considerable alignment was visible within 2 hours (Shapiro and Hsu, 1989). Colonies inoculated as spots did not immediately begin to expand over the agar, as might have been anticipated if expansion resulted only from division of cells at the periphery. Instead, the entire inoculated zone filled in by cell

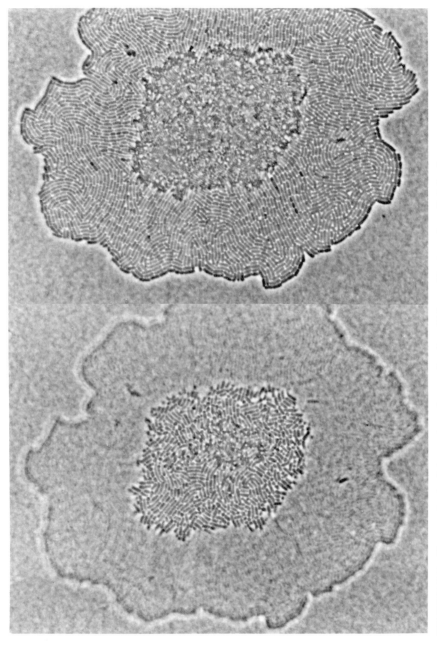

FIGURE 2.5 Phase-contrast, oil immersion micrographs of an 11-hour-old microcolony that developed from a single colony-forming unit on a thin agar layer. These micrographs were taken at different planes of focus to illustrate that the central region is two cell layers thick. (Approximately ×2000)

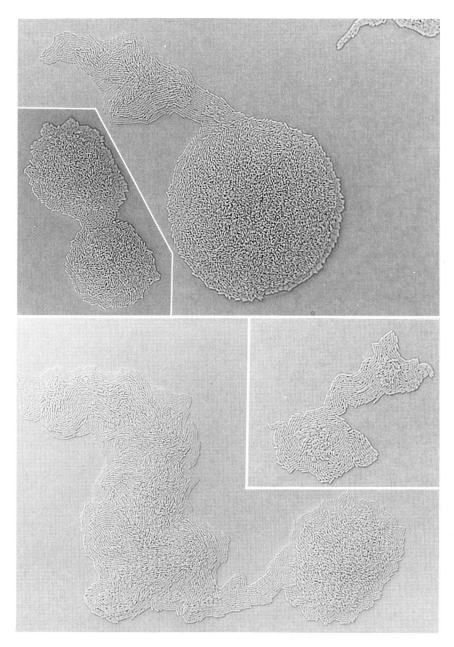

FIGURE 2.6 Initial stages of mergers between *E. coli* microcolonies on an agar surface. These microcolonies had undergone 7 hours of growth on the surface of a normal petri dish when the photographs were taken. Note the bridges formed by palisaded cells during the process of micro-colony fusion.

23

division, and the ring at the edge accumulated several layers so it formed a mound just inside the periphery (Fig. 2.7). Once this consolidated mound structure was formed, the colony began active expansion over the substrate. From high magnification time-lapse videotapes, it appeared that cell divisions among the organized cells in the mound pushed out the palisaded, radially oriented (nondividing) cells at the periphery. Colony expansion driven by a limited zone of exponentially growing cells around the edge would fit quantitative measurements showing a linear increase in colony diameter with time (Pirt, 1967). *E. coli* colonies also elaborated extracellular materials during this preexpansion consolidation phase. These materials were visible as bright halos around the living colonies by light microscopy and as aggregated polymeric structures extending over the bacteria and outward from the colony edge in scanning electron micrographs of fixed material (Fig. 2.8) (Shapiro, 1987). Serendipitous observations on the behavior of young cell populations surrounding glass fibers extending outside the normal colony perimeter confirmed the need for an initial stage of structural preparation for expansion over the substrate (see below).

Once active expansion had begun, the colony assumed a great deal of form visible at low magnification in the scanning electron microscope (Shapiro, 1987). Under the conditions studied, the initial inoculation zone was demarcated by a thin depression ring that remained fixed in size; the outer expansion zone was demarcated by a deep circular groove a few hundred micrometers inside the periphery. Because the position of this groove remained relatively constant with respect to the edge, it grew in diameter as the colony expanded. Additional macroscopic features developed during colony growth, especially distinct zones where the bacteria had different sizes, shapes and patterns of multicellular aggregation. Sometimes the boundaries between these zones were sharp and marked the shift from one phase of bacterial growth to another. It is tempting to think of these zones as constituting ''tissues'' composed of differentiated cell types.

Most photography and microscopy examines only the colony surface and its organization in the two dimensions comprising the plane of the agar surface. Additional structure was revealed by sectioning colonies to examine their internal structure perpendicular to the substrate. Sectioning and staining with traditional dyes or examining XGal-stained colonies uncovered a stratified pattern with multiple differentiated layers (Shapiro, 1994). These layers included zones where the bacteria did not stain for protein, and cell-sorter experiments showed that populations of nonviable cells were regularly produced during colony growth (see Chapter 10). Thus it is likely that some kind of programmed cell death occurred during bacterial colony development (Yarmolinsky, 1995). It should be noted that a histologically stratified structure in bacterial colonies was first reported by Legroux and Magrou in 1920. They inferred (correctly, in my opinion) that these layers must play distinct physiological roles. The exact adaptive functions of all these layers are not yet known, although it may be speculated that zones of empty cells help protect bacterial populations on the bottom of the colony from attack by antagonists, such as disinfectants (Costerton et al., 1987) and bacteriophages (Shapiro, 1994). Important aspects of bacterial physiology, such as transport and gas exchange, must be profoundly influenced by all features of colony structure, including cellular differentiation, stratification and extracellular materials. It will be interesting to investigate whether bacteria have evolved collective transport mechanisms to mobilize substrates and remove wastes. Observing papillae growing on a colony in response to a substrate in the agar below, it is difficult not to wonder how nutrients penetrate a dense mass of bacteria and exopolymer to feed the growing cells on top.

Regulative Phenomena in Colony Morphogenesis

A classic method for studying morphogenesis in higher organisms has been to introduce experimental disruptions into the developmental process and observe the response. The observation of embryonic regulation (that is, the reestablishment of normal form, frequently by a novel sequence of events) was one of the key indications that intercellular communication played a fundamental role in development. The early experiments with developmental regulation in *E. coli* colonies have been simple, but they suggested that the regularity of colony structure may well owe more to communication than to the statistics of large numbers. In particular, they pointed to an important coordinating role for chemical fields in the agar substrate.

One basic question about colonies concerns why they display good circular symmetry when they grow from point inocula. Where does it come from? We know that biological functions are essential for circularity because changes in growth conditions or mutations can lead to the formation of branched colony structures [see Chapters 13 and 14 for *Bacillus subtilis* examples; and see Shapiro (1988), and Shapiro and Trubatch (1991) for *Proteus mirabilis* examples]. Time-lapse video recording of colony development from an uneven toothpick inoculation showed that the bacterial mass grew in such a way that it acquired a nearly uniform distribution before it began expanding (Fig. 2.9). In a more controlled fashion, it was possible to create an asymmetrical starting population by irradiating the edge of a spot inoculum with a near-ultraviolet (UV) microbeam from a fluorescence microscope. This maneuver created a notch in the microcolony perimeter during the initial phase of growth, but the population in the notch expanded faster than the rest of the colony to reestablish a smooth colony outline within a few hours after irradiation (Shapiro, 1994). Such adjustments to smooth out the colony perimeter indicated the existence of factors that actively promote colony symmetry, but we know nothing yet about how such factors operate.

Another system for disrupting the normal geometry of colony development was discovered by chance. Small fibers were frequently deposited on the agar surface from sterile pipettes stored in tins containing glass wool at the bottom. These fibers proved to be excellent tools for creating new colony geometries. Each fiber was surrounded by a meniscus of liquid, which could be observed by transmitted light microscopy as a bright halo around the fiber. Expanding

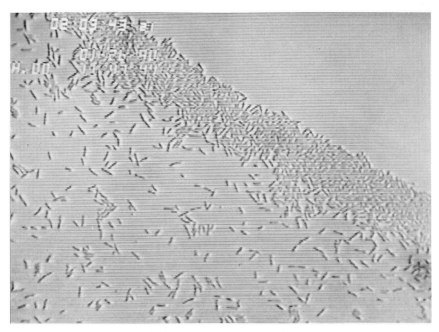

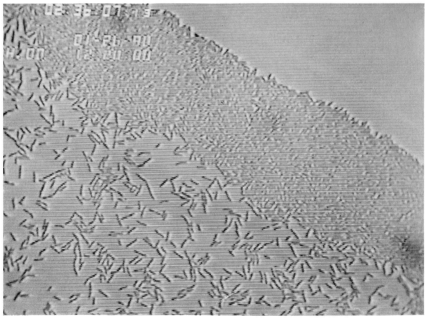

FIGURE 2.7 Sequence of time-lapse video frames illustrating the consolidation of a mound structure at the edge of a spot inoculum preceding *E. coli* colony expansion. The first frame shows the distribution of bacteria after a small drop of culture had dried into the agar medium 2.15 hours after inoculation. The next three frames show how the central region filled in and the edge became thicker (darker) during the next 1.25 hour, with little advance of the colony edge. The thick zone in the last frame corresponds to an early stage in the formation of the mound structure seen a few hours later in the first panel of Figure 2.8. Note also the development of a bright halo around the colony perimeter. The numbers in the upper left corner indicate the time elapsed since inoculation. (Approximately ×2000) Figure continues.

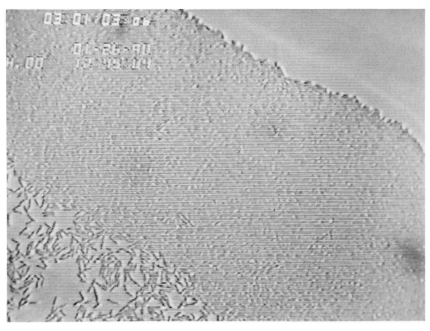

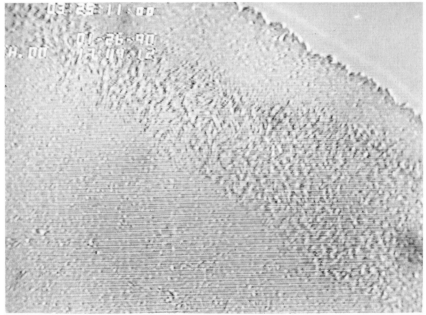

FIGURE 2.7 *Continued*

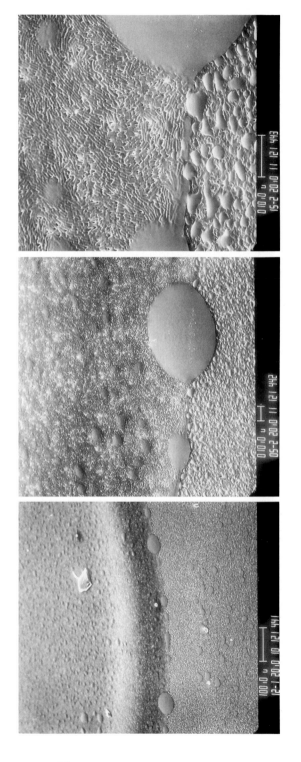

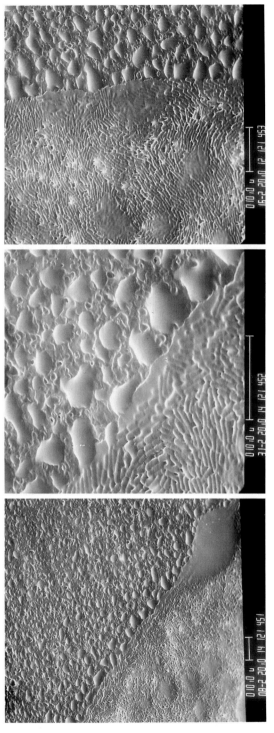

FIGURE 2.8 Scanning electron micrographs showing the edge of a 7.5-hour-old *E. coli* colony (Shapiro, 1987). This colony was just beginning its expansion phase. In frame 441 note the thickened mound just inside the periphery (see Figure 2.7). In the other frames note the presence of extracellular material at the edge and inside the colony. The top row shows successively higher magnifications of the bottom edge of the colony (the magnified region is at the far left of frame 441). The bottom row shows two regions at the top (frames 451 and 452) and right (frame 453) edges. The scale bars indicate 100 or 10 μm. Figure continues.

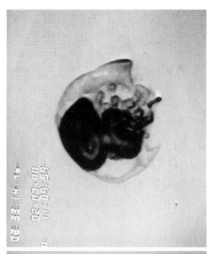

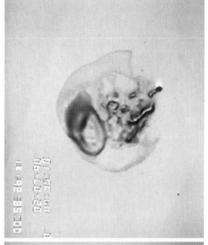

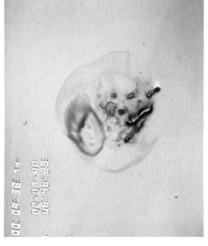

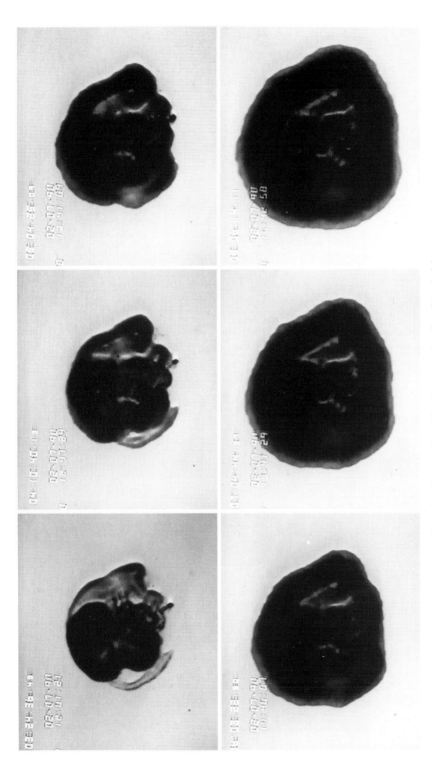

FIGURE 2.9 Time-lapse video frames illustrating the growth and development of an *E. coli* toothpick inoculum on an agar plate. Note the uneven initial distribution of the cell mass and how a reasonably uniform colony density developed during the first 5 to 6 hours of growth preceding active expansion of the colony.
Figure continues.

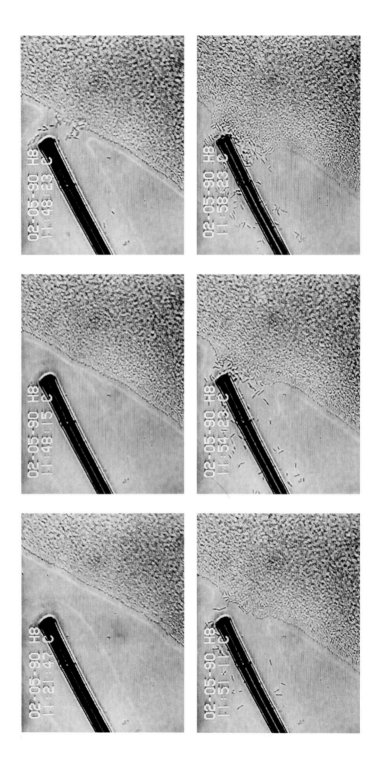

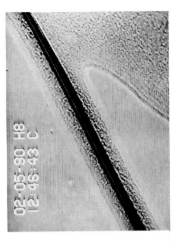

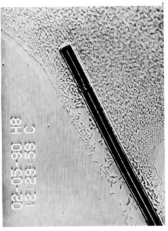

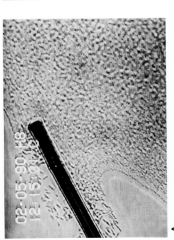

A

FIGURE 2.10 Time-lapse video frames illustrating the encounter between an expanding microcolony and a fragment of glass wool on the surface of a petri dish concluding with engulfment of the fiber. (A) Note how motile cells at the colony perimeter liberated themselves into the liquid surrounding the glass as soon as the halo around the fiber and the halo around the colony came into contact. (Approximately ×1500)
Figure continues.

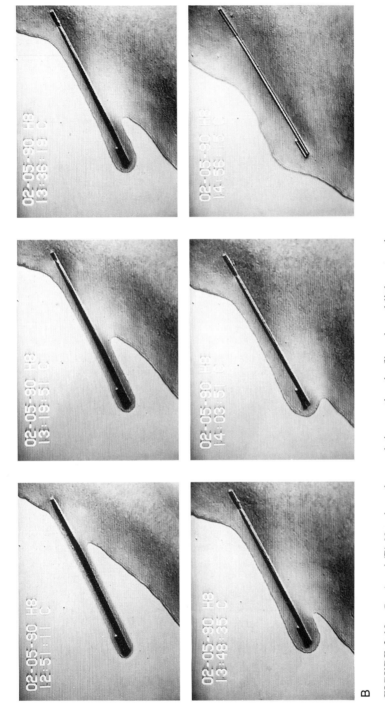

FIGURE 2.10 *Continued* **(B)** Note how the population coating the fiber showed little outward expansion over the agar compared to the colony edge. The colony outline continued to become even more regular after the last frame shown here.

B

34

colonies were also surrounded by a halo, assumed to result from an exopolymer gel over and outside the colony (Shapiro, 1985c, 1987). Time-lapse videotapes of the encounters between colonies and fibers were dramatic. When the halo around a growing colony met the halo around a fiber, cells swam explosively into the meniscus of liquid surrounding the glass, where they circumnavigated until they completely coated the fiber (Fig. 2.10A). This "explosion" of peripheral cells from the colony required active flagellar motility but not chemotaxis. It was not displayed by nonmotile cells, which could migrate a distance into the meniscus around the fiber but could not separate themselves. Apparently, the cells were glued together by exopolymeric material and required flagellar propulsion to detach. The populations that coated the glass fibers did not spread initially because they had not developed the consolidated structure needed for outward expansion. Thus if the fibers were relatively short, they were engulfed by the expanding colony before this structure could form, and the colony perimeter was quickly regulated to a smooth contour (Fig. 2.10B) (Shapiro, 1994). Around longer fibers, however, the coating population developed sufficiently to begin outward expansion before it could be engulfed, so a distorted colony outline formed. Significantly, if the bacteria carried the developmentally regulated *fiu::lacZ* fusion at 18.2 minutes of the chromosome, the XGal staining patterns along the fiber aligned with those around the colony. The patterns emerging from the coated fibers displayed only concentric rings that began at the developmental stage where the colony encountered the fiber (Shapiro and Trubatch, 1991; Shapiro, 1992a, 1994), indicating that the bacterial populations growing out from the fibers were developmentally coordinated with the bacteria at the colony perimeter. Coordination worked through chemical fields in the agar substrate. This activity could be confirmed with extremely long fibers, where the XGal staining patterns did change at 1–2 cm or more from the colony periphery, indicating that the coordinating fields fell below threshold values at these distances (Shapiro, 1992a, 1994).

Another type of colony encounter provided further evidence for long-range coordination through the substrate. Whenever adjacent colonies grew together in crowded fields of XGal-stained bacteria, their ring patterns merged, which was to be expected on purely geometric grounds if all colonies developed more or less synchronously and produced corresponding β-galactosidase rings at the same times. However, it was possible that a more active process of alignment was also occurring. Marc Lavenant, then a graduate student, suggested examining colonies inoculated near each other at different times. Whenever this experiment was performed, the ring patterns still aligned (Fig. 2.11) (Shapiro, 1992a, 1994, 1995b). Moreover, the center of the younger colony always bypassed some of the early stages of pattern formation to come into register with the older colony. The center of a duplicate control colony inoculated far from the older colony underwent all the early stages of colony development, as illustrated in Fig. 2.11 (Shapiro, 1992a). This meant that the chemical field around the older colony induced the bacteria in the younger colony to behave as though it was of the same age. Through these fields, gene expression levels in two colonies inoculated at different times could be synchronized before they

FIGURE 2.11   Communication between colonies of different ages. The double colony on the left grew from two spots inoculated 37.5 hours apart. When the younger colony was inoculated next to the older colony, a sample of the same culture was also inoculated several centimers away to produce the isolated colony shown on the right. Note how the ring patterns in the adjacent colonies aligned and how the centers of the sibling younger colonies differed: The isolated single colony went through all the same stages as the older colony, whereas the adjacent younger colony skipped the earliest stages and synchronized with its older neighbor. The older colony was 12 days old when the photographs were taken. (×3.8)

came into contact with each other. Thus bacterially created chemical fields in the substrate have been observed to play the coordinating role needed to explain the integrity of concentric ring patterns extending through numerous different clonal populations within a single colony.

Beginnings of Genetic Analysis of *E. coli* Colony Patterns

A fundamental question in colony development concerns the genetic control of pattern formation. We know that there is genetic specificity to differential gene expression during colony development because *lacZ* fusions to different regions of the *E. coli* chromosome displayed distinct XGal staining patterns (Fig. 2.3). Hence the signal transduction circuits governing individual genetic loci respond to changing conditions during colony development, and each circuit responds in its own way. The genetic control of colony pattern formation will prove complex because many mutations affect growth and lead to altered colony structures. Some interesting data are currently available, however, from analyzing mini-Tn10 transposon insertions affecting Mu*dlac* patterning.

When Mu*dlac* colonies harboring mini-Tn10 transposon insertions (Way et al., 1984) were screened, a number of distinct new patterns were observed. The pattern differences involved levels and timing of β-galactosidase expression as well as colony outlines and spreading patterns. The inserted transposons were mapped in two ways: (1) by recombination with chromosomal fragments cloned into phage lambda (Kohara et al., 1987), or (2) by cloning the mini-Tn10 into a plasmid vector and matching the restriction map or DNA sequence of the adjacent DNA with known regions of the *E. coli* chromosome. The genetic loci identified in this manner as affecting colony patterning are listed in Table 2.1. They include functions involved in DNA metabolism (*dnaQ*,

TABLE 2.1 Mini-Tn10 Insertion Mutations Affecting Mud/*ac* Colony Patterns

dev Allele[a]	Location[b]	Mapping Technique[c]
2099	polA	Kohara phage transduction (φ547), restriction mapping, sequencing
2103	selB	Sequencing
2105	20.1′ (lrp?)	Restriction mapping, Kohara phage transduction (φ214, φ215)
2131	dnaQ	Restriction mapping (J.-P Bouché), P1 transduction, verifying mutator phenotype
2135	cya	Sequencing
2137	gidA	Kohara phage transduction (φ560); sequencing

[a]Isolation number of mini-Tn10 insertion mutations altering colony structure, Mud*lac* β-galactosidase expression patterns or both (labelled *dev* alleles when first isolated to indicate an effect on colony *dev*elopment).
[b]Chromosomal location of each insertion.
[c]Technique(s) used to determine the chromosomal location.

polA), cellular metabolism (*cya*, *selB*), and the control of cell division (*gidA*). This diversity suggests that there is no simple set of ''pattern'' functions in *E. coli* but, rather, that bacterial physiology and colony morphogenesis are inseparable.

The *polA2099*::mini-Tn10 mutation led to studies that demonstrated differential function and differential expression of DNA polymerase I during colony development (Shapiro, 1992b). Examining patterns of *polA-lacZ* fusion expression revealed that even basic ''housekeeping'' functions were subject to differential multicellular regulation. In other words, developmentally regulated gene expression patterns were not limited to some special set of functions. With *polA-lacZ* fusions, long-distance regulation through the agar substrate could also be observed. Older colonies inhibited fusion expression in colonies inoculated near them, sometimes by distances of more than 1 cm. Accompanying the long-range inhibitory effect on *polA* expression, there was a growth-stimulating (presumably protective) effect diffusing through the agar that compensated for the poor initial growth of *polA* bacteria lacking DNA polymerase I function. The *polA* mutants were particularly sensitive to oxidative DNA damage and produced aberrant microcolonies filled with elongated cells during the first few hours of development due to SOS inhibition of cell division. Despite the abnormal microcolony development, however, *polA* and *pol*[+] colonies were morphologically indistinguishable after 3 days of growth. Thus the *polA*::mini-Tn10 mutation provided an additional example of morphogenetic regulation operating during colony development.

The *selB2103*::mini-Tn10 insertion prevented formate dehydrogenase synthesis and changed the oxidative metabolism of the bacteria. SelB is a translation factor necessary for the incorporation of selenocysteine into formate dehydrogenase (Leinfelder et al., 1988). The *selB*::mini-Tn10 (*dev2103*) mutation not only affected Mud*lac* derepression but, when introduced into other genetic

backgrounds, modified colonial morphology, converting the colony centers produced by the particular strain illustrated from domes to craters (Fig. 2.12). It
is probable that this structural change resulted from poor anaerobic growth
under the colony center due to the formate dehydrogenase deficiency. The
cya2135::mini-Tn10 mutation prevented the synthesis of cyclic adenosine
monophosphate (cAMP), the molecule that *E. coli* cells use as their central
monitor of carbohydrate metabolism (Magasanik and Neidhardt, 1987). The
cya mutant Mu*lac* strains displayed greatly reduced β-galactosidase expression. This observation was confirmed using well characterized mutations in

Dev⁺   *selB*

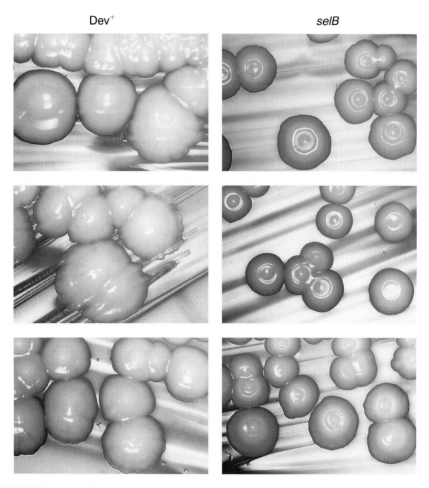

FIGURE 2.12   Effect of a *selB*::miniTn10 mutation on colony morphology. Each row shows
colonies produced after 4 days of incubation on glucose-casmino acids agar by a *galT* mutant of
HfrHayes (Dev⁺) (Adhya and Shapiro, 1969) (**on the left**) and a *selB*::miniTn10 transductant
(*dev*2103) (Table 2.1) (**on the right**). Note the effect on colony size and the structure of the
colony centers.

both *cya* and *crp* (which encodes the cAMP receptor protein, a transcription factor) (Shapiro, 1994).

Perhaps the most intriguing result of the mini-Tn10 mutagenesis studies was isolation of the *gidA2137*::mini-Tn10 mutation. The *gidA* phenotype was reported to involve formation of elongated cells in glucose medium (von Meyerneberg et al., 1982), and GidA function has been implicated in the cellular differentiation of *Proteus mirabilis*, as *gidA* mutants in that species are unable to form swarmer cells (see Chapter 7). The *gidA* defect had a dramatic effect on Mu*dlac* XGal-staining patterns; rather than a light colony center surrounded by dark rings, *gidA* colonies displayed a dark center with alternating concentric light and dark rings. In some way, the GidA function participated in a signal transduction network controlling Mu*dlac* derepression in response to changing cellular physiology during colony development. It is likely that this network includes the cAMP-Crp system because the *gidA2137*::mini-Tn10 mutation also affected the response of the Mu*dlac* element to glucose gradients in the growth medium.

Another component of the signal transduction system regulating Mu*dlac* is the ClpPX protease, which is known to play at least two roles in phage Mu derepression and replication (Geuskens et al., 1992; Mhammedi-Alaoui et al., 1994). The ClpPX protease is a member of a family of two component ATP-dependent proteases that share a catalytic subunit (ClpP) associated with one of a number of regulatory subunits (e.g., ClpX, ClpA), each of which is an Hsp100 stress-regulated protein (Katayama et al., 1988; Gottesman et al., 1990). The ClpP and ClpX subunits are required for Mu repressor inactivation, and ClpX alone is additionally required for some subsequent step in Mu (and Mu*dlac*) replication (Mhammedi-Alaoui et al., 1994). Colonies with a *clpX* mutation were white owing to the double block to derepression and replication (Shapiro and Maenhaut-Michel, unpublished observations). Colonies with a *clpP* mutation had a more complex phenotype (Shapiro, 1993). Streak colonies starting development from a single cell (i.e., colonies always undergoing a regular, highly organized process of development) were uniformly pale owing to the low levels of Mu*dlac* derepression. Stab colonies, however, started development from disorganized cell masses (Fig. 2.10), in which conditions could be highly variable, and the colony centers displayed high levels of XGal staining. The rest of the colony, which developed from an orderly process of expansion, was pale, similar to the streak colonies. Thus, ClpP-independent Mu*dlac* derepression could occur under certain growth conditions. Intriguingly, the conditions for ClpP-independent derepression could also be created genetically because *gidA clpP* double mutant colonies showed XGal staining similar to the *gidA* pattern. In genetic terminology, *gidA* stimulation of Mu*dlac* derepression during normal growth was epistatic to *clpP* inhibition.

Clearly, much analysis remains to be done to understand the architecture and logic of the signal transduction network controlling developmentally specific Mu*dlac* derepression and replication. Nonetheless, the initial results established the feasibility of a genetic dissection of colony pattern formation, and the participation of several classes of interacting regulatory molecules has been established.

Self-Organization of Chemotactic *E. coli* Colonies in Fluid Media

Escherichia coli not only forms colonies on standard laboratory medium containing such a high percentage of agar that the cells are immobile. Some of the most striking bacterial patterns have been found in expanding colonies of chemotactic bacteria in semisolid media where the cells could swim (Budrene and Berg, 1991, 1995). That chemotactic bacteria such as *E. coli* form expanding concentric rings when grown from a point inoculum in soft agar has been known for years (Adler, 1966). As the bacteria multiply and consume nutrients around the central inoculated point, they create gradients of attractants, leading the cells to migrate outward. Sequential uptake of different nutrients in the medium leads to the formation of several gradients of chemoattractants across the petri plate, and different populations chase each gradient, forming distinct rings. In a typical experiment, the swarm rings disappear when substrates are exhausted. Although this behavior could be satisfactorily explained with statistical analysis as reflecting the average movement of many independent cells (Nossal, 1972), the operation of multicellular coordination mechanisms could not be excluded. Dramatically different patterns formed in chemotactic colonies when *E. coli* cells grew on Krebs cycle intermediates or when a stress agent, such as hydrogen peroxide, was incorporated into the growth medium. The active participation of the chemotactic control circuit (the best understood of all molecular neural networks) (Bray et al., 1993) could be demonstrated by showing that genetic or chemical blockage of the Tar chemoreceptor prevented pattern formation. The bacteria were found to emit autoaggregation signals. Amino acid analysis showed that the stressed bacteria began to excrete aspartate and glutamate, both powerful chemoattractants sensed by the Tar chemoreceptor. Once in an aggregate, a high proportion of the bacteria frequently lost their motility and so produced a fixed spot of nonmotile cells. How these bacteria lost their motility so easily and so permanently remains a key mystery in this system. Distinct patterns of these dots form in different soft agar media. Analysis of how pattern dynamics depend on substrate concentration revealed that the motile bacteria migrate between two kinds of multicellular structure: the chemotactic ring and the autoaggregates. Under certain conditions, aggregates are observed to remain intact for long periods and sometimes behave as integrated units (E. Budrene, personal communication). For example, they can migrate, split, and fuse. It is likely that additional forms of intercellular communication are operating within and between the aggregates. Clearly, these patterns show that situations exist in which migrating *E. coli* cells do not move independently but are coordinated by means of secreted signals processed by the chemotaxis system. This mode of behavior in a fluid environment is biologically adaptive because bacteria in multicellular aggregates have a much better chance than isolated cells of surviving exposure to toxic substances (Ma and Eaton, 1992). Comparable patterns have been reported in *Salmonella* swarm colonies (Blat and Eisenbach, 1995).

Multicellularity and Processes of Genetic Change

One very important but frequently overlooked phenomenon is the importance of multicellularity in *E. coli* and other bacteria for processes of genetic change. This oversight is doubly ironic because studies of genetic exchange in bacteria were one of the main research areas leading to the development of molecular biology, and current controversies over adaptive mutation involve multicellular genetic phenomena. DNA transfer is an important form of intercellular communication, and modern studies of conjugation have indicated that mating occurs most frequently in aggregates (Achtman, 1975) rather than in the conveniently bourgeois mating pairs imagined by the pioneers of bacterial genetics during the 1950s (Jacob and Wollman, 1961; Hayes, 1968). Intercellular (and especially interspecific) DNA transfer is a fundamental aspect of the evolution and spread of bacterial antibiotic resistance, a problem of enormous practical significance (Watanabe, 1963). In addition, we are continually learning how many features of bacterial activity in the environment, such as biodegradation and symbiosis, are encoded by transmissible plasmids (Levy et al., 1981). The virtuosity of bacteria in mobilizing DNA molecules has revolutionized our thinking about fundamental aspects of genetics, especially evolution (Shapiro, 1992c). Nonetheless, the strength of traditional views on genetics as a purely intracellular phenomenon have inhibited our appreciation of how the bacterial genome is affected by multicellularity.

In fact, there was strong evidence of multicellular influences on bacterial heredity long before the days of molecular biology. At the end of the nineteenth century and during the early decades of the twentieth, bacteriologists interested in inherited changes in bacterial cultures for vaccine production knew that different growth conditions could influence hereditary variation. They showed that situations such as colony aging could stimulate the process called "microbic dissociation," which meant the occurrence in sectors and papillae of new strains displaying multiple phenotypic changes "dissociating" the inherited properties that identified a particular bacterial culture (Hadley, 1927). In a fundamental way, the old observations on microbic dissociation were analogous to the results described above for Mu*dlac* derepression patterns in colonies. Both results established a connection between changing bacterial physiology during colony development and systems that can reorganize DNA molecules.

This connection between cellular physiology and DNA biochemistry was at the heart of the phenomenon now known as "adaptive mutation" (increased occurrence of genetic changes under selective conditions when they lead to adaptively useful new phenotypes) (Shapiro, 1995a). The basic observation in all adaptive mutation systems concerned the kinetics with which mutant clones appeared on a selective medium where the plated bacteria could not proliferate but remained viable (Shapiro, 1984a; Hall, 1988; Cairns et al., 1988). Few (or no) clones appeared during the first 2 days of incubation, indicating a low frequency of mutational events during normal growth prior to plating. However, mutant clones accumulated in unexpectedly high numbers upon longer incubation, indicating that selection somehow induced (adaptive) genetic changes leading to proliferation. Because the conditions were not lethal, bacteria surviving under selection underwent many physiological changes that altered the biochemical systems acting on the genome, frequently inducing higher levels of mutation. Adaptive mutation experiments differed in this key experimental respect from the classical studies of spontaneous mutagenesis during the 1940s and 1950s in which the killing of nonmutant bacteria by phage and bactericidal antibiotics precluded the possibility of genetic change under selective conditions (Luria and Delbrück, 1943; Newcombe, 1949; Cavalli-Sforza and Lederberg, 1956).

The first carefully documented example of selection-induced mutation involved the ability of a Mu prophage to form an *araB-lacZ* genetic fusion encoding a hybrid β-galactosidase activity (Shapiro, 1984a). In this system there were temporal and spatial features of the process of genetic change that revealed the connections between metabolism, population dynamics, and the genome. The kinetics of colony appearance over time indicated that some kind of triggering event for the fusion process occurred after plating. The addition of small amounts of glucose to the selective medium accelerated this triggering, indicating a link to carbohydrate metabolism. This link was confirmed by finding that fusions occurred at high frequency in response to aerobic starvation in glucose medium (Mittler and Lenski, 1990; Foster and Cairns, 1994; Maenhaut-Michel and Shapiro, 1994). In addition, the first fusion colonies to appear on the selection plates frequently clustered in a small region of the substrate (Shapiro, 1984a). This clustering indicated a response to locally created depletion zones in the substrate, much as nutritional depletion fields stimulated Mu*lac* derepression in colonies (Shapiro, 1994).

Genetic analysis of *araB-lacZ* fusions has indicated additional parallels with the Mu*lac* system. In the *araB-lacZ* system, the Mu prophage and Mu transposition functions played an active role in making the genetic fusions. Mutants defective in Mu transposase and proteins needed for transposase expression or function were blocked in fusion production (Shapiro and Leach, 1990). Thus an important aspect of the "adaptive" aspect of *araB-lacZ* fusion formation was the induction of Mu transposition activities under selective conditions. These activities shared regulatory features with Mu*lac* derepression, such as the requirement for the ClpPX protease (Shapiro, 1993). There is one important difference, however, between the molecular events that gave rise to *araB-lacZ* fusions and to Mu*lac* replication. In both cases Mu functions had to be derepressed (hence the common requirement for ClpPX) and a strand-transfer complex (STC) had to be formed; but the processing of the STC must have differed markedly in the two cases. To create *araB-lacZ* fusions, replication could not occur because it was lethal to cellular survival.

Speculating on the meaning of the Mu*dlac* and *araB-lacZ* systems in the context of the adaptive mutation debate, Higgins (1992) suggested that intercellular genetic exchange may occur far more frequently in the mutational process than is commonly believed. "Altruistic" genetic changes could occur in cells that were terminally differentiated, such as those having undergone Mu*dlac* replication; but DNA carrying these changes could be incorporated into other viable cells. These prescient speculations have been confirmed by genetic analysis of another adaptive mutation system, reversion of the *lacI-Z33* frameshift deletion (Cairns and Foster, 1991). When plated on lactose medium, *lacI-Z33* strains produced few early colonies from revertant bacteria that arose before plating, but the plates continued to sprout "adaptive" Lac$^+$ colonies with time. Analysis of the DNA in these revertant clones revealed that the rare frameshifts that occurred prior to plating had a sequence different from that of the selection-induced frameshifts (Foster and Trimarchi, 1994; Rosenberg et al., 1994). Thus new molecular machinery must have been brought into play on the selection plates. This induction process paralleled the activation of Mu transposition functions in the *araB-lacZ* system. The induced molecular machinery included homologous recombination functions because *recA* and *recBC* mutants did not display selection-induced *lacI-Z33* reversion (Foster and Cairns, 1992; Harris et al., 1994). This finding was particularly significant in light of the recent discovery that carbohydrate starvation could induce the SOS system and that RecA protein synthesis was regulated by the cAMP-Crp system (F. Taddei, I. Matic, and M. Radman, personal communication). The molecular machinery for *lacI-Z33* reversion also included plasmid transfer functions. The *lacI-Z33* strains used in these experiments carried the mutation on a F'*lac* plasmid. Peters and Benson (1995) discovered that F'*lac* plasmid transfer was frequent (even between male bacteria) under conditions similar to those on the lactose selection plates, and three papers have reported that induced *lacI-Z33* reversion depended on plasmid transfer (Foster and Trimarchi, 1995; Galitski and Roth, 1995; Radicella et al., 1995). Many of the "revertant" bacteria were actually exonjugants carrying a transferred and frameshifted F'*lac* plasmid (Radicella et al., 1995). It has also been discovered that *rec* mutations affected plasmid transfer in the same quantitative manner as they affected *lacI-Z33* reversion (Peters et al., 1996). Thus it appears that *lacI-Z33* reversion probably occurs during plasmid transfer, an intercellular event. Indeed, it is possible that the entire "adaptive" aspect of *lacI-Z33* reversion may depend on differential regulation of plasmid replication and transfer, just as the kinetics of *araB-lacZ* fusion depend on differential regulation of Mu functions.

The demonstration of "altruistic" genetic change is highly significant for understanding the multicellular aspect of bacterial population genetics. Intercellular movement of DNA segments by plasmids, bacteriophages, and transformation by naked or vesicle-enclosed DNA is ubiquitous in the prokaryotic world (Hayes, 1968; Shapiro, 1985a). Redfield (1988) had also proposed a kind of genetic altruism based on transformation that in nature involves the uptake of DNA liberated from lysed bacteria. From an adaptive point of view, a useful mutation is of equal value to the population whether it becomes part of the genome by purely intracellular processes within a viable cell or by incorporation of mutated exogenous DNA from another (possibly nonviable) cell, which may have carried multiple genetic alterations. It has been proposed that certain "mutations" arise from Rec-dependent gene conversion events that may involve partially homologous exogenous DNA (Rosenberg, 1994). Some pathogenic species, such as *Neisseria gonorrhoeae*, use gene conversation and DNA-mediated transformation as a mechanism to generate useful antigenic and phase variation of surface proteins (Meyer, 1987). Thus like all other intracellular systems, the genome itself participates in the multicellular life of bacterial populations.

New Paradigms in Prokaryotic Biology—Bacteria Are Not Stupid

From classical microbiology we have known that bacteria are the most versatile chemists and ecologists on the planet. The molecular biology of bacteria has revealed them to be sophisticated and complex beyond our ability to describe. Studies of symbiosis and pathogenesis (Salyers and Whitt, 1994) have shown how talented they are as cell and developmental biologists, able to make efficient use of higher organisms for their own benefit. Taken together, all aspects of bacterial research combine with the results summarized above to generate a new paradigm in prokaryotic biology: a picture of bacteria as sensitive, communicative, decisive organisms integrating information from their environment and from their neighbors in order to carry out the complex tasks of reproduction and survival in organized multicellular populations. This concept is strikingly different from the conventional wisdom about bacteria as isolated, small, primitive organisms with a limited repertoire of automatic responses to changing conditions.

When this new paradigm has been presented to audiences of microbiologists and molecular geneticists who are used to the early concepts of bacteria, heated controversy has sometimes erupted. Objections are launched that "artifactual" and "unnatural" examples of *E. coli* multicellularity in the laboratory are insignificant because bacteria normally grow in suspension. There is a great irony in these objections because the well aerated suspension culture is the true artifact, created in the laboratory to examine populations of physiologically homogeneous cells. Suspension cultures have found their most sophisticated technology in the chemostat, but chemostat experiments routinely come to an end when the *E. coli* cells follow their natural instincts and settle out of suspension to produce "wall growth," or biofilms, on the inside of the culture vessel. Indeed, it can be argued that the tendency of bacteria to resist being maintained in suspension culture is the key result of any chemostat experiment. In nature, of course, most bacteria live as adherent microcolonies and biofilms (Winogradsky, 1949; Costerton et al., 1987), and it is well known to water microbiologists that part of the normal *E. coli* ecological cycle involves survival in water supplies, often as biofilms growing inside the pipes

(Mackerness et al., 1993). If we think about why *E. coli* and other bacteria prefer to adopt a multicellular life style, it is evident that they can derive the same benefits as other organisms: strength in numbers, cellular division of labor, and functional specialization. Far from being an artifact, the colony is a laboratory expression of fundamental capabilities that *E. coli* can use as it confronts the myriad challenges of growth and survival throughout its complex life cycle.

It should not be surprising that the "bacteria as single cells" paradigm has proved strong and resilient despite the facts about bacterial multicellularity presented here and elsewhere in this volume. In this regard it resembles other major concepts in the history of science (Kuhn, 1960). Science as whole, however, is undergoing major conceptual changes, away from Cartesian reductionism toward a more integrated connectionist vision of nature (Waldrop, 1992; Kelly, 1994). Bacteria will occupy an important role in that change. They may even become key protagonists in this scientific revolution, as they were in the molecular biology revolution that is still transforming our lives. In bacterial pattern formation, such as that observed with *E. coli*, the problems of self-organization and complexity merge with the powerful technology of molecular genetics. We can anticipate that fundamental problems in pattern formation will find their first solutions in bacterial systems. Thus although we cannot predict the future in detail, it is possible to state with confidence that *E. coli* and other bacteria have many amazing lessons yet to teach.

Acknowledgments

I thank the late Nancy Cole and my colleagues Pat Higgins, Genevieve Michel, Hewson Swift, David Leach, Jean-Pierre Bouché, Dina Newman, Pamela Westfall, Pamela Brinkley, Robert Alonso, Clara Hsu, David Trubatch, Mike Bayley, Teresa Reyes, and Dawn Holliday, all of whom have contributed in important ways to the research reported here. My research on *E. coli* colony development and the physiological regulation of genetic change has been funded by the National Science Foundation.

References

Achtman, M. (1975) Mating aggregates in *Escherichia coli* conjugation. *J. Bacteriol.* **123**:505–515.

Adhya, S., and Shapiro, J.A. (1969) The galactose operon of *E. coli* K-12. I: structural and pleiotropic mutants of the operon. *Genetics* **62**:231–248.

Adler, J. (1966) Chemotaxis in bacteria. *Science* **153**:708–716.

Ben-Jacob, E., Shmueli, H., Schochet, O., and Tenenbaum, A. (1992) Adaptive self-organization during growth of bacterial colonies. *Physica A* **187**:378–424.

Blat, Y., and Eisenbach, M. (1995) Tar-dependent and -independent pattern formation by *Salmonella typhimurium*. *J. Bacteriol.* **177**:1683–1691.

Bray, D., Bourret, R.B., and Simon, M.I. (1993) Computer simulation of the phosphorylation cascade controlling bacterial chemotaxis. *Mol. Biol. Cell* **4**:469–482.

Bremer, E., Silhavy, T.J., and Weinstock, G.M. (1988) Transposition of λplacMu is mediated by the A protein altered at its carboxy-terminal end. *Gene* **71**:177–186.

Budrene, E.O., and Berg, H.C. (1991) Complex patterns formed by motile cells of *Escherichia coli. Nature* **349**:630–633.

Budrene, E.O., and Berg, H.C. (1995) Dynamics of formation of symmetrical patterns by chemotactic bacteria. *Nature* **376**:49–53.

Cairns, J., and Foster, P.L. (1991) Adaptive reversion of a frameshift mutation in *Escherichia coli. Genetics* **128**:695–701.

Cairns, J., Overbaugh, J., and Miller, S. (1988) The origin of mutants. *Nature* **335**: 142–145.

Castilho, B.A., Olfson, P., and Casadaban, M. (1984) Plasmid insertion mutagenesis and *lac* gene fusion with Mini-Mu bacteriophage transposons. *J. Bacteriol.* **158**: 488–495.

Cavalli-Sforza, L.L., and Lederberg, J. (1956) Isolation of preadaptive mutants in bacteria by sib selection. *Genetics* **41**:367.

Costerton, J.W., Cheng, K.-J., Geesey, G.G., et al. (1987) Bacterial biofilms in nature and disease. *Annu. Rev. Microbiol.* **41**:435–464.

Foster, P.L., and Cairns, J. (1992) Mechanisms of directed mutation. *Genetics* **131**:783.

Foster, P.L., and Cairns, J. (1994) The occurrence of heritable Mu excisions in starving cells of *Escherichia coli. EMBO J.* **13**:5240–5244.

Foster, P.L., and Trimarchi, J.M. (1994) Adaptive reversion of a frameshift mutation in *Escherichia coli* by simple base deletions in homopolymeric runs. *Science* **265**: 407–409.

Foster, P.L., and Trimarchi, J.M. (1995) Adaptive reversion of an episomal frameshift mutation in *Escherichia coli* requires conjugal functions but not actual conjugation. *Proc. Natl. Acad. Sci. USA* **92**:5487–5490.

Fujikawa, H., and Matsushita, M. (1989) Fractal growth of *Bacillus subtilis* on agar plates. *J. Phys. Soc. Jpn.* **58**:3875–3878.

Fuqua, W.C., Winans, S.C., and Greenberg, E.P. (1994) Quorum sensing in bacteria: the LuxR-LuxI family of cell density-responsive transcriptional regulators. *J. Bacteriol.* **176**:269–275.

Galitski, T., and Roth, J.R. (1995) Evidence that F plasmid transfer replication underlies apparent adaptive mutation. *Science* **268**:421–423.

Geuskens, V., Mhammedi-Alaoui, A., Desmet, L., and Toussaint, A. (1992) Virulence in bacteriophage Mu: a case of trans-dominant proteolysis by the *E. coli* Clp serine protease. *EMBO J.* **13**:5121–5127.

Gottesman, S., Squires, C., Pichersky, E., et al. (1990) Conservation of the regulatory subunit for the Clp ATP-dependent protease in prokaryotes and eukaryotes. *Proc. Natl. Acad. Sci. USA* **87**:3513–3517.

Graham-Smith, G.S. (1910) The division and post-fission movements of bacilli when grown on solid media. *Parasitology* **3**:17–53.

Hadley, P. (1927) Microbic dissociation. *J. Infect. Dis.* **40**:1–312.

Hall, B.G. (1988) Adaptive evolution that requires multiple spontaneous mutations. I. Mutations involving an insertion sequence. *Genetics* **120**:887–897.

Harris, R.S., Longerich, S., and Rosenberg, S.M. (1994) Recombination in adaptive mutation. *Science* **264**:258–260.

Hayes, W. (1968) *The Genetics of Bacteria and their Viruses*, 2nd ed. Wiley, New York.

Higgins, N.P. (1992) Death and transfiguration among bacteria. *Trends Biochem. Sci.* **17**:207–211.

Huisman, G.W., and Kolter, R. (1994) Sensing starvation: a homoserine lactone-dependent signalling pathway in *Escherichia coli*. *Science* **265**:537–539.

Jacob, F., and Wollman, E.L. (1961) *Sexuality and the Genetics of Bacteria*. Academic Press, Orlando, FL.

Kaiser, D., and Losick R. (1993) How and why bacteria talk to each other. *Cell* **73**: 873–885.

Katayama, Y., Gottesman, S., Pumphrey, J., Rudikoff, S., Clark, W.P., and Maurizi, M.M. (1988) The two-component, ATP-dependent Clp protease of *Escherichia coli*. *J. Biol. Chem.* **263**:15226–15236.

Kell, D.B., Kaprelyants, A.S., and Grafen, A. (1995) Pheromones, social behaviour and the functions of secondary metabolism in bacteria. *Trends Ecol. Evol.* **10**:126–129.

Kelly, K. (1994) *Out of Control*. Addison-Wesley, Reading, MA.

Kohara, Y., Akiyama, K., and Isono, K. (1987) The physical map of the whole *E. coli* chromosome: application of a new strategy for rapid analysis and sorting of a large genomic library. *Cell* **50**:495–508.

Kuhn, T. (1960) *The Structure of Scientific Revolutions*. University of Chicago Press, Chicago.

Legroux, R., and Magrou, J. (1920) État organisé des colonies bactériennes. *Ann. Inst. Pasteur.* **34**:417–431.

Leinfelder, W., Forchhammer, K., Zinoni, F., Sawers, G., Mandrand-Berthelot, M.-A., and Böck, A. (1988) *Escherichia coli* genes whose products are involved in selenium metabolism. *J. Bacteriol.* **170**:540–546.

Levy, S., Clowes, R.C., and Koenig, E.L. (1981) *Molecular Biology, Pathogenicity and Ecology of Bacterial Plasmids*. Plenum, New York.

Luria, S., and Delbrück, M. (1943) Mutations of bacteria from virus sensitivity to virus resistance. *Genetics* **28**:491.

Ma, M., and Eaton, J.W. (1992) Multicellular oxidant defense in unicellular organisms. *Proc. Natl. Acad. Sci. USA* **89**:7924–7928.

Mackerness, C.W., Colbourne, J.S., Dennis, P.L.J., Rachwal, T., and Keevil, C.W. (1993) Formation and control of coliform biofilms in drinking water distribution systems. In *Microbial Biofilms: Formation and Control*, S.P. Denyer, S.P. Gorman, and M. Sussman, (eds.), pp. 217–226. Society of Applied Bacteriology Technical Series **30**. Blackwell, Oxford.

Maenhaut-Michel, G., and Shapiro, J.A. (1994) The roles of starvation and selective substrates in the emergence of *araB-lacZ* fusion clones. *EMBO J.* **13**:5229–5239.

Magasanik, B., and Neidhardt, F.C. (1987) Regulation of carbon and nitrogen utilization. In *Escherichia coli and Salmonella typhimurium* F.C. Neidhardt et al., (eds.), pp. 1318–1325, American Society for Microbiology, Washington, DC.

Meyer, T.F. (1987) Molecular basis of surface antigen variation in *Neisseria*. *Trends Genet.* **3**:319–324.

Mhammedi-Alaoui, A., Pato, M., Gama, M.-J.,and Toussaint, A. (1994) A new component of bacteriophage Mu replicative transposition machinery: the *E. coli* ClpX protein. *Mol. Microbiol.* **11**:1109–1116.

Mittler, J., and Lenski, R.E. (1990) Further experiments on excisions of Mu from *Escherichia coli* MCS2 cast doubt on directed mutation hypothesis. *Nature* **344**: 173–175.

Newcombe, H.B. (1949) Origin of bacterial variants. *Nature* **164**:150.

Nossal, R. (1972) Growth and movement of rings of chemotactic bacteria. *Exp. Cell Res.* **75**:138–142.

Peters, J.E., Bartoszyk, I.M., Dheer, S., and Benson, S.A. (1996) Redundant homosexual F transfer facilitates selection induced reversion of plasmid mutations. *J. Bacteriol.* **178**:3037–3043.

Peters, J.E., and Benson, S.A. (1995) Redundant transfer of F' plasmids occurs between *Escherichia coli* cells during nonlethal selections. *J. Bacteriol.* **177**:847–849.

Pirt, S.J. (1967) A kinetic study of the mode of growth of surface colonies of bacteria and fungi. *J. Gen. Microbiol.* **47**:181–197.

Radicella, J.P., Park, P.U., and Fox, M.S. (1995) Adaptive mutation in *Escherichia coli*: a role for conjugation. *Science* **268**:418–420.

Redfield, R.J. (1988) Evolution of bacterial transformation: is sex with dead cells ever better than no sex at all? *Genetics* **119**:213.

Rosenberg, S. (1994) In pursuit of a molecular mechanism for adaptive mutation. *Genome* **37**:893–899.

Rosenberg, S.M., Longerich, S., Gee, P., and Harris, R.S. (1994) Adaptive mutation by deletions in small mononucleotide repeats. *Science* **265**:405–407.

Salyers, A.A., and Whitt, D.D. (1994) *Bacterial Pathogenesis.* American Society for Microbiology, Washington, DC.

Shapiro, J.A. (1979) A molecular model for the transposition and replication of bacteriophage Mu and other transposable elements. *Proc. Natl. Acad. Sci. USA* **76**: 1933–1937.

Shapiro, J.A. (1984a) Observations on the formation of clones containing *araB-lacZ* cistron fusions. *Mol. Gen. Genet.* **194**:79–90.

Shapiro, J.A. (1984b) The use of Mu*dlac* transposons as tools for vital staining to visualize clonal and non-clonal patterns of organization in bacterial growth on agar surfaces. *J. Gen. Microbiol.* **130**:1169–1181.

Shapiro, J.A. (1984c) Transposable elements, genome reorganization and cellular differentiation in gram-negative bacteria. *Symp. Soc. Gen. Microbiol.* **36**(Part 2): 169–193.

Shapiro, J.A. (1985a) Intercellular communication and genetic change in bacteria. In *Engineered Organisms in the Environment: Scientific Issues*, H.O. Halvorson, D. Pramer, and M. Rogul (eds.), pp. 63–69, American Society for Microbiology, Washington, DC.

Shapiro, J.A. (1985b) Photographing bacterial colonies. *Am. Soc. Microbiol. News* **51**: 62–69.

Shapiro, J.A. (1985c) Scanning electron microscope study of *Pseudomonas putida* colonies. *J. Bacteriol.* **164**:1171–1181.

Shapiro, J.A. (1987) Organization of developing *Escherichia coli* colonies viewed by scanning electron microscopy. *J. Bacteriol.* **169**:142–156.

Shapiro, J.A. (1988) Bacteria as multicellular organisms. *Sci. Am.* **256**:82–89.

Shapiro, J.A. (1992a) Concentric rings in *E. coli* colonies. In *Oscillations and Morphogenesis*, L. Rensing (ed.), pp. 297–310, Marcel Dekker, New York.

Shapiro, J.A. (1992b) Differential action and differential expression of *E. coli* DNA polymerase I during colony development. *J. Bacteriol.* **174**:7262–7272.

Shapiro, J.A. (1992c) Natural genetic engineering in evolution. *Genetica* **86**:99–111.

Shapiro, J.A. (1993) A role for the Clp protease in activating Mu-mediated DNA rearrangements. *J. Bacteriol.* **175**:2625–2631.

Shapiro, J.A. (1994) Pattern and control in bacterial colonies. *Sci. Progr.* **76**:399–424.

Shapiro, J.A. (1995a) Adaptive mutation: who's really in the garden? *Science* **268**: 373–374.

Shapiro, J.A. (1995b) The significances of bacterial colony patterns. *Bioessays* **17**: 597–607.

Shapiro, J.A., and Higgins, N.P. (1988) Variation of β-galactosidase expression from Mu*dlac* elements during the development of *E. coli* colonies. *Ann. Inst. Pasteur* **139**:79–103.

Shapiro, J.A., and Higgins, N.P. (1989) Differential activity of a transposable element in *E. coli* colonies. *J. Bacteriol.* **171**:5975–5986.

Shapiro, J.A., and Hsu, C. (1989) *E. coli* K-12 cell–cell interactions seen by time-lapse video. *J. Bacteriol.* **171**:5963–5974.

Shapiro, J.A., and Leach, D. (1990) Action of a transposable element in coding sequence fusions. *Genetics* **126**:293–299.

Shapiro, J.A., and Trubatch, D. (1991) Sequential events in bacterial colony morphogenesis. *Physica D* **49**:214–223.

von Meyerneberg, K., Jørgensen, B.B., Neilsen, J., and Hansen, F.G. (1982) Promoters of the *atp* operon coding for the membrane-bound ATP synthase of *Escherichia coli* mapped by Tn*10* insertion mutations. *Mol. Gen. Genet.* **188**:240–248.

Waldrop, M. (1992) *Complexity*. Simon & Schuster, New York.

Watanabe, T. (1963) Infective heredity of multiple drug resistance in bacteria. *Bacteriol. Rev.* **27**:87–115.

Way, J.C., Davis, M.A., Morisato, D., Roberts, D.E., and Kleckner, N. (1984) New Tn10 derivatives for transposon mutagenesis and for construction of *lacZ* operon fusions by transposition. *Gene* **32**:369–379.

Winogradsky, S. (1949) *Microbiologie du Sol: Problemes et Methodes*. Masson, Paris.

Yarmolinsky, M.B (1995) Programmed cell death in bacterial populations. *Science* **267**: 836–837.

II.

Intercellular Communication

3

Pheromone-Inducible Conjugation in *Enterococcus faecalis*

Mating Interactions Mediated by Chemical Signals and Direct Contact

ROBERT E. RUHFEL, BETTINA A. B. LEONARD, & GARY M. DUNNY

The most interesting form of gram-positive conjugation from the standpoint of cell–cell interaction is the transfer of pheromone-inducible conjugative plasmids typified by the well studied plasmids pCF10 (Dunny et al., 1981) and pAD1 (Tomich et al., 1979). *Enterococcus faecalis* cells containing this class of plasmid sample the environment for the presence of potential recipient cells. Plasmid-free cells broadcast their presence by means of an array of small hydrophobic peptide pheromones, typically seven or eight amino acid residues in length (Mori et al., 1988); these substances are produced by the cell and released into the growth medium. Plasmid-containing cells bind these pheromones by means of pheromone receptors, which are highly specific, apparently binding only one pheromone with high affinity (Ruhfel et al., 1993; Tanimoto et al., 1993). Once bound, pheromone is transported into the cell where it interacts with cytoplasmic components of the plasmid's regulatory system. Pheromone binds internally to a putative plasmid replication protein, possibly readying the plasmid for transfer; the pheromone also disables a negative regulatory system (Leonard et al., 1994), an action that allows the positive regulatory machinery to be activated, resulting in the expression of a cell-surface adhesin (Kao et al., 1991; Chung and Dunny, 1992). Induced donor cells bind all *E. faecalis* cells, plasmid-free or not, by the interaction of the surface adhesin and lipotechoic acid-containing receptor of the cell wall (Ehrenfeld et al., 1986; Trotter and Dunny, 1990; Bensing and Dunny, 1993). Additionally, a surface exclusion function is expressed (Dunny et al., 1985). Finally, through an undefined mechanism the plasmid DNA is passed from the donor cell to the recipient. Once this passage has occurred the cell is converted to a donor. As such it expresses several genes, including those that prevent the cell's own endogenous pheromone from inducing this newly acquired plasmid (Clewell et al., 1987; An and Clewell, 1994; Leonard et al., 1994; Nakayama et al., 1994b).

We now describe each of these steps in more detail by following the fate of pheromone as it moves from recipients to pheromone-inducible plasmid-containing donor cells. As we look at this simple form of bacterial conversation we describe how donor cells receive the recipient's pheromone signal and the subsequent interactions with regulatory machinery that lead to plasmid transfer.

Elaboration of Pheromone by Recipient Cells: First Step in Cell–Cell Interaction

Of all the events leading to pheromone-inducible plasmid transfer, the synthesis and export of pheromone by recipients is least well understood. Cells that have presumably never encountered a pheromone-inducible plasmid excrete five known pheromones (probably more). Those that have been characterized have been found to be highly hydrophobic heptameric or octameric peptides. Is not known why these cells produce pheromones, though it has been postulated that they are fragments of leader sequences that have been released from the cell following processing of secreted proteins (Dunny, 1990). Indeed, the leader peptide from the *Staphylococcus aureus* pSK41 encoded lipoprotein TraH contains sequences nearly identical to the pheromone cAD1 and in general is similar to the leader peptide of iAD1 (Firth et al., 1993), the pAD1 encoded competitive inhibitor of cAD1.

Alternatively, these peptides might be synthesized and secreted specifically for the purpose of signaling cellular information to the cell itself as well as within the cell population. There has been a suggestion that the oligopeptide permease system (Opp) of *Bacillus subtilis* may operate as a growth-monitoring link communicating the status of cell wall synthesis to cytoplasm through the transport of cell wall peptides (Perego et al., 1991).

Studies on the synthesis of pheromone have shown that some pheromones are differentially affected by cultural conditions (Weaver and Clewell, 1991). Aeration of cultures containing plasmid-free *E. faecalis* strains resulted in a marked decrease in measurable cAD1 pheromone in filtrates. There was, however, no such decrease in the activity of two other assayed pheromones, cPD1 and cAM373. It could represent a purposeful regulation of cAD1 for communication of information, such as the oxygen tension of the culture; it could also reflect the regulation of some surface protein from which cAD1 represents a cleavage product. One possibility does not exclude the other.

Unlike the *B. subtilis* competence pheromone gene, the sequences responsible for encoding *E. faecalis* sex pheromones have not been found. The chromosomally encoded competence pheromone of *B. subtilis* is a peptide of nine or ten amino acids containing a modified residue (Magnuson et al., 1994). This peptide, ComX, does not have a leader peptide and therefore must be secreted by a pathway different from the major secretory pathway. Interestingly, the competence pheromone of *Streptococcus pneumoniae* is probably transported by an ATP-dependent export system (Hui and Morrison, 1991).

Bacterial chromosomes, phage genomes, and plasmids often incorporate genes from other systems that can be exploited for novel purposes. It may be that the sex pheromone system evolved from a signaling system utilized by plasmid-free *E. faecalis* cells. This type of chromosomal system would be difficult to identify without an easily observable phenotype.

Mutants that fail to produce pheromones have not been found, although they have been sought. One mutant strain that was isolated on the basis of failure to produce pheromone (Ike et al., 1983) appears to be altered in peptide export rather than synthesis (B. Leonard, unpublished observations). These data suggest that pheromones play some essential role in maintenance of the cell.

Pheromone Binding: Receiving the Message

The presence of a plasmid-free recipient is sensed by potential donors through specific pheromone binding by a receptor. Each pheromone-responsive plasmid encodes its own receptor, which imparts a selectivity that results in little or no cross-reactivity between the various pheromones at biologically relevant pheromone concentrations (Ruhfel et al., 1993; Tanimoto et al., 1993). Because the pheromone receptors probably utilize the chromosomally encoded peptide transport machinery other than the binding protein, we briefly discuss what is known about these systems. Several types of peptide transport system are known. Within the enteric bacteria *S. typhimurium* and *E. coli* are three genetically distinct peptide transport systems (Hiles et al., 1987): dipeptide permease (Dpp), tripeptide permease (Tpp), and oligopeptide permease (Opp). The major one is the Opp system, which serves a nutritional role as well as having responsibility for recycling cell wall peptides during growth.

The gram-positive species *B. subtilis* (Perego et al., 1991) and *Lactococcus lactis* (Tynkkynen et al., 1993) contain an Opp system with constituents homologous to those from the enteric bacteria. Here, as in the enteric system, OppB and OppC are membrane-spanning proteins that receive the peptide from the cell surface lipoprotein OppA. Peptide transport is coupled to ATP hydrolysis by the cytoplasmic proteins OppD and OppF. In *L. lactis* it has been shown that Opp is required for growth on milk and can transport oligopeptide up to eight residues in length (Tynkkynen et al., 1993). For *B. subtilis*, in addition to any nutritional role it may serve, Opp is required for regulation of competence and under some conditions sporulation (Magnuson et al., 1994).

The pheromone-binding proteins encoded by pCF10 and pPD1 (Nakayama et al., 1994a) display a strong degree of homology to each other, to pAD1, and to a lesser extent to the oligopeptide-binding proteins of *B. subtilis* (Perego et al., 1991; Rudner et al., 1991); *Escherichia coli* (Kashiwagi et al., 1990), and *Salmonella typhimurium* (Hiles et al., 1987). Like the *B. subtilis* OppA protein, the predicted product encoded by these plasmids contains a lipoprotein signal sequence (LxxCx) 19 residues from the amino-terminus and is therefore likely anchored to the membrane by a lipid moiety. In *B. subtilis* the lipoprotein signal sequence is cleaved, leaving the amino-terminal cysteine residue of the mature protein covalently modified by a thioester linkage of diacylglycerol, which contains an amide-linked fatty acid.

Interruption in the pheromone receptor genes of both pCF10 (Leonard et al., 1994) and pAD1 (Tanimoto et al., 1993)—*prgZ* and *traC*, respectively— still allows the cell to be induced by pheromone, but the amount of pheromone required is increased. The homology between the highly discriminatory pheromone receptor and the nonspecific chromosomal oligopeptide permease binding protein likely explains the ability of receptor mutants to respond to pheromone when supplied in significantly higher concentrations. It can be reasoned that pheromone may enter the cell by an *E. faecalis* encoded oligopeptide permease system, binding to chromosomally encoded OppA protein, which likely shares the broad specificity of the other well studied systems. Indeed, no other peptide transport proteins have been identified on pCF10 or pAD1. The plasmid encoded binding proteins utilize components of the chromosomally encoded peptide transport machinery to import pheromone (Leonard et al., 1996).

The pheromone-binding proteins have been shown to be similar to the OppA protein of the oligopeptide permease transport systems of organisms of various genera. The closely related *B. subtilis* OppA protein shares 35% identical amino acids with a large number of conserved substitutions. The pheromone receptors also show a high degree of identity to *S. typhimurium* OppA. The crystal structure of the oligopeptide-binding protein of *S. typhimurium* (OppA) bound to a peptide ligand has been determined, as have the residues that interact with tripeptide and tetrapeptide ligand substrates (Tame et al., 1994). A comparison of these crucial amino acids on an alignment of PrgZ from pCF10 and the TraC molecules from pPD1 and pAD1 sheds some light on the binding interactions of pheromone and receptor.

The structure of *S. typhimurium* OppA comprises three domains. Domains I and III surround the ligand and are linked by two regions that facilitate opening and closing. Domain II has relatively little contact with the ligand, and no function for this domain has been proposed. The alignment of the receptors PrgZ and the two TraCs, shown in Figure 3.1, contains only the portions of the alignment with *S. typhimurium* OppA that interact with the peptide ligand. The segments corresponding to the *S. typhimurium* OppA domains I, II, and III are indicated. Within domain II, the three pheromone receptors are nearly identical. This region also contains the highest degree of homology between the pheromone receptors and *S. typhimurium* OppA (34%). It is tempting to speculate that this area is the portion of the pheromone receptors that interacts with the *E. faecalis* membrane-spanning constituents of the oligopeptide permease system, homologs of OppB and OppC. If pCF10, pAD1, pPD1, and the other pheromone-inducible plasmids utilize chromosomally encoded OppB and OppC homologs to transport the pheromone into the

```
pCF10 PrgZ    CgtntaTKdsqdateKkVEQVAT1tAgTPVQSLDPATAVD
pPD1  TraC    CgintaTKdsqdvteKkVEQVAT1tAgTPVQSLDPATAVD
pAD1  TraC    CngkegTKn--nsksKeVEQVATfsAmTPVQSLDPATAVD
              ........................(I)...........................

     V34
41   QTSITLLANVMEGLYRLDeKNQPQPAIAAGQPKvSNNGKTYTIVIRDGAKWsDGTqITAsD
     QTSITLLANVMEGLYRLDqKNQPQPAIAAGQPKvSNNGKTYTIVIRDGAKWsDGTqITAsD
     QTSITLLANVMEGLYRLDeKNQPQPAIAAGQPKiSNNGKTYTIVIRDGAKWaDGTdITAdD
     ..........(I)...-------------------------------------(II)-------------------------

102  FVaAWQRVVDPKTvSPnVELFSAIKNAKEIASGKQaKdTLAVKSiGeKTlEIELvePTPYF
     FVaAWQRVVDPKTaSPnVELFSAIKNAKEIASGKQvKdTLAVKSiGeKTlEIELvePTPYF
     FVtAWQRVVDPKTaSPsVELFSAIKNAKEIASGKQsKeTLAVKSnGnKTiEIELekPTPYF
     --------------------------------------------------(II)----------------------

163  TDLLsLTAYYPVQQKAIKEYGKDYGvSqKaIVTNGAFNLTnLEGVGTSDKWTISKNKEYWD
     TDLLsLTAYYPVQQKAIKEYGKDYGtSkKaIVTNGAFNLTiLEGVGTSDKWTISKNKEYWD
     TDLLaLTAYYPVQQKAIKEYGKDYGtSqKsIVTNGAFNLTsLEGVGTSDKWTISKNKEYWD
     --------------------------------------------------------(II)---------------.....(I)....

                                              C271    E276
224  qKdVSMdKINFQVVKEINTGINLYNDGQLDeaPlaGEYAKQYKkdKEySTTLMANTMfLEm
     kKdVSMdKINFQVVKEINTGINLYNDGQLDdaPlaGEYAKQYKkdKEySTTLMANTMfLEm
     qKeVSMeKINFQVVKEINTGINLYNDGQLDdtPvtGEYAKQYKdnKEfSTTLMANTMyLEl
     ................................................(I)...................***(III)*

                                K307
285  NQtgeNklLqNKNVRKAInYaIDReslvkklLDNGSvasVGVVpkemAfnPvnKKDFANek
     NQtgeNKlLkNKNVRKAIsYpIDRdslveklLDNGSipsVGVVpekmAynPktKKDFANek
     NQrekNsiLqNKNVRKAIsYaIDRsnyakniLDNGSisaVGVVakdvAfdPstKKDFANkm
     ************************************************(III)************************

                                 H371
346  LVefnkkqAeeYWdkaKkeidlskntsLdllvsdgEfeKKAgEflQGQLqdsLeGLKvtvT
     LVeynkkqArtYWetvKskdsvsekleLdifvgdgEfeKKAgEflQGQLeenLeGLKvniT
     LVhfdtekAqsYWnkaKkelnikeqvtLniltneeEttKKAaEyiQGQLeenLkGLKitiT
     ************************************************(III)************************

       W397 L401 RH405      R GWC D419
407  PiPANVfmERltKkDFtlSLSGWQADYaDPiSFLaNFEtNSpmNhGGYSNknYDEllKDsS
     PvPANVfmERltKkDFaiSLSGWQADYaDPiSFLaNFEtNSplNhGGYSNknYDEliKDtS
     PvPANVqiERvmKhDFtiSLSGWQADYpDPmSFLgNFEsNSvlNfGGYSNtkYDEylKDtS
     ************************************************(III)************************

468  sKRWqeLKKAEkllinDmGVvPifQVGtakLeKskikNvlmHsiGAkYDYKKMriek
     sKRWqeLKKAEkividDaGViPvfQVGiarLqKntirNlviHpvGArYDYKKMmvqn
     dKRWksLKKAEellleDaGVvPflQVGnskLlKstmrNietHsiGAkYDYKKMkrle
     *********************(III)***********.................(I)....................
```

FIGURE 3.1 Alignment of predicted amino acid sequence of the pheromone receptors from pCF10 (PrgZ), pPD1 (TraC), and pAD1 (TraC). Numbering of pCF10 begins with the presumed amino terminal lipid-modified cysteine. The boldface lowercase letters indicate the residues that are not conserved in all receptor proteins. Underlining indicates the three domains of *Salmonella typhimurium* OppA homolog determined by x-ray crystallography and aligned with the pheromone receptors (domain I ....), (domain II --------), and (domain III ....). The amino acid residues of the *S. typhimurium* OppA that interact with those of the peptide ligands are noted above. Some of the important residues of the pheromone receptor that specifically interact with the side chains of the pheromone peptide may be inferred from the three binding pockets identified in OppA. A valine residue (V^{34}) and the cysteine residues (C^{271}, C^{417}), which form a disulfide bridge in OppA, interact with the first and third side chains of the ligand's β and γ carbons. The pocket for the second side chain in OppA contains two tryptophan residues (W^{397}, W^{416}) and a leucine residue (L^{401}), forming a collar that then widens and contains a glutamic acid residue (E^{32}) in an ion pair with a histidine residue (H^{405}) and a glutamic acid residue (E^{276}) in an ion pair with an arginine residue (R^{404}).

cell, one would expect that the sequences that mediate docking of the phero-mone receptors to the outer surface of this oligopeptide transport channel are constrained to resist genetic drift and must match as closely as possible the relevant sequences on *E. faecalis* OppA homolog. This constraint may also slow the sequence divergence of functionally related proteins between the gen-era and perhaps be the reason for the higher level of homology for *S. typhi-murium* OppA to this portion of the plasmid-determined receptors.

The backbone of the peptide ligand, when bound within the OppA crystal structure, has interactions with residues 32–34, 415, and 417 (indicated in Figure 3.1 as superscript on the alignment). The carboxy-terminal carboxylate group of the various peptide ligands forms salt bridges to an arginine residue (R^{413}) with the tripeptide, a histidine residue (H^{371}) with the tetrapeptide, and possibly a lysine residue (K^{307}) with a pentapeptide. More importantly and perhaps the most interesting feature of the *S. typhimurium* OppA is an aspartic acid residue (D^{419}), which forms a salt bridge with the α-amino group of the peptide ligand. The binding of these ligands to OppA provides some insight into the possible mechanism of pheromone binding to its receptor.

The pheromone cCF10, H_3N-Leu-Val-Thr-Leu-Val-Phe-Val-COOH, can be modified by the addition of a carboxy-terminal tyrosine to yield a fully active eight-residue peptide (Leonard et al., 1994, Leonard, et al., 1996). If this pep-tide is tethered by the carboxy-terminal end to Sephadex 4B, the ability to bind the receptor is not impaired. If it is tethered by the amino-terminus, however, it loses the ability to bind receptor. By analogy to OppA, an important recog-nition feature of the pheromone must be the interaction of the amino-terminal residue with aspartic acid to form an analogous salt bridge. This action pre-sumably establishes the register in which the pheromone binds to the receptor, allowing highly discriminatory interactions of the receptor with the hydropho-bic amino acid side chains of the pheromone.

Although it is not possible at this time to determine the specific interactions between the side chains of the amino acid residues of the pheromone with certain amino acids of the receptor, it does seem that the specificity is contained in the region homologous to domains I and III of *S. typhimurium* OppA. Once they are known, it will be interesting to compare the amino acid sequences of the other pheromone receptors and the *E. faecalis* chromosomal OppA with those aligned here.

Prevention of Self-Induction: Keeping Transfer Functions Turned Off in the Absence of Recipients.

It has been shown that two genes are required to prevent donor cells from responding to internally produced pheromone. One gene encodes a competitive inhibitor peptide (Clewell et al., 1987; Nakayama et al., 1994b), and the other encodes a putative membrane-spanning protein that works in a less obvious way (Ruhfel et al., 1993; An and Clewell, 1994; Leonard et al., 1994). In the pAD1 system, this gene, TraB, is referred to as a ''pheromone shut-down'' gene.

The mechanisms for preventing autoinduction in the pCF10 and pAD1 systems are similar but have some differences. Cells carrying pAD1 do not excrete active pheromone but excrete a pheromone inhibitor, iAD1. Enough inhibitor peptide is found in these pAD1-containing cell filtrates to inhibit the response of indicator cells to the pheromone cAD1. Unfractionated filtrates of cultures of strains carrying pCF10 have no significant inhibitory effect on indicator cells. However, pCF10 does in fact synthesize an inhibitor. Examination of the DNA sequence revealed a gene, *prgQ*, that encodes an inhibitor peptide (Nakayama et al., 1994b). A high performance liquid chromatography (HPLC) assay has been used to show that the lack of pheromone activity in donor pCF10 *E. faecalis* cultures is due to secretion of a mixture of iCF10 and cCF10 rather than abolition of cCF10 secretion (Nakayama et al., 1994b). The lack of pheromone activity can be attributed to the inhibitor's ability to mask the effect of pheromone. Interestingly, the amount of cCF10 excreted by *E. faecalis* is apparently unchanged with the acquisition of pCF10.

A chemically synthesized peptide corresponding to the seven residues of the carboxy-terminal portion deduced from the nucleotide sequence of *prgQ* confers inhibitor activity, whereas neither peptide composed of the six or eight terminal sequences has any activity. By analogy to the studies with tethered pheromone mentioned above, it is likely that the amino-terminal sequences are critical for orienting this peptide in the receptor and cannot be modified.

The sequences of the inhibitor peptide gene are known for pAD1 (Clewell et al., 1990), pCF10 (Nakayama et al., 1994b), and pPD1 (Nakayama et al., 1994a). Each encodes a precursor slightly larger than 20 amino acids in length, of which the carboxy-terminal seven to eight residues are cleaved to yield the active molecule, designated iCF10, iPD1, and iAD1, respectively. Because the precursor molecule resembles the leader sequence of excreted proteins, the active inhibitor peptide may not accumulate within the cell but, rather, associates with the secretory machinery and is processed outside at the cell surface. Indeed, no unusual plasmid-encoded processing is required, as mature peptides of both iCF10 and iAD1 are excreted from *E. coli* cells carrying only these genes cloned onto shuttle vectors.

When the inhibitor peptides are compared with their corresponding pheromones, it can be seen that about 25–50% of the amino acid residues are identical. Hence it can be surmised that these amino acids, of which one or more are always leucine residues, contribute to recognition and binding. It has been shown that iCF10 does bind to the pheromone receptor PrgZ (Nakayama et al., 1994b), and there is no reason to believe that it is not transported into the cell where it may bind other components of the pCF10 regulatory system.

There have been several attempts to clone pheromone genes from the chromosome, all of which have been unsuccessful. A reaction that changed several bases of the *prgQ* sequence to generate one that would encode the pheromone peptide was carried out by the polymerase chain reaction (Fig. 3.2). When a plasmid carrying this mutated gene was inserted into either *E. coli* or *E. fae-*

ACGACACTCGAAGATGTGTTTATTAAGCTATATCCCTTTTTTTTAAAAAAAAAATACATATTTTAGTT
left primer

```
                                          M   K   T   T   L   K   K
GAAAATATAATACTTAGATGTTAAGATGTTTTTTATAGGAGGGGTGTAAATGAAAACCACTCTAAAAA


    L   S   R   Y   I   A   V   V   I   l   v   T   L   v   F   v        EcoRI
AACTATCAAGATATATAGCTGTTGTAATTttAgTAACCTTAgTATTTgTCTGATAGaattcatca
                             right primer
```

FIGURE 3.2 Mutation of the *prgQ* gene of pCF10 encoding the iCF10 inhibitor peptide by polymerase chain reaction gene amplification with a primer designed to alter the sequence to encode cCF10. The processed protein is Leu-Val-Thr-Leu-Val-Phe-Val, cCF10 rather than Ala-Ile-Thr-Leu-Ile-Phe-Ile, iCF10. Altered bases and resultant altered amino acid residues are indicated by boldface lowercase letters.

calis, active pheromone was produced. This maneuver demonstrates that the difficulty of identifying, cloning, and expressing the chromosomal gene for pheromone is not a result of some lethal aspect of unregulated synthesis of this molecule. Our recent unpublished results suggest that a pheromone molecule may be processed to cCF10 by a specific maturation mechanism. However, we were not able to replace the wild-type *prgQ* gene with the mutant derivative in a plasmid containing the other regulatory genes described below without generating large deletions and plasmid instability.

The other genes responsible for preventing self-induction are *prgY* in the pCF10 system (Ruhfel et al., 1993; Leonard et al., 1994) and *traB* in the pAD1 system (An and Clewell, 1994). Inactivation of these components of the regulatory machinery results in a constitutively clumpy phenotype for each system. It has been shown that in the pCF10 system there is essentially the same amount of cCF10 pheromone in the filtrates of *E. faecalis* in cultures regardless of whether they contain pCF10. However, the cancellation of cCF10 activity by iCF10 in plasmid-containing cultures is not sufficient to prevent self-induction. The activity exerted by PrgY appears to be on the level of cCF10 found inside the cell (Leonard et al., 1994). Membrane fractions of cells that do not contain pCF10 can produce at least 100 times more cCF10 than the amount of excreted pheromone. The membrane fractions of cells that do contain pCF10 have no activity at all. It appears that PrgY may regulate or sequester cell-associated pheromone.

The result of the function of these two gene products is that pCF10-containing cells can detect as few as one to five molecules of exogenous cCF10 per donor cell within the background of cCF10 and iCF10 molecules released by these donors. The binding and subsequent internalization of pheromone initiates cessation of negative regulation and allows expression of positive regulatory genes. The tethered pheromone experiments have shown that cCF10 binds to PrgW, several host proteins, and *prgQ* RNA (Leonard et al., 1994; Leonard et al., 1996). These substances, as well as PrgN and PrgX, may be responsible for the negative regulation (Hedberg et al., 1996).

Positive Regulation: *cis*-Activation of Transfer Functions

The pCF10 *prgB* gene, which encodes the aggregation substance protein Asc10, is transcribed in wild-type donor cells when the pheromone signaling process described above abolishes negative control. In *E. faecalis* cells carrying pCF10 derivatives where the negative control region has been disrupted genetically, *prgB* is expressed constitutively. Simple abolition of negative control is insufficient for this expression; expression of a separate positive control region is also required. This region spans genes denoted *prgQ-S* on the map of pCF10 shown in Figure 3.3. Genetic and transcriptional analysis of this region (Chung and Dunny, 1994) has shown that products of the *prgQ* and *prgS* genes (see the map of pCF10 shown in Figure 3.3) are required for positive control. There are several unusual features of this regulatory system. As can be seen on the map, the target gene is a considerable distance from the regulatory genes, suggesting that *trans*-acting positive regulatory gene products from this region might be involved in the activation process. Even though experimental evidence indicates that *prgB* transcription initiates from a promoter just upstream from the start of the open reading frame (ORF) encoding Asc10, the regulatory genes involved in positive control seem to act only when they are in *cis* and in the same orientation relative to the target, although they can be positioned more than 10 kilobases (kb) upstream from the target and still function (Chung and Dunny, 1992). These results suggested the possibility that the mechanism could involve the tracking of regulatory molecules along the DNA. Furthermore, the active positive regulatory molecule encoded by the *prgQ* region seems to be an RNA rather than a protein. As noted above, *prgQ* encodes the iCF10 inhibitor peptide; this peptide is processed from a 22-amino-acid precursor encoded by a 66 basepairs (bp) ORF. The predominant mRNA identified with *prgQ* probes is a constitutively produced 430 nucleotide species called Q_S, which contains the ORF at its 5′ end. Pheromone-induced cells contain a second species, Q_L, which is extended by approximately 130 nucleotides at the 3′ end. Genetic evidence indicates that Q_S is required for *prgS* and *prgA* expression, the latter two genes are transcribed constitutively, and *prgS* is required for Q_L formation or stability. The fact that the nucleotide sequences of the Q_S and Q_L messages contain multiple stop codons in all reading frames and the recent observation that these messages interact with other intracellular regulatory molecules as noted above suggest that these RNA species are directly involved in positive control.

A model illustrating our current idea of the relation between the signaling system and the positive and negative control circuits in the pCF10 system is shown in Figure 3.4. Key experimental results in support of the model include those indicating that internalization of the pheromone is crucial for activation of *prgB* expression and the identification of a number of key players in the regulatory process. Two major questions still unresolved are the reasons for the *cis*-acting nature of positive control and a plausible molecular mechanism by which the Q_L molecule activates *prgB* expression. The positive control mechanism in the pAD1 system has some important differences from that in

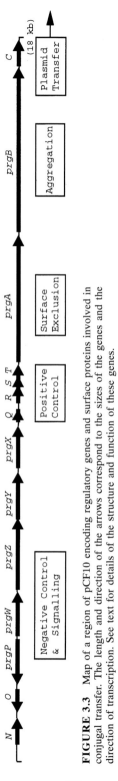

FIGURE 3.3 Map of a region of pCF10 encoding regulatory genes and surface proteins involved in conjugal transfer. The length and direction of the arrows correspond to the sizes of the genes and the direction of transcription. See text for details of the structure and function of these genes.

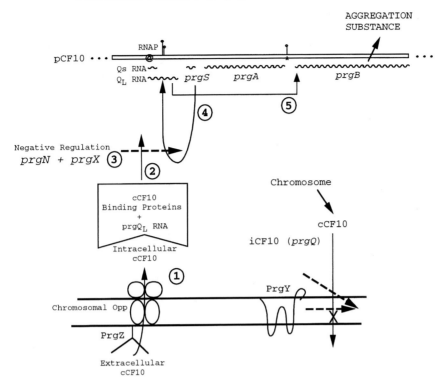

FIGURE 3.4 Model for the regulatory circuits controlling *prgB* (asc10) expression in the pCF10 system. As shown at bottom right, the products of the prgQ and prgY genes act to prevent endogenous active pheromone production by donor cells. Exogenous pheromone from recipients binds to prgZ and is imported by and Opp system (1). Direct interaction with intracellular proteins and RNAs (2) abolishes a negative control system (3), which normally prevents synthesis or stabilization of the Q_L RNA. Once negative control is abolished, the prgS gene product enhances production or stability of Q_L (4), allowing activation of prgB expression (5). See text for details.

pCF10. The sequence of pAD1 contains an inhibitor peptide gene, *iad*; and mRNA species analogous to Q_S and Q_L have been identified (Galli et al., 1992; Tanimoto and Clewell, 1993). There is no counterpart to *prgS* in the pAD1 system, and pAD1 encodes a protein called TraE1, which behaves as a classical *trans*-acting positive control protein (Galli et al., 1992; Tanimoto and Clewell, 1993). The reasons for these differences are not clear, but analysis of the regulatory regions of both plasmids indicates that pAD1 seems to contain additional sequences not found in pCF10. Therefore the former plasmid may be more highly evolved and somewhat more sophisticated in its regulatory mechanism.

Donor–Recipient Interactions: Aggregation, Surface Exclusion, and DNA Transfer

Cell–cell clumping in liquid culture is the hallmark of the pheromone-inducible plasmids typified by pCF10, pPD1, and pAD1. The induction of these high frequency transfer plasmids by pheromone leads to the expression of a surface protein that binds to components of the *E. faecalis* cell wall (Fig. 3.5). The surface adhesin designated aggregation substance (AS) mediates attachment of the plasmid-containing donor cells to the chromosomally encoded receptors called enterococcal binding substance (EBS) found on all *E. faecalis* cell walls. This receptor may be lipoteichoic acid (LTA), a major component of the *E. faecalis* cell wall. Initial studies had shown that purified LTA inhibited aggregation of *E. faecalis* (Ehrenfeld et al., 1986). Genetic analysis was subsequently carried out in which conjugation-negative mutants, defective as recipients in liquid matings, were generated by Tn*916* mutagenesis (Trotter and Dunny, 1990). This analysis showed that multiple insertions were required to inactivate the expression of EBS and resulted in alterations of the cell wall. When each of these mutations were segregated into four strains, it was found that those strains each had a normal recipient phenotype.

Complementation results have shown that at least three of the four regions mutated play a role in expression of EBS on the recipient cell surface. Possibly, then, there are multiple types of binding site for AS, or there may be redun-

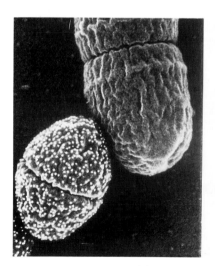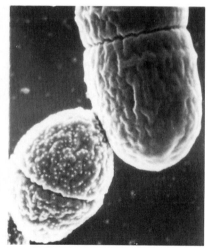

FIGURE 3.5 Low voltage scanning electron microscopy of an *E. faecalis* mating pair. A mating mixture of a plasmid-free recipient strain and a pheromone-induced donor were fixed, stained, and viewed with a Hitachi field emission scanning electron microscope at low accelerating voltages. The donor cells were labeled with a monoclonal antibody directed against Asc10 and conjugated to 12 nm colloidal gold. The image at left is from a backscatter detector using 3.5 kV accelerating voltage; the image at right is from a secondary electron detector at 3 kV accelerating voltage. (Image prepared by Stephen Olmsted.)

dancy in enzymes involved in the metabolism of the cell wall. The inactivated genes have been sequenced and analyzed genetically for one locus (Bensing and Dunny, 1993). This locus contains four ORFs in which the transposon insertion occurred between the two divergently transcribed *ebsA* and *ebsB* genes. The gene *ebsC* appears to be responsible for the complementing activity, as its inactivation interferes with complementation, and the inactivation of *ebsA* or *ebsB* caused stimulation of complementation. EbsB may function as a cell wall hydrolase, with the *ebsA* product perhaps assisting in its localization in the cell wall. EbsC may inhibit the expression or activity of either of these other two genes.

The pheromone-responsive plasmids appear to contain an entry exclusion mechanism. Pheromone induction of "male recipients" decreased their ability to act as recipients by a factor of 10–300 (Dunny et al., 1985). Genes for one of the proteins involved in this phenotype have been identified: In pCF10 *prgA* encodes Sec10, and in pAD1 *sea1* encodes Sea1. Inactivation of the gene or interference with the protein's function by monoclonal antibodies increases the frequency of transfer to plasmid-containing recipient cells from donors containing genetically tagged homologous plasmids. This gene is necessary but not sufficient for entry exclusion, as clones expressing *prgA* did not exhibit the entry exclusion phenotype. Also the phenotype is pheromone-inducible, whereas Sec10 is expressed constitutively.

Most of the genes involved in the physical transfer of pCF10 DNA in mating aggregates have not been sequenced. However, PrgC resembles a surface protein from the pAMβ1 system and may be part of the DNA transfer of pCF10 (Kao et al., 1991). In the pAD1 system, a number of pheromone-inducible Tn917-lac insertions that affect mating aggregate stability and transfer have been isolated 3′ from the *asa1* gene (Clewell, 1993); presumably these mutations affect the transfer apparatus.

High frequency transfer of conjugative plasmids in liquid culture can be achieved in systems (e.g., the *E. faecalis* pheromone-inducible plasmids) that encode an aggregation mechanism. *Bacillus thuringiensis* subsp. *israelensis* (Andrup et al., 1993) and *Lactococcus lactis* (Gasson et al., 1992) are additional examples of this phenomenon, although in neither of these systems is a recipient-derived pheromone signal suspected.

Conclusion

The high frequency transfer of the pheromone-inducible plasmids is the result of the elaboration of surface adhesins. These adhesins mediate aggregation with all other *E. faecalis* cells. Although such action may lock cells together and allow plasmid transfer in difficult mating situations (as in liquid culture), this phenotype would confer a disadvantage to cells if it were expressed constitutively. By encoding inducible expression of the surface adhesins, the plasmid does not confer a generally disadvantageous phenotype on its bacterial host. To express the aggregation phenotype only when recipients are readily available for mating, the pheromone-inducible conjugative plasmids have evolved a sophisticated signaling system, probably from the chromosomally encoded oligopeptide transport system of *Enterococcus*. An intriguing feature is the high degree of specificity of the pheromone system. Detailed analysis of the molecular basis of this specificity is of great importance for understanding pheromone signaling and other forms of oligopeptide transport in bacteria.

Note Added in Proof. It has recently been determined (B.A. Bensing, B.J. Meyer, and G.M. Dunny, 1996. *Proc. Natl. Acad. Sci USA.* **93**:7794–7799) that *prgB* transcription actually initiates from the *prgQ* promoter.

Acknowledgments

The pCF10 research described here was supported by NIH grants 1RO1GM49540 to G.M.D. and grant 1F32AI08742 to B.A.B.L. We thank Stephen Olmsted for preparation of the electron micrograph.

References

An, F.Y., and Clewell, D.B. (1994) Characterization of the determinant (*traB*) encoding sex pheromone shutdown by the hemolysin/bacteriocin plasmid, pAD1 in *Enterococcus faecalis*. Plasmid 31:215–221.

Andrup, L., Damgaard, J., and Wassermann, K. (1993) Mobilization of small plasmids in *Bacillus thuringiensis* subsp. *israelensis* is accompanied by specific aggregation. *J. Bacteriol.* **175**:6530–6536.

Bensing, B.A., and Dunny, G.M. (1993) Cloning and molecular analysis of genes involved in expression of binding substance, the recipient-encoded receptor(s) mediating mating aggregate formation in *Enterococcus faecalis. J. Bacteriol.* **175**: 7421–7429.

Chung, J.W., and Dunny, G.M. (1992) cis-Acting, orientation-dependent, positive control system activates pheromone-inducible conjugation functions at distances greater than 10 kilobases upstream from its target in *Enterococcus faecalis. Proc. Natl. Acad. Sci. USA* **89**:9020–9024.

Chung, J.W., and Dunny, G.M. (1994) Genetic analysis of a region of the *Enterococcus faecalis* plasmid pCF10 involved in positive regulation of conjugative transfer functions. *J. Bacteriol.* **176**:2118–2124.

Clewell, D.B. (1993) Sex pheromones and the plasmid-encoded mating response in *Enterococcus faecalis*. In *Bacterial Conjugation.* D.B. Clewell (ed.) pp. 349–368. Plenum Press, New York.

Clewell, D.B., An, F.Y., Mori, M., Ike, Y., and Suzuki, A. (1987) *Streptococcus faecalis* pheromone cAD1 response: evidence that the peptide inhibitor excreted by pAD1-containing cells may be plasmid determined. *Plasmid* **17**:65–68.

Clewell, D.B., Pontius, L.T., An, F.Y., Ike, Y., Suzuki, A., and Nakayama, J. (1990) Nucleotide sequence of the sex pheromone inhibitor (iAD1) determinant of *Enterococcus faecalis* conjugative plasmid pAD1. *Plasmid* **24**:156–161.

Dunny, G., Funk, C., and Adsit, J. (1981) Direct stimulation of the transfer of antibiotic resistance by sex pheromones in *Streptococcus faecalis. Plasmid* **6**:270–278.

Dunny, G.M. (1990) Genetic functions and cell–cell interactions in the pheromone-inducible plasmid transfer system of *Enterococcus faecalis. Mol. Microbiol.* **4**: 689–696.

Dunny, G.M., Zimmerman, D.L., and Tortorello, M.L. (1985) Induction of surface exclusion (entry exclusion) by *Streptococcus faecalis* sex pheromones: use of monoclonal antibodies to identify an inducible surface antigen involved in the exclusion process. *Proc. Natl. Acad. Sci. USA* **82**:8582–8586.

Ehrenfeld, E.E., Kessler, R.E., and Clewell, D.B. (1986) Identification of pheromone-induced surface proteins in *Streptococcus faecalis* and evidence of a role for lipoteichoic acid in the formation of mating aggregates. *J. Bacteriol.* **168**:6–12.

Firth, N., Ridgway, K.P., Byrne, M.E., et at. (1993) Analysis of a transfer region from the staphylococcal conjugative plasmid pSK41. *Gene* **136**:13–25.

Galli, D., Friesnegger, A., and Wirth, R. (1992) Transcriptional control of sex-pheromone-inducible genes on plasmid pAD1 of *Enterococcus faecalis* and sequence of a third structural gene for (pPD1-encoded) aggregation substance. *Mol. Microbiol.* **6**:1297–1308.

Gasson, M.J., Swindell, S., Maeda, S., and Dodd, H.M. (1992) Molecular rearrangement of lactose plasmid DNA associated with high-frequency transfer and cell aggregation in *Lactococcus lactis* 712. *Mol. Microbiol.* **6**:3231–3223.

Hedberg, P.J., Leonard, B.A.B., Ruhfel, R.E., and Dunny, G.M. (1996) Identification and characterization of the genes of *Enterococcus faecalis* plasmid pCF10 involved in replication and in negative control of pheromone-inducible conjugation. *Plasmid* **35**:46–57.

Hiles, I.D., Gallagher, M.P., Jamieson, D.J., and Higgins, C.F. (1987) Molecular characterization of the oligopeptide permease of *Salmonella typhimurium*. *J. Mol. Biol.* **195**:125–142.

Hui, F.M., and Morrison, D.A. (1991) Genetic transformation in *Streptococcus pneumoniae*: nucleotide sequence analysis shows *comA*, a gene required for competence induction, to be a member of the bacterial ATP-dependent transport protein family. *J. Bacteriol.* **173**:372–381.

Ike, Y., Craig, R.A., White, B.A., Yagi, Y., and Clewell, D.B. (1983) Modification of *Streptococcus faecalis* sex pheromones after acquisition of plasmid DNA. *Proc. Natl. Acad. Sci. USA* **80**:5369–5373.

Kao, S.-M., Olmsted, S.B., Viksnins, A.S., Gallo, J.C., and Dunny, G.M. (1991) Molecular and genetic analysis of a region of plasmid pCF10 containing positive control genes and structural genes encoding surface proteins involved in pheromone-inducible conjugation in *Enterococcus faecalis*. *J. Bacteriol.* **173**:7650–7664.

Kashiwagi, K., Yamaguchi, Y., Sakai, Y., Kobayashi, H., and Igarashi, K. (1990) Identification of the polyamine-induced protein as a periplasmic oligopeptide binding protein. *J. Biol. Chem.* **265**:8387–8391.

Leonard, B.A.B., Bensing, B.A., Hedberg, P.J., and Dunny, G.M. (1994) Pheromone-inducible gene regulation and signalling in the control of aggregation substance expression in the conjugative plasmid, pCF10. In *Genetics of Streptococci, Enterococci, and Lactococci*, J. Ferretti, M. Gilmore, and T. Klaenhammer (eds.), pp. 212–213. International Bureau of Biological Standards, Washington, DC.

Leonard, B.A.B., Podbielski, A., Hedberg, P.J., and Dunny, G.M. (1996) *Enterococcus faecalis* peromone binding protein PrgZ recruits a chromosomal oligopeptide permease system to import sex pheromone cCF10 for induction of conjugation. *Proc. Natl. Acad. Sci. USA* **93**:260–264.

Magnuson, R., Solomon, J., and Grossman, A.D. (1994) Biochemical and genetic characterization of a competence pheromone from *B. subtilis*. *Cell* **77**:207–216.

Mori, M., Sakagami, Y., Ishii, Y., et al. (1988) Structure of cCF10, a peptide sex pheromone which induces conjugative transfer of the *Streptococcus faecalis* tetracycline resistance plasmid, pCF10. *J. Biol. Chem.* **263**:14574–14578.

Nakayama, J., Isogai, A., Clewell, D.B., and Suzuki, A. (1994a) Molecular and genetic analysis of a region of *Enterococcus faecalis* plasmid, pPD1 containing sex pheromone sensitivity (*traC*), pheromone shutdown (*traB*) and pheromone inhibitor (*ipd*) genes. GenBank Accession D28859.

Nakayama, J., Ruhfel, R.E., Dunny, G.M., Isogai, A., and Suzuki, A. (1994b) The *prgQ* gene of the *Enterococcus faecalis* tetracycline resistance plasmid, pCF10, encodes a peptide inhibitor, iCF10. *J. Bacteriol.* **176**:2003–2004.

Perego, M., Higgins, C.F., Pearce, S.R., Gallagher, M.P., and Hoch, J.A. (1991) The oligopeptide transport system of *Bacillus subtilis* plays a role in the initiation of sporulation. *Mol. Microbiol.* **5**:173–185.

Rudner, D.Z., LeDeaux, J.R., Ireton, K., and Grossman, A.D. (1991) The *spoOK* locus of *Bacillus subtilis* is homologous to the oligopeptide permease locus and is required for sporulation and competence. *J. Bacteriol.* **173**:1388–1398.

Ruhfel, R.E., Manias, D.A., and Dunny, G.M. (1993) Cloning and characterization of a region of the *Enterococcus faecalis* conjugative plasmid, pCF10, encoding a sex pheromone binding function. *J. Bacteriol.* **175**:5253–5259.

Tame, J.R.H., Murshudov, G.N., Dodson, E.J., et al. (1994) The structural basis of sequence-independent peptide binding by OppA protein. *Science* **264**:1578–1581.

Tanimoto, K. and Clewell, D.B. (1993). Regulation of the pAD1-encoded sex pheromone response in *Enterococcus faecalis*: expression of the positive regulator TraE1. *J. Bacteriol.* **175**:1008–1018.

Tanimoto, K., An, F.Y., and Clewell, D.B. (1993) Characterization of the traC determinant of the *Enterococcus faecalis* hemolysin-bacteriocin plasmid pAD1: binding of sex pheromone. *J. Bacteriol.* **175**:5260–5264.

Tomich, P.K., An, F.Y., Damle, S.P., and Clewell, D.B. (1979) Plasmid related transmissibility and multiple drug resistance in *Streptococcus faecalis* subspecies *zymogenes* strain DS16. *Antimicrob. Agents Chemothera.* **15**:828–830.

Trotter, K.M., and Dunny, G.M. (1990) Mutants of *Enterococcus faecalis* deficient as recipients in mating with donors carrying pheromone-inducible plasmids. *Plasmid* **24**:57–67.

Tynkkynen, S., Buist, G., Kunji, E., et al. (1993) Genetic and biochemical characterization of the oligopeptide transport system of *Lactococcus lactis*. *J. Bacteriol.* **175**: 7523–7532.

Weaver, K.E., and Clewell, D.B. (1991) Control of *Enterococcus faecalis* sex pheromone cAD1 elaboration: effects of culture aeration and pAD1 plasmid-encoded determinants. *Plasmid* **25**:177–189.

4

N-Acyl-L-Homoserine Lactone Autoinducers in Bacteria

Unity and Diversity

PAUL V. DUNLAP

Autoinduction is an intercellular signaling and gene regulatory mechanism by which bacteria sense and respond to their local population density. It was first characterized as controlling luminescence in certain marine luminous bacteria, and the first chemically and genetically defined autoinduction system was that controlling luminescence in the marine symbiotic bacterium *Vibrio fischeri*. Autoinduction, since identified in a wide range of bacteria, marine and terrestrial, has been shown to control a variety of activities potentially adaptive at high population density or in association with higher organism hosts. Autoinduction of luminescence in *V. fischeri* provides a model for these newly discovered systems, the widespread nature of which indicates that autoinduction is a regulatory mechanism of fundamental importance in microbiology.

General Considerations

Autoinduction and Bacterial Multicellularity

The signal components of autoinduction systems are the autoinducers *N*-acyl-L-homoserine lactones (*N*-acyl-L-HSLs). These self-produced, highly specific, membrane-permeant compounds are released from cells and accumulate in localized environments in a population density-dependent manner. Upon attaining a threshold concentration, autoinducers trigger gene transcription in localized populations of cells, activating biochemical and physiological functions in the

LIVERPOOL JOHN MOORES UNIVERSITY
LEARNING SERVICES

TABLE 4.1 N-Acyl-L-Homoserine Lactone Autoinducers in Bacteria

Proteobacterial Subdivision and Species	Autoinducer-Controlled Activity or Product	Autoinducer Acyl Moiety	Genes[a] Signal Generator	Genes[a] Activator/Sensor/Modulator/Repressor	Comments
Alpha					
Agrobacterium tumefaciens	Ti plasmid conjugal transfer	N-3-Oxooctanoyl	*traI* (30%)	*traR* (18%) *traM* (ns)	Negative modulator
Rhizobium leguminosarum	Rhizome interaction	C_{14} molecule	?	*rhiR* (23%)	
Beta					
Chromobacterium violaceum	Violacein, hemolysin, exoprotease	?	?	?	
Gamma					
Enterobacter agglomerans	?	N-3-Oxohexanoyl	*eagI* (25%)	?	
Erwinia carotovora	Extracellular virulence factors, carbapenem	N-3-Oxohexanoyl	*expI* (30%) *carI* (25%)	*expR* (?%) *carR* (?%)	
Erwinia stewartii	Extracellular capsular, polysaccharide	N-3-Oxohexanoyl	*esaI* (25%)	*esaR* (24%)	
Escherichia coli	Cell division, starvation response	? Unidentified factor	? ?	*sdiA* (24%) ? *rspA* (ns)	HSL degradation?

Pseudomonas aeruginosa	Extracellular virulence factors, rhamnolipid biosurfactant	*N*-3-oxododecanoyl *N*-butanoyl	*lasI* (35%) *rhlI* (?%)	*lasR* (27%) *rhlR* (23%)	
Pseudomonas aureofaciens	Phenazine	?	*phzI* (31%)	*phzR* (23%)	
Vibrio fischeri	Luminescence ?	*N*-3-Oxohexanoyl *N*-Hexanoyl *N*-Octanoyl *N*-Octanoyl	*luxI* *luxI* *ainS* (ns) *ainS*	*luxR* *luxR* *luxR* *ainR* (ns)	*lux* Negative modulation Non-*lux* gene regulation?
Vibrio harveyi	Luminescence, PHB	*N*-3-Hydroxybutanoyl Unidentified factor	*luxLM* (ns) ?	*luxN* (ns) *luxPQ* (ns) *luxO* (ns) *luxR* (ns)	Repressor Activator
Yersinia enterocolitica	?	*N*-3-Oxohexanoyl	*yenI* (20%)	*yenR* (21%)	

PHB, poly-β-hydroxybutyric acid; HSL, homoserine lactone.
[a]Numbers in parentheses indicate the percent identity to *V. fischeri* LuxI or LuxR; (ns) indicates no obvious similarity, and (?%) indicates that the percent identity is not yet reported. A question mark (?) indicates a present lack of information. See text for references.

population in a coordinated manner. This coordination and the possibility that the functions activated are adaptive for cells at high population density indicate that autoinducers mediate a transiently multicellular, cooperative state in bacteria.

Autoinducers as Bacterial Pheromones

Autoinducers are not essential for growth and reproduction (e.g., Kuo et al., 1996) and so can be viewed as secondary metabolites (Vining, 1990). They exhibit functional similarities to pheromones, self-produced compounds released from cells that have specific effects on other cells of that species (Eberhard, 1972; Stephens, 1986). Pheromone-like compounds are used as signals in many bacteria. They range in chemical composition from butyrolactones, which are involved in regulating antibiotic production and the formation of aerial mycelia in *Streptomyces* spp., to single amino acids, peptides, and proteins, which mediate, for example, swarming in *Proteus*, competence in *Bacillus* and *Pneumococcus*, fruiting body formation in *Myxococcus*, and conjugal transfer of certain plasmids in mating types of *Enterococcus* (Dworkin, 1991; Rudner et al., 1991; Clewell, 1993; Kaiser and Losick, 1993; Dunny et al., 1995; Kell et al., 1995). Nonetheless, autoinducers are chemically and genetically distinct from other pheromones. They do not cross react physiologically, for example, with butyrolactones; and the proteins involved in the synthesis of and response to autoinducers show no obvious similarity to proteins involved in the synthesis and response to butyrolactones (Eberhard et al., 1986; Horinouchi et al., 1989; Horinouchi and Beppu, 1992; Okamoto et al., 1992; Chhabra et al., 1993; Kell et al., 1995).

Unity and Diversity of Autoinduction in Bacteria

Autoinduction was for many years thought to be unique to the luminescence system of *V. fischeri* and certain closely related marine bacteria. However, several other species of bacteria recently have been found to produce *N*-acyl-L-HSLs chemically similar or identical to that of *V. fischeri*. In these other species, the autoinducers regulate activities as diverse as conjugative transfer of plasmids, synthesis of antibiotics, and production of extracellular enzymes. In many cases, proteins homologous to those of *V. fischeri* necessary for synthesis of and response to autoinducer (LuxI and LuxR proteins, respectively) have been identified in these other bacteria (Fuqua et al., 1994). The diversity of species using autoinduction and the chemical and genetic similarities of their autoinduction systems (Table 4.1) indicate that autoinduction is an evolutionarily conserved regulatory mechanism, common among bacteria, for sensing and responding to population density.

Evolutionary Origin of Autoinduction in Bacteria

The bacteria presently known to produce autoinducers belong to the proteobacteria, a physiologically diverse group. Within the proteobacteria, species using autoinduction have been found in the alpha, beta, and gamma subdivisions (Table 4.1), which are thought to have a common origin in an early phototroph. Representatives have not been reported, however, for the delta and epsilon subdivisions, which diverged earlier from an ancestral phototroph (Brock et al., 1994). It is possible therefore that autoinduction arose in the early phototroph that gave rise to the alpha, beta, and gamma subdivisions. Furthermore, the bacteria presently known to use autoinducers associate with higher organisms as pathogenic, mutualistic, commensal, or saprophytic symbionts. Autoinduction therefore might be an adaptation to invading and colonizing higher organisms. Possibly, then, autoinduction was retained in lineages derived from the early phototroph that developed associations with plants and animals as these higher organisms arose. These speculative views are subject to revision as more autoinduction systems are characterized. Furthermore, the origin of autoinduction is more complex than this simple view implies. Autoinduction systems evolutionarily distinct from the LuxI/LuxR system of *V. fischeri* have been identified in certain members of the proteobacterial gamma subdivision (Bassler et al., 1994a; Kuo et al., 1994; Gilson et al., 1995).

Autoinduction in Luminous Bacteria

Autoinduction was first described as a regulatory mechanism controlling light production (Fig. 4.1) in the marine luminous bacteria *Vibrio harveyi* and *V. fischeri* (Nealson et al., 1970; Eberhard, 1972). Luminescence and synthesis of luciferase, the light-emitting enzyme, are strongly regulated in these species, inducing dramatically in batch culture during the late exponential phase of

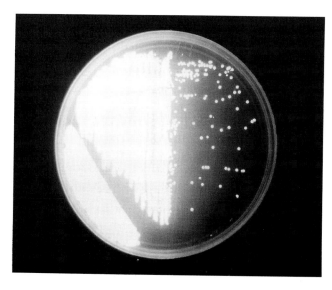

FIGURE 4.1 Bacterial luminescence. Colonies of the marine symbiotic bacterium *V. fischeri* growing on a nutrient seawater agar plate and fully induced for light production were photographed by their own light. (Photo by the author)

growth (Fig. 4.2). Both species can be isolated from coastal seawater and various other marine habitats (Baumann and Baumann, 1981).

Autoinduction of Luminescence in Vibrio fischeri

Vibrio fischeri is one of three identified light-producing species, among approximately 12 known light-emitting bacteria, that forms a specific bioluminescent mutualism, or light-organ symbiosis, with certain marine squid and fish. The animal harbors *V. fischeri* extracellularly within specialized tissues called light organs and uses the bacterial light in various behaviorally related luminescence displays (for references: Dunlap and Greenberg, 1991; Nealson and Hastings, 1992; Ruby and McFall-Ngai, 1992; Haygood, 1993; Meighen and Dunlap, 1993). Autoinduction of luminescence is presently best understood in *V. fischeri*, and the autoinduction system in this species serves as a model for autoinduction systems newly discovered in other bacteria.

Key Elements of *Vibrio fischeri* Luminescence Autoinduction System

Studies of luminescence and its regulation in *V. fischeri* led to two key advances fundamental to understanding the mechanism of luminescence autoinduction: identification of the *V. fischeri* autoinducer and isolation of the luminescence regulatory genes. These advances also provided a biochemical and genetic foundation for the discovery of autoinduction systems in other bacteria.

The *V. fischeri* luminescence autoinducer, the first bacterial autoinducer to be characterized, was extracted and purified from medium conditioned by the growth of *V. fischeri* to high population density and full induction of luminescence. The compound was identified as *N*-3-oxohexanoyl-L-HSL (Eberhard et al., 1981) (Fig. 4.3, Table 4.1). The *V. fischeri* autoinducer has no effect on luminescence in *V. harveyi* (Eberhard, 1972; Nealson, 1977).

The second key advance was the isolation by Engebrecht et al. (1983) of a chromosomal fragment of DNA from *V. fischeri* that conferred on *Escherichia coli* the ability to produce light and to do so in an autoinducible manner. The luminescence (*lux*) genes of *V. fischeri* are organized in two divergently transcribed units, *luxR* and *luxICDABEG* (the *lux* operon), separated by a regulatory region containing the *luxR* and *lux* operon promoters. *luxA* and *luxB* specify the α and β subunits of the light-emitting enzyme luciferase. *luxC*, *luxD*, and *luxE* encode the polypeptides of the *lux*-specific fatty acid reductase complex (reductase, acyltransferase, and acyl protein synthase, respectively) involved in the synthesis and recycling of aldehyde substrate for luciferase (Boylan et al., 1985, 1989). The last *lux* operon gene, *luxG*, encodes a protein with unknown function that is not required for luminescence or its regulation in *E. coli* (for references: Meighen and Dunlap, 1993); *luxG* is followed by a strong transcriptional terminator (Swartzman et al., 1990a). The first gene of the *lux* operon, *luxI*, and the single gene of the second transcriptional unit, *luxR*, are

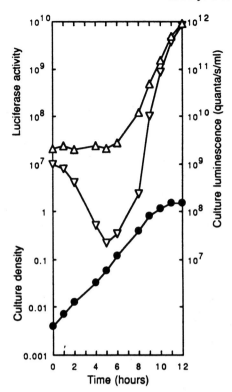

FIGURE 4.2 Autoinduction of luciferase synthesis and luminescence in *V. fischeri*. Cells were inoculated at low density; growth, as density of the culture (●), was monitored over time. Samples of the culture were obtained periodically to measure luciferase in vitro activity (△), in relative units per milliliters of culture, and luminescence (▽), in quanta s^{-1} ml^{-1}. (Reprinted from Dunlap and Greenberg, 1991, with permission)

regulatory. *luxI* is necessary for cells to synthesize *N*-3-oxohexanoyl-L-HSL and is thought therefore to specify autoinducer synthase; *luxR* specifies a protein necessary for cells to activate *lux* operon transcription in response to autoinducer (Engebrecht et al., 1983; Engebrecht and Silverman, 1984).

lux Regulatory Region

The *V. fischeri lux* regulatory region, 219 basepairs (bp) in length, contains the *luxR* promoter (with −10 and −35 regions similar to consensus sequences for *E. coli* "housekeeping" promoters) and the *lux* operon promoter (with a −10 region but no −35 region identifiable by comparison with *E. coli* consensus sequences). At this position instead, centered 40 bp upstream of the *lux* operon transcriptional start, is a 20-bp inverted repeat sequence that is required for activation of *lux* operon transcription and thus is implicated as the binding site for LuxR (Engebrecht and Silverman, 1987; Devine et al., 1988a,b; Baldwin et al., 1989; Shadel et al., 1990; Stevens et al., 1994). A sequence closely similar to this palindromic sequence, or *lux* box, has been identified at the same location in the *lux* regulatory regions of various *V. fischeri* strains as well as in the promoter regions controlling expression of the autoinducible genes *lasB*

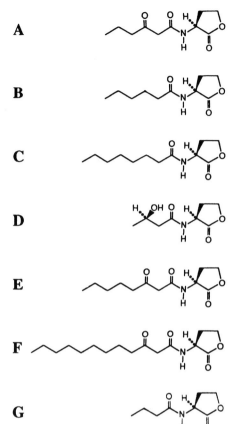

FIGURE 4.3 Chemical structures of bacterial N-acyl-L-homoserine lactone autoinducers. (**A**) N-3-Oxohexanoyl-L-HSL (*V. fischeri, E. carotovora*) (Eberhard et al., 1981; Bainton et al., 1992b). (**B**) N-Hexanoyl-L-HSL (*V. fischeri*) (Kuo et al., 1994). (**C**) N-Octanoyl-L-HSL (*V. fischeri*) (Kuo et al., 1994). (**D**) N-3-Hydroxybutanoyl-L-HSL (*V. harveyi*) (Cao and Meighen, 1989). (**E**) N-3-Oxooctanoyl-L-HSL (*A. tumefaciens*) (Zhang et al., 1993). (**F**) N-3-Oxododecanoyl-L-HSL (*P. aeruginosa*) (Pearson et al., 1994). (**G**) N-Butanoyl-L-HSL (*P. aeruginosa*) (Pearson et al., 1995). (Modified from Fuqua et al., 1994, with permission)

in *Pseudomonas aeruginosa* and *traA* and *traI* of the octopine-type Ti plasmid and *traI* of the nopaline-type plasmid, respectively, of *Agrobacterium tumefaciens* (Fig. 4.4) (Fuqua and Winans, 1994; Fuqua et al., 1994; Gray et al., 1994; Hwang et al., 1994). The *lux* box represents a conserved regulatory sequence; its presence upstream of a bacterial gene would be consistent with autoinducer-mediated control of that gene (Gray et al., 1994).

A consensus cAMP receptor protein (CRP) binding site is present upstream of the *luxR* promoter, at position -59 from the *luxR* transcriptional start (Engebrecht and Silverman, 1987; Devine et al., 1988a). cAMP and CRP are necessary for induction of luminescence in *V. fischeri*, functioning to activate transcription from the *luxR* promoter and thereby potentiating *lux* operon induction by N-3-oxohexanoyl-L-HSL (Dunlap and Greenberg, 1985; 1988; Dunlap, 1989; Dunlap and Ray, 1989; Dunlap and Kuo, 1992). A variety of other factors contribute to luminescence induction in *V. fischeri* and are reviewed elsewhere (Ulitzur and Dunlap, 1995).

V. fischeri ATCC 7744 / MJ1 *luxI*	A	C	C	T	G	T	A	G	G	A	T	C	G	T	A	C	A	G	G	T
V. fischeri ES114 *luxI*	A	G	C	T	G	T	A	G	G	A	T	G	G	T	A	C	A	G	G	T
V. fischeri H905 *luxI*	G	G	C	T	G	T	A	G	G	A	T	A	G	T	A	C	A	G	G	T
V. fischeri EM30 *luxI*	G	C	C	T	G	T	A	G	G	A	T	C	G	T	A	C	A	G	G	T
P. aeruginosa *lasB*	A	C	C	T	G	C	C	A	G	T	T	C	T	G	G	C	A	G	G	T
A. tumefaciens Oct *traA*	A	T	G	T	G	C	A	·	G	A	T	C	·	T	G	C	A	C	A	T
A. tumefaciens Oct *traI*	A	C	G	T	G	C	A	·	G	A	T	C	·	T	G	C	A	C	A	T
A. tumefaciens Nop *traI*	A	G	G	T	G	C	A	·	A	A	T	C	·	T	G	C	A	C	G	T
Consensus	R	N	S	T	G	Y	A	X	G	A	T	N	X	T	R	C	A	S	R	T

FIGURE 4.4   Palindromic *lux* box-like elements. Shown are putative or known LuxR homolog binding sites identified upstream of the *lux* operon in various strains of *V. fischeri*, upstream of *lasB* in *P. aeruginosa*, and upstream of *traA* and *traI* of *A. tumefaciens*. Boxes indicate sites at which five or more nucleotides are identical. Consensus sequence abbreviations: N = A, T, C, or G; R = A or G; S = C or G; Y = T or C; X = N or gap in sequence. See text for details and references. (Reprinted from Gray et al., 1994, with permission)

Model for Luminescence Autoinduction in *Vibrio fischeri*

Autoinduction of luminescence occurs when a threshold level of autoinducer has accumulated (Rosson and Nealson, 1981). When fully induced *V. fischeri* cells are inoculated into fresh medium at low population density, the *N*-3-oxohexanoyl-L-HSL in the cells rapidly diffuses out (Kaplan and Greenberg, 1985), dropping the concentration below the threshold necessary to maintain transcriptional activation of the *lux* operon. Synthesis of autoinducer continues, though at a low rate (Dunlap, 1992), owing to activity of the previously produced LuxI protein, which presumably is diluted as the culture grows, and to the activity of a small amount of newly produced LuxI, the synthesis of which presumably results from basal, inactivated transcription of the *lux* operon. As cells accumulate in physically restricted environments, such as test tubes or light organs, *N*-3-oxohexanoyl-L-HSL also gradually accumulates in the medium and in cells. Once *N*-3-oxohexanoyl-L-HSL attains a threshold level, it interacts with LuxR, forming a complex that facilitates association of RNA polymerase with the *lux* operon promoter, thereby activating transcription of the *lux* operon (Engebrecht and Silverman, 1984; Stevens et al., 1994). The presence of *luxI* as part of the *lux* operon creates a positive feedback circuit for *N*-3-oxohexanoyl-L-HSL synthesis (Engebrecht et al., 1983; Friedrich and Greenberg, 1983; Eberhard et al., 1991), leading to a rapid increase in the autoinducer once induction begins (Dunlap, 1992). This increase presumably rapidly saturates the available LuxR protein in the cell, leading thereby to maximal activation of *lux* operon transcription. Consequently, once the threshold concentration of *N*-3-oxohexanoyl-L-HSL is attained, luciferase synthesis and luminescence increase rapidly (Fig. 4.2). The high rate of synthesis of *N*-

3-oxohexanoyl-L-HSL and its membrane-permeant nature presumably cause local environments to quickly become saturated with the autoinducer.

Ecological rationales for autoinduction of luminescence are based on the bacterial light being visible to higher organisms. In the light-organ symbiosis of *V. fischeri* with monocentrid fish and sepiolid squid, the host is dependent on the bacterial light for various luminous displays (e.g., Dunlap and Greenberg, 1991; Ruby and McFall-Ngai, 1992). During the symbiosis, the bacteria are present at exceptionally high population density, are exposed to a high level of autoinducer, and are highly induced for luminescence (Ruby and Nealson, 1976; Ruby and Asato, 1993; Boettcher and Ruby, 1995). In contrast, at low population densities of *V. fischeri* free in seawater (Ruby and Nealson, 1978; Nealson and Hastings, 1979; Ruby et al., 1980), where light production by individual cells would presumably not confer a benefit, autoinducer would diffuse away and luminescence presumably would not be expressed. Also, light production by commensal or saprophytic *V. fischeri* cells in fecal pellets or colonizing marine snow and decaying tissues of marine animals could lead to these materials being fed upon by marine animals, thereby returning *V. fischeri* cells to the nutrient-rich enteric habitat (Nealson and Hastings, 1992). A physiological rationale for the luminescence reaction, in which the production of visible light would be incidental, is less obvious but might relate to competition for oxygen (for references: Dunlap, 1991).

Autoinduction might be self-limiting at high concentrations of *N*-3-oxohexanoyl-L-HSL and LuxR protein. *N*-3-Oxohexanoyl-L-HSL together with LuxR represses *luxR* expression posttranscriptionally and transcriptionally (Engebrecht and Silverman, 1986; Dunlap and Greenberg, 1988; Dunlap and Ray, 1989). *luxR* transcriptional negative autoregulation is dependent on the 20-bp palindromic *lux* box and a regulatory site in *luxD* (Shadel and Baldwin, 1991, 1992a). Conversely, at low concentrations *N*-3-oxohexanoyl-L-HSL together with LuxR can activate expression from the *luxR* promoter (Shadel and Baldwin, 1991); *luxR* positive autoregulation involves both cAMP-CRP-dependent and cAMP-CRP-independent activation of transcription from the *luxR* promoter (Shadel and Baldwin, 1992b).

LuxR and LuxR Homologs

The *luxR* gene encodes a protein, LuxR, of 250 amino acid residues [relative molecular weight (M_r) 27 kilodaltons (kDa)], which is required for *E. coli* containing the *lux* genes and for *V. fischeri* to respond with luminescence to *N*-3-oxohexanoyl-L-HSL (Engebrecht et al., 1983; Engebrecht and Silverman, 1984, 1987; Devine et al., 1988a; Baldwin et al., 1989; Dunlap and Kuo, 1992). LuxR proteins from different strains of *V. fischeri* show a high degree of sequence conservation; *luxR* gene sequences are 98% identical between MJ-1 and ATCC 7744 and 80% identical between MJ-1 and ES114 (Gray and Greenberg, 1992b). Immunoprecipitation experiments indicate that LuxR, like many other bacterial transcriptional activators that respond to environmental signals, associates with the cytoplasmic membrane (Kolibachuk and Greenberg,

1993); and association of *N*-3-oxohexanoyl-L-HSL with LuxR might be facilitated at the membrane by protein–lipid interactions (Fuqua et al., 1994). Active protein that binds the autoinducer is dependent on the products of the *groE* operon, indicating that chaperonin-mediated folding is necessary for stabilization of the native form of LuxR (Adar et al., 1992; Dolan and Greenberg, 1992; Adar and Ulitzur, 1993; Hanzelka and Greenberg, 1995). Studies describing the modular nature of LuxR, which is composed of regulatory and activator domains (Shadel et al., 1990; Slock et al., 1990; Choi and Greenberg, 1991, 1992a,b; Fuqua et al., 1994; Hanzelka and Greenberg, 1995; Poellinger et al., 1995), and the interactions between LuxR, *N*-3-oxohexanoyl-L-HSL, CRP, and *lux* regulatory DNA (Stevens et al., 1994) are reviewed elsewhere (Ulitzur and Dunlap, 1995).

Proteins with overall sequence similarity to LuxR (approximately 18–27% sequence identity) (Table 4.1, Fig. 4.5) and involved in autoinduction (LuxR homologs) have been identified in various other species of bacteria, as described below. The similarities tend to cluster in regions aligning with two portions of LuxR: a portion of the N-terminal domain thought to be involved in interaction with *N*-3-oxohexanoyl-L-HSL and the helix-turn-helix portion of the C-terminal domain (Fuqua et al., 1994). The C-terminal domain of LuxR exhibits sequence similarity to DNA-binding regions of members of the UphA/FixJ family of two-component (sensor-regulator) proteins, which are involved in cellular responses to various intracellular and environmental stimuli (Henikoff et al., 1990; Kahn and Ditta, 1991; Fuqua et al., 1994).

LuxI and *LuxI* Homologs

The second key protein controlling *V. fischeri* luminescence induction is LuxI, an M_r 25 kDa protein (Engebrecht and Silverman, 1984) necessary and sufficient for *V. fischeri* and *E. coli* to synthesize *N*-3-oxohexanoyl-L-HSL (Eberhard et al., 1981; Engebrecht et al., 1983; Engebrecht and Silverman, 1984; Kuo et al., 1994). LuxI is considered therefore to be autoinducer synthase (Engebrecht and Silverman, 1984). Proteins with overall sequence similarity to LuxI (25–35% identity) (Table 4.1, Fig. 4.5) have been identified in various other species of bacteria, as described below.

The deduced LuxI sequences of *V. fischeri* strains exhibit a high degree of identity. The sequences of two strains, MJ-1 (from the light organ of a monocentrid fish) and ATCC 7744 (from sea water), are identical, whereas that of a third, ES114 (from light organs of the coastal Hawaiian squid *Euprymna scolopes*), diverges somewhat (89% amino acid residue identity). The low synthesis of *N*-3-oxohexanoyl-L-HSL and the unusually low expression of light by ES114 in laboratory culture, but not in the symbiosis, might reflect this divergence through weak translation of the *luxI* mRNA or low specific activity of LuxI (Boettcher and Ruby, 1990; Gray and Greenberg, 1992a,b). Alternatively, ES114 grows exceptionally rapidly in culture (Dunlap et al., 1995), so low availability of substrates for *N*-3-oxohexanoyl-L-HSL, due to their consumption in biosynthetic pathways, might account for low autoinducer synthesis.

```
LuxR  MKNINADDTY  RIINKIKACR  SNNDINQCLS  DMTKMVHC.E  YYLLAIIYPH
LasR  ..MALVDG..  .....FLELE  RSSGKLEWSA  ILQKMASDLG  FSKILFGLLP
EsaR  ..........  ...MFSFFLE  NQTITDTLQT  YIQRKLSPLG  SPDYAYTVVS
TraR  .......MQH  WLDKLTDLAA  IQGDECILKD  GLADLAEHFG  FTGYAYLHI.
PhzR  ..MELGQQLG  WDAYFYSIFA  RTMDMQEFTA  VALRALRELR  FDFFRYGMCS
RhiR  MKEESSAVSN  LVFDFLSESA  SAKSKDDVLL  LFGKISQYFG  FSYFAISGIP


LuxR  SMVKSDIS..  ILDNYPKKWR  QYYDDANLIK  YDPIVDYSN.  SNHSPINWNI
LasR  KDSQDYENAF  IVGNYPAAWR  EHYDRAGYAR  VDPTVSHCTQ  SVLPIFWEPS
EsaR  KK..NPSNVL  IISSYPDEWI  RLYRANNFQL  TDPVILTAFK  RTSPFAWDEN
TraR  ....QHKHTI  AVTNYHRDWR  SAYFENNFDK  LDPVVKRAKS  RKHVFAWSGE
PhzR  VTPFMRPRTY  MYGNYPEDWV  QRYQAANYAV  IDPTVKHSKV  SSSPILASNE
RhiR  SPIERIDSYF  VLGNWSVGWF  DRYRENNYVH  ADPIVHLSKT  CDHAFVWSEA


LuxR  FENNAVNKKS  PNVIKEAKTS  GLITGFSFPI  HTANNGFCML  SFAHSEKD..
LasR  IYQTR.KQH.  .EFFEEASAA  GLVYGLTMPL  HGARGELGAL  SLSVEAENRA
EsaR  ITLMS.DLRF  TKIFSLSKQY  NIVNGFTYVL  HDHMNNLALL  SVIIKGNDQT
TraR  QERSRLSKEE  RAFYAHAADF  GIRSGITIPI  KTANGSMSMF  TLA.SERPAI
PhzR  LF..R.GCP.  .DLWSEANDS  NLRHGLAQPS  FNTQGRVGVL  SLARKDNPIS
RhiR  LRDQKLDRQS  RRVMDEAREF  KLIDGFSVPL  HTAAGFQSIV  SFG.AEKV..


LuxR  ....NYIDS.  ...LFLHACM  NIPLIVPSLV  .DNYRKINIA  NNKSNNDLTK
LasR  ......EANR  FMESVLPTLW  MLKDYAL...  QSGAGLAFEH  PVSKPVVLTS
EsaR  ALEQRLAAEQ  GTMQMLLIDF  NEQMYRLAGT  EGERAPALNQ  SADKTI.FSS
TraR  DLDREIDAAA  AAGAVGQLHA  RISFLQTTPT  VEDAAW....  ......LDP
PhzR  L..QEFEALK  VVTKAFAAAV  HEKISEL...  ESDV.....R  VFNTDVEFSG
RhiR  ....ELSTCD  RSALYLMAAY  AHSLLRAQIG  NDASRKI...  ..QALPMITT


LuxR  REKECLAWAC  EGKSSWDISK  ILGCSERTVT  FHLTNAQMKL  NTTNRCQSIS
LasR  REKEVLQWCA  IGKTSWEISV  ICNCSEANVN  FHMGNIRRKF  GVTSRRVAAI
EsaR  RENEVLYWAS  MGKTYAEIAA  ITGISVSTVK  FHIKNVVVKL  GVSNARQAIR
TraR  KEATYLRWIA  VGMTMEEVAD  VEGVKYNSVR  VKLREAMKRF  DVRSKAHLTA
PhzR  RECDVLRWTA  DGKTSEEIGV  IMGVCTDTVN  YHHRNIQRKI  GASNRVQASR
RhiR  REREIIHWCA  AGKTAIEIAT  ILGRSHRTIQ  NVILNIQRKL  NVVNTPQMIA


LuxR  KAILTGAIDC  PYFKN.
LasR  MAVNLGLITL  ......
EsaR  LGVFLDLIRP  AASAAR
TraR  LAIRRKLI..  ......
PhzR  YAVAMGYI..  ......
RhiR  ESFRLRIIR.  ......


A
```

FIGURE 4.5 Amino acid sequence alignments of proteins homologous to the LuxR and LuxI proteins of *V. fischeri*. (**A**) LuxR homologs. Figure continues.

```
LuxI  MTIMIKKSDF  LAIPSEEYKG  ILSLRYQVFK  QRLEWDL...  VVENNLESDE
LasI  MIVQIGRRE.  .EFDKKLLGE  MHKLRAQVFK  ERKGWDV...  SVIDEMEIDG
TraI  MRILTVSPDQ  YERYRSFLKQ  MHRLRATVFG  GRLEWDV...  SIIAGEERDQ
PhzI  ..MHMEEHTL  NQMSDELKLM  LGRFRHEQFV  EKLGWRLPAH  PSQAGCEWDQ
EsaI  .MLELFDVSY  EELQTTRSEE  LYKLRKKTFS  DRLGWEV...  ICSQGMESDE
ExpI  .MLEIFDVSY  TLLSEKKSEE  LFTLRKETFK  DRLNWAV...  KCINGMEFDQ
CarI  .MLEIFDVNH  TLLSETKSGE  LFTLRKETFK  DRLNWAV...  QCTDGMEFDQ
EagI  .MLEIFDVSY  NDLTERRSED  LYKLRKITFK  DRLDWAV...  NCSNDMEFDE
YenI  .MLKLFNVNF  NNMPERKLDE  IFSLRKITFK  DRLDWKV...  TCIDGKESDQ


LuxI  YDNSNABYIY  ACDDT..ENV  SGCWRLLPTT  GDYMLKSVFP  ELLGQQSAPK
LasI  YDALSPYYML  IQEDTPEAQV  FGCWRILDTT  GPYMLKNTFP  ELLHGKEAPC
TraI  YDNFKPSYLL  AITDS..GRV  AGCVRLLPAC  GPTMLEQTFS  QLLEMGSLAA
PhzI  YDTEHARYLL  AFNEDR..AI  VGCARLIPTT  FPNLLEGVFG  HTCAGAP.PK
EsaI  FDGPGTRYIL  GICE...GQL  VCSVRFTSLD  RPNMITHTFQ  HCFSDVTLP.
ExpI  YDDDNATYLF  GVEG...DQV  ICSSRLIETK  YPNMITGTFF  PYFEKIDIP.
CarI  Y.DNNTTYLF  GIKD...NTV  ICSLRFIETK  YPNMITGTFF  PYFEKEINIP.
EagI  FDNSGTRYML  GIYD...NQL  VCSVRFIDLR  LPNMITHTFQ  HLFGDVKLP.
YenI  YDDENTNYIL  GTID...DTI  VCSVRFIDMK  YPTMITGPFA  PYFSDVSLP.


LuxI  DPNIVELSRF  AVGKN..SSK  INNSASEITM  KLFEAIYKHA  VSQGITEYVT
LasI  SPHIWELSRF  AI.NS..GQK  GSLGFSDCTL  EAMRALARYS  LQNDIQTLVT
TraI  HSGMVESSRF  CVDTSLVSRR  DASQLHLATL  TLFAGIIEWS  MASGYTEIVT
PhzI  HPAIWEMTRF  ..........  .TTREPQLAM  PLFWRSLKTA  SLACADAIVG
EsaI  AYG.TESSRF  FVDKARARAL  LGEHYPISQV  .LFLAMVNWA  QNNAYGNIYT
ExpI  EGKYIESSRF  FVDKARSKTI  LGNSYPVSTM  .FFLATVNYS  KSKGYDGVYT
CarI  EGNYLESSRF  FVDKSRAKDI  LGNEYPISSM  .LFLSMINYS  KDKGYDGIYT
EagI  EGDYIESSRF  FVDKNRAKAL  LGSRYPISYV  .LFLSMINYA  RHHGHTGIYT
YenI  IDGFIESSRF  FVEKALARDM  VGNNSSLSTI  .LFLAMVNYA  RDRGHKGILT


LuxI  VTSTAIERFL  KFIKVPCHRI  GDKEIHVLGD  TKSVVLSMPI  N.EQFKKAVL
LasI  VTTVGVEKMM  IFAGLDVSRF  GPH..LKIGI  ERAVALRIEL  N.AKTQIALY
TraI  ATDLRFERIL  KFAGIPMRRL  GEPT..AIGN  TIAIAGRLPA  DRASFEQVCP
PhzI  IVNSTIERYY  KINGVHYERL  GPVT..VHQN  EKILAIKLSA  HREHHRSAVA
EsaI  IVSRAMLKIL  TRSGIQIKVI  KEA..FLTEK  ERIYLLTLPA  GQDDKQQLGG
ExpI  IVSHPMLTIL  KRSGIKISIV  EQG..MSEKH  ERVYLLFLPV  DNESQDVLVR
CarI  IVSHPMLTIL  KRSGIGIRVV  EQG..LSEKE  ERVYLVFLPV  DDENQEALAR
EagI  IVSRAMLTIA  KRSGIEIEVI  KEG..FVSEN  EPIYLLRLPI  DCHNQHLLAK
YenI  VVSRGIFILL  KPSGINITVL  NQG..ESEKN  EVIYLLHLGI  DNDSQQQLIN


LuxI  N.........  ..........  ..........
LasI  GGVLVEQRLA  VS........  ..........
TraI  PGYYSIPRID  VAAIRSAA..  ..........
PhzI  PSAFMSDTLL  RETA......  ..........
EsaI  DVVSRTGCPP  VAVTTIPLTL  PV......
ExpI  RINHNQEFVE  SKLREIPLSF  EPMTEPVG
CarI  RINRSGTFMS  NELKQIPLRV  PAAIAQA.
EagI  RIRDQSESNI  AALCQIPMSL  TVTPEQV.
YenI  KILRVHQVEP  KTLETIPIIV  PGIIK...
```

B

FIGURE 4.5 *Continued* **(B)** LuxI homologs. Identities are indicated by white letters on a black background. See text for species designations. (Modified from Beck von Bodman and Farrand, 1995, with permission)

Autoinducer Biosynthesis

The enzymatic activity of the LuxI protein or of LuxI homologs has not been reported. Consequently, the pathway by which cells synthesize N-acyl-L-HSLs is not well established. Precursors for N-acyl-L-HSLs are present in *E. coli* and are not likely to be unusual compounds, as *E. coli* cells containing, for example, the *V. fischeri luxI* gene, synthesize autoinducer (Engebrecht et al., 1983). Substrates for the synthesis of N-3-oxohexanoyl-L-HSL may be S-adenosylmethionine (SAM), from methionine and homoserine, and 3-oxohexanoyl-coenzyme A; the addition reaction and the HSL cyclization reaction might be catalyzed by LuxI, or the cyclization reaction might occur spontaneously (Eberhard et al., 1991). Genetic evidence suggests that homoserine phosphate instead of homoserine might serve as the major precursor for HSL synthesis, with homoserine perhaps partially contributing as a substrate (Huisman and Kolter, 1994). Addition of the acyl moiety to HSL might occur in this case as a reaction distinct from the cyclization reaction that converts homoserine phosphate to HSL (Huisman and Kolter, 1994), and presumably either reaction (or both) could be catalyzed by LuxI. With respect to the acyl moiety, its source may be the fatty acid biosynthetic pathway. Because a high concentration of 3-oxohexanoyl-coenzyme A fails to saturate the synthase activity in extracts of *V. fischeri*, this compound might first be converted to 3-oxohexanoyl-acyl carrier protein (Eberhard et al., 1991; Cao and Meighen, 1993). Regardless, the cellular pools of various N-acyl compounds and the substrate specificity of a given autoinducer synthase presumably would determine the autoinducers produced by a species (Kuo et al., 1994). Recently, these issues were resolved in that purified LuxI protein was shown to catalyze the synthesis of N-hexanoyl-L-HSL, a second *luxI*-dependent autoinducer (Kuo et al., 1994; 1996), from the substrates SAM and hexanoyl-ACP (E.P. Greenberg, personal communication).

Regulated Production of Autoinducer

Consistent with the regulatory model for *lux* operon autoinduction described above, production of N-3-oxohexanoyl-L-HSL in *V. fischeri* is regulated in a population-density-dependent manner. In early work, production of autoinducer was hypothesized to be under the same transcriptional control as luciferase (Friedrich and Greenberg, 1983), a hypothesis confirmed by the finding that the putative autoinducer synthase gene, *luxI*, was part of the *lux* operon (Engebrecht et al., 1983; Engebrecht and Silverman, 1984). The autocatalytic nature of autoinducer synthesis was later confirmed (Eberhard et al., 1991). Consistent with these observations, the production of autoinducer (presumably mostly N-3-oxohexanoyl-L-HSL) in batch cultures of *V. fischeri* exhibits a biphasic pattern, a gradual preinduction increase that is followed by a rapid increase after induction begins (Fig. 4.6) (Dunlap, 1992). Enzymatic degradation also might be involved in controlling the cellular levels of N-acyl-L-HSLs and homoserine lactones (Huisman and Kolter, 1994).

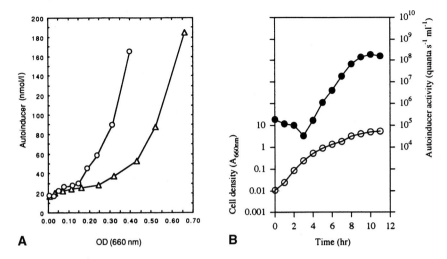

FIGURE 4.6   Regulated production of autoinducers in *V. fischeri*. (**A**) Production of *N*-3-oxo-hexanoyl-L-HSL during growth of *V. fischeri* MJ-1 (wild-type) in the absence (△) and presence (○) of an iron chelator. (**B**) Production of *N*-octanoyl-L-HSL (●) during growth (○) of *V. fischeri* MJ-211 (Δ*luxI*). (**A** Reprinted from Dunlap, 1992; **B** reprinted from Kuo et al., 1994; with permission)

Second Autoinduction System in *Vibrio fischeri*

A second autoinduction system, unrelated to *luxI*, is present in *V. fischeri*. In addition to *N*-3-oxohexanoyl-L-HSL (and *N*-hexanoyl-L-HSL), the synthesis of which is dependent on *luxI*, *V. fischeri* also produces a novel, *luxI*-independent autoinducer, *N*-octanoyl-L-HSL (Kuo et al., 1994) (Table 4.1). Synthesis of *N*-octanoyl-L-HSL is directed by *ainS*, the deduced amino acid sequence of which exhibits no obvious similarity to LuxI or its homologs (Gilson et al., 1995). Like *N*-3-oxohexanoyl-L-HSL, the production of *N*-octanoyl-L-HSL is regulated in a population density-dependent manner (Kuo et al., 1994) (Fig. 4.6), suggesting that expression of *ainS*, like that of *luxI*, is under autoinducer control. Consistent with this suggestion, a possible *lux* box sequence has been identified in the *ainS* promoter region (Gilson et al., 1995). A variety of synthetic autoinducer analogs, including *N*-octanoyl-L-HSL, have been shown to activate *lux* operon expression via LuxR (Eberhard et al., 1986; Schaefer et al., 1996).

Studies indicate that *N*-octanoyl-L-HSL operates as a negative modulator of the *lux* system, competitively inhibiting the interaction of *N*-3-oxohexanoyl-L-HSL with LuxR (Kuo et al., 1996). The effect of *N*-octanoyl-L-HSL may explain how cells prevent induction at low population density. One or two molecules of *N*-3-oxohexanoyl-L-HSL per cell can initiate some induction in *V. fischeri* (Kaplan and Greenberg, 1985); and as a consequence a long-standing question is how premature induction is avoided. *N*-Octanoyl-L-HSL, by interfering with the activity of *N*-3-oxohexanoyl-L-HSL, apparently raises the threshold concentration of N-3-oxohexanoyl-L-HSL necessary to initiate *lux* operon transcriptional activation, and this activity may prevent premature induction at low population density. Also, by delaying induction more *V. fischeri* cells accumulate and contribute to light production, once it is initiated (Kuo et al., 1996). Preliminary evidence indicates that autoinducers control the expression of a variety of genes in *V. fischeri* in addition to the *lux* genes (S.M. Callahan and P.V. Dunlap, unpublished data).

Why N-acyl-L-HSLs?

A possible explanation for the use of N-acyl-L-HSLs as the signal components of autoinduction systems is that the levels of the HSL and acyl groups in the cell may vary with the flow of carbon, nitrogen, and energy through amino acid and fatty acid biosynthesis pathways, increasing with perturbations in the flow associated with the onset of the stationary phase. If so, the levels of HSL and acyl groups could provide a sensitive intracellular signal for changing growth conditions. When combined to form N-acyl-L-HSLs, these metabolites would become membrane-permeant intercellular signals that activate functions adaptive at high population density (Huisman and Kolter, 1994).

Luminescence and Synthesis of Poly-β-hydroxybutyric Acid in Vibrio harveyi

Vibrio harveyi is closely related to *V. fischeri* but differs ecologically, occurring in coastal seawater and sediments at temperatures generally warmer than those in which *V. fischeri* is found. Also, *V. harveyi* is not known to establish light organ symbiosis with marine fish and squids (Nealson and Hastings, 1979; Baumann and Baumann, 1981).

Vibrio harveyi Autoinducer

An autoinducer necessary for population density-responsive expression of luminescence in *V. harveyi* was identified as N-3-hydroxybutanryl-L-HSL (Cao and Meighen, 1989) (Fig. 4.3, Table 4.1). The presence of a chiral center at the 3-carbon of the acyl moiety of N-3-hydroxybutanryl-L-HSL permitted assessment of the pathway by which this compound is synthesized in *V. harveyi*. Chemical synthesis and analysis of the 3-carbon D and L isomers of N-3-hydroxybutanryl-L-HSL indicated that only the D isomer was active as an autoinducer. The D configuration for the 3-hydroxy group implies that the acyl moiety is derived either from fatty acid biosynthesis in the form of an acyl acyl carrier protein (acyl ACP) or via the poly-D-3-hydroxybutyrate biosynthetic pathway in the form of acyl coenzyme A (CoA), not from fatty acid degradation as would be implied by the L configuration. Consistent with the fatty acid biosynthesis pathway, blockage of fatty acid biosynthesis with exogenous fatty acids and with cerulenin, a compound that specifically inhibits 3-keto acyl ACP synthetases, decreased N-3-hydroxybutanryl-L-HSL production in *V. harveyi*

(Cao and Meighen, 1993). These results are consistent with the view, described above, that the acyl moiety of the *V. fischeri* autoinducer *N*-3-oxohexanoyl-L-HSL also is derived from the fatty acid biosynthesis pathway (Eberhard et al., 1991; Cao and Meighen, 1993).

Vibrio harveyi lux Operon

The luminescence enzyme structural genes of the *V. harveyi lux* operon (*luxCDABEGH*) are preceded by a strong inducible promoter, the activity of which is population density-dependent and catabolite-repressible by glucose. Consistent with catabolite repression, the region upstream of the *lux* operon promoter contains a CRP binding site. Regulatory genes with sequence homology to *luxR* and *luxI* of *V. fischeri*, however, are not present in the upstream region or elsewhere on the chromosome (Miyamoto et al., 1988a,b). The presence of a recognizable −10 sequence but no corresponding −35 sequence in the *lux* operon promoter region implied nonetheless that a regulatory protein was necessary to activate transcription by RNA polymerase (Miyamoto et al., 1990; Swartzman et al., 1990b).

Vibrio harveyi LuxR Protein

The gene for the regulatory protein LuxR, a *trans*-acting regulatory factor necessary for stimulation of *V. harveyi lux* operon expression, was identified and found not to be linked to the *lux* operon on the *V. harveyi* chromosome (Table 4.1). Furthermore, *V. harveyi* cells mutagenized in the chromosomal region containing the gene for the regulatory protein continued to produce autoinducer, indicating that a third locus, distinct from that region and from the *lux* operon, was necessary for autoinducer synthesis (Martin et al., 1989; Showalter et al., 1990). Although apparently analogous to the *V. fischeri* LuxR protein, and therefore also designated LuxR, and exhibiting features similar to certain DNA-binding proteins, the *V. harveyi* LuxR protein exhibits no obvious sequence similarity to the *V. fischeri* protein. Furthermore, its stimulation of *lux* operon expression, which is not population density-responsive, does not require autoinducer; conditioned medium (i.e., containing autoinducer) failed to induce *lux* operon expression in the presence of the *V. harveyi luxR* gene in *E. coli* containing the *lux* operon. This apparent lack of autoinducibility suggested that an additional regulatory element was involved in the autoinduction mechanism in this species (Showalter et al., 1990). Nonetheless, primer extension analysis of the *lux* operon promoter and mobility shift assays confirmed the requirement in *lux* operon transcriptional activation for the LuxR protein, an M_r 23-kDa polypeptide, and its *lux*-specific DNA binding at two sites in the regulatory region upstream of the *lux* operon transcription initiation site (Swartzman et al., 1992; Swartzman and Meighen, 1993; Miyamoto et al., 1994). Binding of the protein at both sites was found to be necessary for maximal *lux* operon expression, which was stimulated by autoinducer and repressed by glucose. The *luxC* proximal site, however, is critical for transcriptional activation. Elimination of LuxR binding at this site simultaneously eliminated the autoinducer stimulation and glucose repression, suggesting that these effects operate either indirectly (by controlling the expression of LuxR) or directly (by mediating LuxR binding) (Miyamoto et al., 1994).

Multiple Signaling Systems Controlling Luminescence in *Vibrio harveyi*

Substantial clarification of the autoinduction mechanism in *V. harveyi* was obtained in studies that identified, in addition to LuxR, two separate signal-response systems and a repressor controlling *lux* operon expression. At the protein sequence level, these signal-response systems exhibit no obvious similarity to the *V. fischeri*-LuxI and LuxR proteins (Bassler et al., 1993, 1994a,b). By analyzing recombinant clones that restored luminescence to *V. harveyi* dim and dark mutants, several regulatory components were identified (Table 4.1). One set of mutants, defining one signal-response system, included strains with defects in *luxL* and *luxM*, which are necessary for synthesis of an autoinducer, presumably *N*-3-hydroxybutanryl-L-HSL, and *luxN*, which encodes a protein necessary for activating *lux* operon expression in response to that autoinducer. LuxN, with domains for the histidine kinase sensor and the response regulator, apparently is a member of the two-component family of signal transduction proteins (Bassler et al., 1993). Another regulatory gene, *luxO*, encoding a repressor protein, was also identified. The deduced amino acid residue sequence of LuxO shows similarity to the response regulator domain of the two-component signal transduction proteins, and it contains a DNA-binding motif (Bassler et al., 1994b). The second signaling system includes a putative second autoinducer, as yet not identified at the chemical or genetic level, and the *luxPQ* genes, which direct the synthesis of a sensor protein necessary for response to

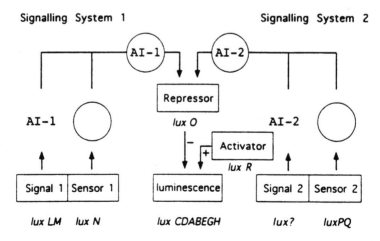

FIGURE 4.7   Regulatory model for control of *lux* gene expression in *V. harveyi*. See text for details. (Reprinted from Bassler et al., 1994a, with permission)

the second autoinducer. Like LuxN, the amino acid residue sequence of LuxQ exhibits regions with similarity to the histidine protein kinase and the response regulator components of the two-component family of signal transduction proteins (Bassler et al., 1994a). These sequence similarities introduce the possibility that interactions of LuxN and LuxQ with the autoinducers initiate phospho-relay reactions leading to *lux* operon transcriptional activation (Bassler et al., 1994a). Specifically, through autophosphorylation of LuxN and LuxQ at the sensor kinase domains in response to autoinducers, and subsequent phosphoryl transfer to the response regulator domain, the autoinducers would mediate interaction of LuxN and LuxQ with the repressor LuxO. That interaction is proposed to inactivate the repressor, thereby potentiating the *lux* operon for transcriptional activation by LuxR (Bassler et al., 1994a) (Fig. 4.7).

Comparison of the *Vibrio fischeri* and *Vibrio harveyi* Autoinduction Systems

Autoinduction of luminescence in *V. fischeri* and *V. harveyi* is in many ways similar, including population density-dependent induction mediated by a diffusible autoinducer, control by cAMP and CRP, involvement of two autoinducers (the synthesis of which is directed by two genes), and requirement for a transcriptional activator protein for *lux* operon expression. (Nealson et al., 1970, 1972; Eberhard, 1972; Ulitzur and Yashphe, 1975; Chen et al., 1985; Martin et al., 1989; Showalter et al., 1990; Bassler et al., 1993, 1994a).

Despite these similarities, the autoinduction mechanisms of these two species differ in fundamental ways, as was implied by the absence of regulatory genes contiguous with the enzyme structural genes in *V. harveyi* (Miyamoto et al., 1988b). The major differences between the two systems include the presence of a repressor protein, LuxO, in the *V. harveyi* system, the role of the *V. harveyi* autoinducers, which via sensor proteins are thought to inactivate the repressor instead of directly effecting *lux* operon transcription by activating LuxR, and the lack of interaction between the *V. harveyi* LuxR protein and the autoinducers (Bassler et al., 1993, 1994a,b). The significance of these differences is not obvious from evolutionary and ecological viewpoints, as *V. fischeri* and *V. harveyi* are closely related and coexist in some habitats; they also both exist in a variety of habitats (Baumann and Baumann, 1981). Nonetheless, the presence of multiple cross-acting autoinduction systems in both species suggests the importance of being able to sense and respond to complex and changing conditions in a variety of habitats.

An intriguing link between the luminescence autoinduction systems of *V. harveyi* and *V. fischeri* has been identified. The C-terminal portion of the *V. fischeri* AinS deduced sequence is 34% identical to *V. harveyi* LuxM, and the N-terminal portion of *V. fischeri* AinR is 38% identical to the amino-terminus of *V. harveyi* LuxN (Gilson et al., 1995). These observations highlight the notions that autoinduction systems are complex, and that as a consequence even evolutionarily related autoinduction systems in closely related bacteria can appear dissimilar in the absence of detailed genetic analysis.

Control of PHB Production by Autoinducer in *Vibrio harveyi*

Production of the fatty acid storage product poly-3-hydroxybutyrate (PHB) was shown to be controlled in a population density-dependent manner by *N*-3-hydroxybutanoyl-L-HSL (Sun et al., 1994) (Table 4.1). These data comprise the first documentation of a non-*lux* function controlled by an autoinducer in a luminous bacterium.

Autoinduction of Luminescence in Other Light-Producing Bacteria

Many bacteria other than *V. fischeri* and *V. harveyi* produce light. Most of them are marine members of the Vibrionaceae. Little is known about the regulation of luminescence in these species (for review: Dunlap, 1991). Analysis of luminescence autoinduction in these little-studied species, in comparison with the autoinduction systems described for *V. fischeri* and *V. harveyi*, could provide substantial insight on the mechanistic similarities and differences within an ecologically and phylogenetically closely related group of bacteria. Nonluminous members of the Vibrionaceae also would interesting subjects for consideration.

Autoinduction in Nonluminous Bacteria

In addition to the light-producing bacteria, several species of proteobacteria have been found to utilize *N*-acyl-L-HSL autoinducers. The functions that are controlled and the bacteria themselves are diverse, indicating that autoinduction is a multifaceted, highly integrated aspect of bacterial life. The themes of the chemical similarity of the autoinducers and homologies to *V. fischeri* LuxI/LuxR aside, the presently identified bacteria using autoinducers associate, like *V. fischeri*, with higher organisms, in many cases as pathogens.

Virulence Determinants in Pseudomonas aeruginosa

Pseudomonas aeruginosa is an opportunistic pathogen of humans, colonizing lung tissues of immunocompromised individuals and those with cystic fibrosis and infecting tissues at wound and burn sites. Several extracellular factors important to pathogenesis are produced by this bacterium, including elastases, an exotoxin, and an alkaline protease (for references: Passador et al., 1993).

Pseudomonas aeruginosa lasR and lasI Genes and Autoinducer

Early indications of autoinducer control of virulence determinants in *P. aeruginosa* include the regulated, late-exponential and stationary phase synthesis of elastase, the weak activity of the promoter for an elastase structural gene *lasB* in *E. coli* compared to that in *P. aeruginosa*, and the identification of a gene (*lasR*) encoding a transcriptional activator necessary for elastase synthesis, with 27% identity to the deduced amino acid sequence of *luxR* (Iglewski et al., 1990; Gambello and Iglewski, 1991) (Fig. 4.5). Expression of *lasR*, which is not regulated by LasR, is low until cells reach the late-exponential phase of growth; *lasR* activation, which is suppressed in the presence of iron and NaCl, requires an unidentified regulatory factor that is present in *P. aeruginosa* but absent from *E. coli* (Albus and Iglewski, 1992). The activity of the LasR protein requires the autoinducer *N*-3-oxododecanoyl-L-HSL, the production of which is dependent on the *P. aeruginosa lasI* gene in *P. aeruginosa* and *E. coli* (Fig. 4.3, Table 4.1). The deduced amino acid sequence of LasI exhibits approximately 35% identity and 56% similarity to that of the *V. fischeri* LuxI protein (Passador et al., 1993; Pearson et al., 1994). A variety of functions are controlled in *P. aeruginosa* by LasR (for references: Passador et al., 1993; Pearson et al., 1994); it is likely that *N*-3-oxododecanoyl-L-HSL also controls their expression.

In *E. coli*, expression of the *lasB* gene can be activated by the *V. fischeri* LuxR protein and *N*-3-oxohexanoyl-L-HSL, as can expression of the *V. fischeri lux* operon by LasR and *N*-3-oxododecanoyl-L-HSL. However, neither LasR nor LuxR exhibits substantial activity with the heterologous autoinducer. These results imply that the various autoinducer–receptor protein complexes are nonspecific in their recognition of the *lux* box sequences preceding *lasB* and the *lux* operon, but that the receptor proteins are specific in their recognition of their cognate autoinducers (Gray et al., 1994). Consistent with this interpretation, LasR and *N*-3-oxododecanoyl-L-HSL activate transcription from the *lasI* promoter, upstream of which a consensus *lux* box is not found, whereas LuxR and *N*-3-oxohexanoyl-L-HSL do not. It is possible that the region upstream of *lasI* contains a unique ''*las* box'' recognizable, like the *lasB lux* box, to LasR but not to LuxR (Seed et al., 1995).

Control of *N*-3-oxododecanoyl-L-HSL synthesis apparently is autocatalytic in *P. aeruginosa*. Both *lasR* and *lasI* on multicopy plasmids are necessary for stimulation of β-galactosidase synthesis in *E. coli* containing a *lasI::lacZ* fusion; the requirement for *lasI* can be eliminated with exogenous *N*-3-oxododecanoyl-L-HSL (Seed et al., 1995). Furthermore, half-maximal expression from the *lasI* promoter requires one-tenth the *N*-3-oxododecanoyl-L-HSL needed for half-maximal expression from the *lasB* promoter. This observation suggests the presence of an autoinduction regulatory hierarchy in which low concentrations of *N*-3-oxododecanoyl-L-HSL via LasR activate *lasI* transcription, leading to an accumulation of *N*-3-oxododecanoyl-L-HSL, which then also via LasR activates transcription of *lasB* and other virulence determinant genes (Seed et al., 1995).

Second Autoinduction System in *Pseudomonas aeruginosa*

A potentially complicating issue is that *P. aeruginosa* also produces a second autoinducer, *N*-butanoyl-L-HSL, involved in *lasB* expression (Fig. 4.3, Table 4.1). A gene other than *lasI* is necessary for synthesis of *N*-butanoyl-L-HSL, which appears to control *lasB* expression indirectly by influencing *lasR* expression but without a direct effect of *lasR* or *lasI* (Pearson et al., 1995).

Autoinducer-mediated regulation of virulence determinants in *P. aeruginosa* therefore is likely to be complex.

Adding to that complexity, a second LuxR homolog, *rhlR*, was identified recently in *P. aeruginosa* (Table 4.1). The *rhlR* gene specifies a 28-kDa protein with 23%, 31%, 29%, and 40% amino acid sequence identity to *V. fischeri* LuxR, *P. aeruginosa* LasR, *Rhizobium melliloti* RhiR, and *E. coli sdiA* (Ochsner et al., 1994) (Fig. 4.5). The RhlR protein controls the production of rhamnolipid biosurfactants. An *rhlR* mutant of *P. aeruginosa*, in addition to not producing rhamnolipids, also failed to produce elastase activity, indicating an apparent overlap in control by RhlR and LasR (Ochsner et al., 1994).

A gene downstream of *rhlR* has been identified that exhibits sequence similarity to *luxI*. This gene, *rhlI*, is necessary for production of a signal that activates *rhlAB* (rhamnolipid biosynthesis structural genes) via RhlR (Ochsner and Reiser, 1995). Two genes possibly equivalent to *rhlR* and *rhlI*, termed *vsmR* and *vsmI*, have been identified in a different strain of *P. aeruginosa*; *vsmI* directs the synthesis of *N*-butanoyl-L-HSL and *N*-hexanoyl-L-HSL in *P. aeruginosa* and *E. coli* (Latifi et al., 1995; Winson et al. 1995). RhlR and *N*-butanoyl-L-HSL indirectly control *lasB* and directly control several genes other than *lasB* (Pearson et al., 1995; Latifi et al., 1995; Winson et al., 1995).

Rationale for Autoinduction in *Pseudomonas aeruginosa*

A rationale for population density-dependent control of gene expression in *P. aeruginosa* and other pathogens is to delay the expression of virulence factors and thereby avoid invoking defense reactions by the host until enough bacteria have accumulated to survive or overcome that reaction. Similarly, delaying expression of extracellular enzymes until a high population density is attained may counteract diffusion of the enzyme and its degradation products (Fuqua et al., 1994).

Conjugal Ti Plasmid Transfer in Agrobacterium tumefaciens

Virulent strains of *Agrobacterium tumefaciens*, a bacterium common in the plant rhizosphere, can elicit the formation of crown galls (neoplasias) in higher plants. The oncogenic interaction is initiated by the transfer of DNA fragments (T-DNA) from large Ti (tumor inducing) bacterial plasmids to the nucleus of the infected plant cell. Proteins encoded by the T-DNA, in addition to eliciting overproduction of the growth hormones auxin and cytokinin, which stimulate rapid plant cell growth, also cause the plant cells to produce and secrete opines, arginine derivatives utilized by *A. tumefaciens* as nutrients for growth. Opines also elicit the conjugal transfer of Ti-plasmids from plasmid-containing to Ti-plasmidless *A. tumefaciens* cells. Specifically, octopine elicits the transfer of the octopine-type Ti plasmids (e.g., pTiR10), and agrocinopines A and B elicit the transfer of nopaline-type Ti plasmids (e.g., TiC58). The opines regulate both the Ti-plasmid transfer (*tra*) genes and the genes for opine catabolism (octopine catabolism, *occ*; agrocinopine catabolism, *acc*) (for references: Fuqua and Winans, 1994; Fuqua et al., 1994; Hwang et al., 1994).

Agrobacterium tumefaciens traI and *traR* Genes and Autoinducer

The octopine-type and nopaline-type systems are controlled by an autoinducer, *N*-3-oxooctanoyl-L-HSL and a LuxR homolog, TraR (Fig. 4.3, Table 4.1) (Piper et al., 1993; Zhang et al., 1993; Fuqua and Winans, 1994; Hwang et al., 1994). In the octopine-type plasmid, pTiR10, OccR, a LysR-type transcriptional activator, in the presence of octopine activates transcription of the *occ* locus and *traR*. TraR, which exhibits 18% sequence identity to *V. fischeri* LuxR (Fig. 4.5), in turn activates *tra* gene expression in the presence of *N*-3-oxooctanoyl-L-HSL, the synthesis of which is dependent on *traI*, a Ti-plasmid gene whose protein product is approximately 30% identical in amino acid residue sequence to LuxI (Table 4.1, Fig. 4.8). The putative *N*-3-oxooctanoyl-L-HSL–TraR complex activates transcription from the *traI* promoter, thereby forming a positive

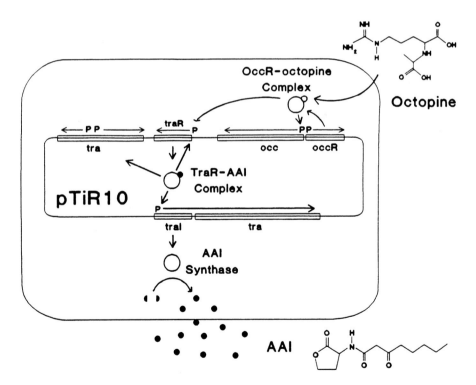

FIGURE 4.8 Regulatory model for control of *tra* gene expression in *A. tumefaciens*. Regulation in the octopine-type Ti-plasmid is shown. AAI refers to *N*-3-oxooctanoyl-L-HSL. In the nopaline-type Ti-plasmid, agrocinopine inactivates a repressor, AccR. See text for details. (Reprinted from Fuqua and Winans, 1994, with permission)

feedback circuit for autoinducer synthesis similar to that in *V. fischeri* (Fuqua and Winans, 1994). In contrast, in the nopaline-type plasmid pTiC58, primary regulation of the system is mediated by a repressor, AccR. The repression of *traR* expression by AccR in pTiC58 is relieved by agrocinopine, permitting *N*-3-oxooctanoyl-L-HSL via TraR to activate expression of *traI* and apparently *acc* (Beck von Bodman et al., 1992; Piper et al., 1993; Hwang et al., 1994). The TraR proteins from the octopine- and nopaline-type Ti plasmids are 81% identical (Fuqua and Winans, 1994). Consistent with *tra* gene control by a putative autoinducer–receptor protein complex in both systems, the region upstream of the octopine *traI* and *traA* genes and of the nopaline *traI* gene contain a *lux* box-like element (Fig. 4.4) (Fuqua and Winans, 1994; Fuqua et al., 1994; Gray et al., 1994; Hwang et al., 1994). Autoinduction of *tra* and *occ* (or *acc*) gene expression might serve as a signal amplification mechanism in each bacterium instead of, or as well as, provoking Ti-plasmid transfer among donor cells (Fuqua and Winans, 1994).

In addition to control by the hierarchy of a host-produced opine and a self-produced autoinducer, an additional element or elements are involved in *A. tumefaciens tra* gene expression. Studies with the octopine-type system indicate that cells in liquid culture induce *tra* gene expression only after the addition of octopine and *N*-3-oxooctanoyl-L-HSL; addition of octopine alone does not lead to *tra* gene induction, even for cells in stationary phase. In contrast, cells grown on solid medium induce *tra* gene expression after the addition of octopine alone. These results implicate an additional control element in *N*-3-oxooctanoyl-L-HSL-TraR-mediated regulation of *tra* gene expression, one associated with solid-phase growth (Fuqua and Winans, 1994).

Negative Modulation of *A. tumefaciens* Autoinduction System

Reports indicate that a novel protein, TraM, negatively modulates the response of TraR to *N*-3-oxooctanoyl-L-HSL in *A. tumefaciens* (Table 4.1) (Fuqua et al., 1995; Hwang et al., 1995). The presence of *traM* at high copy number suppresses *N*-3-oxooctanoyl-L-HSL-TraR-mediated activation of *tra* gene expression, an effect that can be overcome with high levels of TraR but not *N*-3-oxooctanoyl-L-HSL. Furthermore, mutations in *traM* result in strong increases in *tra* gene expression, synthesis of *N*-3-oxooctanoyl-L-HSL, and conjugation. Expression of *traM* is dependent on TraR. TraM, however, has no effect on the expression of *traR*, indicating that the modulatory effect of TraM is not due to *traR* repression. Apparently, TraM functions to prevent *tra* gene induction by low levels of *N*-3-oxooctanoyl-L-HSL, possibly by sequestering TraR or *N*-3-oxooctanoyl-L-HSL (Fuqua et al., 1995; Hwang et al., 1995). TraM of the octopine-type Ti-plasmid is 77% identical to that of the nopaline-type Ti-plasmid (Fuqua et al., 1995). It is not obvious that the activity of TraM would account for the solid-phase dependent *tra* gene regulation mentioned above.

Exoenzyme and Antibiotic Production in Erwinia carotovora

Erwinia carotovora is a plant pathogen causing soft-rot disease through the activity of plant cell wall-degrading exoenzymes. Production of these enzymes is regulated, occurring at the onset of the stationary phase (Saarilahti et al., 1990; Pirhonen et al., 1993). *E. carotovora*, along with actinomycetes and certain other terrestrial enteric bacteria, also produces carbapenems, β-lactam antibiotics with antibacterial and β-lactamase-inhibitory properties. Production in *E. carotovora* of both the exoenzyme virulence determinants and carbapenem is regulated by autoinduction.

Autoinduction System of *Erwinia carotovora*

Studies of the growth phase-dependent production of exoenzymes in *E. carotovora* identified the *expI* gene (exoenzyme production inducer) as necessary for transcriptional activation of the genes encoding polygalacturonase, pectate lyase, pectin lyase, cellulase, and protease and for plant virulence (Pirhonen et al., 1993). The ExpI amino acid sequence is 30% identical to that of *V. fischeri* LuxI (Table 4.1) (Pirhonen et al., 1993). In an *expI* mutant of *E. carotovora*, the *expI* gene or addition of *N*-3-oxohexanoyl-L-HSL complemented exoenzyme production; and addition of *N*-3-oxohexanoyl-L-HSL reversed the avirulent phenotype of the mutant. The functional similarity of ExpI to LuxI and the ability of *N*-3-oxohexanoyl-L-HSL to complement the exoenzyme and virulence defects of the mutant suggest that ExpI directs the synthesis of *N*-3-oxohexanoyl-L-HSL (Pirhonen et al., 1993) or possibly a similar compound.

Initial insight into control of carbapenem production by autoinducer was obtained though cross-feeding studies between two groups of *E. carotovora* mutants unable to produce carbapenem. One group was found to release a compound that restored carbapenem production in the second group. The membrane-permeant compound was purified and identified as *N*-3-oxohexanoyl-L-HSL (Fig. 4.3); the chemically synthesized compound restored carbapemen production in the second group of mutants (Bainton et al., 1992a,b). An *E. carotovora* gene, *carI*, also complemented the ability of a member of the second group to produce carbapenem; CarI exhibits approximately 25% identity to *V. fischeri* LuxI and is similar to ExpI (Table 4.1, Fig. 4.5) (Swift et al., 1993). Consistent with the *expI* mutant studies mentioned above (Pirhonen et al., 1993), the *carI* mutants were found also to produce lower levels of the exoenzymes pectinase, cellulase, and protease than the wild-type, indicating a possible coordinate regulation of carbapenem and exoenzyme production by autoinducer. Exogenous *N*-3-oxohexanoyl-L-HSL restored the wild-type levels of expression of these enzymes and wild-type virulence (Jones et al., 1993). A second gene, *carR*, encoding a protein with sequence homology to *V. fischeri* LuxR, was isolated from *E. carotovora*. CarR functions in *trans* to activate carbapenem production and might play a role in exoenzyme regulation (McGowan et al., 1995).

One can surmise from these observations that two separate but potentially similar autoinduction systems, ExpI (and presumably ExpR) (Pirhonen et al., 1993) and CarI/CarR, are involved in regulating the production of exoenzyme virulence determinants and carbapenem, respectively, in *E. carotovora*, with some cross-talk between the two systems. Why antibiotic production is under population density-dependent control is not obvious. Possibly it is a way by which *E. carotovora* prevents other bacteria from colonizing a plant it has infected and competing with it for nutrients released by its exoenzymes.

Capsular Polysaccharide Production in Erwinia stewartii

Erwinia stewartii is the agent that causes Stewart's wilt of sweet corn. A key virulence factor is the production of an extracellular polysaccharide (EPS) capsule, which provides protection against host defenses and promotes plant necrosis (for references: Beck von Bodman and Farrand, 1995). Production of EPS and virulence are dependent on an autoinducer, identified as *N*-3-oxohexanoyl-L-HSL, the synthesis of which is directed by a gene, *esaI*, encoding a protein with 25% amino acid sequence identity to LuxI of *V. fischeri* (Table 4.1). The deduced amino acid sequence of a linked and convergently expressed gene, *esaR*, exhibits 24% identity to *V. fischeri* LuxR (Beck von Bodman and Farrand, 1995) (Fig. 4.5).

Starvation Survival and Cell Division in Escherichia coli

An *N*-acyl-L-HSL might be involved in the ability of *E. coli* to survive starvation. In *E. coli*, starvation survival requires the expression of several genes, a process regulated in part by the starvation phase specific sigma factor (σ^s), a product of the *rpoS* gene, the expression of which increases at the onset of the stationary phase (for references: Huisman and Kolter, 1994). Another gene, *rspA* (regulatory in stationary phase A), was identified whose product at high copy number blocked *rpoS* expression at the onset of the stationary phase. Sequence similarities of RspA to certain catabolic enzymes suggested that the effect of RspA on *rpoS* expression might involve degradation of a compound involved in signaling starvation, and the similarity of RspA to the lactonizing enzyme chlormuconate cycloisomerase suggested that the signal compound might be a lactone. Consistent with these possibilities, mutants defective in the threonine biosynthesis pathway at steps prior to the synthesis of homoserine fail to induce *rpoS* expression, except upon addition of homoserine lactone. Furthermore, overproduction of RspA interferes with the HSL-mediated stimulation of σ^s accumulation and with *N*-3-oxohexanoyl-L-HSL-mediated *lux* operon induction in *E. coli*. RspA presumably enzymatically destroys HSL and *N*-3-oxohexanoyl-L-HSL (Table 4.1) (Huisman and Kolter, 1994). Thus the accumulation of homoserine lactone at the onset of the stationary phase may be an important intracellular signal that activates the σ^s response under starvation conditions regardless of population density; acylation of this signal, which would confer membrane permeability, converts it into an intercellular signal for inducing functions whose activities are advantageous at high population density (Huisman and Kolter, 1994).

Cell division in *E. coli* also apparently is under an autoinducer-mediated population density-dependent control (Table 4.1). The SdiA protein (suppressor of division inhibition A, previously designated UvrC-28 kDa) of *E. coli*, which exhibits 24% amino acid sequence identity to LuxR of *V. fischeri* (Henikoff et al., 1990), positively controls a cluster of genes (*ftsQ*, *ftsA*, and *ftsZ*) whose products are involved in cell septation. The *ftzQAZ* genes are controlled by two promoters, P1 and P2, upstream of *ftsQ*; *ftsZ* is further controlled by four promoters within *ftsA* (Aldea et al., 1990; Wang et al., 1991). Expression from the P1 promoter, which is enhanced as the growth rate decreases, is dependent on σ^s, whereas the P2 promoter functions primarily in actively growing cells (Aldea et al., 1990). Therefore by mediating *rpoS* expression or the expression and activity of SdiA, *N*-acyl-L-HSL could control both starvation survival and cell division in the stationary phase. Recently, evidence indicating production of an extracellular factor regulating *SdiA* expression in *E. coli* has been obtained (Garcia-Lara et al., 1996).

Phenazine Synthesis in Pseudomonas aureofaciens

Fluorescent pseudomonads produce phenazine antibiotics important to the suppression of fungal pathogenesis of wheat and to long-term survival of the bacteria in the wheat rhizosphere. A gene, *phzR* (phenazine regulator), for a positive regulator of phenazine antibiotic production, has been characterized from *Pseudomonas aureofaciens* 30–84. The PhzR deduced amino acid residue sequence (Fig. 4.5) exhibits 45% or more similarity and 23% or more identity to LuxR, LasR, TraR, and SdiA of *V. fischeri*, *P. aeruginosa*, *A. tumefaciens*, and *E. coli*, respectively, suggesting that phenazine synthesis is controlled by autoinduction. PhzR apparently positively regulates expression from the *phzR* promoter, which is transcribed divergently from the *phz* structural genes (Pierson et al., 1994). Also consistent with autoinduction control, phenazine antibiotics are produced only during the late exponential and stationary phases of batch culture growth (Turner and Messenger, 1986); and culture supernatants of *P. aureofaciens* 30–84 stimulate expression of *phzB*, a phenazine synthesis gene (Pierson et al., 1994). Most recently, a gene (*phzI*) from *P. aureofaciens* 30–84 was characterized that is necessary for *P. aureofaciens* to produce an autoinducer-like compound that induces expression of the phenazine synthesis genes; the PhzI deduced amino acid residue sequence exhibits 31% identity and 49% similarity to LuxI (Table 4.1, Fig. 4.5) (Wood and Pierson, 1996). The *phzI* promoter region contains a 20-bp palindromic sequence similar to the consensus *lux* box (Wood and Pierson, 1996), so transcriptional activation by the putative autoinducer and PhzR in a manner analogous to *lux* operon expression in *V. fischeri* is a reasonable expectation. It is intriguing to note that medium conditioned by the growth of a different soil bacterium, *P. fluorescens*, also activates expression of the phenazine synthesis genes in *P. aureofaciens*, which suggests that the putative *P. aureofaciens* phenazine biosynthesis autoinducer is produced by other rhizosphere bacteria (D. Wood and L.S. Pierson, personal communication). A rationale for population density-dependent control of phenazine synthesis may relate to antagonistic interactions between *P. aureofaciens* and plant pathogenic fungi in the rhizosphere.

Extracellular Signals in Pseudomonas solanacearum

The phytopathogenic bacterium *Pseudomonas solanacearum* uses a multicomponent system to regulate virulence determinants (e.g., extracellular polysaccharide and endoglucanase) involved in a lethal wilting disease of plants. Evidence has been obtained indicating involvement of a volatile extracellular factor, probably 3-hydroxy-hexadecanoic acid methyl ester, and an *N*-acyl-L-homoserine lactone in the expression of virulence determinants (Clough et al., 1994; T.P. Denny, personal communication).

Rhizosphere Interactions in Rhizobium leguminosarum

As indicated by studies of *A. tumefaciens* and *P. aureofaciens* (see above), the rhizosphere habitat is a productive area for research into autoinducers as signal-response mediators of interactions between bacteria and higher organisms. For example, rhizobia, because of their association with legumes and the importance of chemical signaling in the rhizobium-legume symbiosis (Roth and Stacey, 1991), seem to be prime candidates for utilizing autoinducers to control the expression of genes in a host-responsive, population density-dependent manner. It is intriguing also that certain pea-modulating biovars of *Rhizobium leguminosarum* are able to utilize homoserine, a compound excreted in substantial quantity by pea roots, as a carbon and nitrogen source (van Egeraat, 1975; Johnston et al., 1988).

A regulator of rhizosphere-expressed genes (*rhiR*), a gene with homology to the *V. fischeri* LuxR protein (approximately 23% identity) and to the *E. coli sdiA* product, was identified in *R. leguminosarum* biovar viciae (Table 4.1). RhiR positively controls expression of the *rhiABC* genes, the products of which may be involved in permitting *R. leguminosarum* to utilize a plant metabolite (Cubo et al., 1992). The *rhiA* gene product, rhizosphere protein A, is a prominent protein in stationary-phase cultures and is synthesized by the bacteria in the pea rhizosphere but not by bacteroids in pea nodules (Dibb et al., 1984). Expression of *rhiR* is repressed by *nod* gene-inducing flavonoids via the regulatory protein NodD (Cubo et al., 1992). Consistent with these observations, a C_{14}-*N*-acyl-L-HSL has been obtained from *R. leguminosarum* that together with *rhiR* activates *rhiABC* expression as well as a growth inhibiting compound (Gray et al., 1996; Schripsema et al., 1996).

Autoinduction Systems in Other Bacterial Species

Many other bacterial species produce autoinducers and contain LuxI and LuxR homologs; information on them is accumulating rapidly. *Enterobacter agglom-*

erans produces autoinducer activity, possibly *N*-3-oxohexanoyl-L-HSL; and a gene, *eagI*, has been isolated whose deduced amino acid sequence is approximately 25% identical to LuxI (Table 4.1) (Swift et al., 1993). The psychrotropic bacterium *Yersinia enterocolitica*, a causative agent of food poisoning in humans, produces *N*-3-oxohexanoyl-L-HSL, the synthesis of which is directed by a gene, *yenI*, whose protein product is approximately 20% identical in amino acid sequence to LuxI homologs. In contrast to the *V. fischeri* system, the promoter region for *yenI* lacks a *lux* box, and *yenI* expression does not exhibit autoinduction. Furthermore, the deduced amino acid sequence of an open reading frame (ORF) immediately downstream of *yenI* shows 21% identity to *V. fischeri* LuxR (Table 4.1, Fig. 4.5). Analysis of a *yenI* mutant suggests the possibility also of a second signal compound in *Y. enterocolitica* (Throup et al., 1995). *N*-Acyl-L-HSLs also appear to be involved in controlling production of the pigment violacein, a hemolysin and an extracellular protease in *Chromobacterium violaceum* (Table 4.1) through a repressor-mediated mechanism (Winson et al., 1994). Other enteric bacteria (e.g., *Citrobacter freundii, Enterobacter herbicola, Hafnia alvei, Rahnella aquatilis, Proteus mirabilis*, and *Serratia marcescens*) also produce an *N*-acyl-L-HSL (Swift et al., 1993).

Perspectives

Over the next few years, the number of bacterial species known to use *N*-acyl-L-HSLs is likely to increase sharply as more species are examined. Such an increase is particularly likely for members of the Vibrionaceae, in which autoinduction was discovered and first characterized. It is likely also that many bacteria will be found to control multiple activities with autoinducers, using proteins homologous and nonhomologous to *V. fischeri* LuxI and LuxR to produce and respond to multiple autoinducers, and to use more than one autoinducer to control individual activities. A rationale for this proposal is that the conditions of high population density are physiologically complex in any given habitat, and those conditions change with time, presenting the cell with various and changing environmental stimuli. This environmental complexity is likely to require numerous mechanisms for sensing and responding to these stimuli, as well as substantial flexibility in positive and negative regulation to integrate the inputs and coordinate the responses (Bassler et al., 1994a). The use of autoinducers to mediate cellular interactions between species is another possibility (McKenney et al., 1995), especially in host–symbiont interactions and mixed microbial assemblages. For example, a higher organism (plant or animal) might synthesize and release autoinducers to prevent colonization by a potential pathogen by altering gene expression of the pathogen in a deleterious way. Alternatively, activities appropriate to mutualism (i.e., light production or production of antibiotics that block a potential pathogen from infecting the host) could be turned on by a host through release of a symbiont-specific autoinducer. Furthermore, the ability to rapidly degrade the homoserine or HSL signal in the extracellular environment, and possibly to use it as a nitrogen and carbon source, might be a virulence or mutualism determinant. Alternatively, the possibility exists that the bacterial pathogen or mutualist alters host gene expression to benefit the bacterium through the release of an autoinducer in a manner analogous to induction of opine synthesis in plant cells by T-DNA from *A. tumefaciens*. At present, however, there are no reports of higher organisms releasing *N*-acyl-L-HSLs or of eukaryotic gene regulation by these compounds. Similarly, in mixed bacterial assemblages, such as in the plant rhizosphere, the potential exists for one type of bacterium to release an autoinducer that inappropriately alters gene expression in, and lowers the fitness of, a competitor, an interaction in some ways similar to the release of antibiotics. The need for defense against this kind of antagonistic interaction could drive specificity in sensing and responding to *N*-acyl-L-HSLs signals. From the perspective of multicellularity and intercellular signaling, studies of autoinduction are likely to bring forth many insights into cooperative behavior in bacteria and its adaptive significance.

Acknowledgments

I thank S. Beck von Bodman, E.P. Greenberg, M. Silverman, and S.C. Winans for permission to use previously published figures; M. Pascual for comments on the manuscript; and T.P. Denny, E.P. Greenberg, D.W. Wood, L.S. Pierson, and P. Williams for information prior to publication. Work in the author's laboratory is supported in part by National Science Foundation grant MCB 94-08266. This work is contribution 8980 from the Woods Hole Oceanographic Institution.

References

Adar, Y.Y., and Ulitzur, S. (1993) GroESL proteins facilitate binding of externally added inducer by LuxR protein-containing *E. coli* cells. *J. Biolumin. Chemilumin.* **8**: 261–266.

Adar, Y.Y., Simaan, M., and Ulitzur, S. (1992) Formation of the LuxR protein in the *Vibrio fischeri lux* system is controlled by HtpR through the GroESL proteins. *J. Bacteriol.* **174**:7138–7143.

Albus, A.M., and Iglewski, B.H. (1992) Transcriptional activation of LasR. In *Abstracts of the General Meeting, 1992, of the American Society for Microbiology*, abstract D-46, p. 103. American Society for Microbiology, Washington, DC.

Aldea, M., Garrido, T., Pla, J., and Vicente, M. (1990) Division genes in *Escherichia coli* are expressed coordinately to cell septum requirements by gearbox promoters. *EMBO J.* **9**:787–3794.

Bainton, N.J., Bycroft, B.W., Chhabra, S.R., et al. (1992a) General role for the *lux* autoinducer in bacterial cell signalling, control of antibiotic biosynthesis in *Erwinia*. *Gene* **116**:87–91.

Bainton, N.J., Stead, P., Chhabra, S.R., et al. (1992b) *N*-(3-Oxohexanoyl)-L-homoserine lactone regulates carbapenem antibiotic production in *Erwinia carotova*. *Biochem. J.* **288**:997–1004.

Baldwin, T.O., Devine, J.H., Heckel, R.C., Lin, J.-W., and Shadel, G.S. (1989) The complete nucleotide sequence of the *lux* regulon of *Vibrio fischeri* and the *luxABN* region of *Photobacterium leiognathi* and the mechanism of control of bacterial bioluminescence. *J. Biolumin. Chemilumin.* **4**:326–341.

Bassler, B.L., Wright, M., and Silverman, M.R. (1994a) Multiple signalling systems controlling expression of luminescence in *Vibrio harveyi*, sequence and function of genes encoding a second sensory pathway. *Mol. Microbiol.* **13**:273–286.

Bassler, B.L., Wright, M., and Silverman, M.R. (1994b) Sequence and function of LuxO, a negative regulator of luminescence in *Vibrio harveyi*. *Mol. Microbiol.* **12**: 403–412.

Bassler, B.L., Wright, M., Showalter, R.E., and Silverman, M.R. (1993) Intercellular signalling in *Vibrio harveyi*, sequence and function of genes regulating expression of luminescence. *Mol. Microbiol.* **9**:773–786.

Baumann, P., and Baumann, L. (1981) The marine gram-negative eubacteria: genera *Photobacterium, Beneckea, Alteromonas, Pseudomonas,* and *Alcaligenes*. In *The Prokaryotes. A Handbook on Habitats, Isolation, and Identification of Bacteria*, M.P. Starr, H. Stolp, H.G. Trüper, A. Balows, and H.G. Schlegel (eds.), pp. 1302–1331. Springer-Verlag, Berlin.

Beck von Bodman, S., and Farrand, S.K. (1995) Capsular polysaccharide biosynthesis and pathogenicity in *Erwinia stewartii* require induction by an *N*-acylhomoserine lactone autoinducer. *J. Bacteriol.* **177**:5000–5008.

Beck von Bodman, S., Hayman, G.T., and Farrand, S.K. (1992) Opine catabolism and conjugal transfer of the nopaline Ti plasmid pTiC58 are coordinately regulated by a single repressor. *Proc. Natl. Acad. Sci. USA* **89**:643–647.

Boettcher, K.J., and Ruby, E.G. (1990) Depressed light emission by symbiotic *Vibrio fischeri* of the sepiolid squid *Euprymna scolopes*. *J. Bacteriol.* **172**:3701–3706.

Boettcher, K.J., and Ruby, E.G. (1995) Detection and quantification of *Vibrio fischeri* autoinducer from symbiotic squid light organs. *J. Bacteriol.* **177**:1053–1058.

Boylan, M., Graham, A.F., and Meighen, E.A. (1985) Functional identification of the fatty acid reductase components encoded in the luminescence operon of *Vibrio fischeri*. *J. Bacteriol.* **163**:1186–1190.

Boylan, M., Miyamoto, C., Wall, L., Graham, A.F., and Meighen, E.A. (1989) Lux C, D and E genes of the *Vibrio fischeri* luminescence operon code for the reductase, transferase, and synthetase enzymes involved in aldehyde biosynthesis. *Photochem. Photobiol.* **49**:681–688.

Brock, T.D., Madigan, M.T., Martinko, J.M., and Parker, J. (1994) *Biology of Microorganisms*, 7th ed. Prentice Hall, Englewood Cliffs, NJ.

Cao, J.-G., and Meighen, E.A. (1989) Purification and structural identification of an autoinducer for the luminescence system of *Vibrio harveyi*. *J. Biol. Chem.* **264**: 21670–21676.

Cao, J.-G., and Meighen, E.A. (1993) Biosynthesis and stereochemistry of the autoinducer controlling luminescence in *Vibrio harveyi*. *J. Bacteriol.* **175**:3856–3862.

Chen, P.-F., Tu, S.-C, Hagag, N., Wu, F.Y.-H., and Wu, C.-W. (1985) Isolation and characterization of a cyclic AMP receptor protein from luminous *Vibrio harveyi* cells. *Arch. Biochem. Biophys.* **241**:425–431.

Chhabra, S.R., Stead, P., Bainton, N.J., Salmond, G.P.C., Stewart, G.S.A.B., Williams, P., and Bycroft, B.W. (1993) Autoregulation of carbapenem biosynthesis in *Erwinia carotovora* by analogues of *N*-(3-oxohexanoyl)-homoserine lactone. *J. Antibiot. (Tokyo)* **46**:441–454.

Choi, S.H., and Greenberg, E.P. (1991) The C-terminal region of the *Vibrio fischeri* LuxR protein contains an inducer-independent *lux* gene activating domain. *Proc. Natl. Acad. Sci. USA* **88**:11115–11119.

Choi, S.H., and Greenberg, E.P. (1992a) Genetic dissection of DNA binding and luminescence gene activation by the *Vibrio fischeri* LuxR protein. *J. Bacteriol.* **174**: 4064–4069.

Choi, S.H., and Greenberg, E.P. (1992b) Genetic evidence for multimerization of LuxR, the transcriptional activator of *Vibrio fischeri* luminescence. *Mol. Marine Biol. Biotechnol.* **1**:408–413.

Clewell, D.B. (1993) Bacterial sex pheromone-induced plasmid transfer. *Cell* **73**:9–12.

Clough, S.J., Schell, M.A., and Denny, T.P. (1994) Evidence for involvement of a volatile extracellular factor in *Pseudomonas solanacearum* virulence gene expression. *Mol. Plant Microbe Interact.* **7**:621–630.

Cubo, M.T., Economou, A., Murphy, G., Johnston, A.W.B., and Downie, J.A. (1992) Molecular characterization and regulation of the rhizosphere-expressed genes *rhiABCR* that can influence modulation by *Rhizobium leguminosarum* biovar viciae. *J. Bacteriol.* **174**:4026–4035.

Devine, J.H., Countryman, C., and Baldwin, T.O. (1988a) Nucleotide sequence of the *luxR* and *luxI* genes and structure of the primary regulatory region of the *lux* regulon of *Vibrio fischeri* ATCC 7744. *Biochemistry* **27**:837–842.

Devine, J.H., Shadel, G.S., and Baldwin, T.O. (1988b) Identification of the operator of the *lux* regulon from the *Vibrio fischeri* strain ATCC 7744. *Proc. Natl. Acad. Sci. USA* **86**:5688–5692.

Dibb, N.J., Downie, J.A., and Brewin, N.J. (1984) Identification of a rhizosphere protein encoded by the symbiotic plasmid of *Rhizobium leguminosarum*. *J. Bacteriol.* **158**: 621–627.

Dolan, K.M., and Greenberg, E.P. (1992) Evidence that GroEL, not σ^{32}, is involved in transcription regulation of the *Vibrio fischeri* luminescence genes in *Escherichia coli*. *J. Bacteriol.* **174**:5132–5135.

Dunlap, P.V. (1989) Regulation of luminescence by cyclic AMP in *cya*-like and *crp*-like mutants of *Vibrio fischeri*. *J. Bacteriol.* **171**:1199–1202.

Dunlap, P.V. (1991) Organization and regulation of bacterial luminescence genes. *Photochem. Photobiol.* **54**:1157–1170.

Dunlap, P.V. (1992) Mechanism for iron control of the *Vibrio fischeri* luminescence system, involvement of cyclic AMP and cyclic AMP receptor protein and modulation of DNA level. *J. Biolumin. Chemilumin.* **7**:203–214.

Dunlap, P.V., and Greenberg, E.P. (1985) Control of *Vibrio fischeri* luminescence gene expression in *Escherichia coli* by cyclic AMP and cyclic AMP receptor protein. *J. Bacteriol.* **164**:45–50.

Dunlap, P.V., and Greenberg, E.P. (1988) Control of *Vibrio fischeri lux* gene transcription by a cyclic AMP receptor protein-LuxR protein regulatory circuit. *J. Bacteriol.* **170**:4040–4046.

Dunlap, P.V., and Greenberg, E.P. (1991) Role of intercellular chemical communication in the *Vibrio fischeri*-monocentrid fish symbiosis. In *Microbial Cell–Cell Interactions*, M. Dworkin (ed.), pp. 219–253. American Society for Microbiology, Washington, DC.

Dunlap, P.V., and Kuo, A. (1992) Cell density-dependent modulation of the *Vibrio fischeri* luminescence system in the absence of autoinducer and LuxR protein. *J. Bacteriol.* **174**:2440–2448.

Dunlap, P.V., and Ray, J.M. (1989) Requirement for autoinducer in transcriptional negative autoregulation of the *Vibrio fischeri luxR* gene in *Escherichia coli*. *J. Bacteriol.* **171**:3549–3552.

Dunlap, P.V., Kita-Tsukamoto, K., Waterbury, J., and Callahan, S.M. (1995) Isolation and characterization of a visibly luminous variant of *Vibrio fischeri* strain ES114 from the sepiolid squid *Euprymna scolopes*. *Arch. Microbiol.* **164**:194–202.

Dunny, G.M., Leonard, B.A.B., and Hedberg, P. (1995) Pheromone-inducible conjugation in *Enterococcus faecalis*: interbacterial and host–parasite chemical communication. *J. Bacteriol.* **177**:871–876.

Dworkin, M. (1991) Cell–cell interactions in Myxobacteria. In *Microbial Cell–Cell Interactions*, M. Dworkin (ed.), pp. 179–216. American Society for Microbiology, Washington, DC.

Eberhard, A. (1972) Inhibition and activation of bacterial luciferase synthesis. *J. Bacteriol.* **109**:1101–1105.

Eberhard, A., Burlingame, A.L., Eberhard, C., Kenyon, G.L., Nealson, K.H., and Oppenheimer, N.J. (1981) Structural identification of autoinducer of *Photobacterium fischeri* luciferase. *Biochemistry* **20**:2444–2449.

Eberhard, A., Longin, T., Widrig, C.A., and Stranick, S.J. (1991) Synthesis of the *lux* gene autoinducer in *Vibrio fischeri* is positively autoregulated. *Arch. Microbiol.* **155**:294–297.

Eberhard, A., Widrig, C.A., McBath, P., and Schineller, J.B. (1986) Analogs of the autoinducer of bioluminescence in *Vibrio fischeri*. *Arch. Microbiol.* **146**:35–40.

Engebrecht, J., and Silverman, M. (1984) Identification of genes and gene products necessary for bacterial bioluminescence. *Proc. Natl. Acad. Sci. USA* **81**:4154–4158.

Engebrecht, J., and Silverman, M. (1986) Regulation and expression of bacterial genes for bioluminescence. *Genet. Eng.* **8**:31–44.

Engebrecht, J., and Silverman, M. (1987) Nucleotide sequence of the regulatory locus controlling expression of bacterial genes for bioluminescence. *Nucleic Acids Res.* **15**:10455–10467.

Engebrecht, J., Nealson, K., and Silverman, M. (1983) Bacterial bioluminescence, isolation and genetic analysis of functions from *Vibrio fischeri*. *Cell* **32**:773–781.

Friedrich, W.F., and Greenberg, E.P. (1983) Glucose repression of luminescence and luciferase in *Vibrio fischeri*. *Arch. Microbiol.* **134**:87–91.

Fuqua, C., Burbea, M., and Winans, S.C. (1995) Activity of the *Agrobacterium* Ti plasmid conjugal transfer regulator TraR is inhibited by the product of the *traM* gene. *J. Bacteriol.* **177**:1367–1373.

Fuqua, C., and Winans, S.C. (1996) Conserved *cis*-acting promoter elements are required for density-dependent transcription of *Agrobacterium tumefaciens* conjugal transfer genes. *J. Bacteriol.* **178**:435–440.

Fuqua, W.C., and Winans S.C. (1994) A LuxR-LuxI type regulatory system activates *Agrobacterium* Ti plasmid conjugal transfer in the presence of a plant tumor metabolite. *J. Bacteriol.* **176**:2796–2806.

Fuqua, W.C., Winans, S.C., and Greenberg, E.P. (1994) Quorum sensing in bacteria, the LuxR-LuxI family of cell density-responsive transcriptional regulators. *J. Bacteriol.* **176**:269–275.

Garcia-Lara, J., Shang, L.H., and Rothfield, L.I. (1996) An extracellular factor regulates expression of *sidA*, a transcriptional activator of cell division genes in *Escherichia coli*. *J. Bacteriol.* **178**:2742–2748.

Gambello, M.J., and Iglewski, B.H. (1991) Cloning and characterization of the *Pseudomonas aeruginosa lasR* gene, a transcriptional activator of elastase expression. *J. Bacteriol.* **173**:3000–3009.

Gilson, L., Kuo, A., and Dunlap, P.V. (1995) AinS and a new family of autoinducer synthesis proteins. *J. Bacteriol.* **177**:6946–6951.

Gray, K.M., and Greenberg, E.P. (1992a) Physical and functional maps of the luminescence gene cluster in an autoinducer-deficient *Vibrio fischeri* strain isolated from a squid light organ. *J. Bacteriol.* **174**:4384–4390.

Gray, K.M., and Greenberg, E.P. (1992b) Sequencing and analysis of *luxR* and *luxI*, the luminescence regulatory genes from the squid light organ symbiont *Vibrio fischeri* ES114. *Mol. Marine Biol. Biotechnol.* **1**:414–419.

Gray, K.M., Passador, L., Iglewski, B.H., and Greenberg, E.P. (1994) Interchangeability and specificity of components from the quorum-sensing regulatory systems of *Vibrio fischeri* and *Pseudomonas aeruginosa*. *J. Bacteriol.* **176**:3076–3080.

Gray, K.M., Pearson, J.P., Downie, J.A., Boboye, B.E.A., and Greenberg, E.P. (1996) Cell-to-cell signaling in the symbiotic nitrogen-fixing bacterium *Rhizobium leguminosarum*: autoinduction of a stationary phase and rhizosphere-expressed genes. *J. Bacteriol.* **178**:372–376.

Hanzelka, B.L., and Greenberg, E.P. (1995) Evidence that the N-terminal region of the *Vibrio fischeri* LuxR protein constitutes an autoinducer-binding domain. *J. Bacteriol.* **177**:815–817.

Haygood, M.G. (1993) Light organ symbiosis in fishes. *Crit. Rev. Microbiol.* **19**:191–216.

Henikoff, S., Wallace, J.C., and Brown, J.P. (1990) Finding protein sequence similarities with nucleotide sequence databases. *Methods Enzymol.* **183**:111–132.

Horinouchi, S., and Beppu, T. (1992) Autoregulatory factors and communication in actinomycetes. *Annu. Rev. Microbiol.* **46**:377–398.

Horinouchi, S., Suzuki, H., Nishiyama, M., and Beppu, T. (1989) Nucleotide sequence and transcriptional analysis of the *Streptomyces griseus* gene (*afsA*) responsible for A-factor biosynthesis. *J. Bacteriol.* **171**:1206–1210.

Huisman, G.W., and Kolter, R. (1994) Sensing starvation, a homoserine lactone-dependent signalling pathway in *Escherichia coli*. *Science* **265**:537–539.

Hwang, I., Cook, D.M., and Farrand, S.K. (1995) A new regulatory element modulates homoserine lactone-mediated autoinduction of Ti plasmid conjugal transfer. *J. Bacteriol.* **177**:449–458.

Hwang, I., Li, P.-L., Zhang, L., et al. (1994) TraI, a LuxI homologue, is responsible for production of conjugation factor, the Ti plasmid *N*-acylhomoserine lactone autoinducer. *Proc. Natl. Acad. Sci. USA* **91**:4639–4643.

Iglewski, B.H., Rust, L., and Bever, R. (1990) Molecular analysis of *Pseudomonas* elastase. In *Pseudomonas: Biotransformations, Pathogenesis, and Evolving Biotechnology*. S. Silver, A.M. Chakrabarty, B. Iglewski, and S. Kaplan (eds.), pp. 36–43. American Society for Microbiology, Washington, DC.

Johnston, A.W.B., Burn, J.E., Economou, A., Davis, E.O., Hawkins, F.K.L., and Bibb, M.J. (1988) Genetic factors affecting host range in *Rhizobium leguminosarum*. In *Molecular Genetics of Plant–Microbe Interactions*, R. Palacios and D.P.S. Verma (eds.), pp. 378–384. APS Press, St. Paul, MN.

Jones, S., Yu, B., Bainton, N.J., et al. (1993) The *lux* autoinducer regulates the production of exoenzyme virulence determinants in *Erwinia cartovora* and *Pseudomonas aeruginosa*. *EMBO J.* **12**:2477–2482.

Kahn, K., and Ditta, G. (1991) Modular structure of FixJ, homology of the transcriptional activator domain with the −35 binding domain of sigma factors. *Mol. Microbiol.* **5**:987–997.

Kaiser, D., and Losick, R. (1993) How and why bacteria talk to each other. *Cell* **73**: 873–885.

Kaplan, H.B., and Greenberg, E.P. (1985) Diffusion of autoinducer is involved in regulation of the *Vibrio fischeri* luminescence system. *J. Bacteriol.* **163**:1210–1214.

Kell, D.B., Kaprelyants, A.S., and Grafen, A. (1995) Pheromones, social behavior and the functions of secondary metabolism in bacteria. *Trends Ecol. Evol.* **10**:126–129.

Kolibachuk, D., and Greenberg, E.P. (1993) The *Vibrio fischeri* luminescence gene activator LuxR is a membrane-associated protein. *J. Bacteriol.* **175**:7307–7312.

Kuo, A., Blough, N.V., and Dunlap, P.V. (1994) Multiple *N*-acyl-homoserine lactone autoinducers of luminescence in the marine symbiotic bacterium *Vibrio fischeri*. *J. Bacteriol.* **176**:7558–7565.

Kuo, A., Callahan, S.M., and Dunlap, P.V. (1996) Modulation of luminescence operon expression by *N*-octanoyl-homoserine lactone in *ainS* mutants of *Vibrio fischeri*. *J. Bacteriol.* **178**:971–976.

Latifi, A., Winson, M.K., Foglino, M., et al. (1995) Multiple homologues of LuxR and LuxI control expression of virulence determinants and secondary metabolites through quorum sensing in *Pseudomonas aeruginosa* PAO1. *Molec. Microbiol.* **17**: 333–343.

Martin, M., Showalter, R., and Silverman, M. (1989) Identification of a locus controlling expression of the luminescence genes in *Vibrio harveyi*. *J. Bacteriol.* **171**:2406–2414.

McGowan, S., Sebaihia, M., Jones, S., et al. (1995) Carbpenem antibiotic production in *Erwinia carotovora* is regulated by CarR, a homologue of the LuxR transcriptional activator. *Microbiology* **141**:541–550.

McKenney, D., Brown, K.E., and Allison, D.G. (1995) Influence of *Pseudomonas aeruginosa* exoproducts on virulence factor production in *Burkholderia cepacia*: evidence of interspecies communication. *J. Bacteriol.* **177**:6989–6992.

Meighen, E.A., and Dunlap, P.V. (1993) Physiological, biochemical and genetic control of bacterial bioluminescence. *Adv. Microbial Physiol.* **34**:1–67.

Miyamoto, C.M., Boylan, M., Graham, A.F., and Meighen, E.A. (1988a) Organization of the *lux* structural genes of *Vibrio harveyi*. *J. Biol. Chem.* **263**:13393–13399.

Miyamoto, C.M., Graham, A.F., and Meighen, E.A. (1988b) Nucleotide sequence of the *luxC* gene and the upstream DNA from the bioluminescent system of *Vibrio harveyi*. *Nucleic Acids Res.* **16**:1551–1562.

Miyamoto, C.M., Meighen, E.A., and Graham, A.F. (1990) Transcriptional regulation of *lux* genes transferred into *Vibrio harveyi*. *J. Bacteriol.* **172**:2046–2054.

Miyamoto, C., Smith, E.E., Swartzman, E., Cao, J.-G., Graham, A.F., and Meighen, E.A. (1994) Proximal and distal sites bind LuxR independently and activate expression of the *Vibrio harveyi lux* operon. *Mol. Microbiol.* **14**:255–262.

Nealson, K.H. (1977) Autoinduction of bacterial luciferase: occurrence, mechanism and significance. *Arch. Microbiol.* **112**:73–79.

Nealson, K.H., and Hastings, J.W. (1979) Bacterial bioluminescence: its control and ecological significance. *Microbiol. Rev.* **43**:496–518.

Nealson, K.H., and Hastings, J.W. (1992) The luminous bacteria. In *The Prokaryotes*, 2nd ed., A. Balows, H.G. Trüper, M. Dworkin, W. Harder, and K.-H. Schleifer (eds.), pp. 625–639. Springer-Verlag, Berlin.

Nealson, K.H., Eberhard, A., and Hastings, J.W. (1972) Catabolite repression of bacterial bioluminescence, functional implications. *Proc. Natl. Acad. Sci. USA* **69**: 1073–1076.

Nealson, K.H., Platt, T., and Hastings, J.W. (1970) Cellular control of the synthesis and activity of the bacterial luminescent system. *J. Bacteriol.* **104**:313–322.

Ochsner, U.A., Koch, A.K., Fiechter, A., and Reiser, J. (1994) Isolation and characterization of a regulatory gene affecting rhamnolipid biosurfactant synthesis in *Pseudomonas aeruginosa. J. Bacteriol.* **176**:2044–2054.

Ochsner, U.A., and Reiser, J. (1995) Autoinducer-mediated regulation of rhamnolipid biosurfactant synthesis in *Pseudomonas aeruginosa. Proc. Natl. Acad. Sci. USA* **92**:6424–6428.

Okamoto, S., Nihira, T., Kataoka, H., Suzuki, A., and Yamada. Y. (1992) Purification and molecular cloning of a butyrolactone autoregulator receptor from *Streptomyces virginiae. J. Biol. Chem.* **267**:1093–1098.

Passador, L., Cook, J.M., Gambello, M.J., Rust, L., and Iglewski, B.H. (1993) Expression of *Pseudomonas aeruginosa* virulence genes requires cell-to-cell communication. *Science* **260**:1127–1130.

Pearson, J.P., Gray, K.M., Passador, L., et al. (1994) Structure of the autoinducer required for expression of *Pseudomonas aeruginosa* virulence genes. *Proc. Natl. Acad. Sci. USA* **91**:197–201.

Pearson, J.P., Passador, L., Iglewski, B.H., and Greenberg, E.P. (1995) A second *N*-acylhomoserine lactone signal produced by *Pseudomonas aeruginosa. Proc. Natl. Acad. Sci. USA* **92**:1490–1494.

Pierson, L.S., Keppenne, V.D., and Wood, D.W. (1994) Phenazine antibiotic biosynthesis in *Pseudomonas aureofaciens* 30–84 is regulated by PhzR in response to cell density. *J. Bacteriol.* **176**:3966–3974.

Piper, K.R., von Bodman, S.B., and Farrand, S.K. (1993) Conjugation factor of *Agrobacterium tumefaciens* regulates Ti plasmid transfer by autoinduction. *Nature* **362**:448–450.

Pirhonen, M., Flego, D., Heikinheimo, R., and Palva, E.T. (1993) A small diffusible signal molecule is responsible for the global control of virulence and exoenzyme production in the plant pathogen *Erwinia carotovora. EMBO J.* **12**:2467–2476.

Poellinger, K.A., Lee, J.P., Parales, J.V., and Greenberg, E.P. (1995) Intragenic suppression of a *luxR* mutation: characterization of an autoinducer-independent LuxR. *FEMS Microbiol. Lett.* **129**:97–102.

Rosson, R.A., and Nealson, K.H. (1981) Autoinduction of bacterial bioluminescence in a carbon-limited chemostat. *Arch. Microbiol.* **129**:299–304.

Roth, L.E., and Stacey, G. (1991) Rhizobium-legume symbiosis. In *Microbial Cell–Cell Interactions*, M. Dworkin (ed.), pp. 255–301. American Society for Microbiology, Washington, DC.

Ruby, E.G., and Asato, L.M. (1993) Growth and flagellation of *Vibrio fischeri* during initiation of the sepiolid squid light organ symbiosis. *Arch. Microbiol.* **159**:160–167.

Ruby, E.G., and McFall-Ngai, M.J. (1992) A squid that glows in the night, development of an animal-bacterial mutualism. *J. Bacteriol.* **174**:4865–4870.

Ruby, E.G., and Nealson, K.H. (1976) Symbiotic association of *Photobacterium fischeri* with the marine luminous fish *Monocentris japonica*, a model of symbiosis based on bacterial studies. *Biol. Bull.* **141**:574–5867.

Ruby, E.G., and Nealson, K.H. (1978) Seasonal changes in the species composition of luminous bacteria in nearshore seawater. *Limnol. Oceanogr.* **23**:530–533.

Ruby, E.G., Greenberg, E.P., and Hastings, J.W. (1980) Planktonic marine luminous bacteria: species distribution in the water column. *Appl. Environ. Microbiol.* **39**:302–306.

Rudner, D.Z., LeDeaux, J.R., Ireton, K., and Grossman, A.D. (1991) The *spoOK* locus of *Bacillus subtilis* is homologous to the oligopeptide permease locus and is required for sporulation and competence. *J. Bacteriol.* **173**:1388–1398.

Saarilahti, H.T., Heino, P., Pakkanen, R., Kalkkinen, N., Palva, I., and Palva, E.T. (1990) Structural analysis of the *pehA* gene and characterization of its protein product endopolygalacturonase of *Erwinia carotovora* ssp. *carotovora. Mol. Microbiol.* **4**: 1037–1044.

Schaefer, A.L., Hanzelka, B.L., Eberhard, A., and Greenberg, E.P. (1996) Quorum sensing in *Vibrio fischeri*: probing autoinducer-LuxR interactions with autoinducer analogs. *J. Bacteriol.* **178**:2897–2901.

Schripsema, J., de Rudder, K.E., van Vliet, T.B., et al. (1996) Bacaateriocin *small* of *Rhizobium leguminosarum* belongs to the class of *N*-acyl-L-homoserine lactone molecules, known as autoinducers and as quorum sensing co-transcription factors. *J. Bacteriol.* **178**:366–371.

Seed, P.C., Passador, L., and Iglewski, B.H. (1995) Activation of the *Pseudomonas aeruginosa lasI* gene by LasR and the *Pseudomonas* autoinducer PAI, an autoinduction regulatory hierarchy. *J. Bacteriol.* **177**:654–659.

Shadel, G.S., and Baldwin, T.O. (1991) The *Vibrio fischeri* LuxR protein is capable of bidirectional stimulation of transcription and both positive and negative regulation of the *luxR* gene. *J. Bacteriol.* **173**:568–574.

Shadel, G.S., and Baldwin, T.O. (1992a) Identification of a distantly located regulatory element in the *luxD* gene required for negative autoregulation of the *Vibrio fischeri luxR* gene. *J. Biol. Chem.* **267**:7690–7695.

Shadel, G.S., and Baldwin, T.O. (1992b) Positive autoregulation of the *Vibrio fischeri luxR* gene. *J. Biol. Chem.* **267**:7696–7702.

Shadel, G.S., Young, R., and Baldwin, T.O. (1990) Use of regulated cell lysis in a lethal genetic selection in *Escherichia coli*, identification of the autoinducer-binding region of the LuxR protein from *Vibrio fischeri* ATCC 7744. *J. Bacteriol.* **172**: 3980–3987.

Showalter, R.E., Martin, M.O., and Silverman, M.R. (1990) Cloning and nucleotide sequence of *luxR*, a regulatory gene controlling bioluminescence in *Vibrio harveyi*. *J. Bacteriol.* **172**:2946–2954.

Slock, J., VanReit, D., Kolibachuk, D., and Greenberg, E.P. (1990) Critical regions of the *Vibrio fischeri* LuxR protein defined by mutational analysis. *J. Bacteriol.* **172**: 3974-3979.

Stephens, K. (1986) Pheromones among the prokaryotes. *CRC Crit. Rev. Microbiol.* **13**: 309–334.

Stevens, A.M., Dolan, K.M., and Greenberg, E.P. (1994) Synergistic binding of the *Vibrio fischeri* transcriptional activator domain and RNA polymerase to the *lux* promoter region. *Proc. Natl. Acad. Sci. USA* **91**:12619–12623.

Sun, W., Cao, J.-G., Teng, K., and Meighen, E.A. (1994) Biosynthesis of poly-3-hydroxybutyrate in the luminescent bacterium, *Vibrio harveyi*, and regulation by the *lux* autoinducer, *N*-(3-hydroxybutanoyl) homoserine lactone. *J. Biol. Chem.* **269**: 20785–20790.

Swartzman, E., and Meighen, E.A. (1993) Purification and characterization of a poly(dA-dT) *lux*-specific DNA-binding protein from *Vibrio harveyi* and identification as LuxR. *J. Biol. Chem.* **268**:16706–16716.

Swartzman, E., Kapoor, S., Graham, A.F., and Meighen, E.A. (1990a) A new *Vibrio fischeri lux* gene precedes a bidirectional termination site for the *lux* operon. *J. Bacteriol.* **172**:6797–6802.

Swartzman, E., Miyamoto, C., Graham, A., and Meighen, E. (1990b) Delineation of the transcriptional boundaries of the *lux* operon of *Vibrio harveyi* demonstrates the presence of two new *lux* genes. *J. Biol. Chem.* **265**:3513–3517.

Swartzman, E., Silverman, M., and Meighen, E.A. (1992) The *luxR* gene product of *Vibrio harveyi* is a transcriptional activator of the *lux* promoter. *J. Bacteriol.* **174**: 7490–7493.

Swift, S., Winson, M.K., Chan, P.F., et al. (1993) A novel strategy for the isolation of *luxI* homologues, evidence for the widespread distribution of a LuxR, LuxI super-family in enteric bacteria. *Mol. Microbiol.* **10**:511–520.

Throup, J.P., Camara, M., Briggs, G.S., Winson, M.K., Chhabra, S.R., Bycroft, B.W., Williams, P., and Stewart, G.S.A.B. (1995) Characterization of the *yenI/yenR* locus from *Yersinia enterocolitica* mediating the synthesis of two *N*-acylhomoserine lactone signal molecules. *Molec. Microbiol.* **17**:345–356.

Turner, J.M., and Messenger, A.J. (1986) Occurrence, biochemistry and physiology of phenazine pigment production. *Adv. Microb. Physiol.* **27**:211–275.

Ulitzur, S., and Dunlap, P.V. (1995) Regulatory circuitry controlling luminescence autoinduction in *Vibrio fischeri*. *Photochem. Photobiol.* **62**:625–632.

Ulitzur, S., and Yashphe, J. (1975) An adenosine 3′,5′-monophosphate-requiring mutant of the luminous bacteria *Beneckea harveyi*. *Biochim. Biophys. Acta* **404**:321–328.

Van Egeraat, A.W.S.M. (1975) The possible role of homoserine in the development of *Rhizobium leguminosarum* in the rhizosphere of pea seedlings. *Plant Soil* **42**: 381–386.

Vining, L.C. (1990) Functions of secondary metabolites. *Annu. Rev. Microbiol.* **44**: 395–427.

Wang, X., de Boer, P.A.J., and Rothfield, L.I. (1991) A factor that positively regulates cell division by activating transcription of the major cluster of essential cell division genes of *Escherichia coli*. *EMBO J.* **10**:3363–3372.

Winson, M.K., Bainton, N.J., Chhabra, S.R., et al. (1994) Control of *N*-acyl homoserine lactone-regulated expression of multiple phenotypes in *Chromobacterium violaceum*. In *Proceedings of the 94th General Meeting, American Society for Microbiology, Las Vegas*, abstract H-71. American Society for Microbiology, Washington, DC.

Winson, M.K., Camara, M., Briggs, G.S. et al. (1995) Multiple *N*-acyl-L-homoserine lactone signal molecules regulated production of virulence determinants and secondary metabolites in *Pseudomonas aeruginosa*. *Proc. Natl. Acad. Sci. USA* **92**: 9427–9431.

Wood, D.W., and Pierson, L.S., III (1996) The *phzI* gene of *Pseudomonas aureofaciens* 30-84 is responsible for the production of a diffusible signal required for phenazine antibiotic production. *Gene* **168**:49–53.

Zhang, L., Murphy, P.J., Kerr, A., and Tate, M.E. (1993) *Agrobacterium* conjugation and gene regulation by *N*-acyl-homoserine lactones. *Nature* **362**:446–448.

III.

Multicellular
Life Styles

5

Cyanobacteria

DAVID G. ADAMS

Although still frequently referred to as blue-green algae, cyanobacteria are in fact an enormously diverse group of gram-negative prokaryotes. Their diversity is reflected in their guanine + cystosine (G + C) contents, which cover the same range as all the rest of the prokaryotes (Herdman et al., 1979). They vary in diameter from less than 1 μm to more than 100 μm, and their large size facilitated the numerous early studies of morphology. Indeed, the first description of cyanobacterial akinetes (Carter, 1856), which are a form of resting cell, was published more than 20 years before the discovery of the bacterial endospore (Cohn, 1877; Koch, 1877).

All cyanobacteria are photoautotrophs that employ oxygenic photosynthesis of the type found in eukaryotic algae and higher plants (Castenholz and Waterbury, 1989; Tandeau de Marsac and Houmard, 1993; Barry et al., 1994; Golbeck, 1994). Many can also grow photoheterotrophically, employing light for energy and organic compounds as a source of carbon. Some are also capable of chemoheterotrophic growth in the dark, when both energy and carbon are derived from organic compounds, although under such conditions growth is usually much slower than in the light (Castenholz and Waterbury, 1989; Tandeau de Marsac and Houmard, 1993). A small number of cyanobacteria can also employ anoxygenic photosynthesis in which H_2S, rather than H_2O, serves as a supply of reducing power (Padan and Cohen, 1982; Schmidt, 1988).

Many cyanobacteria show a remarkable capacity to differentiate a variety of morphological forms in response to changes in their environment (Fig. 5.1). The best known is the heterocyst, which is specialized for dinitrogen fixation;

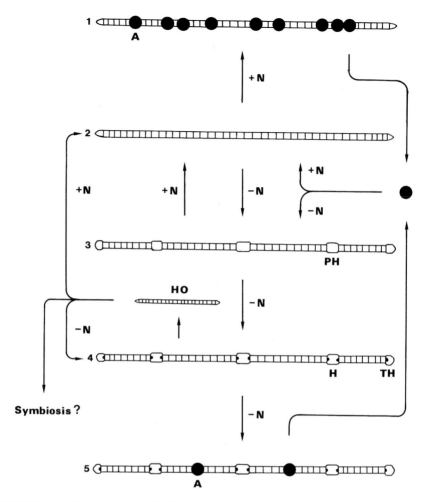

FIGURE 5.1 Some of the possibilities for morphologic development of filamentous cyano-bacteria. When grown in the presence of a source of fixed nitrogen, the filament consists entirely of undifferentiated vegetative cells (2). At the end of the exponential growth phase when light (energy) becomes limiting, some vegetative cells can differentiate into the spore-like cells called akinetes (A), which in the absence of heterocysts are randomly placed within the filament (1). In the absence of fixed nitrogen the vegetative filament differentiates highly specialized dinitrogen-fixing cells (heterocysts) at regular intervals within the filament (H) and in terminal positions (TH). Heterocysts are characterized by their thickened cell walls, relatively agranular cytoplasm, and polar bodies at the point of attachment to vegetative cells, there being two in heterocysts within the filament but only one in terminal heterocysts (4). During their development heterocysts pass through an intermediate stage, the proheterocyst (PH), which unlike the mature cell does not have the thickened cell wall and polar bodies and is able to dedifferentiate in the presence of fixed nitrogen (3). When akinetes develop in dinitrogen-fixing cultures, they do so at locations with a precise spatial relationship to the heterocysts (see text for details) such as midway between (5). Akinetes can germinate and give rise to filaments with or without heterocysts depending on the availability of combined nitrogen. Hormogonia (HO) are short, motile, undifferentiated fila-ments that develop as a result of a variety of stimuli (see text for details). Their formation usually involves the rapid division of vegetative cells without concomitant growth, followed by fragmen-tation of the filament to release heterocysts and motile hormogonia. The latter can give rise to heterocystous or nonheterocystous filaments depending on the availability of combined nitrogen. Hormogonia can also serve as the infective agents in the establishment of symbiotic associations with a wide variety of plant groups (see text for details). (Adapted from Adams, 1992)

other forms include akinetes, hormogonia, baeocytes, and hairs, all of which are discussed herein. Many cyanobacteria are therefore multicellular organisms that can develop a variety of highly specialized cell types that often display a precise spatial relationship to each other.

The purpose of this chapter is not to present a comprehensive review of all that is known about the morphological forms and development of cyanobacteria. There are many excellent reviews that can be consulted for such detailed information, and they are indicated in the text. Indeed, in many instances I have referred to recent review articles rather than, or in addition to, the original literature, as they can provide the reader with access to a greater range of information. My wish is to convey to those unfamiliar with these fascinating organisms their remarkable complexity and developmental capabilities. Unlike most bacteria, much can be seen by viewing cyanobacteria with even low power light microscopy. Those who have worked with cyanobacteria for many years tend to endow them with human attributes. They can be stubborn, unpredictable, and intriguing. Perhaps the only consolation is that, unlike most bacteria, you can readily see the source of your frustration.

Cell Structure

The cell wall of cyanobacteria is of the gram-negative type, consisting of two unit membranes: the cytoplasmic membrane and the outer membrane, separated

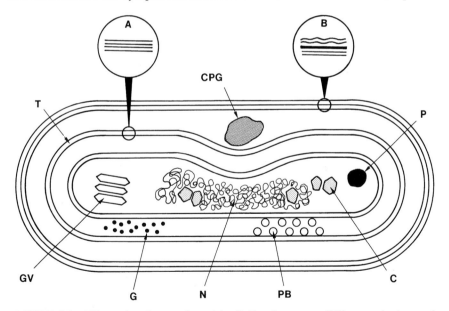

FIGURE 5.2   Thin section of a cyanobacterial cell. C, carboxysome; CPG, cyanophycin granule; T, thylakoid; P, polyphosphate granule; N, nucleoplasmic region; G, glycogen granules; PB, phycobilisome; GV, gas vesicle. (**Inset A**) Enlarged view of a thylakoid, showing paired unit membranes. (**Inset B**) Enlarged view of the cell envelope, showing the outer membrane, peptidoglycan, and cytoplasmic membrane. (Adapted from Stanier and Cohen-Bazire, 1977)

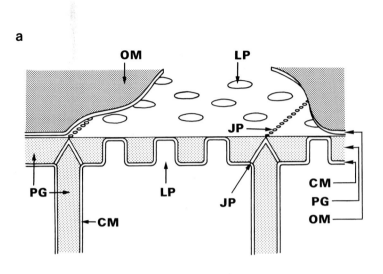

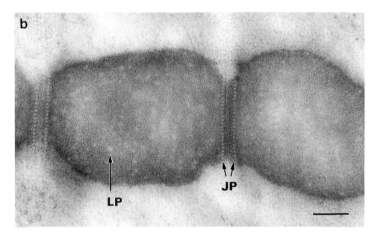

FIGURE 5.3 (a) Cell wall of *Oscillatoria* sp. The peptidoglycan layer (PG) is perforated by large pore pits (LP) up to 70 μm in diameter, which bring the cytoplasmic membrane (CM) into contact with the outer membrane (OM). The large pore pits are seen in cross section in the lower half of the diagram and appear as circles when viewed from above in the top half of the diagram. Junctional pores (JP) traverse the peptidoglycan between cells and meet at the surface, appearing as a single row of small circles when viewed from above in the top half of the diagram. (b) Electron micrograph of a thin, longitudinal section of an unidentified member of the order *Oscillatoriales*. The large pore pits can be seen as bright circles scattered throughout the peptidoglycan. The plane of the section is within the peptidoglycan layer, which means that the junctional pores appear as parallel rows of small, closely positioned circles at the cell cross walls, rather than the single rows that would be seen at the surface of the peptidoglycan. Osmium tetroxide fixation; uranyl acetate stain. Bar = 0.2 μm. (Electron microscopy by Denise Ashworth. **A**, adapted from Halfen, 1979)

112

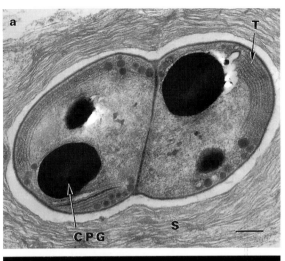

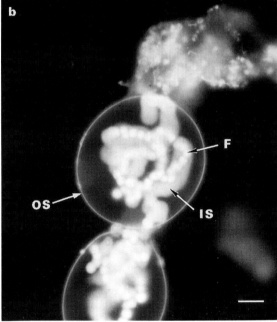

FIGURE 5.4 Variations in the polysaccharide sheaths of filamentous cyanobacteria. (a) Electron micrograph of a thin, longitudinal section of *Nostoc* sp., showing the extensive, laminated sheath material (S) external to the cell envelope. CPG, cyanophycin granule; T, thylakoid. Osmium tetroxide fixation; uranyl acetate stain. Bar = 0.5 μm. (Electron microscopy by Denise Ashworth). (b) Epifluorescence micrograph of *Nostoc* sp. This cyanobacterium frequently grows within a well defined outer sheath (OS), which encloses the whole filament (F); it is also surrounded by a diffuse inner sheath (IS). The polysaccharide of the sheath layers has been stained with the fluorescent dye calcofluor. The cells of the filament also exhibit autofluorescence owing to their photosynthetic pigments. Bar = 10 μm.

by the electron-opaque peptidoglycan layer (Fig. 5.2) (Stanier, 1988; Weckesser and Jurgens, 1988; Castenholz and Waterbury, 1989; Gantt, 1994). The peptidoglycan layer is usually 1–10 nm thick but can reach 200 nm in *Oscillatoria princeps*, which has cells 100 μm in diameter. The thickness of the peptido-

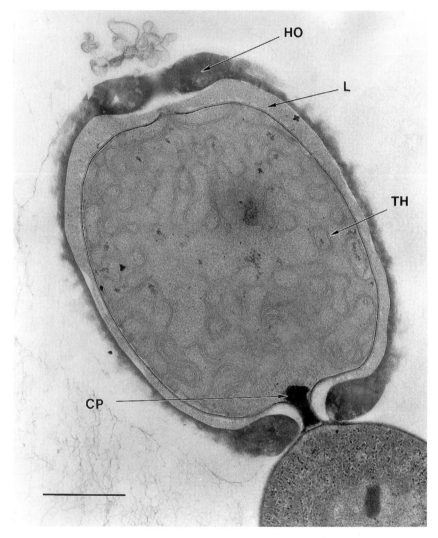

FIGURE 5.5   Electron mirograph of a thin, longitudinal section of a filament of *Nostoc* sp. strain LBG1. The cell is a mature heterocyst showing the presence of the laminated (L) and homogeneous (HO) layers external to the normal cell wall. The junction between the heterocyst and the neighboring vegetative cell is narrow, and this polar region shows the characteristic deposition of a plug of cyanophycin (CP). The relatively agranular nature of the cytoplasm of the heterocyst can be seen when compared with the vegetative cell at the bottom of the picture. TH, thylakoids. Osmium tetroxide fixation; uranyl acetate stain. Bar = 1 μm. (Electron microscopy by D. Ashworth. Reprinted from Adams, 1992, with permission)

glycan, together with its degree of cross-linking and the presence of covalently linked polysaccharide, are properties more characteristic of gram-positive than gram-negative cell walls (Weckesser and Jurgens, 1988). In cyanobacteria such as *O. princeps* the peptidoglycan is usually traversed by large pores (approximately 70 nm in diameter), which carry the cytoplasmic membrane outward and into contact with the outer membrane (Fig. 5.3). The walls of all cyanobacteria contain small diameter (5–13 nm) pores, the distribution of which varies greatly. In the longitudinal wall of filamentous strains the pores can be distributed over the whole surface or in a ring around the cross walls, when they are referred to as junctional pores (Fig. 5.3). The cell septum may contain a single tiny central pore or many pores (microplasmodesmata).

Many cyanobacteria have an additional layer, external to the outer membrane, that is predominantly polysaccharide (Castenholz and Waterbury, 1989; Gantt, 1994). The consistency of this outer envelope varies considerably, resulting in a variety of names being applied to it, including glycocalyx, sheath, capsule, mucilage, and slime. Many sheaths show a fibrillar structure (Fig. 5.4A) and can be extremely tough, often enclosing more than one cell or filament (Fig. 5.4B).

The photosynthetic membranes of cyanobacteria, the thylakoids, are the most conspicuous structures within the cell (Stanier, 1988; Castenholz and Waterbury, 1989; Gantt, 1994). They consist of two unit membranes separated by an electron-transparent space of approximately 3–5 nm and are usually 60–70 nm from adjacent thylakoids (Fig. 5.2). In many unicellular and filamentous cyanobacteria the thylakoids are arranged in a row of three to six, which run parallel to each other and to the cell wall. However, less ordered arrangements also occur, and during heterocyst development the thylakoids become more convoluted (Fig. 5.5). The thylakoids contain the chlorophyll *a*-protein complexes, photosynthetic reaction centers, carotenoids, and the electron transport system. Attached to the membranes are rows of hemidiscoidal structures, 10–12 nm across, known as phycobilisomes (Fig. 5.2), which contain the major light-harvesting pigments, the phycobiliproteins (Bryant, 1991; Sidler, 1994). These proteins are largely responsible for the characteristic colors of cyanobacteria, which range from blue-green, red, and brown to almost black.

Classification

The classification of cyanobacteria has relied heavily on their morphological complexity. Detailed descriptions can be found elsewhere (Rippka et al., 1979, 1981; Castenholz and Waterbury, 1989) and are not considered here. However, the presently used taxonomic system is briefly explained, as it provides a framework for discussion of the developmental capabilities of cyanobacteria.

The classification system of Rippka et al. (1979, 1981) divided the cyanobacteria into five sections, subsequently referred to as subsections and given ordinal rank (Castenholz and Waterbury, 1989). Subsections I and II (Orders Chroococcales and Pleurocapsales) contain cyanobacteria that are unicellular,

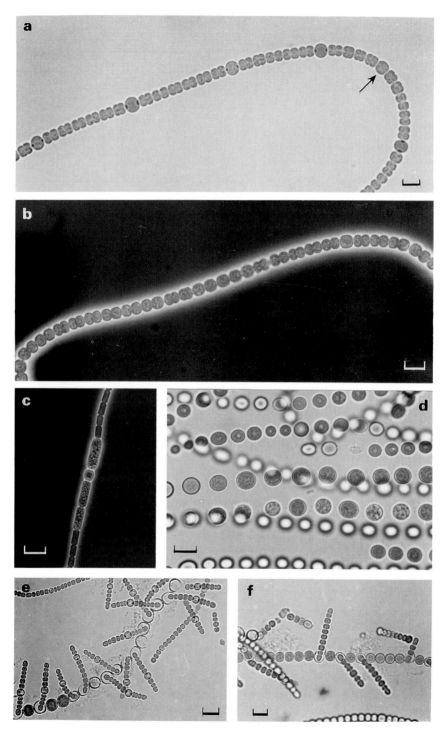

although they may exist as nonfilamentous aggregates of cells. In the Chroococcales the cells are spherical, cylindrical, or oval and reproduce by symmetrical or asymmetrical binary fission in one, two, or three planes, or by budding. In a structural sense they are the simplest of the cyanobacteria. The Pleurocapsales reproduce by internal multiple fission (which is unique among the cyanobacteria), or by multiple and binary fission (see Figs. 5.11–5.14 and the section on pleurocapsalean cyanobacteria).

The remaining three subsections consist of filamentous cyanobacteria. Subsection III (Order Oscillatoriales) contains cyanobacteria in which cell division is in one plane only (at right angles to the filament axis) and that are incapable of forming heterocysts. In subsection IV (Order Nostocales) cell division is in one plane only, and filaments are capable of producing heterocysts if the concentration of combined nitrogen is low (Castenholz, 1989a) (Figs. 5.1, 5.6). Subsection V (Order Stigonematales) contains the most morphologically and developmentally complex cyanobacteria in which heterocysts are produced in the absence of combined nitrogen, and cell division is commonly in more than one plane, giving rise to multiseriate trichomes (filaments with more than one row of cells), trichomes with branches, or both (Castenholz, 1989b).

Heterocysts

The ability of many cyanobacteria to fix atmospheric nitrogen is remarkable, as these organisms generate a by-product of photosynthesis that is highly toxic to nitrogenase, namely oxygen. Cyanobacteria protect nitrogenase from inactivation by oxygen in a number of ways (Fay, 1992; Gallon, 1992), the most fascinating of which is the heterocyst, a highly differentiated cell the sole purpose of which is to provide a suitable anaerobic environment for the functioning of nitrogenase in an otherwise aerobic filament of vegetative cells (Adams and Carr, 1981; Wolk, 1982; Adams, 1992; Wolk et al., 1994). Heterocysts develop in the filamentous cyanobacteria of subsections IV (Order Nostocales) and V (Order Stigonematales). In many cases, notably in members of the Order Nos-

---

FIGURE 5.6   Photomicrographs illustrating some of the differentiated cell types of cyanobacteria. (**a**) *Anabaena* sp. strain CA showing the regular spacing of heterocysts, which are the sites of dinitrogen fixation, and a developing proheterocyst (arrow). (**b**) *Anabaena* sp. strain CA grown in the presence of nitrate, which completely suppresses heterocyst development. (**c**) *Anabaena cylindrica*. Note the large, granular akinetes developing immediately adjacent to a heterocyst. The two akinetes below the heterocyst show a characteristic gradient of maturity, with the largest and oldest being closer to the heterocyst. (**d**) Old culture of nitrate-grown *Anabaena* sp. strain CA in which all vegetative cells have transformed into spherical akinetes. Although the akinetes have become separated as a result of pressure created by the coverslip being placed on the sample, the line of the original filaments can still be seen. Dilution of such a culture leads to germination of the akinetes. If the medium used for dilution of the akinetes does not contain a source of fixed nitrogen, each short filament that emerges from the akinete coat contains a heterocyst (**e**). However, if the medium contains nitrate, the filaments remain undifferentiated (**f**). Bars = 10 μm. (**b, c**) Phase contrast optics. (a, b, d–f reprinted from Adams, 1992; **c** reprinted from Nichols and Adams, 1982; with permission)

tocales, heterocysts develop singly at regular, spaced intervals, forming a one-dimensional pattern (Fig. 5.6).

When heterocysts develop from vegetative cells they undergo a series of major modifications that ensure suitability for their future role as sites of dinitrogen fixation. They acquire additional cell wall layers (see below) that reduce the diffusion of gases, including oxygen, and help to generate an anaerobic interior to the cell. This process is aided by the loss of photosystem II activity, which eliminates the photosynthetic generation of oxygen. Any remaining oxygen, and any that diffuses into the cell along with dinitrogen via the pores linking the heterocyst with the neighboring cells, is removed by several mechanisms, the most important of which is probably respiration. Heterocysts lack ribulose bisphosphate carboxylase and so do not fix CO_2. They must therefore rely on vegetative cells for a supply of fixed carbon to support dinitrogen fixation. Dinitrogen is fixed to NH_3, which is combined with glutamate to form glutamine and is exported to the vegetative cells (Fay, 1992; Gallon, 1992; Wolk et al., 1994).

Structure

Heterocysts are readily discernible by light microscopy, as they are usually larger than vegetative cells with less granular cytoplasm, thickened cell walls, and polar bodies at either end of the cell at the point of attachment to neighboring cells (Figs. 5.5–5.7). The characteristic thickened cell wall is a result of the deposition of three extra cell wall layers external to the normal cell envelope (Wolk, 1982). The innermost is the laminated layer, which consists of glycolipid. The next is the homogeneous layer, consisting of polysaccharide, and the outermost is the fibrous layer, which is probably uncompacted strands of the same polysaccharide. The fibrous layer is the first to be synthesized during heterocyst development and is one of the first changes visible on electron micrographs (Fig. 5.7). This phase is followed by deposition of the homogeneous and laminated layers (Fig. 5.5). These extra wall layers, particularly the glycolipid, help to reduce gas diffusion into the heterocyst and so maintain an anaerobic interior (Walsby, 1985).

Development

In the presence of ammonium or nitrate, cyanobacteria such as *Anabaena* and *Nostoc* grow as undifferentiated vegetative filaments (Fig. 5.6b). Heterocysts develop in such cultures within approximately one generation time of their transfer to medium lacking combined nitrogen. Although the heterocyst is terminally differentiated, it passes through an intermediate stage, the proheterocyst, which is capable of returning to vegetative growth under the appropriate conditions (Adams, 1992). Although clearly differentiated from the vegetative cell, the proheterocyst lacks the thickened walls and polar bodies typical of the fully mature heterocyst (Figs. 5.6a, 5.7), and it may be the lack of thickened walls that permits the proheterocyst to dedifferentiate.

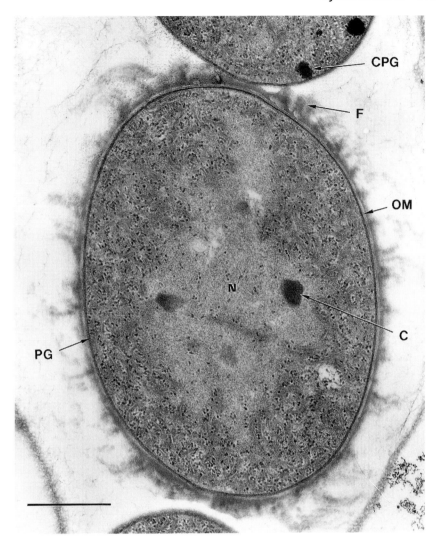

FIGURE 5.7   Electron micrograph of a thin, longitudinal section of a filament of *Nostoc* sp. strain LBG1. The cell is at an early stage of heterocyst development, showing little change from the neighboring vegetative cells, other than the deposition of loose, fibrous material (F) external to the normal cell envelope. N, nucleoplasmic region; C, carboxysome; CPG, cyanophycin granule; PG, peptidoglycan layer; OM, outer membrane. Osmium tetroxide fixation; uranyl acetate stain. Bar = 1 μm. (Electron microscopy by D. Ashworth. Reprinted from Adams, 1992, with permission)

In cultures such as those described above, the stimulus to develop heterocysts is clearly nitrogen starvation. Considerable new protein synthesis is required for each heterocyst, and the nitrogen for this synthesis must come from the hydrolysis of internal reserves and the turnover of vegetative cell proteins.

The reserves consist of phycocyanin, one of the major light harvesting pigments (Glazer, 1987; Tandeau de Marsac and Houmard, 1993), and cyanophycin, a copolymer of arginine and aspartic acid that is unique to cyanobacteria (Simon, 1987). Both of these nitrogen reserves are degraded by specific enzymes (Adams, 1992; Tandeau de Marsac and Houmard, 1993). In actively growing, dinitrogen-fixing cultures, new heterocysts develop midway between existing ones, thereby maintaining the regular pattern. In this situation, in which nitrogen starvation of the culture as a whole does not occur, it is not yet clear if localized depletion of intracellular combined nitrogen or some other trigger provides the stimulus for heterocyst development.

Under aerobic conditions dinitrogen fixation is confined to mature heterocysts, and the appearance of nitrogenase activity therefore coincides with the development of these cells. However, if heterocyst differentiation is induced anaerobically using an argon atmosphere and adding dichlorophenyl dimethylurea (DCMU) to inhibit the photosynthetic production of oxygen, heterocysts do not acquire their characteristic thickened walls (Rippka and Stanier, 1978). Although it might be imagined that under these anaerobic conditions nitrogen fixation may also occur in vegetative cells, the use of immunoferritin labeling to detect the molybdenum-iron (Mo-Fe) protein of nitrogenase has shown that the enzyme is localized to the incompletely formed heterocysts (Murry et al., 1984). These experiments imply that the expression of *nif* genes in heterocystous cyanobacteria is not simply under environmental regulation but has a developmental component, which means that their expression requires nitrogen starvation, anaerobiosis, and completion (or at least partial completion) of heterocyst development. This developmental control of *nif* gene expression has been well demonstrated by Elhai and Wolk (1990; Wolk et al., 1994) using the *luxAB* genes (which encode bacterial luciferase) as transcriptional reporters. This elegant technique has allowed the temporal and spatial expression of specific genes to be followed within single cells during heterocyst development (Wolk et al., 1994).

Pattern

Heterocysts do not occur in random positions in vegetative filaments but at precise locations that are characteristic for each cyanobacterium (Rippka et al., 1979, 1981; Castenholz and Waterbury, 1989). In the genus *Nostoc* heterocysts occur mostly within the filament (at intercalary locations), whereas in the genus *Anabaena* they can occur at both intercalary and terminal positions, and in *Cylindrospermum* they are found only terminally. Intercalary heterocysts are spaced at regular intervals along the filament (Fig. 5.6a). This pattern permits the most even distribution of fixed nitrogen to the vegetative cells and in turn allows the heterocysts to receive adequate supplies of fixed carbon. During growth of a dinitrogen-fixing culture, cell division increases the number of vegetative cells between heterocysts (referred to as the interheterocyst interval). However, the regular spacing and frequency of heterocysts is maintained by differentiation of a new cell at the center of each interval once the number of

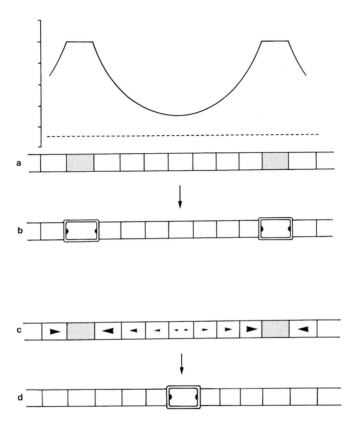

FIGURE 5.8 Alternative models (Wolk, 1989) to explain the development of a spaced pattern of heterocysts in a filamentous cyanobacterium after removal of fixed nitrogen from the medium. (**a, b**) In the first model the first cells to respond to nitrogen deprivation (shaded cells) arise at random locations and produce an inhibitor of heterocyst development, which diffuses along the filament. The inhibitor is degraded as it diffuses, establishing a decreasing concentration gradient, indicated by the solid line above the filament in (**a**). The horizontal dashed line represents a threshold intracellular concentration of inhibitor, below which a cell can differentiate and above which it cannot. The production of inhibitor prevents neighboring cells from differentiating, and the first cells to respond to nitrogen starvation therefore develop into heterocysts (**b**). Cell division eventually increases the number of vegetative cells between the heterocysts and results in a cell at the center of the interval falling below threshold and being able to differentiate. (**c, d**) In the second model the first cells to respond to nitrogen starvation (shaded cells) activate amino acid pumps, which draw the products of proteolysis (which is associated with heterocyst development) from adjacent cells, which in turn activate pumps and drain their neighbors of fixed nitrogen (**c**). In this way pump activation is propagated along the filament, but the magnitude of the pump decreases with the distance from the initial cell. The arrowheads in each cell indicate the direction of flow of fixed nitrogen, and their size indicates the magnitude of the pump drawing fixed nitrogen from that cell. In this model the cells that differentiate (**d**) are not the ones that first respond to nitrogen deprivation but are those that are being pumped from both directions (middle cell of the interval in **c**) and are therefore severely nitrogen-starved. (Adapted from Adams, 1992)

vegetative cells in that interval has doubled. In such cultures the existing pattern of heterocysts can serve to regulate the position of newly developing cells. However, a regular heterocyst pattern can be formed de novo when a culture grown in the presence of ammonium chloride, and therefore lacking any heterocysts, is transferred to a medium without a source of combined nitrogen.

How do cyanobacteria achieve the de novo establishment of a regular pattern of heterocysts? Two models have been proposed to explain it (Fig. 5.8) (Wolk, 1989, 1991; Adams, 1992; Wolk et al., 1994). In the first model, after removal of fixed nitrogen, developing heterocysts establish zones of inhibition around themselves by producing a diffusible inhibitor (Fig. 5.8a). A vegetative cell can commence heterocyst development only if the concentration of inhibitor within it is below a threshold level.

The triggering of heterocyst development by removal of combined nitrogen from the medium results in extensive proteolysis of many cell proteins. Haselkorn (1978) suggested that the amino acids liberated by this proteolysis might serve as diffusible inhibitors of differentiation; the second model of pattern control is based on this idea (Fig. 5.8). Because proteolysis occurs in all cells, Haselkorn (1978) proposed that the establishment of gradients of inhibition of the type needed to generate a spaced pattern of heterocysts would require amino acid pumps. In this model the first cells to respond to nitrogen starvation would establish such pumps to scavenge fixed nitrogen from adjacent cells (Wolk, 1989, 1991) (Fig. 5.8c). These cells would themselves respond by activating their own pumps from the next cell, and so on along the filament.

In the first model the cells most likely to differentiate are those that were the first to respond to nitrogen starvation (Fig. 5.8b). However, the second model predicts that the cells that will eventually form heterocysts are not the first to respond to nitrogen starvation but are those midway between, which are being drained of fixed nitrogen from both sides and therefore have no alternative but to generate their own fixed nitrogen (Fig. 5.8d). There is no conclusive evidence to support either model, although there are reasons to believe that the second one is unlikely to be correct (Adams, 1992).

Cell Selection During Differentiation and Pattern Formation

The de novo establishment of a regular pattern of heterocysts from a filament of apparently similar vegetative cells requires a flexible system of cell selection because as far as we know all cells have the genetic capacity to become heterocysts. How is it achieved? At present we have no answer, but there is evidence that cell division, cell cycle/DNA replication, and competition between developing cells may play a part.

Cell division in at least some *Anabaena* spp. is asymmetrical, and heterocysts can develop from only the smaller daughter cell (Wilcox et al., 1975b; Adams, 1992). Incubation of *Anabaena cylindrica* at a high light intensity for a short period following the induction of heterocyst development results in a three- to fourfold increase in the frequency of symmetrical cell divisions and leads to the production of adjacent (double) heterocysts (Fig. 5.9a,b) (Adams,

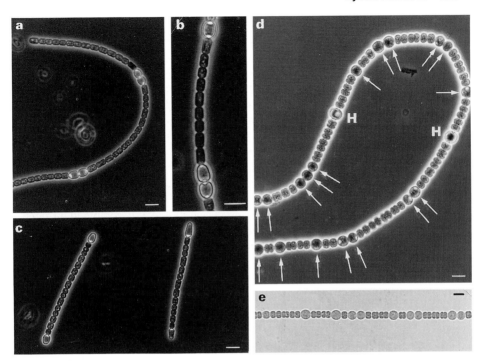

FIGURE 5.9   Photomicrographs of *Anabaena* spp. showing alterations to the normal heterocyst pattern. (**a, b**) *Anabaena cylindrica*, showing double heterocysts induced by incubation at high light intensity for a short period following removal of fixed nitrogen. (**c**) *A. cylindrica*, showing two short filaments with terminal heterocysts at both ends; they were induced by fragmentation of filaments after transfer to a medium lacking fixed nitrogen. (**d, e**) *Anabaena* sp. strain CA grown in the absence of nitrate but with 7-azatryptophan (10^{-5} M). The effect of the analog on heterocyst pattern can be clearly seen when this figure is compared with Figure 5.6a. (**d**) mature heterocysts (H) and proheterocysts (arrows) are indicated. (**e**) Proheterocysts are clearly visible. **a–d**: Phase contrast microscopy. **e** Bright field microscopy. Bars = 10 μm. (Reprinted from Adams and Carr, 1981, with permission)

1992). Asymmetrical cell division may therefore provide a means of selecting cells for development. It is unclear, however, if this mechanism applies to all filamentous cyanobacteria that produce intercalary heterocysts, as there are reports that in some cases, such as *Anabaena* PCC 7120, cell division is symmetrical (Buikema and Haselkorn, 1993).

The use of specific inhibitors has shown that DNA, RNA, and protein synthesis are required for heterocyst development, and that a heterocyst fulfills the requirement for each of these processes at different stages of its formation (Adams, 1992). The requirement for DNA synthesis has led to the suggestion that DNA replication serves as a timer that ties heterocyst differentiation to the cell cycle (Adams, 1992).

Although selection mechanisms based on asymmetrical cell division and position in the cell cycle may help limit the number of vegetative cells that can differentiate, they cannot be sufficient to generate the final regular pattern of heterocysts. This is because the first cells to begin development are likely to be randomly spaced in the filament and are therefore sometimes too close to other developing cells. A competitive mechanism must therefore operate to generate the regular pattern, which is apparent even when the first prohetero-cysts are becoming visible by light microscopy. Evidence for such a competitive mechanism was obtained by Wilcox et al. (1975b; Adams and Carr, 1981). Competition is probably especially important between daughter cells because of their inherent similarities. Treatments that disrupt heterocyst spacing, such as incubation with 7-azatryptophan or at high light intensity, probably do so by interfering with competition (see next section).

Disruption of Heterocyst Pattern

Under normal growth conditions the regular intercalary pattern of single het-erocysts in cyanobacteria such as *Anabaena* is remarkably consistent. Exposure to high light intensity during heterocyst development can lead to the production of double (adjacent) heterocysts (Fig. 5.9a,b), although the number of vegetative cells between heterocysts is little changed, implying that the inhibitory zones around heterocysts are not affected (Adams, 1992). There are several ways in which both the frequency of heterocysts and the regularity of their spacing can be altered. The first of these methods is the isolation of mutants, some of which produce highly irregular patterns with as many as six to eight heterocysts adjacent to each other (Wilcox et al., 1975a; Adams, 1992). A remarkably similar alteration of the pattern can be produced by incubating cyanobacteria with the tryptophan analog 7-azatryptophan (Fig. 5.9d,e) (Mitch-ison and Wilcox, 1973; Adams, 1992). Finally, when a plasmid carrying the *hetR* gene is transferred to wild-type *Anabaena* PCC 7120, a large number of heterocysts are produced, usually in pairs but sometimes with up to five in a row (Buikema and Haselkorn, 1993).

Genome Rearrangements Associated with Heterocyst Development

During the late stages of heterocyst development in *Anabaena* PCC 7120, two developmentally regulated DNA rearrangements occur in the *nifD* and *fdxN* genes (Haselkorn, 1992). The first is excision of an 11-kilobase (kb) element from within the coding region of the *nifD* gene, restoring the *nifHDK* operon, which can then be transcribed from the *nifH* promoter as a polycistronic mes-sage. Whereas excision is essential for *nifK* expression (from the *nifH* pro-moter), it has no role in heterocyst development or spacing (Tandeau de Marsac and Houmard, 1993). This conclusion is supported by the observation that the *nifHDK* genes are contiguous in the heterocystous strain *Fischerella* ATCC 27929, as they are in all nonheterocystous strains (Tandeau de Marsac and Houmard, 1993).

The second gene rearrangement that occurs during heterocyst development in *Anabaena* PCC 7120 involves excision of a 55-kb element that interrupts the *fdxN* gene. The recombination restores the *fdxN* gene and results in formation of the *nifB-fdxN-nifS-nifU* operon. Excision of the 55-kb element is required for dinitrogen fixation, yet as with the 11-kb element it is not necessary for normal heterocyst differentiation and pattern (Carrasco et al., 1994). This finding is confirmed by the absence of the 55-kb element from the genomes of a number of *Anabaena* and *Nostoc* strains (Haselkorn, 1992).

What advantage does possession of the two interrupting elements confer on *Anabaena* and *Nostoc*? We presently have no answer to this question. Neither is required for heterocyst development. The 55-kb element is missing from many strains, and although the 11-kb element is present in most strains there are exceptions. A possibility proposed by Haselkorn (1992) is that they code for a DNA repair enzyme. This situation is analogous to that in *Escherichia coli* carrying Tn5, which confers an advantage to the cell by virtue of the DNA repair activity coded for by the bleomycin resistance gene (Blot et al., 1991).

*Identification and Analysis of Genes Involved in
Heterocyst Development*

A number of genes involved in heterocyst development have been cloned and sequenced (Wolk et al., 1994). The genes *hepA* (originally called *hetA*) and *hglK* are involved in production of the heterocyst envelope polysaccharide layer (Wolk et al., 1988, 1993; Holland and Wolk, 1990; Buikema and Haselkorn, 1993) and the glycolipid layer (Buikema and Haselkorn, 1993), respectively. The *hetR* gene was cloned by complementation of a mutant of *Anabaena* PCC 7120, which failed to produce heterocysts (Buikema and Haselkorn, 1991b, 1993). Expression of *hetR* is low in the presence of combined nitrogen but increases three- to fivefold early during heterocyst development. Wild-type cells containing *hetR* on a multicopy plasmid produce increased numbers of heterocysts, often as pairs or sometimes higher multiples (Buikema and Haselkorn, 1991a, 1993). *hetR* is essential for heterocyst development, for which it is a positive control element (Buikema and Haselkorn, 1991a; Black et al., 1993); and its expression is autoregulated (Black et al., 1993; Buikema and Haselkorn, 1993). To regulate the heterocyst pattern, *hetR* seems to interact with a second gene, *patA* (Liang et al., 1992, 1993).

Akinetes

Akinetes (Greek: *akinetos*, motionless) are thick-walled cells produced by many strains of subsections IV (Order Nostocales) and V (Order Stigonematales), usually as cultures approach the stationary phase. These cells are a resting form that are generally resistant to cold and desiccation but not to heat; and they can germinate to produce new filaments when conditions are suitable (Nichols and Adams, 1982; Herdman, 1987, 1988; Adams, 1992). Although frequently referred to as spores, particularly in the early literature, the term is inappropriate as they clearly differ from the variety of resistance and resting structures, including bacterial endospores, produced by other microorganisms. Akinetes were first described (Carter, 1856; de Bary, 1863; Rabenhorst, 1865) long before the discovery of the bacterial endospore (Cohn, 1877; Koch, 1877).

Akinetes provide cyanobacteria with a means of overwintering and surviving dry periods (Herdman, 1987). For example, akinetes of *Anabaena cylindrica* can survive for at least 5 years in the dark in the dry state, whereas vegetative cells can survive no longer than 2 weeks under similar conditions (Yamamoto, 1975). Akinetes of *Nostoc* PCC 7524 can survive in the dark at 4°C for 15 months, whereas vegetative cells cannot survive more than 7 days at this temperature (Sutherland et al., 1979). An even clearer indication of their survival capabilities in the environment is the isolation of viable akinetes from sediments as old as 64 years (Livingstone and Jaworski, 1980).

Structure and Composition

Most akinetes are readily distinguished from vegetative cells by light microscopy, being generally larger than vegetative cells; they have a thickened cell wall, a multilayer extracellular envelope, and more granular cytoplasm (Figs. 5.6, 5.10) (Nichols and Adams, 1982; Herdman, 1987, 1988). Akinetes accumulate cyanophycin, a nitrogen reserve material unique to cyanobacteria that consists of a polymer of arginine and aspartic acid (Simon, 1987) and that can be seen as granules in the cytoplasm (Figs. 5.6c,d).

In *Nostoc* PCC 7524 the mean cellular contents of RNA, DNA, and protein are similar in vegetative cells and akinetes (Sutherland et al., 1979). In contrast, the much larger akinetes of *A. cylindrica* contain the same amount of RNA, more than twice as much DNA, and 10 times as much protein as vegetative cells (Simon, 1977).

Relation to Heterocysts

Although akinetes are produced only by heterocystous cyanobacteria, the presence of heterocysts is not essential, as they can develop in some strains when heterocyst development has been repressed by the presence of fixed nitrogen. For example, in *Anabaena* sp. strain CA grown in the presence of sodium nitrate, and therefore lacking heterocysts, every cell in a filament can become an akinete (Fig. 5.6d). However, when heterocysts are present they exert an influence on the positions at which akinetes can develop, and this trait is characteristic of the cyanobacterial strain. For example, in *Anabaena* sp. strain CA akinetes develop randomly in the absence of heterocysts but midway between heterocysts when the latter are present. Alternative patterns are possible, as akinetes develop immediately adjacent to heterocysts in *A. cylindrica* (Fig. 5.6c) and three cells away in *Anabaena circinalis* and some other planktonic species (Fay et al., 1984). In most cases akinetes develop in strings that show a clear gradient of decreasing maturity away from the first to form (Fig. 5.6c).

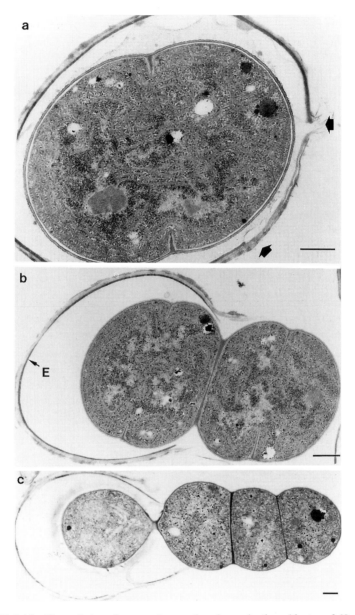

FIGURE 5.10 Transmission electron micrographs of germinating akinetes of *Nostoc* PCC 7524. (a) Germinating akinete showing the early stage of cross wall formation of the first cell division. A wide break in the laminated layer of the envelope is indicated by the arrows. (b) Germinating akinete at the second division stage. The cell division is well synchronized in both cells, and the short filament is emerging from the still rigid envelope (E). (c) Germination has reached the four-cell stage. The terminal cell is still within the akinete envelope and has begun to develop into a heterocyst with the characteristic neck-like connection with the neighboring vegetative cell. Bars = 550 nm. (Reprinted from Sutherland et al., 1985b, with permission)

It might be thought that because akinetes accumulate large amounts of fixed nitrogen in the form of cyanophycin they would develop close to a source of fixed nitrogen. It would explain their occurrence next to heterocysts in *A. cylindrica* but not their presence midway between heterocysts in *Anabaena* sp. strain CA and *Nostoc* PCC 7524. These two extremes of akinete pattern suggest at least two regulatory mechanisms, of which we know nothing at present. It is known, however, that placement of akinetes can be altered chemically: by incubation with the tryptophan analog 7-azatryptophan (Sutherland et al., 1979) and the arginine analog canavanine (Nichols et al., 1980; Adams, 1992).

Factors That Influence Akinete Development

In common with bacterial endospores, akinetes develop at the end of the exponential growth phase, even in the presence of excess inorganic nutrients in the medium (Fay, 1969; Nichols and Adams, 1982; Herdman, 1987, 1988). In such cultures the only factor to become limiting is light—as a result of self-shading when the culture becomes dense. Indeed, there is a direct correlation between the light intensity at which *A. cylindrica* is grown and the cell density at which akinete formation begins (Nichols et al., 1980). The involvement of light energy limitation can be shown with a facultative photoheterotroph, such as *Nostoc* PCC 7524, in which the exponential phase can be prolonged and akinete formation delayed by the addition of a utilizable carbon source such as sucrose (Sutherland et al., 1979).

Although light energy limitation is the clearest trigger for akinete differentiation, limitation of a wide variety of nutrients, notably phosphate, has also been implicated (Nichols and Adams, 1982; Herdman, 1987, 1988). It may be that these diverse stimuli induce a common physiological trigger that results in akinete development, but nothing is known of this possibility at present.

Cell-free supernatants of akinete-containing cultures of *Cylindrospermum licheniforme* contain a compound (formula: C_7H_5OSN) that stimulates akinete formation in young cultures of the same cyanobacterium (Fisher and Wolk, 1976; Hirosawa and Wolk, 1979a,b). It is not clear whether this compound functions as an interfilamentous or an intrafilamentous signal, although there is the interesting possibility that it may become concentrated in a natural body of water during desiccation or stagnation and trigger the development of akinetes to survive the impending drought (Fisher and Wolk, 1976). Attempts to detect similar activities in other cyanobacteria have proved unsuccessful (Nichols and Adams, 1982).

Akinete Germination

The major stimulus for akinete germination is increased light intensity, which is commonly achieved by diluting an akinete-containing culture with fresh medium (Herdman, 1987, 1988). Germination requires respiration and photophosphorylation but is most efficient when both photosystems I and II are active. No germination occurs in the dark, even when strains capable of chemoheterotrophic growth in the dark are supplied with the appropriate sugar (Chauvat et al., 1982).

During akinete germination the envelope usually remains intact, apart from a pore at one end, through which the germling emerges, having first undergone one or two cell divisions (Fig. 5.10) (Nichols and Adams, 1982; Herdman, 1987). Empty akinete envelopes are frequently seen after germination (Fig. 5.6e). In the absence of a source of combined nitrogen the germinating akinete must form heterocysts to supply its demand for fixed nitrogen. Usually a single heterocyst develops at a time and position that are characteristic of the cyanobacterium from which the akinete developed. For example, in *Nostoc* PCC 7524 the first heterocyst develops in a terminal position when the germling is three cells long (Fig. 5.10c) (Sutherland et al., 1985b), whereas in *Anabaena* CA a single heterocyst forms at approximately the center of the germling when it reaches six or seven cells in length (Fig. 5.6e) (Nichols and Adams, 1982). Once the newly formed filament becomes long enough, additional heterocysts develop at intercalary locations in both of these cyanobacteria.

In *Nostoc* PCC 7524 akinetes, protein synthesis begins immediately on initiation of germination and continues for 11 hours (Sutherland et al., 1985a). RNA synthesis continues throughout germination, whereas DNA synthesis does not begin until 80 minutes after initiation but is continuous thereafter. Germination continues in the presence of phenethyl alcohol, an inhibitor of DNA synthesis (Herdman, 1987), resulting in the production of short filaments of 10 vegetative cells. This is possible because the akinetes contain 10 genome equivalents of DNA (Sutherland et al., 1979), thereby allowing cell division to continue in the absence of DNA replication until each newly formed vegetative cell contains one genome copy.

Hormogonia

Hormogonia are short, undifferentiated filaments that develop as a result of fragmentation of the parent trichome; they are often motile by gliding (Fig. 5.1) (Rippka et al., 1979, 1981; Adams, 1992; Tandeau de Marsac and Houmard, 1993; Tandeau de Marsac, 1994). In the Order Oscillatoriales, which do not develop heterocysts or akinetes, the hormogonia are simply short, motile fragments of the parent trichome, which is itself motile (Rippka et al., 1979, 1981). Fragmentation of the trichome results from death and lysis of cells, which in some cases appear to be sacrificed specifically for this purpose. These cells are referred to as necridia (Rippka et al., 1979, 1981). It is unclear whether the formation of this type of hormogonium requires any specific changes in gene expression, as it resembles the parent trichome in all but length, and it may simply be an indirect result of cell lysis and consequent fragmentation of the parent trichome. However, in the heterocyst-forming cyanobacteria of subsections IV (the Nostocales) and V (the Stigonematales), hormogonia are different from the parent trichomes, as they are not just shorter but their cells are much smaller, they show gliding motility (the parent trichomes are immotile), and many develop gas vesicles. Therefore their formation clearly requires differential gene expression and seems to result from changes in the environment.

Function

The formation of hormogonia provides a means of dispersal that is not otherwise available because, with the exception of the *Oscillatoriales*, the parent trichomes of all other hormogonia-forming cyanobacteria are immotile. Cyanobacteria that do not form hormogonia are all motile. Thus in the Order Nostocales, the genera *Anabaena*, *Nodularia*, *Cylindrospermum*, and *Aphanizomenon* exhibit permanent gliding motility but do not form hormogonia, whereas the genera *Nostoc*, *Scytonema*, and *Calothrix* have immotile trichomes but do form hormogonia. The effectiveness of hormogonia as a means of dispersal is enhanced by their motility, their generally small size, and the presence of gas vesicles, which give them buoyancy (Tandeau de Marsac and Houmard, 1993; Tandeau de Marsac, 1994). Using these means they can glide or float away from the original colony. These characteristic properties are only transient, as the hormogonia must eventually come to rest and form vegetative trichomes once more. The latter is essential, as hormogonia lack heterocysts and therefore become nitrogen-starved after a relatively short period.

Hormogonia Development

The development of the small-cell hormogonia, typical of Nostocales and Stigonematales, is preceded by a round of rapid, relatively synchronous cell division (Tandeau de Marsac et al., 1988; Campbell and Meeks, 1989; Adams, 1992; Tandeau de Marsac, 1994). The latter occurs in the absence of cell growth and therefore results in a characteristic decrease in cell size. This phenomenon has an interesting parallel in the formation of baeocytes by multiple fission in the pleurocapsalean cyanobacteria (see below). The enforcement of synchronized cell divisions on an asynchronous population of cells clearly means that many cells divide prior to completion of DNA replication. Indeed, the synchronous cell divisions that initiate hormogonia formation occur with little or no DNA replication (Herdman and Rippka, 1988; Damerval et al., 1991; S. Babic and D.G. Adams, unpublished observations).

Why does it not result in cells lacking DNA? Cyanobacteria commonly have multiple genome copies (Waterbury and Stanier, 1978; Sutherland et al., 1979; Herdman and Rippka, 1988; Haselkorn, 1991), which permits several rounds of cell division without DNA replication, as can be seen with germination of akinetes. After a short motile period, hormogonia lose motility, develop heterocysts at positions characteristic of the particular genus, and eventually return to the typical mature filament morphology (Rippka et al., 1979, 1981; Tandeau de Marsac, 1994).

Factors That Trigger Hormogonia Formation

Because hormogonia provide a means of dispersal it is no surprise that environmental factors influence their formation, although the significance of some of the stimuli is clearer than that of others. It might be thought that the greatest need for dispersal occurs when the conditions for active growth are poorest, yet these conditions are the ones that trigger development of the cyanobacterial resting cell, the akinete, whereas hormogonia usually develop in response to improved conditions. For example, the addition of phosphate to phosphate-starved *Calothrix* spp. growing on agar can trigger hormogonia formation (Whitton, 1989). An interesting example occurs with *Mastigocladus laminosus* (a member of the order Stigonematales) in which hormogonia formation is triggered by transfer of filaments from liquid medium to the surface of medium solidified with agar (Hernandez-Muniz and Stevens, 1987). The environmental significance of this action may at first seem obscure without the additional observation that, in its native habitat, *Mastigocladus laminosus* can rapidly colonize new areas of the rock on which it is growing (W. Hernandez-Muniz, personal communication). Coverage of fresh regions of substratum, created for example by the scouring effect of fast-flowing water, can be rapid when permanently motile cyanobacteria move in from adjoining areas of growth. This phenomenon is often seen in fish tanks in which the glass is covered with a rich blue-green or brown growth of cyanobacteria, such as *Oscillatoria* spp. When a small area of this growth is cleared, it begins to be recolonized within hours by the movement of filaments from surrounding areas. The only way in which nonmotile cyanobacteria can compete in such circumstances is to be able to rapidly produce motile hormogonia.

The triggering of hormogonia formation as a result of sudden clearing of the substratum has similarities with that occurring in response to the transfer of old cultures of some *Nostoc* spp. to fresh medium (Rippka et al., 1979; Herdman and Rippka, 1988). In both cases it can be speculated that a factor inhibitory to hormogonia formation is being diluted by either the fresh medium or the sudden removal of neighboring organisms. Indeed, it has been shown that late exponential, or stationary phase, cultures of *Nostoc* PCC 7119 excrete a compound that inhibits hormogonia formation even when diluted as much as 20-fold (Herdman and Rippka, 1988). The identity of the inhibitor is unknown, although it is destroyed by autoclaving and is dialyzable.

Interestingly, factors that stimulate hormogonia formation have also been reported. Hormogonia-containing cultures of *Nostoc muscorum* A and *Nostoc commune* 584 excrete a heat-labile substance that induces hormogonia formation when added to dark-grown cultures (Robinson and Miller, 1970). Such stimulatory factors may play a role in coordinating hormogonia formation by enabling extracellular communication within large colonies of cyanobacteria. A similar situation arises in the establishment of cyanobacteria–plant symbioses, when extracellular factors produced by the plant trigger hormogonia formation. It increases the chance of the plant becoming infected, as it is the hormogonia that are the infective agents (Meeks, 1990).

Another major factor in the triggering of hormogonia formation is light intensity and wavelength. For example, red light (640–650 nm) stimulates the differentiation of hormogonia in *Nostoc muscorum* A (Lazaroff, 1973) and *Nostoc commune* 584 (Robinson and Miller, 1970). Green light reverses the effect. A similar response is seen in *Calothrix* PCC 7601 in which hormogonia formation is stimulated by red light and inhibited by green light (Damerval et al., 1991; Tandeau de Marsac and Houmard, 1993; Tandeau de Marsac, 1994; Campbell et al., 1993).

Pleurocapsalean Cyanobacteria

The pleurocapsalean cyanobacteria are distinguished from all others by their ability to undergo multiple fission, forming small, spherical cells known as baeocytes (Greek: small cell), which serve as a means of dispersal (Waterbury and Stanier, 1978; Waterbury, 1989).

In pleurocapsalean cyanobacteria the vegetative cells are surrounded by a fibrous layer (Figs. 5.12, 5.13) that resembles the sheaths of other cyanobacteria. The fibrous layer becomes thicker as the cells enlarge and ruptures to release the baeocytes. Unicellular forms divide exclusively by multiple fission, whereas in the remaining forms binary fission produces cell aggregates in which some or all of the cells undergo multiple fission (Fig. 5.14). The feature that distinguishes multiple fission from repeated binary fission is the reduction in cell size that occurs following multiple fission. With binary fission cell growth follows each cell division, whereas with multiple fission rapidly repeated rounds of cell division occur without cell growth, resulting in the production of at least four, and often many hundreds of, daughter cells that are considerably smaller than the parent cell (Waterbury and Stanier, 1978; Waterbury, 1989). This rapid cell division without cell growth has an interesting parallel in hormogonia formation in filamentous cyanobacteria. In common with the synchronous cell divisions involved in akinete germination and hormogonia formation, multiple fission probably does not require de novo DNA synthesis, as the vegetative cells have many genome copies (Waterbury and Stanier, 1978).

In the genera *Dermocarpa* and *Xenococcus* cell division always involves multiple fission, the baeocyte enlarging until it is ready to undergo multiple fission once more (Fig. 5.14). However, in *Dermocarpella* one to three binary fissions occur before multiple fission (Figs. 5.13, 5.14). In other cases binary fission in three planes gives rise to complex aggregates of cells that may be regular cubes, as in *Myxosarcina* (Figs. 5.11, 5.14), or irregular masses with filamentous extensions, as with the *Pleurocapsa* group (Figs. 5.11, 5.14). In either case, some or all of the vegetative cells in the aggregates undergo multiple fission and release baeocytes (Figs. 5.11–5.14). Although multicellular masses give the appearance of being filamentous, they differ from the truly filamentous cyanobacteria, as the cells are separated from one another by the fibrous outer wall layers, which also surround the whole cell mass and keep the cells together (Figs. 5.12, 5.13). The baeocytes produced by most pleuro-capsalean cyanobacteria lack the fibrous layer (Fig. 5.12b, 5.13b), although there are exceptions (Fig. 5.13a) (Waterbury and Stanier, 1978; Waterbury, 1989). As baeocytes increase their cell volume during growth, they cause the fibrous outer wall layer of the parental cell to rupture.

The vegetative cells of all pleurocapsalean cyanobacteria are immotile. Baeocytes surrounded by the fibrous layer are also immotile, whereas those produced without this layer exhibit gliding motility for a short period following their release. They are phototactic, moving toward or away from light depending on its intensity. As baeocytes begin to enlarge into vegetative cells, they synthesize the fibrous layer, which coincides with loss of motility.

The pleurocapsalean cyanobacteria are a remarkably understudied group that pose many fascinating problems. How, for example, is multiple fission regulated? What is its relation to DNA replication? How is chromosome partition achieved when a single cell divides in three dimensions to produce hundreds of baeocytes?

Hairs

Members of the Rivulariaceae (Family III of subsection IV) (Whitton, 1989) are characterized by trichomes that taper from a broad base to a narrow apex, often ending in a long, multicellular hair (Whitton, 1989). The degree of tapering is variable, and a better diagnostic feature of this family is the presence of a single terminal heterocyst that develops at the wide end of the trichome (and is therefore often referred to as a basal heterocyst) in medium deficient in combined nitrogen. The availability of combined nitrogen usually completely inhibits heterocyst development, often reduces the degree of tapering of trichomes, and can result in reduction in the frequency and length of hair formation (Sinclair and Whitton, 1977a,b). Indeed, the morphological appearance of members of the Rivulariaceae in laboratory culture is often considerably different from that in nature. An interesting feature of the Rivulariaceae is that growth is typically meristematic, being restricted to a particular region of the trichome, often either close to the basal heterocyst or to the hair, depending on the genus or species (Whitton, 1987).

The hair can be defined as the region of the trichome where the cells are narrow, elongated, highly vacuolated, and usually apparently colorless (Whitton, 1987). This region can account for two-thirds of a total trichome length of 800 μm or more. Not all members of the Rivulariaceae produce hairs, although it is not clear if this characteristic represents a true genetic difference or an inability to identify the correct environmental trigger. In support of the former is the observation that there are some members of the Rivulariaceae that have never been seen to produce hairs in either the laboratory or the field (Whitton, 1989).

The formation of hairs is stimulated most commonly by phosphate deficiency, less commonly by iron deficiency, and only rarely by a lack of mag-

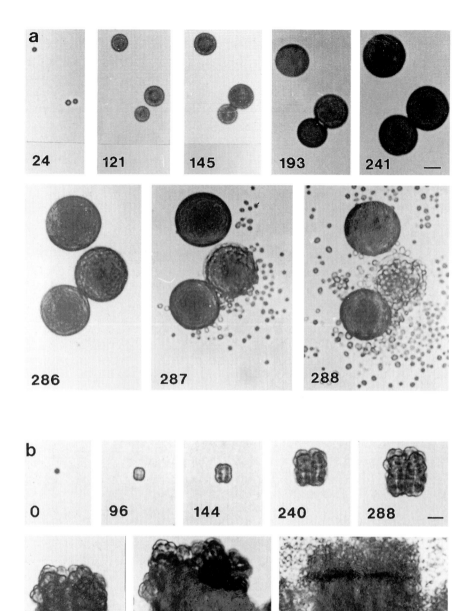

FIGURE 5.11 Development of individual baeocytes of *Dermocarpa* PCC 7302 (**a**), *Myxosarcina* PCC 7312 (**b**), and *Pleurocapsa* PCC 7516 (**c**) on agar medium. The numbers in the photomicrographs indicate the elapsed time, in hours, after the initial observation. The baeocytes follow the developmental cycles shown in Figure 5.14, culminating in multiple fission and the production of more baeocytes. The magnification is the same for all photomicrographs. Bars = 10 μm. (Reprinted from Waterbury and Stanier, 1978, with permission) Figure continues.

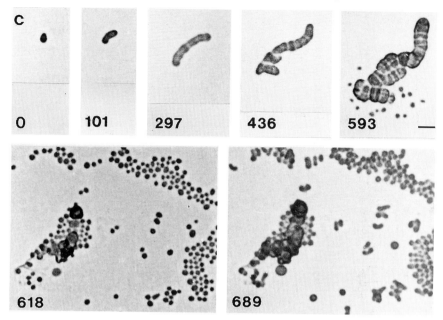

FIGURE 5.11 *Continued*

nesium (Whitton, 1987). Their development does not involve the production of new cells, only the narrowing and considerable elongation of existing ones, resulting in a great increase in surface area. As the hair develops, the cells lose their pigmentation, granular inclusions, and much of the cytoplasm; and they form large intrathylakoidal vacuoles (Whitton, 1989).

What is the function of hairs? Although this question remains controversial, the most promising theory is that they play a role in maximizing phosphate uptake (Whitton, 1989). In support of this are the observations that the production of hairs is stimulated by phosphate deficiency, and that they possess the ability to hydrolyze organic phosphate. For example, in *Calothrix parietina* D184, inducible phosphatase activity commences at about the same time as hair formation in batch cultures in which phosphate concentration is allowed to decrease (Livingstone et al., 1983; Whitton, 1987). Phosphomonoesterase activity is present on the walls of hair cells, the chlorophyll-containing cells adjacent to these, and the part of the sheath furthest from the heterocyst.

Concluding Remarks

It should be clear from the preceding discussion that, for prokaryotes, cyanobacteria possess remarkably complex developmental capabilities. Why should this be the case? In this final section I will attempt to answer this question by drawing on a few facts and a good deal of speculation.

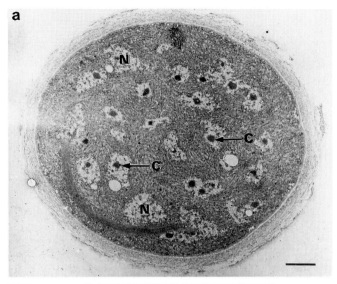

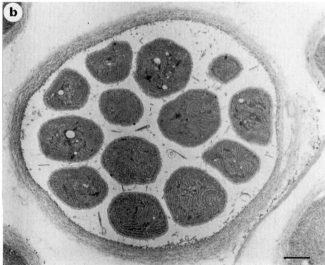

FIGURE 5.12   Electron micrographs of thin sections of *Dermocarpa* PCC 7304 grown in liquid medium. (**a**) Cell immediately prior to multiple fission, containing many nucleoids (N) frequently associated with carboxysomes (C). (**b**) Cell that has completed multiple fission and is filled with baeocytes, each of which is surrounded by a peptidoglycan layer and outer membrane, but not by a fibrous layer. Bars = 1 μm. (Reprinted from Waterbury and Stanier, 1978, with permission)

Why Be Multicellular?

Growth of cyanobacteria as filamentous clumps or colonies has advantages for protection against predation (long filaments are difficult to engulf), the colo-

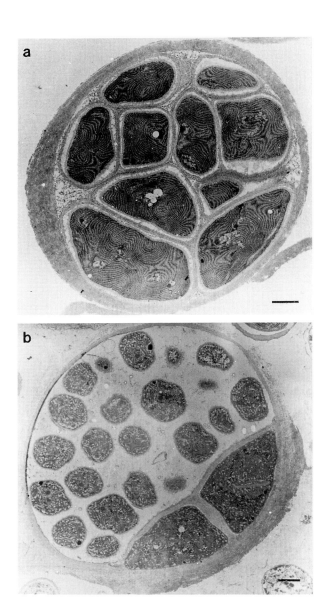

FIGURE 5.13 (a) Electron micrograph of a thin section of a cell of *Xenococcus* PCC 7307 that has divided into numerous baeocytes, each of which is surrounded by a distinct fibrous layer. (b) Electron micrograph of a thin section of a cell of *Dermocarpella* PCC 7326. The two basal cells, which are surrounded by a clear fibrous layer, are the products of previous binary fission, whereas the apical cell has undergone multiple fission to produce many baeocytes, which do not have a fibrous layer. Bars = 1 μm. (Reprinted from Waterbury and Stanier, 1978, with permission)

Developmental cycle

Genus	A	B	C	D	E	F
Dermocarpa and *Xenococcus*	Symmetric baeocyte enlargement	Multiple fission	Release of baeocytes (i)	—	—	—
Dermocarpella	Asymmetric baeocyte enlargement	Binary fission	Multiple fission of apical cell	Baeocyte release	Baeocyte motility	Basal cell enlargement
Myxosarcina and *Chroococcidiopsis*	Symmetric baeocyte enlargement	Repeated binary fission in 3 planes	Multiple fission of nearly all cells and baeocyte release	Baeocyte motility (ii)	—	—
Pleurocapsa groups I and II	Baeocyte enlargement (iii)	Binary fission in many irregular planes	Multiple fission of some cells	Baeocyte release	Baeocyte motility	—

(i) Baeocytes are motile in *Dermocarpa* and non-motile in *Xenococcus*.
(ii) Baeocytes are motile in *Myxosarcina* and non-motile in *Chroococcidiopsis*.
(iii) Symmetric in subgroup I; asymmetric in subgroup II.

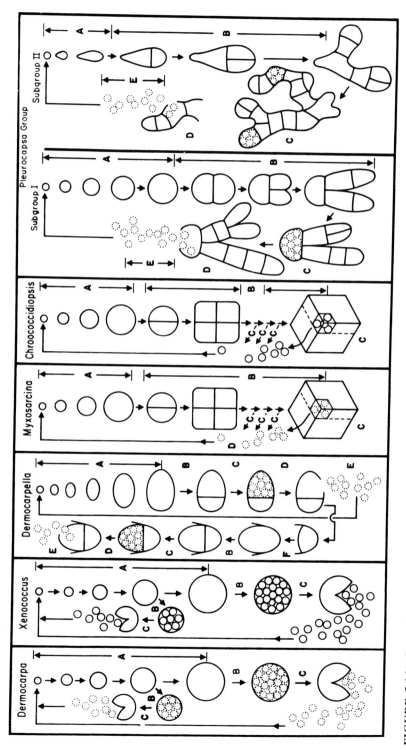

FIGURE 5.14 Comparison of the developmental cycles of pleurocapsalean cyanobacteria. Small, dotted circles represent baeocytes which are not surrounded by a fibrous outer wall layer at the time of release and are consequently motile. Small, solid circles represent baeocytes that are surrounded by a fibrous layer and are immotile. (Reprinted from Waterbury and Stanier, 1978, with permission)

nization of substrates, and the survival of desiccation. In a dense clump of unicellular cyanobacteria, those at the center suffer through lack of light or other nutrients. In filamentous forms, whereas one end of the filament may be buried at the center of the colony, the other may extend toward the outside. Intercellular communication within the filament then permits redistribution of nutrients, which is not possible in the separate cells of a unicellular form. Colonial growth is common in the genus *Nostoc*, in which trichomes are often embedded in gelatinous polysaccharide to produce colonies that vary in size from microscopic (Fig. 5.4b) to 20 cm in diameter (Castenholz, 1989a).

Members of the family Rivulariaceae also frequently grow in colonial form. These cyanobacteria are characterized by trichomes that are broad at the base and become narrower toward the apex (Whitton, 1987, 1989). When deprived of certain nutrients, notably phosphate, the tapering becomes more extreme and the trichome ends in a long, multicellular hair. Single heterocysts are formed at the broad end of the trichome, and in colonial growth it is this end of the filament that is attached to the substrate, forming hemispherical ''cushions'' (Whitton, 1989). In the case of planktonic strains the trichomes are radially arranged with the heterocysts at the center, producing spherical colonies. Here again the use of colonial growth presumably helps to protect against predation and to improve adherence to substrates. In addition, the heterocysts at the center of the colony can supply each trichome with fixed nitrogen, and the other end is free to form a hair, which may aid in phosphate scavenging or in the release of hormogonia for dispersal.

Another benefit of being filamentous is the greatly enhanced motility. Cyanobacteria do not possess flagella and, with one possible exception (Waterbury et al., 1985), cannot swim. Instead, they move by gliding, the mechanism of which is unknown, as indeed it is for all bacterial gliding. This form of motility requires filaments to be attached to a substrate and is ideally suited to movement through sediments or microbial mats. Long, streamlined filaments are likely to move much more efficiently than unicells in these situations, and this form of locomation is particularly common in the Order Oscillatoriales (non-heterocystous cyanobacteria), which can glide at rates of up to 10 μm per second. Fossil evidence has shown that counterparts to the modern-day members of this order were alive 3.5 billion years ago and have changed remarkably little over the intervening period (Schopf, 1994). At the time these cyanobacteria evolved, their environment would have been oxygen-deficient and lacking an effective ultraviolet (UV)-absorbing ozone layer. The ability to glide would have given filamentous cyanobacteria a great advantage in coping with high UV flux in the type of shallow-water settings where they thrive today and where gliding enables them to regulate their exposure to light and oxygen.

Although many filamentous cyanobacteria have permanent gliding motility, others have evolved an alternative strategy, in which they are motile for only part of their life cycle. For example, members of the genus *Nostoc* produce hormogonia as their motile phase, and pleurocapsalean cyanobacteria produce baeocytes. It is interesting that in both cases production of the motile form is preceded by highly synchronized, rapid cell division without accompanying cell growth, resulting in a decrease in cell size. Pleurocapsalean cyanobacteria cells undergo multiple fission, and in the genus *Nostoc* every cell in a filament divides within a fraction of the normal doubling time.

Being filamentous, then, has advantages and is known to be an ancient property of cyanobacteria. Such multicellularity and the associated intercellular communication are also essential prerequisites for the evolution of highly specialized cells because the cell that specializes invariably loses its autonomy and must rely on other cells to provide what it can no longer provide for itself. This situation is clearly the case with the heterocyst, which supplies vegetative cells with fixed nitrogen but cannot fix its own carbon and must receive it from its neighbors. Such cell specialization is possible only in multicellular cyanobacteria.

Why Develop Specialized Cells?

There are many unicellular cyanobacteria that survive and compete successfully, and there are filamentous forms that are incapable of cellular differentiation, yet have changed little in more than 3 billion years. Why then have some cyanobacteria evolved such complex multicellularity? One obvious driving force is the incompatible nature of dinitrogen fixation and oxygenic photosynthesis. Although cyanobacteria have evolved several means to circumvent this problem, most of them present limitations (Fay, 1992; Gallon, 1992). The most flexible solution is production of a specialized cell, dedicated to nitrogen fixation, which of course is the heterocyst.

It is likely that the ability to develop heterocysts arose relatively late in the evolution of cyanobacteria (Giovannoni et al., 1988), although probably still more than 3 billion years ago. How might such a complex cell have evolved? There are interesting similarities in the cell wall glycolipids that help reduce gas diffusion into the heterocyst and those in the thickened walls of the akinete, a cell specialized to survive cold and desiccation. This similarity might imply an akinete-like ancestor for today's heterocysts. Thus the heterocyst may have evolved in a filamentous cyanobacterium that produced akinetes but was capable of dinitrogen fixation only under low oxygen tension. Modifications to the glycolipid-enclosed akinete would provide a new cell type capable of protecting nitrogenase from both externally and internally derived oxygen thereby increasing the range of environments available to the new heterocystous strain. Perhaps the deposition of a cyanophycin plug at the end of each heterocyst (Fig. 5.5) is a remnant of its akinete ancestry, as akinetes contain large numbers of cyanophycin granules as a nitrogen reserve (Fig. 5.6c). Two pieces of genetic evidence support a link between heterocyst and akinete differentiation. It is now known that *hetR*, a regulatory gene in heterocyst development, is also essential for akinete formation (Leganes et al., 1994). In addition, the gene *hepA*, which is required for normal deposition of the polysaccharide layer of the heterocyst, is also required for formation of the akinete envelope (Leganes, 1994).

Because heterocysts represent a considerable energy investment and a loss of division capacity, there would be strong selective pressure to evolve mechanisms for ensuring that only the correct numbers of heterocysts are produced to provide for the filament's combined nitrogen needs. This action in turn leads to further refinement of the developmental system, the proheterocyst (Figs. 5.1, 5.6, 5.7). Because of its ability to cease differentiation and return to vegetative growth, the proheterocyst permits the flexibility required to regulate heterocyst frequency. Again there would be strong selective pressure in favor of strains that develop heterocysts that can dedifferentiate during the early stages over those that become fully committed at the beginning of the developmental process.

Why Is a Spatial Pattern Necessary?

Heterocysts supply neighboring vegetative cells with combined nitrogen and in turn receive the fixed carbon they are unable to provide for themselves. This interchange of nutrients leads to further necessary refinement of the differentiation process. An efficient distribution of both carbon and nitrogen within a long filament requires that heterocysts, which usually constitute only 5–10% of cells, are evenly distributed, and a mechanism must evolve to achieve this state. There are occasions when heterocyst development is initiated in cells that are too close to each other, so there must be a mechanism to permit ''fine tuning'' of the initial crude pattern, and there must be an opportunity for cells to dedifferentiate during the early stages of the process. Again, the proheterocyst provides this flexibility of response.

Why Communicate with Your Neighbors?

Although we know little of the mechanisms involved in the regulation of heterocyst spacing, it is clear that intercellular communication is vital to the process. In turn, the development of akinetes is influenced by the presence of heterocysts. When akinetes develop in filaments in which heterocyst formation has been suppressed by the presence of combined nitrogen, they do so in random locations. When heterocysts are present, however, akinetes develop in positions with a precise spatial relationship to the heterocysts, implying that intercellular communication is involved.

It is clear, then, that communication between the cells of a filament is vital for the interchange of nutrients and the establishment of heterocyst and akinete patterns. Is there communication between filaments? Although extracellular exchange of nutrients from one filament to another is unlikely, signals certainly are exchanged. Akinete-producing cultures excrete compounds into the medium that stimulate akinete development when added to another culture (Fisher and Wolk, 1976; Hirosawa and Wolk, 1979a,b). Such compounds may become concentrated in a body of water as it dries out and may presage an impending drought, thereby helping to coordinate akinete formation in the local population of cyanobacteria and maximizing the chances of survival. Communication between filaments can also be demonstrated during hormogonia formation, in which cultures have been shown to excrete compounds that either inhibit (Herdman and Rippka, 1988) or stimulate (Robinson and Miller, 1970) hormogonia formation. Such communication may serve to synchronize hormogonia formation in a cyanobacterial population and so achieve the optimum response to improved environmental conditions.

An even clearer example of extracellular communication is seen with the triggering of hormogonia formation by plant-derived products, which is the initial phase of the establishment of many plant–cyanobacteria symbioses. Both the hornwort *Anthoceros* (Meeks, 1990) and the angiosperm *Gunnera* (Rasmussen et al., 1994) produce extracellular signals that trigger hormogonia formation in *Nostoc* spp.

Many cyanobacteria, then, are multicellular organisms that can adapt to their environment by producing specialized cells and that can communicate with each other and with bacteria, plants, and animals in a wide range of symbioses. With the application of molecular genetic techniques, we are beginning to make real progress in understanding their complexities; with increasing knowledge we can now appreciate how much we still have to learn.

Acknowledgments

Thanks go to Mike Herdman (Fig. 5.10) and John Waterbury (Figs. 5.11–5.14) for kindly supplying original photographs for use in these figures.

References

Adams, D.G. (1992) Multicellularity in cyanobacteria. In *Prokaryotic Structure and Function: A New Perspective. Society for General Microbiology Symposium*, Vol. 47, S. Mohan, C. Dow, and J.A. Cole (eds.), pp. 341–384. Cambridge University Press, Cambridge.

Adams, D.G., and Carr, N.G. (1981) The developmental biology of heterocyst and akinete formation in cyanobacteria. *CRC Crit. Rev. Microbiol.* **9**:45–100.

Barry, B.A., Boerner, R.J., and de Paula, J.C. (1994) The use of cyanobacteria in the study of the structure and function of photosystem II. In *The Molecular Biology of Cyanobacteria*, D.A. Bryant (ed.), pp. 217–257. Kluwer, Dordrecht.

Black, T.A., Cai, Y., and Wolk, C.P. (1993) Spatial expression and autoregulation of *hetR*, a gene involved in the control of heterocyst development in *Anabaena*. *Mol. Microbiol.* **9**:77–84.

Blot, M., Meyer, J., and Arber, W. (1991) Bleomycin-resistance gene derived from the transposon Tn5 confers selective advantage to *Escherichia coli* K-12. *Proc. Nat. Acad. Sci. USA* **88**:9112–9116.

Bryant, D.A. (1991) Cyanobacterial phycobilisomes: progress towards complete structural and functional analysis via molecular genetics. In *Cell Culture and Somatic Cell Genetics of Plants*, L. Bogorad and I.K. Vasil (eds.), pp. 257–300. Academic Press, Orlando, FL.

Buikema, W.J., and Haselkorn, R. (1991a) Characterization of a gene controlling heterocyst differentiation in the cyanobacterium *Anabaena* 7120. *Genes Dev.* **5**:321–330.

Buikema, W.J., and Haselkorn, R. (1991b) Isolation and complementation of nitrogen fixation mutants of the cyanobacterium *Anabaena* sp. strain PCC 7120. *J. Bacteriol.* **173**:1879–1885.

Buikema, W.J., and Haselkorn, R. (1993) Molecular genetics of cyanobacterial development. *Annu. Rev. Plant Physiol. Plant Mol. Biol.* **44**:33–52.

Campbell, D., Houmard, J., and Tandeau de Marsac, N. (1993) Electron transport regulates cellular differentiation in the filamentous cyanobacterium *Calothrix*. *Plant Cell* **5**:451–463.

Campbell, E.L., and Meeks, J.C. (1989) Characteristics of hormogonia formation by symbiotic *Nostoc* spp. in response to the presence of *Anthoceros punctatus* or its extracellular products. *Appl. Environ. Microbiol.* **55**:125–131.

Carrasco, C.D., Ramaswamy, K.S., Ramasubramanian, T.S., and Golden, J.W. (1994) *Anabaena xisF* gene encodes a developmentally regulated site-specific recombinase. *Genes Dev.* **8**:74–83.

Carter, H.J. (1856) Notes on the freshwater infusoria of the island of Bombay. 1. Organisation. *Ann. Magazine Natural Hist.* **18**(second series):115–132, 221–249.

Castenholz, R.W. (1989a) Subsection IV. Order Nostocales. In *Bergey's Manual of Systematic Bacteriology*, Vol. 3, J.T. Staley, M.P. Bryant, N. Pfennig, and J.G. Holt (eds.), pp. 1780–1793. Williams & Wilkins, Baltimore.

Castenholz, R.W. (1989b) Subsection V. Order Stigonematales. In *Bergey's Manual of Systematic Bacteriology*, Vol. 3, J.T. Staley, M.P. Bryant, N. Pfennig, and J.G. Holt (eds.), pp. 1794–1799. Williams & Wilkins, Baltimore.

Castenholz, R.W., and Waterbury, J.B. (1989) Preface. In *Bergey's Manual of Systematic Bacteriology*, Vol. 3, J.T. Staley, M.P. Bryant, N. Pfennig, and J.G. Holt (eds.), pp. 1710–1727. Williams & Wilkins, Baltimore.

Chauvat, F., Corre, B., Herdman, M., and Joset-Espardellier, F. (1982) Energetic and metabolic requirements for the germination of akinetes of the cyanobacterium *Nostoc* PCC 7524. *Arch. Microbiol.* **133**:44–49.

Cohn, F. (1877) Untersuchungen uber Bakterien. IV. Beitrage zur Biologie der Bacillen. *Beitr. Biol. Pflanz.* **2**:249–276.

Damerval, T., Guglielmi, G. Houmard, J., and Tandeau de Marsac, N. (1991) Hormogonium differentiation in the cyanobacterium *Calothrix*: a photoregulated developmental process. *Plant Cell* **3**:191–201.

De Bary, A. (1863) Beitrag zur Kenntnis der Nostocaceen insbesondere der *Rivularien*. *Flora (Jena)* **35**:553–560.

Elhai, J., and Wolk, C.P. (1990) Developmental regulation and spatial pattern of expression of the structural genes for nitrogenase in the cyanobacterium *Anabaena*. *EMBO J.* **9**:3379–3388.

Fay, P. (1969) Cell differentiation and pigment composition in *Anabaena cylindrica*. *Arch. Mikrobiol.* **67**:62–70.

Fay, P. (1992) Oxygen relations of nitrogen fixation in cyanobacteria. *Microbiol. Rev.* **56**:340–373.

Fay, P., Lynn, J.A., and Majer, S.C. (1984) Akinete development in the planktonic blue-green alga *Anabaena circinalis*. *Br. Phycol. J.* **19**:163–173.

Fisher, R.W., and Wolk, C.P. (1976) Substance stimulating the differentiation of spores of the blue-green alga *Cylindrospermum licheniforme*. *Nature* **259**:394–395.

Gallon, J.R. (1992). Tansley review no. 44; reconciling the incompatible: N_2 fixation and O_2. *N. Phytologist* **122**:571–609.

Gantt, E. (1994) Supramolecular membrane organization. In *The Molecular Biology of Cyanobacteria*, D.A. Bryant (ed.), pp. 119–138, Kluwer, Dordrecht.

Giovannoni, S.J., Turner, S., Olsen, G.J., Barns, D., Lane, D.J., and Pace, N.R. (1988) Evolutionary relationships among cyanobacteria and green chloroplasts. *J. Bacteriol.* **170**:3584–3592.

Glazer, A.N. (1987) Phycobilisomes: assembly and attachment. In *The Cyanobacteria*, P. Fay and C. Van Baalen (eds.), pp. 69–94. Elsevier, Amsterdam.

Golbeck, J.H. (1994) Photosystem I in cyanobacteria. In *The Molecular Biology of Cyanobacteria*, D.A. Bryant (ed.), pp. 319–360. Kluwer, Dordrecht.

Halfen, L.N. (1979). Gliding movements. In *Encyclopedia of Plant Physiology, New Series*, Vol. 7, W. Haupt and M.E. Feinleib (eds.), pp. 250–267. Springer-Verlag, New York.

Haselkorn, R. (1978) Heterocysts. *Annu. Rev. Plant Physiol.* **29**:319–344.

Haselkorn, R. (1991) Genetic systems in cyanobacteria. *Methods Enzymol.* **204**:418–430.

Haselkorn, R. (1992) Developmentally regulated gene rearrangements in prokaryotes. *Annu. Rev. Genet.* **26**:113–130.

Herdman, M. (1987) Akinetes: structure and function. In *The Cyanobacteria*, P. Fay and C. Van Baalen (eds.), pp. 227–250. Elsevier, Amsterdam.

Herdman, M. (1988) Cellular differentiation: akinetes. *Methods Enzymol.* **167**:222–232.

Herdman, M., and Rippka, R. (1988) Cellular differentiation: hormogonia and baeocytes. *Methods Enzymol.* **167**:232–242.

Herdman, M., Janvier, M., Waterbury, J.B., Rippka, R., Stanier, R., and Mandel, M. (1979) Deoxyribonucleic acid base composition of cyanobacteria. *J. Gen. Microbiol.* **111**:63–71.

Hernandez-Muniz, W., and Stevens, S.E., Jr. (1987) Characterization of the motile hormogonia of *Mastigocladus laminosus*. *J. Bacteriol.* **169**:218–223.

Hirosawa, T., and Wolk, C.P. (1979a) Factors controlling the formation of akinetes adjacent to heterocysts in the cyanobacterium *Cylindrospermum licheniforme* Kutz. *J. Gen. Microbiol.* **114**:423–432.

Hirosawa, T., and Wolk, C.P. (1979b) Isolation and characterization of a substance which stimulates the formation of akinetes in the cyanobacterium *Cylindrospermum licheniforme* Kutz. *J. Gen. Microbiol.* **114**:433–441.

Holland, D., and Wolk, C.P. (1990) Identification and characterization of *hetA*, a gene that acts early in the process of morphological differentiation of heterocysts. *J. Bacteriol.* **172**:3131–3137.

Koch, R. (1877) Untersuchungen uber Bakterien. V. Die Aetiologie der Milzbrand-Krankheit, begrundet auf der Entwicklungsgeschichte des *Bacillus anthracis*. *Beitr. Biol. Pflanz.* **2**:227–310.

Lazaroff, N. (1973) Photomorphogenesis and nostocacean development. In *The Biology of Blue-Green Algae*, N.G. Carr and B.A. Whitton (eds.), pp. 279–319. Blackwell, Oxford.

Leganes, F. (1994) Genetic evidence that *hepA* gene is involved in the normal deposition of the envelope of both heterocysts and akinetes in *Anabaena variabilis* ATCC 29413. *FEMS Microbiol. Lett.* **123**:63–68.

Leganes, F., Fernandez-Pinas, F., and Wolk, C.P. (1994) Two mutations that block heterocyst differentiation have no effect on akinete differentiation in *Nostoc ellipsosporum*. *Mol. Microbiol.* **12**:679–684.

Liang, J., Scappino, L., and Haselkorn, R. (1993) The *patB* gene product, required for growth of the cyanobacterium *Anabaena* sp. strain PCC 7120 under nitrogen-limiting conditions, contains ferredoxin and helix-turn-helix domains. *J. Bacteriol.* **175**:1697–1704.

Livingstone, D., and Jaworski, G.H.M. (1980) The viability of akinetes of blue-green algae recovered from the sediments of Rostherne Mere. *Br. Phycol. J.* **15**:357–364.

Livingstone, D., Kohja, T.M., and Whitton, B.A. (1983) Influence of phosphorus on physiology of a hair-forming blue-green alga (*Calothrix parietina*) from an upland stream. *Phycologia* **22**:345–350.

Meeks, J.C. (1990) Cyanobacterial-bryophyte associations. In *CRC Handbook of Symbiotic Cyanobacteria*, A.N. Rai (ed.), pp. 43–63. CRC Press, Boca Raton, FL.

Mitchison, G.J., and Wilcox, M. (1973) Alteration in heterocyst pattern of *Anabaena* produced by 7-azatryptophan. *Nature* **246**:229–233.

Murry, M.A., Hallenbeck, P.C., and Benemann, J.R. (1984) Immunochemical evidence that nitrogenase is restricted to the heterocysts of *Anabaena cylindrica*. *Arch. Microbiol.* **137**:194–199.

Nichols, J.M., and Adams, D.G. (1982) Akinetes. In *The Biology of Cyanobacteria*, N.G. Carr and B.A. Whitton (eds.), pp. 387–412. Blackwell, Oxford.

Nichols, J.M., Adams, D.G., and Carr, N.G. (1980) Effect of canavanine and other amino acid analogues on akinete formation in the cyanobacterium *Anabaena cylindrica*. *Arch. Microbiol.* **127**:67–75.

Padan, E., and Cohen, Y. (1982) Anoxygenic photosynthesis. In *The Biology of Cyanobacteria*, N.G. Carr and B.A. Whitton (eds.), pp. 215–235. Blackwell, Oxford.

Rabenhorst, L. (1865) *Flora Europaea Algarum*, Vol. 2. Leipzig.

Rasmussen, U., Johansson, C., and Bergman, B. (1994) Early communication in the *Gunnera-Nostoc* symbiosis: plant-induced cell differentiation and protein synthesis in the cyanobacterium. *Mol. Plant Microbe Interact.* **7**:696–702.

Rippka, R., and Stanier, R.Y. (1978) The effects of anaerobiosis on nitrogenase synthesis and heterocyst development by Nostocacean cyanobacteria. *J. Gen. Microbiol.* **105**:83–94.

Rippka, R., Deruelles, J., Waterbury, J.B., Herdman, M., and Stanier, R.Y. (1979) Generic assignments, strain histories and properties of pure cultures of cyanobacteria. *J. Gen. Microbiol.* **111**:1–61.

Rippka, R., Waterbury, J.B., and Stanier, R.Y. (1981) Provisional generic assignments for cyanobacteria in pure culture. In *The Prokaryotes*, Vol. I, M.P. Starr, H. Stolp, H.G. Truper, A. Balows, and H.G. Schlegel (eds.), pp. 247–256, Springer-Verlag, Berlin.

Robinson, B.L., and Miller, J.H. (1970) Photomorphogenesis in the blue-green alga *Nostoc commune* 584. *Physiol. Plantarum* **23**:461–472.

Schmidt, A. (1988) Sulfur metabolism in cyanobacteria. **167**:572–583.

Schopf, J.W. (1994) Disparate rates, differing fates: tempo and mode of evolution changed from the Precambrian to the Phanerozoic. *Proc. Nat. Acad. Sci. USA* **91**:6735–6742.

Sidler, W.A. (1994) Phycobilisome and phycobiliprotein structures. In *The Molecular Biology of Cyanobacteria*, D.A. Bryant (ed.), pp. 139–216. Kluwer, Dordrecht.

Simon, R.D. (1977) Macromolecular composition of spores from the filamentous cyanobacterium *Anabaena cylindrica*. *J. Bacteriol.* **129**:1154–1155.

Simon, R.D. (1987) Inclusion bodies in the cyanobacteria: cyanophycin, polyphosphate, polyhedral bodies. In *The Cyanobacteria*, P. Fay, and C. Van Baalen (eds.), pp. 199–225. Elsevier, Amsterdam.

Sinclair, C., and Whitton, B.A. (1977a) Influence of nitrogen source on morphology of Rivulariaceae (Cyanophyta). *J. Phycol.* **13**:335–340.

Sinclair, C., and Whitton, B.A. (1977b) Influence of nutrient deficiency on hair formation in the Rivulariaceae. *Br. Phycol. J.* **12**:297–313.

Stanier, G. (1988) Fine structure of cyanobacteria. *Methods Enzymol.* **167**:157–172.

Stanier, R.Y., and Cohen-Bazire, G. (1977) Phototrophic prokaryotes: the cyanobacteria. *Annu. Rev. Microbiol.* **31**:225–274.

Sutherland, J.M., Herdman, M., and Stewart, W.D.P. (1979) Akinetes of the cyanobacterium *Nostoc* PCC 7524: macromolecular composition, structure and control of differentiation. *J. Gen. Microbiol.* **115**:273–287.

Sutherland, J.M., Reaston, J., Stewart, W.D.P., and Herdman, M. (1985a) Akinetes of the cyanobacterium *Nostoc* PCC 7524: macromolecular and biochemical changes during synchronous germination. *J. Gen. Microbiol.* **131**:2855–2863.

Sutherland, J.M., Stewart, W.D.P., and Herdman, M. (1985b) Akinetes of the cyanobacterium *Nostoc* PCC 7524: morphological changes during synchronous germination. *Arch. Microbiol.* **142**:269–274.

Tandeau de Marsac, N. (1994) Differentiation of hormogonia and relationships with other biological processes. In *The Molecular Biology of Cyanobacteria*, pp. 825–842. Kluwer, Dordrecht.

Tandeau de Marsac, N., and Houmard, J. (1993) Adaptation of cyanobacteria to environmental stimuli: new steps towards molecular mechanisms. *FEMS Microbiol. Rev.* **104**:119–190.

Tandeau de Marsac, N., Mazel, D., Damerval, T., Guglielmi, G., Capuano, V., and Houmard, J. (1988) Photoregulation of gene expression in the filamentous cyanobacterium *Calothrix* sp. PCC 7601: light-harvesting complexes and cell differentiation. *Photosyn. Res.* **18**:99–132.

Walsby, A.E. (1985) The permeability of heterocysts to the gases nitrogen and oxygen. *Proc. R. Soc. Lond. B Biol. Sci.*, **226**:345–366.

Waterbury, J.B. (1989) Subsection II. Order Pleurocapsales Geitler 1925, emend. Waterbury and Stanier 1978. In *Bergey's Manual of Systematic Bacteriology*, Vol. 3, J.T. Staley, M.P. Bryant, N. Pfennig, and J.G. Holt (eds.), pp. 1746–1770. Williams & Wilkins, Baltimore.

Waterbury, J.B., and Stanier, R.Y. (1978) Patterns of growth and development in pleurocapsalean cyanobacteria. *Microbiol. Rev.* **42**:2–44.

Waterbury, J.B., Willey, J.M., Franks, D.G., Valois, F.W., and Watson, S.W. (1985) A cyanobacterium capable of swimming motility. *Science* **230**:74–76.

Weckesser, J., and Jurgens, U.J. (1988) Cell walls and external layers. *Methods Enzymol.* **167**:173–188.

Whitton, B.A. (1987) The biology of Rivulariaceae. In *The Cyanobacteria*, P. Fay and C. Van Baalen (eds.), pp. 513–534. Elsevier, Amsterdam.

Whitton, B.A. (1989) Genus I. *Calothrix Agardh* 1824. In *Bergey's Manual of Systematic Bacteriology*, Vol. 3, J.T. Staley, M.P. Bryant, N. Pfennig, and J.G. Holt (eds.), pp. 1791–1793. Williams & Wilkins, Baltimore.

Wilcox, M., Mitchison, G.J., and Smith, R.J. (1975a) Mutants of *Anabaena cylindrica* altered in heterocyst spacing. *Arch. Microbiol.* **103**:219–223.

Wilcox, M., Mitchison, G.J., and Smith, R.J. (1975b) Spatial control of differentiation in the blue-green alga *Anabaena*. In *Microbiology 1975*, D. Schlessinger (ed.), pp. 453–463. American Society for Microbiology, Washington, DC.

Wolk, C.P. (1982) Heterocysts. In *The Biology of Cyanobacteria*, N.G. Carr and B.A. Whitton (eds.), pp. 359–386. Blackwell, Oxford.

Wolk, C.P. (1989) Alternative models for the development of the pattern of spaced heterocysts in *Anabaena (Cyanophyta)*. *Plant Systemat. Evol.* **164**:27–31.

Wolk, C.P. (1991) Genetic analysis of cyanobacterial development. *Curr. Opin. Genet. Dev.* **1**:336–341.

Wolk, C.P., Cai, Y., Cardemil, L., et al. (1988) Isolation and complementation of mutants of *Anabaena* sp. strain PCC 7120 unable to grow aerobically on dinitrogen. *J. Bacteriol.* **170**:1239–1244.

Wolk, C.P., Elhai, J., Kuritz, T., and Holland, D. (1993) Amplified expression of a transcriptional pattern formed during development of *Anabaena*. *Mol. Microbiol.* **7**:441–445.

Wolk, C.P., Ernst, A., and Elhai, J. (1994) Heterocyst metabolism and development. In *The Molecular Biology of Cyanobacteria*, pp. 769–823. Kluwer, Dordrecht.

Yamamoto, Y. (1975) Effect of desiccation on the germination of akinetes of *Anabaena cylindrica*. *Plant Cell Physiol. (Tokyo)* **16**:749–752.

6

Mycelial Life Style of *Streptomyces coelicolor* A3(2) and Its Relatives

KEITH F. CHATER & RICHARD LOSICK

For most of the known cultivable bacteria any increase in biomass is proportional to an increase in the number of organisms because rods, spheres, or spirals divide by binary fission. The cell cycle and life cycle are often the same thing. *Escherichia coli* provides a fine model for this phenomenon, notwithstanding the ability of cells within *E. coli* populations to show cooperation or differentiation. Remarkably, some mutants of *E. coli* can form branched, septate filaments (Åkerlund et al., 1993), which is an unambiguous multicellularity. So far as we know, *E. coli* has not found any adaptive benefit in this potential during its evolution. However, many members of the Actinomycetes (the section of the gram-positive lineage with DNA of high G + C content) have done so, and this chapter is concerned with their most well known genus, *Streptomyces*.

Multicellular Differentiated Organization of *Streptomyces* Colonies

Streptomyces spp. are numerous inhabitants of most soils. Their success in this habitat arises from their especially effective aptitude for digesting and utilizing abundant but relatively stable and insoluble organic polymers such as cellulose and chitin. To achieve this process the organism grows as a mycelium of branching hyphae (Fig. 6.1). The hyphal tips penetrate solid matrices with the help of secreted or cell surface-bound hydrolytic enzymes, producing a pool of nutrients that are taken up and converted to mycelial biomass.

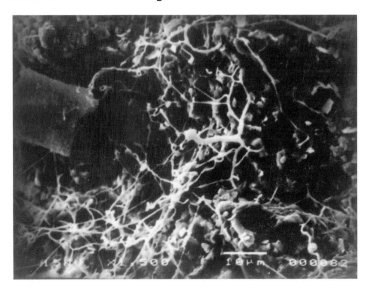

FIGURE 6.1   Web-like growth of *Streptomyces lividans* in soil. Note the comparatively open mycelium, infrequent branching, and occurrence of hyphae which seem to stretch through the air across crevices in the substrate. Some coiled aerial hyphae are also visible. Micrograph is of a 3-day culture. Bar = 10 μm. (Courtesy of E.M.H. Wellington)

In the laboratory, when growing on a suitable agar medium, *Streptomyces* colonies grow fairly rapidly and form a thick mass of mycelium, probably much more compact and luxuriant than is typical of growth in soil (Fig. 6.2). Within a couple of days, obvious changes take place. Various secondary metabolites such as antibiotics and pigments are produced, and the mycelial surface facing the air differentiates and forms aerial branches (except at the edge of the colony), giving a furry, white, mold-like appearance. The edges of the colony continue to grow, but much of the substrate mycelium in the differentiating central region undergoes lysis (Hopwood, 1960; Wildermuth, 1970; Allan and Prosser, 1985); radiolabeling experiments have shown that at least some of the substrate mycelial biomass is cannibalized by the aerial mycelium (Méndez et al., 1985). Aerial hyphae may be considerably more than 100 μm in length before their tips stop growing and turn into long strings of hydrophobic unigenomic spores (Fig. 6.3), which on maturing are easily separated from each other. The spores allow dispersal of what is otherwise an intrinsically immobile organism.

The solubilization of vegetative hyphae to provide food for the aerial mycelium could be rather indiscriminate: other soil bacteria might potentially benefit from the local *Streptomyces* soup kitchen. However, *Streptomyces* colonies appear able to defend the resources they have made available, by poisoning them with antibiotics to which the producing organism is itself resistant (or can become resistant) (Chater and Merrick, 1979; Chater, 1992). Typically, each *Streptomyces* soil isolate can produce an idiosyncratic range of such compounds

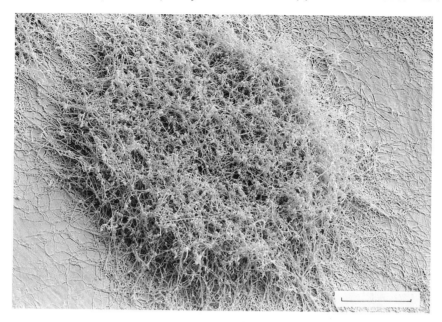

FIGURE 6.2   Luxuriant colony of *Streptomyces coelicolor* growing on a rich agar medium. Most of the colony is densely covered with aerial hyphae, but at the colony edge the foraging substrate mycelium protrudes for a considerable distance. The micrograph is of a 4-day culture of a *whiI* mutant of *S. coelicolor*. Bar = 100 μm. (Courtesy of K. Findlay and N.J. Ryding)

active against various bacteria and other competitors in the soil. This diversity of antibiotic production profiles may reflect both the variety of unrelated coinhabitants in different soil microniches and the need to compete with other streptomycetes or members of other genera of actinomycetes. Possibly, nearby motile bacteria attracted by nutrients released from colonies are first inactivated by the antibiotics and then consumed by the action of lytic enzymes. [An early example dating from 1947 was reviewed by Welsch (1962). Shi and Zusman (1993) documented a comparable situation, which they called ''fatal attraction,'' in *Myxococcus xanthus*.]

In this chapter we summarize results and speculation concerning: (1) the fundamental attributes and mechanisms of vegetative growth as a branching mycelium; (2) the perception of the appropriateness of developmental changes; (3) the mechanism and nutritional underpinning of aerial mycelial development; and (4) the special genetic devices and opportunities arising from multicellularity. Two model species have provided much of this information: *Streptomyces coelicolor* A3(2), which has been extensively studied and has experimentally convenient genetics (Chater and Hopwood, 1993); and the streptomycin producer *Streptomyces griseus*, whose development involves an extracellular signaling molecule, A-factor.

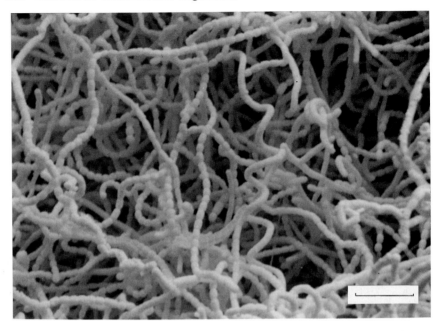

FIGURE 6.3   Long spore chains on the surface of a colony of *Streptomyces coelicolor*. This scanning electron micrograph is of a 4-day colony. Bar = 10 μm. (Courtesy of K. Findlay and N.J. Ryding)

Mycelial Growth

Spores and Their Germination

Mycelial growth is difficult to study because of the rapidity with which the young mycelium becomes physiologically heterogeneous. Most of the available information comes from studies of either growth tips at the advancing colony perimeter or the young hyphae that develop during the first few hours after spore germination. Spores are the only unigenomic, unicellular stage in the life cycle of *Streptomyces* spp. We here briefly summarize our knowledge of spores and their germination (for details and references: Hodgson, 1992).

Streptomyces spores form in a different way from the well known endo-spores of *Bacillus* spp. and other low G + C gram-positive bacteria, and they do not show such extreme resistance to physical and chemical insults as en-dospores. They are more or less spherical and have multilayered, thick cell walls encased in a markedly hydrophobic surface layer, which is often patterned or ornamented with hairs or spines. They are often colored with pigments that differ among species. They are rather inactive metabolically, although much less dormant than endospores. The spore cytoplasm is dehydrated and contains abundant trehalose, which may assist in the considerable desiccation-resistance and slight heat-resistance of spores compared with vegetative hyphae. They also contain trehalase enzyme in a form or subcellular compartment that becomes active (by an unknown mechanism) only during germination.

Germination requirements vary. Some but not all species germinate more rapidly when stressed by mild heat or sonication treatment. Divalent cations (often Ca^{2+}) and CO_2 are usually required. Germination is seen as a rapid loss of phase brightness followed by swelling, which coincides with rapid trehalose degradation, so it may be caused by a resultant increase in osmotic pressure. RNA and protein synthesis are initiated within a few minutes of phase-darkening, with DNA replication beginning about 60–120 minutes later. Germ tube emergence approximately coincides with completion of the first round of DNA replication. The thin cell wall of the newly forming hypha appears to be continuous with the innermost wall layer of the germinated spore. Interestingly, this layer is laid down inside the old hyphal wall during sporulation (Wildermuth and Hopwood, 1970; Manzanal and Hardisson, 1978), so each completed *Streptomyces* life cycle from spore to spore involves the sloughing of a cell wall layer, whereas the mycelial wall retains continuity for the duration of vegetative growth. Early observations suggested that germination or outgrowth (or both) are associated with high levels of cAMP (Gersch and Strunk, 1980).

Tip Growth and the Necessities of Branching

Studies of the first few hours of growth after spore germination (Kretschmer, 1988, 1991; Miguélez et al., 1988) have shown that the germ tube, as it emerges, contains two genomes resulting from the early replication of DNA. Soon these genomes begin to replicate synchronously, and the rate of hyphal extension doubles. After about 45 minutes, the four partially replicated genomes initiate another round of replication, and the extension rate doubles again (Fig. 6.4). Cell wall growth at this and later stages of growth is highly concentrated at hyphal tips (though a lower level of incorporation of cell wall precursors can also be detected throughout the mycelium) (Braña et al., 1982; Gray et al., 1990; Miguélez et al., 1992, 1993a,b).

By the time the germ tube contains 32 genomes, the tip is extending by the length of an *E. coli* cell every 2–3 minutes (i.e., approaching 40 µm per hour) on rich medium, and the physical capacity of the tip to support such an intensity of cell wall construction must become limiting. The problem is solved by the appearance of a new tip, usually in the form of a second germ tube (Kretschmer, 1991). When the problem recurs after each additional cycle of replication, it is relieved by additional new tips emerging as branches from the older part of the growing hypha. Thus the rate of tip extension does not increase; but because the number of branches increases exponentially for a time, a more or less exponential rate of biomass accumulation is achieved during the earliest stages of colony formation.

What signal informs an apical hyphal compartment that its tip is growing at the maximum speed, how does it lead to branching, and why are branches far from tips? Kretschmer (1991) suggested that the critical signal may be an accumulation of cell wall precursors beyond levels that can be consumed by

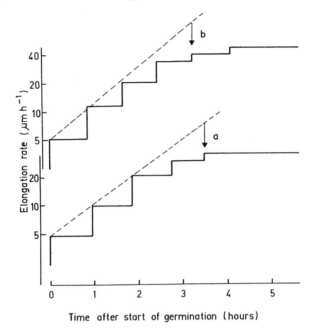

Time after start of germination (hours)

FIGURE 6.4   Stepwise increases in the elongation rates of two germ tubes of *Streptomyces granaticolor*. Arrows indicate the time at which the first branch was visible. The departure from the dashed lines illustrates a gradual deviation from quasiexponential kinetics of elongation after the third growth rate increment. (Reprinted from Kretschmer, 1991, with permission)

the tip. These excess precursors should be most abundant at the farthest distance from a growing tip; and this high local abundance of precursors may stimulate branching.

The Occasional Septa Present in Vegetative Hyphae Are not Essential for Mycelium Formation

The substrate mycelium contains occasional septa, which are more frequent in older parts of the mycelium (Kretschmer, 1982). Often branch points are close to septa. Typically, neither septa nor branch points are formed within 20–30 μm of a growth tip during rapid growth, though this interval decreases at low growth rates (Kretschmer, 1989). It is not known what controls the position of the single septum formed in a hyphal tip compartment after each round of DNA replication. The septation of a hyphal tip results in two kinds of cell: an actively extending apical cell and a slightly shorter nonextending subapical cell (Kretschmer, 1982). In the apical cell replicated chromosomes separate as the cell elongates, whereas in the subapical cell it seems that chromosomes replicate without separating. A "normal" genome/cytoplasm ratio is reestablished in subapical cells by the emergence of a new branch, which becomes populated with the progeny of one of the replicated chromosome pairs (Kretschmer,

1987). Because of repetition of this situation in the newly branched subapical compartment, branches and cross-walls are more frequent in older mycelium (Fig. 6.5).

Septation in bacteria involves several dedicated gene products, notably that of *ftsZ*. FtsZ, a highly abundant protein, somewhat resembles eukaryotic tubulin (Lutkenhaus, 1993). FtsZ is essential for *E. coli* or *Bacillus subtilis* cells to generate colonies (Beall and Lutkenhaus, 1991; Dai and Lutkenhaus, 1991) and is present in streptomycetes (Dharmatilake and Kendrick, 1994; McCormick et al., 1994). Deletion of *ftsZ* in *S. coelicolor* leads to the elimination of cross-walls but remarkably does not prevent formation of a more or less normal colony consisting of a branched mycelium (McCormick et al., 1994). A deletion mutant of another cell division gene, *ftsQ*, has been shown to be viable and to produce a branched mycelium (J. McCormick and R. Losick, unpublished results). Thus the complete development of septa, with the resultant production of two cell types as described above, is not integral and indispensable for branching, though it may well play an accessory role.

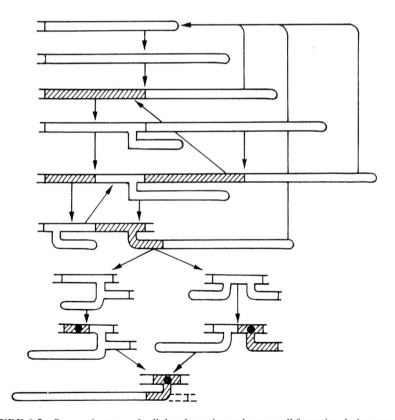

FIGURE 6.5 Proposed pattern of cellular elongation and cross-wall formation during mycelial growth of *Streptomyces granaticolor*. Black dots indicate cells that are no longer able to elongate, and hatched cells temporarily lack elongation. (Modified from Kretschmer, 1982, with permission)

Production of Extracellular Enzymes and Their Inhibitors

Regulation in Relation to Colony Development and
Glucose Repression

Secreted extracellular degradative enzymes provide nutrients to cell populations, whereas intracellular metabolic enzymes serve only a single cell. Consistent with their multicellularity, streptomycetes produce many such enzymes (Molnár, 1994). Diffusible extracellular enzymes are generally most efficiently exploited by their producer when its cell density is high; so they should be most abundantly produced at later stages in colony development rather than at the first ''foraging'' stage. Indeed, production by laboratory cultures is generally maximal in relatively old cultures (e.g., Trigo and Ball, 1994).

Even in the presence of their substrate-related inducers, many extracellular enzymes are strongly repressed when the organism is supplied with a soluble, efficient carbon and energy source such as glucose [e.g., chitinase (Miyashita et al., 1991; Robbins et al., 1992), agarase (Hodgson, 1980; Servín-González et al., 1994), and α-amylases (Virolle et al., 1988)]. Such repression may delay production of exoenzymes by young cultures growing under conditions of glucose sufficiency until readily available soluble carbon sources have become limiting. Carbon limitation is most likely to be encountered when the biomass is relatively high, and in the most densely populated parts of colonies. Thus catabolite repression may contribute to the appropriate temporal and spatial control of extracellular enzyme production.

Unlike the situation in *E. coli*, glucose repression in *S. coelicolor* does not appear to involve glucose transport, cyclic adenosine monophosphate (cAMP), or a cAMP receptor protein, even though an adenylate cyclase gene has been cloned (Danchin et al., 1993) and cAMP is present at perceptible levels (Botsford and Harman, 1992). Instead, the existing evidence (Hodgson, 1982; Angell et al., 1994) links phosphorylation of glucose by an intracellular, presumptively cytoplasmic, glucose kinase to the repression of many genes.

There are several examples of exported proteins that inhibit the activity of extracellular enzymes, including inhibitors of proteases, α-amylases, and β-lactamases (Molnár, 1994). Presumably these proteins serve either to inhibit enzymes produced by other soil coinhabitants or to control the activity of self-produced enzymes. The second possibility could be relevant for enzymes, such as proteases (Ginther, 1979; Gibb and Strohl, 1988), that play a part in the lysis of substrate mycelium to provide some of the nutrients for aerial growth. In a strain of *Streptomyces exfoliatus* cultured in liquid medium (a condition in which sporulation does not take place), an extracellular trypsin-like protease accumulates as glucose is exhausted at the end of the main growth period coinciding with extensive mycelial lysis (Kim and Lee, 1995). The activity of the protease and lysis of the mycelium can both be inhibited by the trypsin inhibitor leupeptin, suggesting that the protease plays a part in lysis. The strain itself produces leupeptin early in growth, providing a potential limitation on premature activity of the trypsin-like protease. The protease is eventually released from inhibition because leupeptin is inactivated by a leupeptin-inactivating enzyme that accumulates during the stationary phase. It will be most interesting to learn whether comparable cascades are widespread among streptomycetes and to expand these physiological investigations by parallel genetic approaches.

Possible Posttranscriptional Influences on Production of
Extracellular Enzymes

Among the many characterized *Streptomyces* genes encoding hydrolytic exo-enzymes, a few contain the rare codon TTA (e.g., an *S. rochei* gene encoding an endoglucanase) (Perito et al., 1994); and one, encoding a lipase, appears to be positively regulated by the product of a TTA-containing gene (Pérez et al., 1994). Because this codon appears to be associated with genes involved in postexponential phase events (see below), its presence in genes associated with exoenzyme production may provide an additional limitation on their expression in young cells. Two other posttranscriptional processes provide potential control points for ensuring the appropriate production and activity of exoenzymes. The first is the secretion machinery itself, which is as yet little studied. The second, unusual among prokaryotes, is glycosylation: some cellulases of streptomycetes and other actinomycetes are glycosylated, presumably affecting their stability and their interactions with their substrate; and *S. lividans*, carrying the relevant genes, can glycosylate the cellulases of some other *Streptomyces* spp. (Wilson, 1992; Ong et al., 1994; and references therein).

Some secreted cellulases and chitinases have specific binding domains, separate from their catalytic domain, that anchor them to their immobile substrate (e.g., Perito et al., 1994). One consequence of this situation is that the enzyme remains close to the producing organism, making the capture of released nutrients as efficient as possible.

Role of A-Factor and Other Extracellular Regulatory Molecules in *Streptomyces* Colony Development

More than 25 years ago, in a ground-breaking study, certain mutants of *Streptomyces griseus* pleiotropically defective in streptomycin production and sporulation were found to be restored to the parental phenotype by growth near a wild-type colony. Using correction of the mutant phenotype as an assay, the agent, termed A-factor, was purified from huge volumes of culture fluid (it is typically present in nanomolar concentrations); it was identified as 2-(6'-methyl-heptanoyl)-3R-hydroxymethyl-4-butanolide (Fig. 6.6) and then chemically synthesized (Khokhlov et al., 1967; Hara and Beppu, 1982; Khokhlov, 1985).

Streptomyces griseus contains a cytoplasmic 37-kilodalton (kDa) protein with a high affinity ($K_D \sim 10^{-9}$ M) for A-factor (Miyake et al., 1989). In the absence of A-factor, this protein, which is present in only 30–40 molecules per chromosome, appears to act as a repressor of streptomycin production and

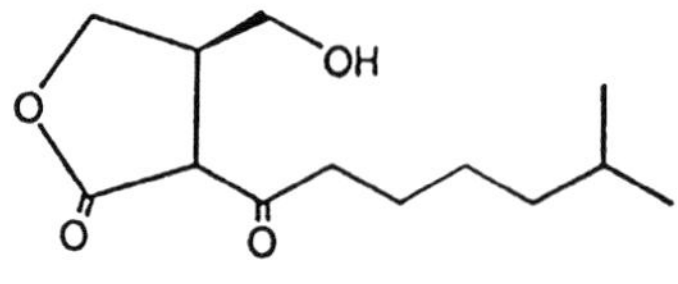

FIGURE 6.6   Structure of A-factor.

sporulation, as elimination of the binding protein caused an A-factor-defective mutant to "revert" to quasinormal development (Miyake et al., 1990). Mutants that lack binding protein sporulate significantly earlier than the wild type but nevertheless undergo significant vegetative growth before sporulating, indicating that A-factor-binding protein is not the sole means by which sporulation is delayed until late in growth. The regulatory pathways leading from A-factor recognition to streptomycin production and sporulation diverge at some point: introducing extra copies of the streptomycin biosynthetic pathway-specific activator gene *strR* into an A-factor-defective mutant leads to streptomycin production and resistance without activating sporulation (Horinouchi and Beppu, 1992), whereas other DNA fragments activate sporulation but not antibiotic production (see below). In a minimal model, A-factor-binding protein represses the promoter of a gene whose product, protein X, activates divergent regulatory cascades, leading separately to sporulation and, via *strR*, to streptomycin production. When A-factor is bound, the binding protein loses its repressor activity (Vujaklija et al., 1993).

A-factor reaches its threshold extracellular concentration of 10^{-9} M rather suddenly, late during culture (Hara and Beppu, 1982). This is reminiscent of the relatively well understood pattern of accumulation of the homoserine lactone (OHHL) involved in activating luminescence in *Vibrio fischeri* (Fuqua et al., 1994). A single gene, *luxI*, is sufficient to stimulate OHHL biosynthesis from general metabolism. This gene is regulated by LuxR, an OHHL-binding protein located in the cytoplasm, where it can monitor extracellular OHHL levels because OHHL, it is thought, diffuses freely through the cell membrane. The LuxR–OHHL complex activates transcription of *luxI* and genes directing light emission. Perhaps an analog of this self-amplifying regulatory loop causes the rapid increase of A-factor in *S. griseus*. Dependence on extracellular autoregulators has also been described for extracellular enzyme and antibiotic production by *Erwinia carotovora* and Ti plasmid transfer by *Agrobacterium tumefaciens* (Fuqua et al., 1994; Williams, 1994). In all these cases the extracellular molecule resembles A-factor in that it consists of a lactone ring with an acyl side chain that probably permits it to pass through membranes; hence cytoplasmic factor-binding proteins can communicate the extracellular information to the gene expression apparatus. In contrast, sporulation, competence, and antibiotic synthesis in *Bacillus subtilis* depend on the accumulation of extracellular oligopeptides, which do not diffuse through membranes. The signal molecules are sensed by means of membrane-bound receptors that appear to communicate indirectly with the gene expression apparatus via protein kinase cascades. One of the *B. subtilis* receptors (referred to below) is an ATP-dependent transport system specified by the *spoOK* genes, and another is a transmembrane protein with a cytoplasmic histidine protein kinase domain typical of two-component regulatory systems (Magnuson et al., 1994; Solomon et al., 1995).

A cluster of three genes (*amfR, A, B*) has been implicated in the sporulation branch of the A-factor-responsive pathway of *S. griseus* because its cloning at high copy number restores sporulation to an A-factor-deficient mutant without restoring streptomycin production (Ueda et al., 1993). The deduced *amf* gene products all resemble known proteins involved directly or indirectly in transport: AmfA and AmfB resemble the Hly family of ATP-dependent secretory proteins, and AmfR resembles a family of response regulators that are subject to phosphorylation by sensor kinases in two-component systems (the closest resemblance of AmfR is to UhpA, which regulates uptake of sugar phosphates in *E. coli*). This finding led Ueda et al. (1993) to propose that AmfR might be involved in a phosphorylation cascade analogous to the SpoOA phospho-relay (Hoch, 1993) that (partially in response to an extracellular signal) initiates sporulation in *B. subtilis*; and that AmfA and AmfB might be more directly involved in detecting such signals in much the same way as described above for the SpoOK proteins in *B. subtilis*. Thus the *amf* genes may be involved in another (i.e., A-factor-independent) mode of sporulation activation by an unknown extracellular factor. Such a process may be general: a related set of genes (*ram*) from *S. coelicolor*, cloned on a low copy-number plasmid, induced accelerated aerial mycelium formation in the closely related *S. lividans*, and disruption of the homologous *ram* genes of *S. lividans* caused a striking aerial mycelium deficiency (Ma and Kendall, 1994).

Substances structurally similar to A-factor are produced by many *Streptomyces* spp., though their function is often unknown. For example, *S. coelicolor* A3(2) produces several such molecules, at least one of which can stimulate streptomycin production in an A-factor-deficient *S. griseus* mutant (Anisova et al., 1984; Efremenkova et al., 1985). *S. coelicolor afs* mutants fail to bring about this stimulation but are not themselves deficient in normal morphological differentiation (Hara et al., 1983). One class, *afsA*, produces antibiotics normally, but the other, *afsB*, is defective in antibiotic production. Antibiotic production was not restored to *afsB* mutants by adding authentic A-factor. These results suggest that the *S. coelicolor*-produced substance that can substitute for A-factor in *S. griseus* is not needed for morphological or physiological differentiation in *S. coelicolor*, and that the product of the *afsB* gene is required for synthesis of this factor and antibiotic production.

Streptomyces virginiae produces ''virginiae butanolides,'' A-factor-like substances that are required at nanomolar concentration for virginiamycin biosynthesis (Yamada et al., 1987). These compounds and A-factor do not substitute for each other *in vivo*, and virginiae butanolides are not bound significantly to the A-factor binding protein (Miyake et al., 1989). There is evidently considerable species specificity for these substances. Similar observations have been reported for the various homoserine lactones that induce luciferase in marine *Vibrio* spp. (Williams, 1994).

Because nutrients become exhausted more rapidly at high cell density, extracellular cell density-related signals such as A-factor that stimulate more or less synchronous sporulation of a colony can be seen as a means of anticipating nutrient exhaustion and ensuring a high degree of biomass conversion into spore progeny. Such a ''safe'' strategy may also confer a numerical disadvantage compared to an alternative but higher risk strategy, in which the decision to sporulate is taken by individual hyphae in a more direct response to nutrient exhaustion. Colonies of the latter kind would probably extract more of the last available nutrients from their environment before sporulating but spend longer in the comparatively vulnerable mycelial state. The pros and cons of these alternative strategies may well be important factors in driving speciation in *Streptomyces*.

Genetic Analysis of Aerial Mycelium Formation in *Streptomyces coelicolor*

Mutations: Prevention of Aerial Mycelium Formation and Simultaneous Interference with Secondary Metabolism

Many *Streptomyces* mutants lacking both aerial mycelium and secondary metabolism have been described. Indeed, among the *bld* (i.e., ''bald'') or Amy⁻ (aerial mycelium defective) mutants of various species, only a few retain normal secondary metabolism (McCue et al., 1992; Champness and Chater, 1994). These observations reinforce the idea of complex coregulation of morphological differentiation and secondary metabolism (at least seven *bld* genes—*bldA*, *B*, *C*, *D*, *G*, *H*, and *I*—are required for normal antibiotic production in *S. coelicolor*).

Few *bld* genes have been characterized at the molecular level, and perhaps the only major lesson yet to emerge from sequencing *bld* genes concerns *bldA*, which specifies the only tRNA that can translate UUA (leucine) codons efficiently (Lawlor et al., 1987; Leskiw et al., 1991b). Corresponding TTA codons are rare in *Streptomyces* genes (Wright and Bibb, 1992) and have not been found in any genes needed for growth; *bldA* mutants have vigorous vegetative growth. Therefore *bldA* may regulate genes for postgrowth functions at the translation level by permitting UUA codons to be translated more efficiently in nongrowing parts of the mycelium (Leskiw et al., 1991a). Evidence bearing on this hypothesis remains inconclusive, as experiments using two different experimental systems gave somewhat different results. Inserting or removing TTA codons in an antibiotic pathway regulatory gene, *act*II-ORF4, had no effect on the time course of synthesis of the cognate antibiotic actinorhodin (Gramajo et al., 1993), whereas translation of mRNA of a reporter gene (*ampC*) containing seven TTA codons was more efficient in older cultures (Leskiw et al., 1993).

Probably experimental conditions significantly influence the regulatory importance of *bldA*. Thus under the conditions of the study by Gramajo et al. (1993), the *bldA*-specified tRNA was present in similar amounts at all time points, whereas with the conditions used by Leskiw et al. (1993) the mature tRNA was much less abundant in young cultures, perhaps allowing its potential regulatory role to be manifested. In both analyses, the 5′ end of the primary, unprocessed *bldA* transcript was much more readily detected in the youngest cultures, in apparent contradiction to expectations and tending to rule out developmental regulation of *bldA* itself at the transcriptional level. Instead, there seems to be an increase with time in the rate of processing the primary to the mature transcript (Leskiw et al., 1993).

If *bldA* fulfills a regulatory role, TTA codons should be present in appropriate target genes, which might be expected to include other *bld* genes and perhaps genes involved in antibiotic production. An *S. griseus* gene termed ORF 1590, which complements or suppresses the sporulation defect of some *bld* mutants (Babcock and Kendrick, 1988, 1990) including A-factor-defective mutants (Ueda et al., 1993), contains a TTA codon, as does its homolog in *S. coelicolor* (Babcock and Kendrick, 1990; McCue et al., 1992). A TTA codon is also present in the *amfR* gene of *S. griseus* (Ueda et al., 1993; see above). Most of the other occurrences of the codon are in pathway-specific regulatory genes for antibiotic biosynthesis, several associated resistance genes, a few genes encoding antibiotic biosynthetic enzymes, and some genes for extracellular enzymes (Leskiw et al., 1991b; Distler et al., 1992; Champness and Chater, 1994).

Apparently normal morphological development can be restored to many, but not all, *bld* mutants by changing the medium: for example, *bldA, C, D, G,* and *H* mutants of *S. coelicolor* produce a normal aerial mycelium on minimal medium when mannitol replaces glucose as the carbon source. This finding suggests that there are alternative routes to aerial mycelium formation: one dependent on *bldA* but independent of carbon source, and the other independent of *bldA* but repressed by glucose.

Intercellular Signaling During Aerial Mycelium Formation in
Streptomyces coelicolor

Although A-factor apparently has no role in the morphological development of *S. coelicolor*, extracellular complementation experiments suggest that other exported molecules may do so. When certain *bld* mutants are grown close to aerial mycelium-producing colonies, aerial mycelium is observed on the nearest edge of the *bld* mutant colony (Willey et al., 1991). Evidently, aerial mycelium formation can be restored by extracellular diffusible molecules. One such molecule is SapB, a proteinaceous substance of low molecular weight, whose appearance coincides temporally and spatially with aerial mycelium formation, and which is absent in *bld* mutants (Willey et al., 1991). Transient aerial mycelium formation is restored by the application of purified SapB to colonies of *bld* mutants. SapB seems to be produced nonribosomally (Willey et al., 1993),

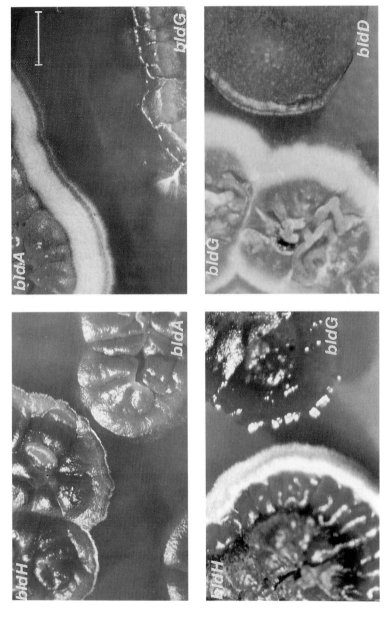

FIGURE 6.7 Extracellular complementation between *bld* mutants. The indicated *bld* mutants were streaked onto R2YE agar plates and grown for 8 days at 30°C. Bar = 500 μm. (Reprinted from Willey et al. 1993, with permission)

but the genes involved are unknown. It is abundant around differentiating colonies and on the surface of aerial hyphae and spores, so it may facilitate the erection of aerial hyphae, perhaps by breaking the surface tension at the colony surface. These observations imply that cooperation between *S. coelicolor* cells leads to a high local concentration of the protein to bring about morphogenesis. The role and behavior of SapB seem to have features in common with the hydrophobins that coat the aerial parts of fungal hyphae, although the hydrophobins are not homologous with SapB and are ribosomally synthesized peptides (Chater, 1991; Wessels, 1993).

Interestingly, pairs of *bld* mutants growing in mixed culture regain the capacity to produce SapB and erect aerial hyphae, indicating that additional extracellularly acting compounds are involved in morphological differentiation (Fig. 6.7) (Willey et al., 1993). A striking feature of extracellular complementation between *bld* mutants is its directionality; only certain pairs of *bld* mutants complement each other and usually with one mutant serving as a donor and the other as respondent. This finding led Willey et al. (1993) to propose a cascade of extracellular signals culminating in SapB-mediated aerial growth. None of the factors involved has been purified, and their chemical nature is unknown.

Why should such a complex exchange of signals have evolved? Perhaps population responsiveness to a single extracellular signal carries the risk that competing organisms could produce analogs of the signal to induce premature sporulation. This risk would be reduced with a more complex signal pathway.

Spores seem to be coated with a variety of other development-specific proteins, some of which may also contribute to hydrophobicity (Guijarro et al., 1988). The spores of *S. coelicolor* strains bearing the plasmid SCP1 carry on their surface several proteins (SapC, D, and E) encoded within the 80-kilobase (kb) terminally repeated region of this 350-kb linear plasmid (J.R. McCormick and R. Losick, unpublished observations). Because the morphology of fully differentiated colonies of SCP1[+] and SCP1[-] strains is similar, the function, if any, of these plasmid-encoded proteins is uncertain. Perhaps they play a part in circumventing self-inhibition by the SCP1-specified antibiotic methylenomycin A, which appears to be most active against the aerial mycelium (Vivian, 1971; Kirby and Hopwood, 1977).

Differentiation in the Aerial Mycelium

The spore-bearing aerial mycelium is the most obvious manifestation of multicellularity in streptomycetes. Unlike substrate hyphae, aerial hyphae grow away from, rather than toward, available nutrients and the water needed to provide turgor pressure. The nutrients therefore must be transported to the top of the aerial mycelium, where growth appears to be localized (Miguélez et al., 1994). A different set of requirements comes into play when the aerial hyphal tips turn into spore chains: continued elongation of hyphae and chromosome replication must stop; genes or proteins involved in septation must be activated or derepressed; and metabolism must be redirected toward the synthesis of spore components before eventually being virtually shut down. The subdivision of aerial hyphae into spore compartments contrasts strikingly with the occasional septation events seen during hyphal growth. Sporulation septa, which differ in detailed structure from vegetative septa, develop synchronously within an aerial hypha to produce unigenomic compartments right up to the tip itself (Wildermuth and Hopwood, 1970; Williams and Sharples, 1970; McVittie, 1974; Hardisson and Manzanal, 1976). It is as if a (hypothetical) inhibitor of tip-proximal septation active during growth becomes inactive at incipiently sporulating aerial hyphal tips, permitting the use of every potential division site.

Provision of Nutrients for Aerial Growth

Hyphae from different parts of *Streptomyces* colonies vary in their content of storage materials. For example, glycogen-like polysaccharides are detectable at two locations in colonies of several species: (1) in hyphae in the central zone of maturing colonies, roughly corresponding to the interface of the substrate and aerial mycelium (phase I deposition), and (2) in developing spore chains (phase II deposition) (Braña et al., 1986; Plaskitt and Chater, 1995). Young vegetative hyphae and the nonsporulating upper parts of aerial hyphae contain no glycogen. A similar spatial pattern of oil droplets has been observed (Plaskitt and Chater, 1995). Probably phase I storage depots supply carbon and energy to growing aerial hyphae. It has been suggested that the soluble materials released from phase I storage macromolecules cause increased osmotic potential, and therefore water uptake, contributing to the development of sufficient turgor pressure to drive extension of hyphae into the air (Chater, 1989). [Osmotic changes might also accompany the condensation and subsequent degradation of phase II storage deposits and contribute to changes in the partitioning of water between the cytoplasm of developing spores and adjacent compartments, which might be the hyphal stalk or an extracellular compartment enclosed within the hydrophobic layer that coats the aerial hyphae (Plaskitt and Chater, 1995).]

Evidence of a developmental association was provided by the near absence of phase I glycogen deposits from various representative *bld* mutants (including *bldA*), even though phase II deposits were present in developing spore chains when sporulation was induced by growing the mutants with mannitol as the carbon source (Plaskitt and Chater, 1995). Remarkably, *S. coelicolor* contains two glycogen branching isoenzymes, encoded by *glgBI* and *glgBII* (Bruton et al., 1995). Disruption of *glgBI* causes abnormalities in phase I glycogen deposits but has no effect on phase II, whereas disruption of *glgBII* has the opposite effect (Fig. 6.8). Hence the spatially separate glycogen deposits are synthesized by pathways that, at least in part, use genetically distinct enzymes that may be subject to developmentally distinctive regulation; and indeed *whiG*-induced ectopic sporulation in the substrate mycelium is accompanied by gly-

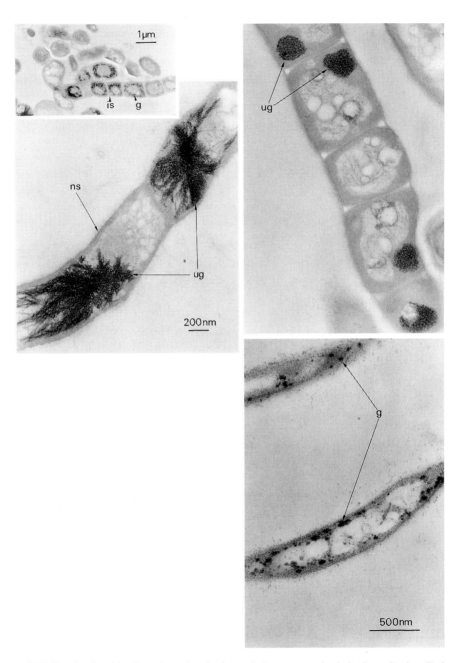

FIGURE 6.8   Specific disruption of each phase of glycogen synthesis in *S. coelicolor*. (**Left panel**) *glgB*I mutant producing unbranched glycogen (ug) specifically in phase I [*inset*: immature spore chain (IS) of the same colony producing normal glycogen (g)]. (**Right panel**) *glgB*II mutant producing unbranched glycogen specifically in phase II (**top**) but with normal glycogen synthesis in phase I (**bottom**). (Reprinted from Bruton et al., 1995, with permission)

cogen deposition in the prespore compartments (Plaskitt and Chater, 1995), indicating that sporulation-associated glycogen biosynthesis is *whiG*-dependent.

Initiation of Sporulation: Involvement of a Specialized σ Factor

Continued hyphal tip extension and sporulation do not take place at the same time in any one aerial hypha. There may therefore be a close, even causal, association between the cessation of growth and the onset of sporulation. This concept is supported by studies of *S. griseus* strains that can sporulate in submerged culture; a sharp nutritional downshift can cause fairly rapid mass sporulation (Kendrick and Ensign, 1983). To investigate the control of sporulation genetically, most effort has been given to studies of *S. coelicolor* mutants that fail to make gray spore pigment, on the assumption that pigment does not accumulate in the absence of spores. This assumption is evidently not wholly correct, as an *ftsZ* mutant makes no spores yet produces a gray aerial mycelium (McCormick et al., 1994). Nevertheless, white colony (*whi*) mutants have provided access to several genes that are important for sporulation (Hopwood et al., 1970; Chater, 1972).

The most readily understood of these genes, *whiG*, encodes a member of a specialized subgroup of σ factors, the other known members of which direct the transcription of genes for flagellar biosynthesis and chemotaxis in gram-negative and gram-positive bacteria (Chater et al., 1989). So similar is σ^{WhiG} to its homologs it appears able to substitute for σ^{FliA} in vivo in activating the motility of *Salmonella typhimurium* (J. Nodwell and R. Losick, unpublished results); and in a three-way heterologous reconstitution experiment, σ^{WhiG} could direct core RNA polymerase of *E. coli* to transcribe the promoter of a motility gene from *B. subtilis* (L. Chamberlin and M.J. Buttner, personal communication). However, the immobility of *S. coelicolor* seems to indicate a divergence of physiological context between σ^{WhiG} and its homologs from non-actinomycetes.

Several observations place σ^{WhiG} in a pivotal position for the initiation of sporulation. The *whiG* mutants have long, virtually undifferentiated aerial hyphae, whereas most other *whi* mutants (*whiA, B, H*) unable to produce sporulation septa show marked coiling of aerial hyphal tips (Fig. 6.9) (Chater, 1972). Double mutants in which a *whiG* mutation is combined with *whiA, whiB*, or *whiH* mutation have the same phenotype as the *whiG* single mutant (Chater, 1975). In addition, multiple copies of *whiG* can elicit early and ectopic sporulation, implying that a limitation for the σ^{WhiG} form of RNA polymerase is the reason sporulation generally occurs nowhere but at the tips of aerial hyphae (Chater et al., 1989).

What limits the amount of the σ^{WhiG} form of RNA polymerase? Evidence does not point to marked developmental regulation of *whiG* transcription [at least from the single promoter detected in *S. coelicolor*; but in *Streptomyces aureofaciens* the situation appears to be more complex (J. Kormanec, personal communication)]: *whiG* mRNA is detectable by S1 nuclease protection in young surface-grown cultures before aerial mycelium is visible; and the abun-

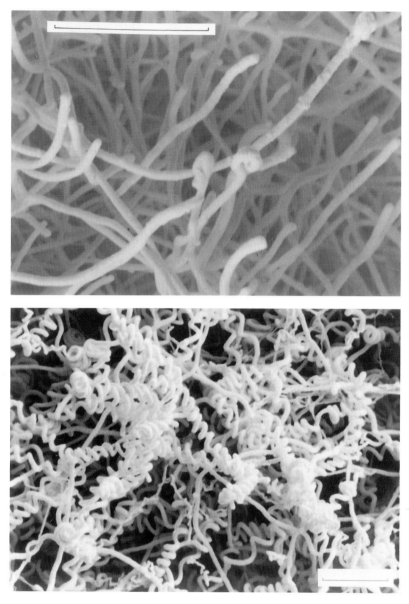

FIGURE 6.9 Scanning electron micrographs of aerial mycelium of *whiG* and *whiA* mutants of *S. coelicolor* (4-day cultures). Note the straight hyphae of the *whiG* mutant (C71) (**top**) and the long helical hyphae of the *whiA* mutant (C72) (**bottom**). Bars = 10 μm. (Courtesy of K. Findlay and N.J. Ryding)

dance of the mRNA changes little as colonies develop, either in the wild-type or in representative early-blocked *whi* mutants; and *whiG* is transcribed in liquid cultures that do not sporulate (G.H. Kelemen et al., 1996). [Multicopy promoter-probing experiments done in *whiA*, *whiB*, and *whiH* mutants had led to an earlier tentative model involving partial *whiG* dependence on these three sporulation genes (Chater, 1993), but this model can no longer be regarded as meaningful because the reporter system used in these experiments has since been found to give inconsistent results (J.M. White, M.J. Bibb, and C.J. Thompson, personal communication).]

Homologs of σ^{WhiG} from other bacteria, notably σ^{FliA} of *Salmonella typhimurium*, are regulated mainly posttranslationally by interaction with a specific anti-σ protein (Ohnishi et al., 1992). In *S. typhimurium*, this protein, the product of the *flgM* gene, responds to information about cellular structure: it is pumped out of the cell by a flagellar hook-basal body structure when the latter has been fully assembled (Hughes et al., 1993). There is as yet no direct evidence of a *flgM*-like gene in *Streptomyces*, but σ^{WhiG} shows noticeable similarities with a region of σ^{FliA} known to interact with FlgM protein (Kutsukake et al., 1994). This finding reinforces earlier tentative suggestions that a FlgM-like protein may regulate σ^{WhiG} activity, perhaps being eliminated from, unexpressed in, or inactivated in aerial hyphae (Chater, 1993; Losick and Shapiro, 1993).

Presumably, several genes are switched on by σ^{WhiG} RNA polymerase to activate sporulation, and several candidates can be imagined: (1) that (or those) for the immediately following step in a supposed regulatory cascade (which might include *whiA*, *whiB*, or *whiH*); (2) genes whose products bring about an early step in morphogenesis of aerial hyphae into spore chains, for example by involvement in sporulation septation (e.g., *ftsZ*); and (3) genes involved in the metabolic changes (e.g., pigment formation or storage metabolism) associated with sporulation. The next three sections deal with examples of potentially *whiG*-dependent genes.

Do Other Regulatory Genes Required for Sporulation Septation Depend on whiG?

Two other *whi* genes, *whiB* and *whiH*, known to be required for sporulation septation, have been characterized in some detail. The *whiH* gene product significantly resembles a family of repressors, and evidence is emerging that the *whiH* promoter is transcribed by the σ^{WhiG} form of RNA polymerase (N.J. Ryding, C. Whatling, and K.F. Chater, unpublished observations). This is the first demonstration of a direct effect of one *whi* gene on another. Because σ^{WhiG} is related to σ factors that are regulated by morphological coupling, and the WhiH-like repressors of other bacteria are mostly regulated by intermediary metabolites, the regulatory connection of *whiG* and *whiH* may be part of a control system that coordinates morphogenesis with metabolism.

The *whiB* product does not show obvious similarities to proteins of known function, though its overall features suggest that it is cytoplasmic (it is highly hydrophilic) and potentially capable of several interactions: with DNA via a positively charged, α-helical C-terminal region; with other proteins through its negatively charged, somewhat glutamate-rich N-terminal region; and with metal ions through four cysteine residues (Davis and Chater, 1992). [These features are also found in some other evidently related, but functionally uncharacterized, hypothetical gene products encountered in database searches (Chater 1993).] Transcription of *whiB* is from two promoters: one weakly constitutive and the other stronger and activated during aerial mycelium development (Soliveri et al., 1992). Both promoters are active in a *whiG* mutant, and neither contains sequences resembling the tentative consensus sequence for σ^{WhiG}-dependent promoters (instead, the developmentally regulated *whiB* promoter contains −10 and −35 regions corresponding to the consensus sequence for principal σ factor recognition). Hence the control of sporulation may involve parallel regulatory steps as well as consecutive ones.

Genes with Unknown Functions That Depend on σ^{WhiG} for Transcription

Tan and Chater (1993) described two short fragments of *S. coelicolor* DNA that seem to be transcribed from σ^{WhiG}-dependent promoters. The fragments were isolated by their induction, at multiple copies, of a partial *whiG* pheno-copy in a morphologically wild-type host, as if they were sequestering the limited amount of available σ^{WhiG}-form of RNA polymerase. Both fragments activated the *xylE* reporter gene when cloned in the correct orientation into a high copy-number promoter-probe plasmid in a *whiG^{+}* host but not in a *whiG* mutant host; and they contained −10 and −35 sequences resembling those of motility σ-factor-dependent genes in other bacteria. The functions of the two *whiG*-dependent transcription units are unknown, but DNA sequence analysis suggests that both the σ^{WhiG}-dependent promoters are used to transcribe the last gene in a larger operon that, in some as yet undiscovered circumstances, may be transcribed from another (uncharacterized) promoter farther upstream (Tan and Chater, 1993; H. Yang, personal communication).

Second σ Factor Required Later in Sporulation

As a result of a general screen with an oligonucleotide designed to correspond to a conserved region of σ factors, Potúčková et al. (1995) identified a gene (*sigF*) that when disrupted in the *S. coelicolor* chromosome was revealed to be important for spore development. The *sigF* mutant makes thin-walled, un-pigmented spores that do not separate readily. Thus *sigF* is the first defined late regulatory gene for *Streptomyces* sporulation. Analysis of *sigF* expression is likely to help in elucidating the roles of earlier acting *whi* genes.

The closest relatives of *sigF* product σ^{F} are σ factors from *B. subtilis*: σ^{B} (a stress response σ factor) and, strikingly, the two forespore-specific sigma factors σ^{F} and σ^{G}. This observation is the first documentation of the occurrence of a member of the so-called sporulation branch of σ factors outside of endospore-forming bacteria (Potúčková et al., 1995). Conservation of an ancient role of such σ factors in the physiology of specialized resting cells is suggested by this observation, in sharp contrast to the distinct lineages of σ factors that initiate sporulation in the two genera [i.e., σ^{WhiG} of *S. coelicolor*, which is related to motility σ factors, and σ^{H} of *B. subtilis*, which is different from all of its cousins (Lonetto et al., 1992)].

Spatial Localization of Gene Expression

Heterogeneity of σ Factors

The complexity of *Streptomyces* colonies on laboratory media suggests that there could be numerous cell types, depending on the location in the colony. In an obvious scenario, each cell type might contain RNA polymerase holoenzyme forms with a particular profile of σ factor content, and each gene needing to be expressed in that cell would have a promoter recognizable by one of these forms. Thus a gene expressed only in a single cell type might have a promoter recognized by a σ factor present only in that cell-type. Late sporulation genes, perhaps including σ^{F}-dependent genes, may well be of this class. However, many genes probably need to be expressed in more than one cell type; if so, either their promoters might be recognized by a σ factor present in all of these cell types, or they could have multiple promoters, each a target for a different σ factor.

Streptomyces coelicolor contains at least eight σ factors, two of which (σ^{WhiG} and σ^{F}) play important roles during sporulation (Buttner, 1989; Lonetto et al., 1994; Potúčková et al., 1995). The four σ factors encoded by *hrdA, B, C*, and *D* are closely related to each other and to primary σ factors (Shiina et al., 1991; Tanaka et al., 1991). Only one of them, σ^{HrdB}, is essential for growth; an *hrdACD* triple mutant showed no obvious phenotypic difference from the wild-type parental strain (Buttner et al., 1990; Buttner and Lewis, 1992). Most streptomycetes contain similar sets of *hrd* genes (Tanaka et al., 1988; Kormanec et al., 1992), so they are likely to have adaptive significance. For example, if *hrdB* were poorly expressed in some parts of the life cycle in the soil, its function could perhaps be taken over in the relevant cell types by another *hrd* gene-specified σ factor; but laboratory conditions may blur the distinction between cell types, allowing sufficient σ^{HrdB} to be present in all cells to permit their effective function without significant dependence on other σ^{HrdB} homologs.

Streptomyces genes often contain multiple promoters. For example, each of four *dagA* (agarase gene) promoters is transcribed by a different form of RNA polymerase (Buttner et al., 1988). However, no differences in relative use have been observed among these promoters under different physiological conditions (Servín-González et al., 1994). The striking diversity in the sequences of promoters of *Streptomyces* genes (Strohl, 1992) may indicate that the existing known σ factors are the tip of an iceberg, although some of the sequence diversity must also reflect differences in promoter strength.

As an additional way of ensuring that a particular biochemical attribute is expressed in several, but not all, cell types, there may be tissue-specific enzyme isoforms encoded by differentially regulated genes. Glycogen metabolism provides an example (see above): the branching enzyme genes *glgBI* and *glgBII* encode similar proteins that are active in different cell types (Bruton et al., 1995). Another example is provided by genes involved in polyketide biosynthesis. Related gene sets in the *act* and *whiE* clusters determine the assembly of the polyketide skeletons of a blue antibiotic (actinorhodin) and the gray spore pigment, respectively (Davis and Chater, 1990; Hopwood et al., 1995). Forced expression of the ketosynthase (KS) component of the *whiE*-encoded polyketide synthase complex resulted in complementation of a mutation in the corresponding *act*-encoded component, leading to actinorhodin synthesis, yet KS-deficient *act* mutants fail to produce actinorhodin (Kim et al., 1994). Probably differential cell-type-specific transcription of the homologous genes is responsible for the normal absence of *whiE−act* interaction. Manifest differences in transcriptional regulation are shown by the dependence of *act*, but not *whiE*, transcription on some of the *abs*, *aba*, *afs*, and *bld* genes described below (Guthrie and Chater, 1990; Bruton et al., 1991; Chater et al., 1993; Champness and Chater, 1994; W. Champness, personal communication).

Integration of Information Relevant to Determination of Cell Type

The determination of cell type is likely to involve signaling molecules, which could be external (as described earlier) or internal, arising either from nutrient limitation or from spatially localized sources that give rise to gradients inside hyphae. Among internally generated molecules, ppGpp, which is produced at the onset of nitrogen limitation, is probably an important signal for the initiation of antibiotic production (Takano and Bibb, 1994). Hyphae in the center of the surface layer of a colony may be the earliest to be nutrient-limited and therefore to contain elevated ppGpp levels. Extracellular signaling molecules should also be most abundant here (because the surface is a diffusion boundary). This location therefore tends to be marked in a way that could lead to differential gene expression and different morphological features. Other, still unknown influences on the initiation of secondary metabolism or morphological differentiation may arise from the interplay of streptomycetes with their environment, whether with chemical or physical effectors or with other organisms such as plants and microorganisms. It is therefore not surprising to find that multiple genes (in addition to the *bld* genes already discussed) regulate antibiotic production; this topic is reviewed in detail by Champness and Chater (1994) and Chater and Bibb (1996). Here we briefly consider the example of actinorhodin synthesis in *S. coelicolor* A3(2).

The relevant pathway genes (*act*) (Hopwood et al., 1995) form an approximately 22-kb cluster that includes one pathway-specific regulatory gene (*actII-ORF4*), which is required for transcription of most *act* genes. ActII-ORF4 protein does not resemble known transcription factors, and its mechanism of action is unknown. At present, we know of no reason to reject the simple hypothesis that all regulation of *act* structural gene transcription operates via ActII-ORF4 protein (Chater and Bibb, 1995). The regulatory input seems to involve the two genes of the *absA* locus, the *absB* locus, *afsR*, *afsB*, *abaA*, and several *bld* genes (*A,B,C,D,G,H,I*), as mutation in any of these genes reduces or eliminates synthesis of actinorhodin (and of other secondary metabolites); and whenever this deficiency has been studied, it has been attributed to a failure to transcribe *act* genes (Chater and Bibb, 1995; W. Champness, personal communication). This multiplicity of regulatory genes can be compared to the situation in *B. subtilis*: various sensory pathways (involving diverse gene products) may contribute information to a central regulatory cascade analogous to the sporulation phospho-relay (Hoch, 1993), each step in the cascade being associated with the collection of a particular kind of sensory input. The pathway-specific regulatory elements of each antibiotic pathway would integrate signals from different, probably overlapping, suites of sources, some of which may well contribute positional information. Some of the regulatory genes involved in this peripheral sensory input might remain cryptic in the laboratory because of the impossibility of mimicking particular ecological situations under artificial conditions.

Genetic Consequences of Multicellularity

In unicellular bacteria, competition between siblings leads to strong selection for conservation of the chromosome at every cell division. In *Streptomyces* hyphae many chromosomes are contained in each cell compartment, so defective chromosomes may be "carried" during colony development, at least until their progeny have replicated sufficiently to populate entire hyphal compartments. The eventual formation of unigenomic spore compartments from multigenomic aerial hyphae might be expected to trap defective chromosomes that would impair competitiveness. Defective chromosomes might also be detected when protoplasts with one or a few chromosomes are produced artificially and then regenerated. Indeed, both situations often reveal defects that affect spore pigmentation, aerial mycelium formation, production of secondary metabolites, ultraviolet sensitivity, or arginine biosynthesis (Leblond and Decaris, 1994). Such mutants are viable despite sometimes harboring deletions of up to 2000 kb (i.e., about 25% of the chromosome). Evidence suggests that many protein-coding genes are present in these unstable regions (Leblond and Dercaris, 1994), which correspond to genetically "silent" regions near the ends of the linear chromosome (Lin et al., 1993; Chen et al., 1994). It is noteworthy that the regions of the genome rich in housekeeping genes, on the other hand, seem as stable as those of any other bacteria.

Susceptibility of Multicellular Colonies to Infectious Agents

Because hyphal compartments form a coherent multicellular organism, *Streptomyces* colonies are potentially vulnerable to infectious agents, such as phages or plasmids, that gain entry initially into just a single compartment. Most *Strep-*

tomyces plasmids are conjugative and appear to be efficiently transferred between mycelia. After intimate mixed growth of plasmid-containing and plasmid-free derivatives of any particular strain, most of the spore progeny contain the plasmid. This phenomenon probably occurs in part by spread of the plasmid from compartment to compartment within hyphae (Kieser et al., 1982). Intriguingly, in several otherwise unrelated plasmids, the single gene required for transfer between strains, which together with one or a few other genes is needed for the spread function, encodes a protein with perceptible homology to the SpoIIIE protein of *B. subtilis* (Tomura et al., 1993; Wu et al., 1995). SpoIIIE is implicated in the completion of transfer of a copy of the chromosome from the spore mother cell compartment across the newly formed asymmetrically located sporulation septum and into the prespore compartment (Wu and Errington, 1994).

The most well known intracellular barrier to bacteriophage infection is cleavage of incoming DNA by restriction enzymes, which are often found in *Streptomyces* spp. If, as sometimes happens, the cognate site-specific modification methylase protects a phage genome against cleavage before the restriction enzyme has acted, the result could be catastrophic for a multicellular organism because the restriction-resistant phage progeny of such an infection could go on to infect and lyse all the cells in the colony. *S. coelicolor* A3(2) has a protective mechanism against some phages that does not suffer from this disadvantage. The system, termed Pgl (for phage growth limitation), is known to be active only against temperate phages related to φC31 (Laity et al., 1993). Pgl$^+$ strains are fully susceptible to a first infection by φC31 coming from a different (i.e., Pgl$^-$) last host; but the phages released after one round of infection are inefficient during the next round. The mechanism of this attenuation is not understood, even though at least some of the *pgl* genes have been sequenced (Bedford et al., 1995).

Prospects for Further Study

Understanding the mechanism and implications of multicellularity in *Streptomyces* spp. depends on research into the physiology and developmental biology of *Streptomyces* spp., especially of *S. coelicolor*, which is in turn heavily dependent on genetics. This era is an exciting time for *Streptomyces* researchers because genes have been cloned for many significant processes. Such cloned genes include many of those needed to take carbon sources through central metabolism and on into secondary metabolites and storage compounds (I.S. Hunter and L. Dijkhuisen, personal communication), so investigators will soon be able to analyze and manipulate metabolic flux. The availability of replication origins, chromosome ends, and many of the genes for DNA, RNA, and protein synthesis (van Wezel et al., 1995) will contribute to studies of chromosome partitioning during growth, branching, and sporulation, as well as providing access to some of the key elements in gene regulation and biomass accumulation. All of the loci that are known from "classical" genetic analysis to be essential for sporulation septation of aerial hyphae have now been cloned, as have some of the genes directly involved in cell division (see above) (N.J. Ryding, H. Parry, and K.F. Chater, unpublished observations). The near future should see progress toward a clearer understanding of how their interactions lead to localized and locally synchronized multiple septated events. Progress in the study of intra- and extracellular signaling is promised particularly by the recognition that A-factor-like molecules do exert significant effects on secondary metabolism in *S. coelicolor* (Takano and Bibb, 1994, personal communication): this finding is of particular significance because of the large number of regulatory and biosynthetic genes concerned with secondary metabolism that are known for this strain. The filling of many of the gaps in the catalog of important cloned genes from *S. coelicolor* A3(2) is proceeding rapidly because of the recent development of an ordered cosmid library (M. Redenbach, et al., 1996), which allows gene cloning by chromosome walking and presents the possibility of extensive genome sequencing. This optimistic situation is somewhat tempered by a continuing deficiency in widely applicable reporter systems of the type and sensitivity needed for easy colony screening and for localizing gene expression in colonies [though the *luxAB* reporter genes have found some applications (Schauer et al., 1988; Sohaskey et al., 1992)]. Nevertheless, our appreciation of the *Streptomyces* colony as a remarkable prokaryotic example of a multicellular organism can only increase.

Acknowledgments

We thank Mervyn Bibb, Mark Buttner, David Hopwood, and Tobias Kieser for helpful comments on the manuscript and Kim Findlay, Jamie Ryding, and Elizabeth Wellington for providing micrographs.

References

Åkerlund, T., Nordström, K., and Bernander, R. (1993) Branched *Escherichia coli* cells. *Mol. Microbiol.* **10**:849–858.

Allan, E.J., and Prosser, J.I. (1985) A kinetic study of the colony growth of *Streptomyces coelicolor* A3(2) and J802 on solid medium. *J. Gen. Microbiol.* **131**:2521–2532.

Angell, S., Lewis, C.G., Buttner, M.J., and Bibb, M.J. (1994) Glucose repression in *Streptomyces coelicolor* A3(2): a likely regulatory role for glucose kinase. *Mol. Gen. Genet.* **244**:135–143.

Anisova, L.N., Blinova, I.N., Efremenkova, O.V., et al. (1984) Regulators of the development of *Streptomyces coelicolor* A3(2). *Izv. Akad. Nauk. SSSR. Ser. Biol.* **1**: 98–108.

Babcock, M.J., and Kendrick, K.E. (1988) Cloning of DNA involved in sporulation of *Streptomyces griseus*. *J. Bacteriol.* **170**:2802–2808.

Babcock, M.J., and Kendrick, K.E. (1990) Unusual transcriptional and translational features of a developmental gene of *Streptomyces griseus*. *Gene* **95**:57–63.

Beall, B., and Lutkenhaus, J. (1991) FtsZ in *Bacillus subtilis* is required for vegetative septation and for asymmetric septation during sporulation. *Genes Dev.* **5**:447–455.

Bedford, D., Laity, C., and Buttner, M.J. (1995) Two genes involved in the phase-variable φC31 resistance mechanism of *Streptomyces coelicolor* A3(2). *J. Bacteriol.* **177**:4681–4689.

Botsford, J.L., and Harman, J.G. (1992) Cyclic AMP in prokaryotes. *Microbiol. Rev.* **56**:100–122.

Braña, A.F., Manzanal, M.-B., and Hardisson, C. (1982) Mode of cell wall growth of *Streptomyces antibioticus*. *FEMS Microbiol. Lett.* **13**:231–235.

Braña, A.F., Méndez, C., Díaz, L.A., Manzanal, M.B., and Hardisson, C. (1986) Glycogen and trehalose accumulation during colony development in *Streptomyces antibioticus*. *J. Gen. Microbiol.* **132**:1319–1326.

Bruton, C.J., Guthrie, E.P., and Chater, K.F. (1991) Phage vectors that allow monitoring of secondary metabolism genes in *Streptomyces*. *Biotechnology* **9**:652–656.

Bruton, C.J., Plaskit, K.A., and Chater, K.F. (1995) Tissue-specific glycogen branching isoenzymes in a multicellular prokaryote, *Streptomyces coelicolor* A3(2). *Mol. Microbiol.* **18**:89–99.

Buttner, M.J. (1989) RNA polymerase heterogeneity in *Streptomyces coelicolor* A3(2). *Mol. Microbiol.* **3**:1653–1659.

Buttner, M.J., and Lewis, C.G. (1992) Construction and characterization of *Streptomyces coelicolor* A3(2) mutants that are multiply deficient in the non-essential *hrd*-encoded RNA polymerase sigma factors. *J. Bacteriol.* **174**:5156–5167.

Buttner, M.J., Chater, K.F., and Bibb, M.J. (1990) Cloning, disruption and transcriptional analysis of three RNA polymerase sigma factor genes of *Streptomyces coelicolor* A3(2). *J. Bacteriol.* **172**:3367–3378.

Buttner, M.J., Smith, A.M., and Bibb, M.J. (1988) At least three RNA polymerase holoenzymes direct transcription of the agarase gene (*dagA*) of *Streptomyces coelicolor* A3(2). *Cell* **52**:599–607.

Champness, W.C., and Chater, K.F. (1994) Regulation and integration of antibiotic production and morphological differentiation in *Streptomyces* spp. In *Regulation of Bacterial Differentiation*, P. Piggot, C. Moran, and P. Youngman (eds.), pp. 61–93. American Society for Microbiology, Washington, DC.

Chater, K.F. (1972) A morphological and genetic mapping study of white colony mutants of *Streptomyces coelicolor*. *J. Gen. Microbiol.* **72**:9–28.

Chater, K.F. (1975) Construction and phenotypes of double sporulation deficient mutants in *Streptomyces coelicolor* A3(2). *J. Gen. Microbiol.* **87**:312–325.

Chater, K.F. (1989) Multilevel regulation of *Streptomyces* differentiation. *Trends Genet.* **5**:372–376.

Chater, K.F. (1991) Saps, hydrophobins, and aerial growth. *Curr. Biol.* **1**:318–320.

Chater, K.F. (1992) *Streptomyces coelicolor*: a mycelial, spore-bearing prokaryote. In *Molecular Genetics of Development*, V.A. Russo, S. Brody, D. Cove, and S. Ottolenghi (eds.), pp. 61–74. Blackwell, Oxford.

Chater, K.F. (1993) Genetics of differentiation in *Streptomyces*. *Annu. Rev. Microbiol.* **47**:685–713.

Chater, K.F., and Bibb, M.J. (1996) Regulation of bacterial antibiotic production. In *Biotechnology*, Vol. 7: *Products of Secondary Metabolism*, H. Kleinkauf, and H. Von Döhren (eds.). VCH, Weinheim (in press).

Chater, K.F., and Hopwood, D.A. (1993) *Streptomyces*. In *Bacillus subtilis and Other Gram-Positive Bacteria: Biochemistry, Physiology and Molecular Genetics*, A.L. Sonenshein, J.A. Hoch, and R. Losick (eds.), pp. 83–99, American Society for Microbiology, Washington, DC.

Chater, K.F., and Merrick, M.J. (1979) *Streptomyces*. In *Developmental Biology of Prokaryotes*, J.H. Parish (ed.), pp. 93–114, Blackwell, Oxford.

Chater, K.F., Brian, P., Brown, G.L., et al. (1993) Problems and progress in the interactions between morphological and physiological differentiation in *Streptomyces coelicolor*. In *Genetics and Molecular Biology of Industrial Microorganisms*, C.L. Hershberger, P. Skatrud, and G. Hegeman (eds.), pp. 151–157. American Society for Microbiology, Washington, DC.

Chater, K.F., Bruton, C.J., Plaskitt, K.A., Buttner, M.J., Méndez, C., and Helmann, J. (1989) The developmental fate of *S. coelicolor* hyphae depends crucially on a gene product homologous with the motility sigma factor of *B. subtilis*. *Cell* **59**:133–143.

Chen, C.W., Lin, Y.S., Yang, Y.L., et al. (1994) The linear chromosomes of *Streptomyces*: structure and dynamics. *Actinomycetologica* **8**:103–112.

Dai, K., and Lutkenhaus, J. (1991) *ftsZ* is an essential cell division gene in *Escherichia coli*. *J. Bacteriol.* **173**:3500–3506.

Danchin, A., Pidoux, J., Krin, E., Thompson, C.J., and Ullman, A. (1993) The adenylate kinase catalytic domain of *Streptomyces coelicolor* is carboxy-terminal. *FEMS Microbiol. Lett.* **114**:145–152.

Davis, N.K., and Chater, K.F. (1990) Spore colour in *Streptomyces coelicolor* A3(2) involves the developmentally regulated synthesis of a compound biosynthetically related to polyketide antibiotics. *Mol. Microbiol.* **4**:1679–1691.

Davis, N.K., and Chater, K.F. (1992) The *Streptomyces coelicolor whiB* gene encodes a small transcription factor-like protein dispensable for growth but essential for sporulation. *Mol. Gen. Genet.* **232**:351–358.

Dharmatilake, A.J., and Kendrick, K.E. (1994) Expression of the division-controlling gene *ftsZ* during growth and sporulation of the filamentous bacterium *Streptomyces griseus*. *Gene* **147**:21–28.

Distler, J., Mansouri, K., Mayer, G., Stockmann, M., and Piepersberg, W. (1992) Streptomycin biosynthesis and its regulation in streptomycetes. *Gene* **115**:105–111.

Efremenkova, O.V., Anisova, L.N., and Bartoshevich, Y.E. (1985) Regulators of differentiation in actinomycetes. *Antibiot. Med. Biotekhnol.* **9**:687–707.

Fuqua, W.C., Winans, S.C., and Greenberg, E.P. (1994) Quorum sensing in bacteria: the LuxR-LuxI family of cell density-responsive transcriptional regulators. *J. Bacteriol.* **176**:269–275.

Gersch, D., and Strunk, C. (1980) Cyclic adenosine 3′,5′-monophosphate as "first messenger" in *Streptomyces hygroscopicus*: bimodal regulation of germination and growth. *Curr. Microbiol.* **4**:271–275.

Gibb, G.D., and Strohl, W.R. (1988) Physiological regulation of protease activity in *Streptomyces peucetius*. *Can. J. Microbiol.* **34**:187–190.

Ginther, C.L. (1979) Sporulation and the production of serine protease and cephamycin C by *Streptomyces lactamdurans*. *Antimicrob. Agents Chemother.* **15**:522–526.

Gramajo, H., Takano, E., and Bibb, M.J. (1993) Stationary phase production of the antibiotic actinorhodin is transcriptionally regulated. *Mol. Microbiol.* **7**:837–845.

Gray, D.I., Gooday, G.W., and Prosser, J.I. (1990) Apical hyphal extension in *Streptomyces coelicolor* A3(2). *J. Gen. Microbiol.* **136**:1077–1084.

Guijarro, J., Santamaría, R., Schauer, A., and Losick, R. (1988) Promoter determining the timing and spatial localization of transcription of a cloned *Streptomyces coelicolor* gene encoding a spore-associated polypeptide. *J. Bacteriol.* **170**:1895–1901.

Guthrie, E.P., and Chater, K.F. (1990) The level of a transcript required for production of a *Streptomyces coelicolor* antibiotic is conditionally dependent on a tRNA gene. *J. Bacteriol.* **172**:6189–6193.

Hara, O., and Beppu, T. (1982) Mutants blocked in streptomycin production in *Streptomyces griseus*: the role of A-factor. *J. Antibiot.* **35**:349–358.

Hara, O., Horinouchi, S., Uozumi, T., and Beppu, T. (1983) Genetic analysis of A-factor synthesis in *Streptomyces coelicolor* A3(2) and *Streptomyces griseus*. *J. Gen. Microbiol.* **129**:2939–2944.

Hardisson, C., and Manzanal, M.B. (1976) Ultrastructural studies of sporulation in *Streptomyces*. *J. Bacteriol.* **127**:1443–1454.

Hoch, J.A. (1993) *spo0* genes, the phosphorelay, and the initiation of sporulation. In *Bacillus subtilis and Other Gram Positive Bacteria: Biochemistry, Physiology and Molecular Genetics*, A.L. Sonenshein, J.A. Hoch, and R. Losick (eds.), pp. 747–755. American Society for Microbiology, Washington, DC.

Hodgson, D.A. (1980) Carbohydrate utilization in *Streptomyces coelicolor* A3(2). *PhD thesis,* University of East Anglia, Norwich, UK.

Hodgson, D.A. (1982) Glucose repression of carbon source uptake and metabolism in *Streptomyces coelicolor* A3(2) and its perturbation in mutants resistant to 2-deoxy-glucose. *J. Gen. Microbiol.* **128**:2417–2430.

Hodgson, D.A. (1992) Differentiation in actinomycetes. In *Prokaryotic Structure and Function: A New Perspective, Symposium of the Society of General Microbiology*, Vol. 47, S. Mohan, C. Dow, and J.A. Cole (eds.), pp. 407–440. Cambridge University Press, Cambridge.

Hopwood, D.A. (1960) Phase-contrast observations on *Streptomyces coelicolor*. *J. Gen. Microbiol.* **22**:295–302.

Hopwood, D.A., Chater, K.F., and Bibb, M.J. (1995) Genetics of antibiotic production in *Streptomyces coelicolor* A3(2), a model streptomycete. In *Genetics and Biochemistry of Antibiotic Production*, L.C. Vining and C. Stuttard (eds.), pp. 65–102. Butterworth-Heinemann, Newton, MA.

Hopwood, D.A., Wildermuth, H., and Palmer, H.M. (1970) Mutants of *Streptomyces coelicolor* defective in sporulation. *J. Gen. Microbiol.* **61**:397–408.

Horinouchi, S., and Beppu, T. (1992) Autoregulatory factors and communication in actinomycetes. *Annu. Rev. Microbiol.* **46**:377–398.

Hughes, K.T., Gillen, K.L., Semon, M.J., and Karlinskey, J.E. (1993) Sensing structural intermediates in bacterial flagellar assembly by export of a negative regulator. *Science* **262**:1277–1280.

Kelemen, G.H., Brown, G.L., Kormanec, J., Potúčková, L., Chater, K.F., and Buttner, M.J. (1996) The positions of the sigma-factor genes, *whiG* and *sigF*, in the hierarchy controlling the development of spore chains in the aerial hyphae of *Streptomyces coelicolor* A3(2). *Mol. Microbiol.* **21**. (in press).

Kendrick, K.E., and Ensign, J.C. (1983) Sporulation of *Streptomyces griseus* in submerged culture. *J. Bacteriol.* **155**:357–366.

Khokhlov, A.S. (1985) Actinomycete autoregulators. In *Proceedings of Sixth International Symposium on Actinomycetes Biology*, G. Szabó, S. Biró, and M. Goodfellow (eds.), pp. 791–798. Akadémiai Kiadó, Budapest.

Khokhlov, A.S., Tovarova, I.I., Borisova, L.N., et al. (1967) A-factor responsible for the biosynthesis of streptomycin by a mutant strain of *Actinomyces streptomycini*. *Dokl. Akad. Nauk. SSSR* **177**:232–235.

Kieser, T., Hopwood, D.A., Wright, H.M., and Thompson, C.J. (1982) pIJ101, a multicopy broad host-range *Streptomyces* plasmid: functional analysis and development of DNA cloning vectors. *Mol. Gen. Genet.* **185**:223–238.

Kim, E.S., Hopwood, D.A., and Sherman, D.H. (1994) Analysis of type II polyketide β-ketoacyl synthase specificity in *Streptomyces coelicolor* A3(2) by *trans* complementation of actinorhodin synthase mutants. *J. Bacteriol.* **176**:1801–1804.

Kim, I.S., and Lee, K.J. (1995) Physiological roles of leupeptin and extracellular protease in mycelium development of *Streptomyces exfoliatus* SMF13. *Microbiology* **141**:1017–1025.

Kirby, R., and Hopwood, D.A. (1977) Genetic determination of methylenomycin synthesis by the SCP1 plasmid of *Streptomyces coelicolor* A3(2). *J. Gen. Microbiol.* **98**:239–252.

Kormanec, J., Farkašovský, M., and Potúčková, L. (1992) Four genes in *Streptomcyes aureofaciens* containing a domain characteristic of principal sigma factors. *Gene* **122**:63–70.

Kretschmer, S. (1982) Dependence of the mycelial growth pattern on the individually regulated cell cycle in *Streptomyces granaticolor. Z. Allg. Mikrobiol.* **22**:325–347.

Kretschmer, S. (1987) Nucleoid segregation pattern during branching in *Streptomyces granaticolor* mycelia. *J. Basic Microbiol.* **27**:203–206.

Kretschmer, S. (1988) Stepwise increase of elongation rate in individual hyphae of *Streptomyces granaticolor* during outgrowth. *J. Basic Microbiol.* **28**:35–43.

Kretschmer, S. (1989) Septation behaviour of the apical cell in *Streptomyces granaticolor* mycelia. *J. Basic Microbiol.* **29**:587–595.

Kretschmer, S. (1991) Correlation between branching and elongation in germ tubes of *Streptomyces granaticolor. J. Basic Microbiol.* **31**:259–264.

Kutsukake, K., Iydo, S., Ohnishi, K., and Iino, T. (1994) Genetic and molecular analyses of the interaction between the flagellum-specific sigma and anti-sigma factors in *Salmonella typhimurium. EMBO J.* **13**:4568–4576.

Laity, C., Chater, K.F., Lewis, C.G., and Buttner, M.J. (1993) Genetic analysis of the ΦC31-specific phage growth limitation (Pgl) system of *Streptomyces coelicolor* A3(2). *Mol. Microbiol.* **7**:329–336.

Lawlor, E.J., Baylis, H.A., and Chater, K.F. (1987) Pleiotropic morphological and antibiotic deficiencies result from mutations in a gene encoding a tRNA-like product in *Streptomyces coelicolor* A3(2) *Genes Dev.* **1**:1305–1310.

Leblond, P., and Decaris, B. (1994) New insights into the genetic instability of *Streptomyces. FEMS Microbiol. Lett.* **123**:225–232.

Leskiw, B.K., Bibb, M.J., and Chater, K.F. (1991a) The use of a rare codon specifically during development? *Mol. Microbiol.* **5**:2861–2867.

Leskiw, B.K., Lawlor, E.J., Fernandez-Abalos, J.M., and Chater, K.F. (1991b) TTA codons in some genes prevent their expression in a class of developmental, antibiotic-negative *Streptomyces* mutants. *Proc. Natl. Acad. Sci. USA* **88**:2461–2465.

Leskiw, B.K., Mah, R., Lawlor, E.J., and Chater, K.F. (1993) Temporal regulation of accumulation of the *bldA*-specified tRNA in *Streptomyces coelicolor* A3(2). *J. Bacteriol.* **175**:1995–2005.

Lin, Y.-S., Kieser, H.M., Hopwood, D.A., and Chen, C.W. (1993) The chromosomal DNA of *Streptomyces lividans* 66 is linear. *Mol. Microbiol.* **10**:923–933.

Lonetto, M., Gribskov, M., and Gross, C.A. (1992) The σ^{70} family: sequence conservation and evolutionary relationships. *J. Bacteriol.* **174**:3843–3849.

Lonetto, M.A., Brown, K.L., Rudd, K.E., and Buttner, M.J. (1994) Analysis of the *Streptomyces coelicolor sigE* gene reveals the existence of a subfamily of eubacterial RNA polymerase σ factors involved in the regulation of extracytoplasmic functions. *Proc. Natl. Acad. Sci. USA* **91**:7573–7577.

Losick, R., and Shapiro, L. (1993) Checkpoints that couple gene expression to morphogenesis. *Science* **262**:1227–1228.

Lutkenhaus, J. (1993) FtsZ ring in bacterial cytokinesis. *Mol. Microbiol.* **9**:403–409.

Ma, H., and Kendall, K. (1994) Cloning and analysis of a gene cluster from *Streptomyces coelicolor* that causes accelerated aerial mycelium formation in *Streptomyces lividans*. *J. Bacteriol.* **176**:3800–3811.

Magnuson, R., Solomon, J., and Grossman, A.D. (1994) Biochemical and genetic characterization of a competence pheromone from *B. subtilis. Cell* **77**:207–216.

Manzanal, M.B., and Hardisson, C. (1978) Early stages of arthospore maturation in *Streptomyces. J. Bacteriol.* **133**:293–297.

McCormick, J., Su, E.P., Driks, A., and Losick, R. (1994) Growth and viability of *Streptomyces coelicolor* mutant for the cell division gene *ftsZ. Mol. Microbiol.* **14**: 243–254.

McCue, L.A., Kwak, J., Babcock, M.J., and Kendrick, K.E. (1992) Molecular analysis of sporulation in *Streptomyces griseus. Gene* **115**:173–179.

McVittie, A.M. (1974) Ultrastructural studies on sporulation in wild-type and white colony mutants of *Streptomyces coelicolor. J. Gen. Microbiol.* **81**:291–302.

Méndez, C., Braña, A.F., Manzanal, M.B., and Hardisson, C. (1985) Role of substrate mycelium in colony development in *Streptomyces. Can. J. Microbiol.* **31**:446–450.

Miguélez, E.M., García, M., Hardisson, C., and Manzanal, M.B. (1994) Autoradiographic study of hyphal growth during aerial mycelium development in *Streptomyces antibioticus. J. Bacteriol.* **176**:2105–2107.

Miguélez, E.M., Hardisson, C., and Manzanal, M.B. (1993a) Incorporation and fate of *N*-acetyl-D-glucosamine during hyphal growth in *Streptomyces. J. Gen. Microbiol.* **139**:1915–1920.

Miguélez, E.M., Hardisson, C., and Manzanal, M.B. (1993b) Peptidoglycan synthesis during hyphal elongation in *Streptomyces antibioticus*. In *Bacterial Growth and Lysis*, M.A. de Pedro, J.V. Höltje, and W. Löffelhardt (eds.), pp. 189–196. Plenum Press, New York.

Miguélez, E.M., Martín, M.C., Manzanal, M.B., and Hardisson, C. (1988) Hyphal growth in *Streptomyces*. In *Biology of Actinomycetes '88*, Y. Okami, T. Beppu, and H. Ogawara (eds.), pp. 490–495. Japan Scientific Societies Press, Tokyo.

Miguélez, E.M., Martín, C., Manzanal, M.B., and Hardisson, C. (1992) Growth and morphogenesis in *Streptomyces. FEMS Microbiol. Lett.* **100**:351–360.

Miyake, K., Kuzuyama, T., Horinouchi, S., and Beppu, T. (1990) The A-factor-binding protein of *Streptomyces griseus* negatively controls streptomycin production and sporulation. *J. Bacteriol.* **172**:3003–3008.

Miyake, K., Yoshida, M., Chiba, N., et al. (1989) Detection and properties of the A-factor-binding protein from *Streptomyces griseus. J. Bacteriol.* **171**:4298–4302.

Miyashita, K., Fijii, T., and Sawada, Y. (1991) Molecular cloning and characterisation of chitinase genes from *Streptomyces lividans* 66. *Gene* **137**:2065–2072.

Molnár, I. (1994) Secretory production of homologous and heterologous proteins by recombinant *Streptomyces*: what has been accomplished? In *Recombinant Microbes for Industrial and Agricultural Applications*, Y. Murooka and T. Imanaka (eds.), pp. 81–104. Marcel Dekker, New York.

Ohnishi, K., Kutsukake, K., Suzuki, H., and Iino, T. (1992) A novel transcriptional regulation mechanism in the flagellar regulon of *Salmonella typhimurium*: an anti-sigma factor inhibits the activity of the flagellum-specific sigma factor, σ^F. *Mol. Microbiol.* **6**:3149–3157.

Ong, E., Kilburn, D.G., Miller, R.C., and Warren, R.A.J. (1994) *Streptomyces lividans* glycosylates the linker region of a β-1,4-glycanase from *Cellulomonas fimi*. *J. Bacteriol.* **176**:999–1008.

Pérez, C., Castro, C., Cruz, H., and Servín-González, L. (1994) Expression of the *Streptomyces exfoliatus* extracellular lipase gene requires a *bldA*-dependent transcriptional activator. In *Abstract Book, 7th International Symposium on the Genetics of Industrial Microorganisms*, Poster P111. National Research Council of Canada, Ottawa.

Perito, B., Hanhart, E., Irdani, T., Iqbal, M., McCarthy, A.J., and Mastromei, G. (1994) Characterization and sequence analysis of a *Streptomyces rochei* A2 endoglucanase-encoding gene. *Gene* **148**:119–124.

Plaskitt, K.A., and Chater, K.F. (1995) Influences of developmental genes on localised glycogen deposition in colonies of a mycelial prokaryote, *Streptomyces coelicolor* A(3)2: a possible interface between metabolism and morphogenesis. *Philos. Trans. R. Soc. Lond. Biol. Sci.* **347**:105–121.

Potúčková, L., Kelemen, G.H., Findlay, K.C., Lonetto, M.A., Buttner, M.J., and Kormanec, J. (1995) A new RNA polymerase sigma factor, σ^F, is required for the late stages of morphological differentiation in *Streptomyces* sp. *Mol. Microbiol.* **17**:37–48.

Redenbach, M., Kieser, H.M., Denapaite, D., Eichner, A., Cullum, J., Kinashi, H., and Hopwood, D.A. (1996) A set of ordered cosmids and a detailed genetic and physical map for the 8 Mb *Streptomyces coelicolor* A3(2) chromosome. *Mol. Microbiol.* **21**:77–96.

Robbins, P.W., Overbye, K., Albright, C., Benfield, B., and Pero, J. (1992) Cloning and high-level expression of chitinase-encoding gene of *Streptomyces plicatus*. *Gene* **111**:69–76.

Schauer, A., Ranes, M., Santamaría, R., et al. (1988) Visualizing gene expression in time and space in the morphologically complex filamentous bacterium *Streptomyces coelicolor*. *Science* **240**:768–772.

Servín-González, L., Jensen, M.R., White, J., and Bibb, M. (1994) Transcriptional regulation of the four promoters of the agarase gene (*dagA*) of *Streptomyces coelicolor* A3(2). *Microbiology* **140**:2555–2565.

Shi, W., and Zusman, D.R. (1993) Fatal attraction. *Nature* **366**:414–415.

Shiina, T., Tanaka, K., and Takahashi, H. (1991) Sequence of *hrdB*, an essential gene encoding sigma-like transcription factor of *Streptomyces coelicolor* A3(2): homology to principal sigma factors. *Gene* **107**:145–148.

Sohaskey, C.D., Im, H., Nelson, A.D., and Schauer, A.T. (1992) Tn*4556* and luciferase: synergistic tools for visualizing transcription in *Streptomyces*. *Gene* **115**:67–71.

Soliveri, J., Brown, K.L., Buttner, M.J., and Chater, K.F. (1992) Two promoters for the *whiB* sporulation gene of *Streptomyces coelicolor* A3(2), and their activities in relation to development. *J. Bacteriol.* **174**:6215–6220.

Solomon, A.D., Magnuson, R., Srivastava, A., and Grossman, A.D. (1995) Convergent sensing pathways mediate response to two extracellular competence factors in *Bacillus subtilis*. *Genes Dev.* **9**:547–558.

Strohl, W.R. (1992) Compilation and analysis of DNA sequences associated with apparent streptomycete promoters. *Nucleic Acids Res.* **5**:961–974.

Takano, E., and Bibb, M.J. (1994) The stringent response, ppGpp and antibiotic production in *Streptomyces coelicolor* A3(2). *Actinomycetologica* **8**:1–16.

Tan, H., and Chater, K.F. (1993) Two developmentally controlled promoters of *Streptomyces coelicolor* A3(2) that resemble the major class of motility-related promoters in other bacteria. *J. Bacteriol.* **175**:933–940.

Tanaka, K., Shiina, T., and Takahashi, H. (1988) Multiple principal sigma factor homologs in eubacteria: identification of the "rpoD box." *Science* **242**:1040–1042.

Tanaka, K., Shiina, T., and Takahashi, H. (1991) Nucleotide sequence of genes *hrdA*, *hrdC* and *hrdD* from *Streptomyces coelicolor* A3(2) having similarity to *rpoD* genes. *Mol. Gen. Genet.* **229**:234–240.

Tomura, T., Kishino, H., Doi, K., Hara, T., Kuhara, S., and Ogata, S. (1993) Sporulation-inhibitory gene in pock-forming plasmid pSA1–1 of *Streptomyces azureus*. *Biosci. Biotechnol. Biochem.* **57**:438–443.

Trigo, C., and Ball, A.S. (1994) Is the solubilized product from the degradation of lignocellulose by actinomycetes a precursor of humic substances? *Microbiology* **140**:3145–3152.

Ueda, K., Miyake, K., Horinouchi, S., and Beppu, T. (1993) A gene cluster involved in aerial mycelium formation in *Streptomyces griseus* encodes proteins similar to the response regulators of two-component regulatory systems and membrane translocators. *J. Bacteriol.* **175**:2006–2016.

Van Wezel, G.P., Buttner, M.J., Vijgenboom, E., Bosch, L., Hopwood, D.A., and Kieser, H.M. (1995) Mapping of genes involved in macromolecular biosynthesis on the chromosome of *Streptomyces coelicolor* A3(2). *J. Bacteriol.* **177**:473–476.

Virolle, M.-J., Long, C.M., Chang, S., and Bibb, M.J. (1988) Cloning, characterisation and regulation of an α-amylase gene from *Streptomyces venezuelae*. *Gene* **74**: 321–334.

Vivian, A. (1971) Genetic control of fertility in *Streptomyces coelicolor* A3(2): plasmid involvement in the interconversion of UF and IF strains. *J. Gen. Microbiol.* **69**: 353–364.

Vujaklija, D., Horinouchi, S., and Beppu, T. (1993) Detection of an A-factor-responsive protein that binds to the upstream activation sequence of *strR*, a regulatory gene for streptomycin biosynthesis in *Streptomyces griseus*. *J. Bacteriol.* **175**:2652–2661.

Welsch, M. (1962) Bacteriolytic enzymes from Streptomycetes. A review. *J. Gen. Physiol.* **45**:115–124.

Wessels, J.G.H. (1993) Wall growth, protein excretion and morphogenesis in fungi. *N. Phytol.* **123**:397–413.

Wildermuth, H. (1970) Development and organisation of the aerial mycelium in *Streptomyces coelicolor*. *J. Gen. Microbiol.* **60**:43–50.

Wildermuth, H., and Hopwood, D.A. (1970) Septation during sporulation in *Streptomyces coelicolor*. *J. Gen. Microbiol.* **60**:51–59.

Willey, J., Santamaría, R., Guijarro, J., Geistlich, M., and Losick, R. (1991) Extracellular complementation of a developmental mutation implicates a small sporulation protein in aerial mycelium formation by *Streptomyces coelicolor*. *Cell* **65**:641–650.

Willey, J., Schwedock, J., and Losick, R. (1993) Multiple extracellular signals govern the production of a morphogenetic protein involved in aerial mycelium formation by *Streptomyces coelicolor*. *Genes Dev.* **7**:895–903.

Williams, P. (1994) Compromising bacterial communications skills. *J. Pharmacol.* **46**: 1–10.

Williams, S.T., and Sharples, G.P. (1970) A comparative study of spore formation in two *Streptomyces* species. *Microbios* **5**:17–26.

Wilson, D.B. (1992) Biochemistry and genetics of actinomycete cellulases. *Crit. Rev. Biotechnol.* **12**:45–63.

Wright, F., and Bibb, M.J. (1992) Codon usage in the G+C-rich *Streptomyces* genome. *Gene* **113**:55–65.

Wu, L.J., and Errington, J. (1994) *Bacillus subtilis* SpoIIIE protein required for DNA segregation during asymmetric cell division. *Science* **264**:572–575.

Wu, L.J., Lewis, P.J., Allmansberger, R., Hauser, P.M., and Errington, J. (1995) A conjugation-like mechanism for prespore chromosome positioning during sporulation in *Bacillus subtilis*. *Genes Dev.* **9**:1316–1326.

Yamada, Y., Sugamura, K., Kondo, K., Yanagimoto, M., and Okada, H. (1987) The structure of inducing factors for virginiamycin production in *Streptomyces virginiae*. *J. Antibiot.* **40**:496–504.

7

Proteus mirabilis and Other Swarming Bacteria

ROBERT BELAS

The writer Victor Hugo wrote, "Where the telescope ends, the microscope begins. Which of the two has the grander view?" It is perhaps because of that "grand view" that many young students are drawn into the field of microbiology. Indeed, much has been learned by observing bacteria as they move in and through their environment. Often one of the "views" that so captivates the imagination is the multicellular behavioral phenomenon referred to as swarming, a form of bacterial translocation that has held the interest of scientists for more than a century. Unlike swimming behavior, which occurs in liquid environments, swarming is the organized movement over a surface, dependent on a specialized cell with unique characteristics that provide an adaptive advantage for living on such surfaces (Henrichsen, 1972). Swarming motility has been demonstrated in many bacterial genera, both gram-negative and gram-positive. It is the objective of this chapter to describe the current state of knowledge regarding swarming behavior, both as a form of bacterial differentiation and in terms of a prokaryotic multicellular behavior. To do so, *Proteus mirabilis* is used as the model swarming cell, as it exemplifies many of the features common to the various swarming bacteria. Although I am biased in favor of the *P. mirabilis* model system, I do emphasize the strengths of the other model of swarming behavior and highlight the differences between these bacteria.

Swarming Behavior

Swarming Life of Proteus mirabilis

Proteus mirabilis is a motile gram-negative bacterium, similar in many aspects of its physiology to other members of the family Enterobacteriaceae, such as

Escherichia coli and *Salmonella typhimurium*. It was originally described and named by Hauser in 1885 for the character in Homer's *Odyssey*, who "has the power of assuming different shapes in order to escape being questioned" (Hoeniger, 1964). *P. mirabilis* is considered an opportunistic pathogen and one of the principal causes of urinary infections in hospital patients with indwelling urinary catheters (Story, 1954; Senior, 1983; Mobley and Warren, 1987). It is believed that the ability of *P. mirabilis* to colonize the surfaces of catheters and the urinary tract may be aided by the characteristic first described more than a century ago and currently referred to as swarmer cell differentiation.

When grown in suitable liquid medium *P. mirabilis* exists as a 1.5- to 2.0-µm motile cell with 6–10 peritrichous flagella. These bacteria, called swimmer cells, display characteristic swimming and chemotactic behavior, moving toward nutrients and away from repellents (Adler, 1983). However, a dramatic change in cell morphology takes place when cells grown in liquid are transferred to a nutrient medium solidified with agar. Shortly after encountering an agar surface, the cells begin to elongate (Hughes, 1956; Hoeniger, 1964, 1965, 1966; Hoeniger and Cinits, 1969). This change is the first step in the production of a morphologically and biochemically differentiated cell, referred to as the swarmer cell. The process of elongation, which takes place with only a slight increase in cell width, is due to inhibition in the normal septation mechanism (Belas et al., 1995). Elongation of the swarmer cell can give rise to cells 60–80 µm in length. During this process, DNA replication proceeds without significant change in rate compared to that in the swimmer cell (Gmeiner et al., 1985). Not surprisingly, the rate of synthesis of certain proteins (i.e., flagellin) is altered markedly in the swarmer cell (Armitage and Smith, 1978; Armitage et al., 1979; Armitage, 1981; Hoffman and Falkinham, 1981; Falkinham and Hoffman, 1984; Jin and Murray, 1988). The result of this process is a long, nonseptate, polyploid cell, with 10^3–10^4 flagella per cell (Fig. 7.1A). The number of chromosomes in the swarmer cell is roughly proportional to the increase in length, such that a 40 µm swarmer cell has about 20 chromosomes (Belas, 1992). Eventually septation and division do take place at the ends of the long swarmer cells, producing a microcolony of differentiated cells. Table 7.1 compares the physical features of swimmer and swarmer cells.

Concurrent with cellular elongation, changes take place in the rate of synthesis of flagella on the swarmer cell (Fig. 7.2). Whereas swimmer cells have only a few flagella, the elongated swarmer cells are profusely covered by hundreds to thousands of new flagella synthesized specifically as a consequence of growth on the surface (Houwink and van Iterson, 1950; Hoeniger, 1965, 1966; Armitage and Smith, 1978). The term "flagellin factories" was first used by Hoeniger (1965) to describe the marked synthesis of new flagella (composed of the protein flagellin) in swarmer cell differentiation. The newly synthesized surface-induced flagella are composed of the same flagellin subunit as the swimmer cell flagella, indicating that the same flagellar species is overproduced upon surface induction. The result of the surface-induced differentiation process is a swarmer cell, which differs from the swimmer cells by having the unique ability to move over solid media in a translocation process referred to as swarm-

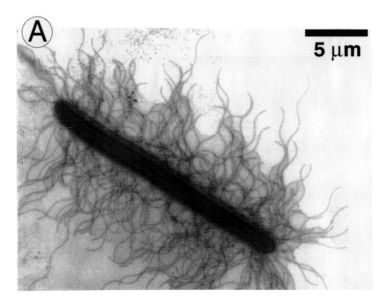

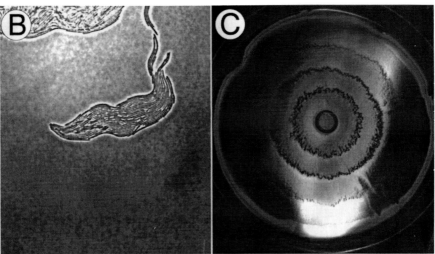

FIGURE 7.1 *P. mirabilis* swarming colony formation and swarmer cell morphology. (**A**) Fully differentiated swarmer cell of a negative-stained specimen taken from the swarming periphery as seen by electron microscopy. The cell shown is approximately 25 μm in length. (**B**) Movement of a mass of cells from the swarming periphery onto uncolonized agar. The swarming periphery is shown at bottom right. (**C**) Characteristic bulls-eye colony of latticed zones of bacteria produced by the cyclic events of differentiation, movement, cessation of movement, and dedifferentiation.

TABLE 7.1 Physical Features of Swimmer and Swarmer Cells of Swarming Bacteria

Characteristic	Swimmer Cells	Swarmer Cells	Reference
Flagella			
No. per cell	1–10	500–5000	Hoeniger (1965)
Length (μm)	0.75	5.25	Hoeniger (1965)
No. per unit cell length	3	150	Hoeniger (1965)
Motility	Tumbling	Smooth	Dick et al. (1985); Hoeniger (1965)
Cell dimensions (μm)	0.7 × 1–2	0.7 × 10 to >80	Hoeniger (1965)
Chromosome no.	1–2	Multiple; correlated with cell length	Hoeniger (1966)

ing (Henrichsen, 1972). However, individual swarmer cells by themselves do not have the ability to swarm (Brogan et al., 1971; Bisset, 1973a,b; Bisset and Douglas, 1976; Douglas and Bisset, 1976; Douglas, 1979). As seen in Figure 7.1B, swarming is the result of a coordinated, multicellular effort of groups of differentiated swarmer cells (Dienes, 1946; Bisset, 1973a,b). The process be-

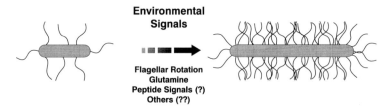

Swimmer Cell	Characteristic	Swarmer Cell
1.5 to 2.0 μm	*Length*	10 to >80 μm
4 to 10	*Flagella*	10^3 to 10^4
1 to 2	*Genomes*	Polyploid
Swimming & Chemotaxis	*Motile Behavior*	Swarming, Chemotaxis & Coordinated Cell-to-Cell Communication

FIGURE 7.2   Characteristics of *P. mirabilis* swarmer cell differentiation and swarming motility. Swarmer cell differentiation is controlled through the combined sensing of environmental conditions that reduce wild-type flagellar filament rotation and through sensing a specific chemical stimulus, the amino acid glutamine. Other signals may play a role in this process (e.g., peptide signals and perhaps other compounds). The differentiated swarmer cell is an elongated, polyploid cell that synthesizes numerous flagella in response to the aforementioned signals.

gins when a group of differentiated swarmer cells move outward as a mass, and it continues until the swarming mass of bacteria is reduced in number owing to loss of constituent cells, which fall behind on the surface, or when the mass reverses direction.

Swarming of *P. mirabilis* is cyclic in nature (Fig. 7.3). Once swarmer cells have fully differentiated, the swarming colony moves outward in unison from all points along the periphery for a period of several hours and then stops (Bisset, 1973a,b; Bisset and Douglas, 1976; Douglas and Bisset, 1976; Douglas, 1979). This cessation of movement is accompanied by a dedifferentiation of the swarmer cell back to swimmer cell morphology, a process referred to as consolidation (Hoeniger, 1964, 1965, 1966; Hoeniger and Cinits, 1969; Bisset, 1973a,b; Williams and Schwarzhoff, 1978). The cycle of swarming and consolidation is then repeated several times until the agar surface is covered by concentric rings formed by the swarming mass of bacteria (Douglas and Bisset, 1976). This cycle of events gives rise to the characteristic bulls-eye appearance of *P. mirabilis* colonies (Fig. 7.1C). The swarmer cell requires contact at all times with the surface to maintain the differentiated state. When removed from the surface of an agar plate and suspended in liquid medium, cells quickly begin to septate and divide into short cells, and the synthesis of flagella returns

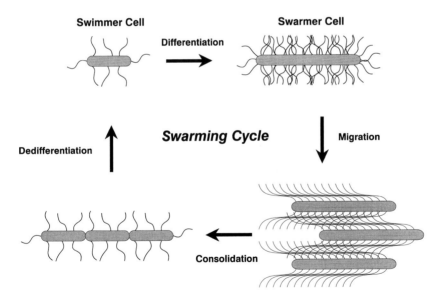

FIGURE 7.3 Temporal cycling of *P. mirabilis* swarmer cell differentiation. Swarmer cell differentiation and swarming motility are cyclic in nature. Swimmer cells differentiate into swarmer cells upon sensing conditions that induce the expression of swarmer cell-specific genes. Once fully differentiated, the swarmer cells then move en masse and migrate away from the initial point of inoculation. Such migration is punctuated by cyclic events of dedifferentiation (referred to as consolidation), whereupon migration stops and the cells revert to a morphology similar to that of the swimmer cell. This process of differentiation and dedifferentiation then conditions until the agar plate is covered by the bacterial mass.

to the level observed in swimmer cells (Jeffries and Rogers, 1968). Thus the differentiation process is reversible as a result of the consolidation process and as a consequence of removing the inducing stimulus from the surface (Fig. 7.4).

Swarming Life of Serratia and Vibrio

Strains of *Serratia marcescens* (Alberti and Harshey, 1990) and of *Vibrio alginolyticus* and *Vibrio parahaemolyticus* (Shinoda et al., 1970; Miwatani and Shinoda, 1971; de Boer et al., 1975; Ulitzur, 1975a,b; Belas et al., 1986; McCarter et al., 1988) have also been reported to swarm as a result of a surface-induced differentiation from swimmer to swarmer cell. The regulation of swarmer cell differentiation was first understood in *V. parahaemolyticus*, which is an estuarine gram-negative bacterium that is often found associated with chitinaceous shells of crustaceans (Kaneko and Colwell, 1973). It is motile by means of a single polar flagellum, which is sheathed by a continuation of the outer membrane (Shinoda et al., 1970). The constitutively synthesized polar flagellum rotates like the flagella in other bacteria (McCarter et al., 1988; McCarter and Silverman, 1990; Atsumi et al., 1992). The polar flagellum of *V. parahaemolyticus* also serves in a sensory capacity by monitoring external conditions that cause the inhibition of flagellar rotation (Belas et al., 1986; McCarter et al., 1988; McCarter and Silverman, 1990). When such conditions are encountered, inhibition of the rotation of the polar flagellum sends a signal into the cell, triggering a chain of events that ultimately produce a differentiated swarmer cell. The differentiated cell swarms by means of hundreds of newly synthesized surface-induced lateral flagella. These flagella, which are not sheathed, are composed of a protein subunit distinct from that of the polar flagellum (McCarter et al., 1988). Whereas the polar flagellum is required for swimming motility, only the force of the lateral flagella is required for movement across surfaces (Sar et al., 1990). Patterns of cyclic consolidation are not manifested in the swarming colony of *V. parahaemolyticus*, although dedifferentiation does occur, as cells in the center of the swarming colony do not have lateral flagella, nor are they elongated (McCarter et al., 1988). In addition to the control exerted by monitoring the inhibition of polar flagellum rotation, a secondary signal is evidently sensed when iron becomes limiting (McCarter and Silverman, 1989). The combination of conditions that inhibit the rotation of the polar flagellum and limit iron concentration are essential to begin transcription of the *laf* genes (McCarter et al., 1988).

Determinants of Swarming

The advent of the techniques of modern molecular biology has brought forth a renewed interest in understanding *P. mirabilis* swarmer cell differentiation as it affects the invasiveness and pathogenicity of this organism. My laboratory has chosen to employ Tn5 transposon mutagenesis as a means of constructing mutations in the swarmer cell genes (De Lorenzo et al., 1990; Belas et al.,

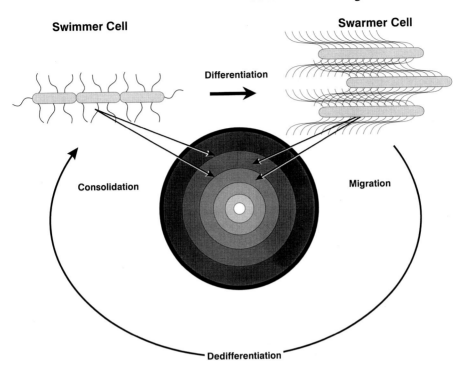

Swimmer Cell

Differentiation

Swarmer Cell

Consolidation

Migration

Dedifferentiation

FIGURE 7.4 Spatial cycling of *P. mirabilis* swarmer cell differentiation. On agar medium swarmer cell differentiation is observed as a series of ever-expanding concentric rings (see Figure 7.1A). These rings are composed of morphologically and biochemically distinct cell types. During swarming the leading edge of the swarming zone is composed of completely differentiated swarmer cells. At consolidation swarming motility stops, and the cells revert to a swimmer cell morphology. This cycle is then repeated as described in Figure 7.3.

1991a,b). A similar methodology has been employed in other laboratories (Allison and Hughes, 1991b). We found that Tn5 transposition is random; and once integrated into the *P. mirabilis* chromosome, the transposon is stably maintained (Belas et al., 1991b). Characterization of the subsequent swarming defective mutants (Swr⁻) has revealed several phenotypic classes including mutants that are defective in (1) swimming motility (Swm⁻, including defects in flagellin synthesis, Fla⁻; motor rotation, Mot⁻; and chemotaxis, Che⁻); (2) swarmer cell elongation [both null mutants (Elo⁻) and constitutively elongated cells (Eloᶜ)]; and (3) a large group of mutants with an impaired swarming (Swrᶜʳ) caused by an unknown defect (Belas et al., 1991a). Although this last group does manifest (nonwild-type) swarming motility, many of these mutants are altered in their ability to form discrete consolidation zones. It has been speculated that these mutants have arisen from transposon insertions in genes coding for production of extracellular signal molecules or the receptors of those signals (Belas, 1992). Localization of the various Tn5 insertions on the chromosomes of swarming mutants using pulse gel electrophoresis has revealed that the genes necessary for swarmer cell differentiation and swarming behavior occur on closely linked genetic loci (Allison and Hughes, 1991b).

The genetic lesions that result in the various mutant phenotypes are now being characterized and the defective genes analyzed for regulation and expression during swarmer cell differentiation and multicellular motility and behavior. Swarmer cell differentiation and swarming motility of *P. mirabilis* are the result of at least four phenomena: (1) the ability to sense cues from the environment; (2) production of an elongated swarmer cell; (3) synthesis of vastly increased amounts of flagellin; and (4) coordinated multicellular interactions resulting in cyclic waves of cellular differentiation and dedifferentiation (Belas, 1992). A defect in any of these events results in either abnormal swarming behavior or the complete lack of differentiation or motility (or both). The following section deals with the genetic aspects of the first three phenomena. The section Multicellular Migration then focuses on what is known concerning the fourth aspect of swarming.

Surface Signal

Flagellar Rotational Tethering

Studies measuring gene expression in *P. mirabilis*, *S. marcescens*, and *V. parahaemolyticus* have convincingly demonstrated that the central environmental stimulus sensed by the cell is physical in nature (Belas et al., 1986, 1995; McCarter et al., 1988; Alberti and Harshey, 1990; Allison et al., 1993). Swarmer cell genes are induced when swimmer cells are transferred to the surface of solidified media, suspended in media of high viscosity, or tethered by antibody specific to the bacterial cell surface. Growth in media amended with polymers such as Ficoll 400 or polyvinylpyrrolidone 360 is as effective for inducing swarmer gene expression as growth on an agar surface (Belas et al., 1986, 1995; McCarter et al., 1988; Alberti and Harshey, 1990; Allison et al., 1993). These branched polymers increase the microviscosity of the medium, thereby creating a matrix that interferes with the movement of swimming bacteria. An even more effective inducing condition is the addition of antibody (but not preimmune serum) to the growth medium, which results in cell tethering and agglutination (McCarter et al., 1988). McCarter et al. (1988) employed an antibody raised against the cell surface of *V. parahaemolyticus* that recognized primarily the polar flagellum sheath. Agglutination with this antibody produced abnormal expression of swarmer cell differentiation in *V. parahaemolyticus*. What these conditions have in common is that they inhibit normal rotation of the flagellar filament(s). This observation has led to the hypothesis that the flagellum functions as a tactile sensor of external conditions and directly transfers that information into the cell where the signal is transduced to transcriptionally control expression of the genes associated with swarmer cell differentiation.

Genetic techniques have strengthened the hypothesis that the rotation of the flagellar filament is the main factor controlling expression of swarmer cell differentiation. McCarter and Silverman (1990) showed that genetic defects that impair the structure of the polar flagellum of *V. parahaemolyticus* affect the regulation of swarmer cell genes. The composition of the *V. parahaemolyticus* polar filament is complex, being encoded by separate genes: *flaA*, *flaB*, *flaC*, and *flaD*. The genes are organized in two loci, with *flaA-flaB* and *flaC-flaD* being physically paired, respectively. Mutants with defects in *flaA*, *flaB*, or both are indistinguishable from the wild-type cell with respect to swimming motility. Thus FlaA and FlaB flagellins are not critical to polar flagellar function. In contrast, all mutants with lesions in *flaC* (i.e., *flaC*, Δ *flaAB flaC*, and Δ *flaAB* Δ*flaCD*) have a defective polar flagellum. All strains with *flaC* mutations move poorly in semisolid motility medium, and by light microscopy they are seen to swim aberrantly with a slow, wobbly movement. Electron microscopic examination of strains with *flaC* defects grown in liquid medium showed ''hairy'' bacteria with many unsheathed flagella. Flagellar preparations confirmed that the unsheathed flagella were lateral flagella. The lateral filaments are polymerized from a single protein, LafA, which is of a different molecular size from the polar flagellins. Thus the FlaC flagellin is important for filament structure, and regulation of *laf* genes is released from surface dependence in mutants with *flaC* defects.

Genetic analyses of the flagellin-encoding genes of *P. mirabilis* has also demonstrated a complex genetic system that is unlike that of *V. parahaemolyticus*. Swimmer cells and swarmer cells of *P. mirabilis* produce the same flagellin protein encoded by the same gene. Belas and Flaherty (1994) showed that there are three copies of flagellin-encoding genes in *P. mirabilis* (Belas, 1994; Belas and Flaherty, 1994). Only one of these flagellin-encoding genes, *flaA*, is expressed in wild-type cells, whether in the swimmer or swarmer cell phase (Belas and Flaherty, 1994). Mutations that disrupt the function of *flaA* result in abnormal expression of swarmer cells, demonstrating the central role of the flagellar filament in signal transduction (Belas and Flaherty, 1994). Adding to this information, defects in other essential flagellar genes of *P. mirabilis* also result in abnormal swarmer cell differentiation (see below). Harshey described a type of swarmer cell differentiation that occurs in *E. coli* and *S. typhimurium* (Harshey and Matsuyama, 1994). In these studies, mutations that affected flagellar filament function also produced abnormal expression of the swarmer cell phenotype, demonstrating the universal nature of sensing via monitoring flagellar rotation to produce the differentiated swarmer cell.

Glutamine

Many attempts have been made to determine if chemotactic behavior plays a role in swarmer cell differentiation and swarming behavior. Evidence both supporting and discounting chemotaxis in swarming behavior has been presented over the years; but the most recent evidence, mostly acquired through careful analysis of swarming mutants that have clear defects in genes encoding protein components of the chemotaxis mechanism, has demonstrated the importance of chemotaxis in swarming motility and differentiation. (A complete accounting of chemotaxis as it relates to swarmer cell behavior is given elsewhere in this chapter.)

Although chemotaxis is now considered essential for swarmer cell differentiation and motility in all of the swarming bacteria thus far studied, only for *P. mirabilis* has a specific interaction been described linking the amino acid glutamine and swarmer cell differentiation (Allison et al., 1993). By supplementing minimal agar medium unable to support swarming migration, Allison et al. (1993) identified a single amino acid, glutamine, as sufficient to signal initiation of cell differentiation and migration. In contrast, addition of the other 19 common amino acids (excluding glutamine) individually or in combination did not initiate differentiation even after prolonged incubation. In liquid minimal medium amended with polyvinylpyrrolidone to increase viscosity and inhibit flagellar rotation (as described above), addition of glutamine also induces swarmer cell differentiation. Furthermore, the induction can be completely inhibited by glutamine analogs, indicating the specificity of the glutamine signal in swarmer cell differentiation. Interestingly, glutamine is chemotactic only to the differentiated swarmer cell, not to the swimmer cell (Allison et al., 1993). These data suggest that glutamine functions in a dual role: initiating differentiation and directing the migration of swarming cells.

This study indicates that *P. mirabilis* swarmer cells are chemotactic but respond to a more limited range of attractants than swimmer cells. Moreover, the discovery that amino acid chemoattractants that are mutually exclusive for either swimming (e.g., glutamate) or swarming (glutamine) cells may indicate the presence of separate sensory components coupled with the two forms of motility. These results are tantalizing and suggest that a combination of signals, including inhibition of flagellar rotation and the presence of glutamine, are necessary for the induction of *P. mirabilis* swarmer cell differentiation.

Iron Limitation in *Vibrio parahemolyticus*

Swarmer cell differentiation of *V. parahaemolyticus* is controlled by a second signal: limitation for iron in the growth medium (McCarter and Silverman, 1989, 1990). Mutants with defects in *flaC* (one of four genes encoding polar flagellin and required for polar flagellum synthesis) produce lateral flagella when grown in 2216 marine broth but failed to synthesize lateral flagella when grown in heart infusion broth. One pertinent difference in the composition of these two media is the availability of iron (McCarter and Silverman, 1989, 1990). Addition of iron chelators to heart infusion broth results in the synthesis of lateral flagella by strains with *flaC* mutations. Furthermore, growth of the wild-type strain in viscous, iron-rich broth did not result in production of lateral flagella, whereas growth of the wild-type strain in viscous, iron-poor broth did elicit lateral flagella synthesis. Therefore expression of *V. parahaemolyticus* swarmer cell genes requires both iron limitation and interference with polar flagellar rotation. Iron deficiency alone is not sufficient to induce swarmer cell differentiation, as lateral flagella synthesis is not induced when the wild type is grown in iron-limited broth (McCarter and Silverman, 1989, 1990). Regulation by the availability of iron affects transcription of *laf* (lateral flagellin) genes. Direct measurements of mRNA encoding lateral flagellin subunit (i.e., *lafA* mRNA) and light production of *laf*::*lux* transcriptional fusion strains revealed that transcription of *laf* genes requires perturbation of polar flagellar function and iron-limiting growth conditions (McCarter and Silverman, 1989, 1990). The effect of iron limitation on the regulation of swarmer cell differentiation has not been described for *P. mirabilis* or *S. marcescens* 274, but it would not come as a surprise if either or both of these bacteria have specialized requirements for iron that may tie into the regulation of swarmer cell differentiation.

Genes and Proteins Involved in Differentiation

flaA Locus of *Proteus mirabilis*

During differentiation from swimmer to swarmer cell, the rate of synthesis of certain proteins undergoes a dramatic increase. The most evident of these overproduced proteins is flagellin (FlaA, the subunit of the flagellar filament). Although swimmer cells have only a few flagella, the elongated swarmer cells are profusely covered by thousands of newly synthesized flagella (Hoeniger, 1965). In many ways the increase in flagellin expression is the hallmark of the differentiated swarmer cell, as has been shown not just for *P. mirabilis* but also for *S. marcescens* (Alberti and Harshey, 1990; O'Rear et al., 1992) and *V. parahaemolyticus* (Belas et al., 1986; McCarter et al., 1988; McCarter and Silverman, 1990).

Biochemical studies of flagella isolated from *P. mirabilis* swimmer and swarmer cells (Bahrani et al., 1991; Belas et al., 1991a), along with genetic data from Tn5 mutageneses (Belas et al., 1991a,b), suggest that *P. mirabilis* synthesizes only a single flagellin species. Belas and Flaherty reported on the cloning of a region of *P. mirabilis* chromosomal DNA capable of complementing *E. coli* FliC⁻ mutants (Belas and Flaherty, 1994). Nucleotide and deduced amino acid sequence analyses revealed that this region contains three open reading frames (ORFs) that were identified based on their homology to other known flagellar genes. The region included the 5'-end of a putative homolog of *E. coli fliD* (an essential flagellar gene responsible for filament assembly) and two nearly identical copies of flagellin-encoding genes called *flaA* and *flaB*. (A third copy of a possible flagellin-encoding gene, referred to as *flaC*, was also found in this study by DNA:DNA homology.) The *flaA* gene encodes a protein that complements *E. coli* FliC⁻ mutants, but *flaB* fails to complement any *E. coli* defects (Belas and Flaherty, 1994). Moreover, *flaB* encodes a protein that is larger than predicted from its deduced amino acid sequence when the gene was used in *E. coli* minicell protein programming experiments. These data suggest that although *flaA* is functional in *E. coli flaB* is not.

Three methods have been used to determine the transcriptional regulation of *flaA* and *flaB*: plasmid-borne transcriptional fusions, Northern blot DNA: RNA hybridization, and primer extension experiments (Belas, 1994). The results from each confirm that *flaA* is the sole flagellin gene transcribed by wild-type cells. Furthermore, *flaA* is transcribed as a monocistronic message and regulated coordinately with swarmer cell differentiation such that *flaA* transcription increases about eightfold during induction. The transcriptional analyses support the data from previous work using Tn5 mutants, which demonstrated that swimmer and swarmer cell flagellin synthesis was interconnected and suggested that only one flagellin gene was involved in the synthesis of both organelles (Belas et al., 1991a).

The function of the gene product of *flaD*, *flaA*, and *flaB* has been established by constructing mutations in *P. mirabilis* (Table 7.2). Mutations in *flaA* completely abolished all motility as well as swarmer cell differentiation, whereas *flaB* mutations have no demonstrable change in wild-type phenotype. These observations emphasize the central role of FlaA flagellin in swarmer cell differentiation and behavior. Presumably, the cell responds to environmental conditions that prevent normal rotation of the FlaA flagella (Allison and Hughes, 1991a; Belas et al., 1991a; Belas, 1992). Loss of FlaA filaments resulted inappropriately in an undifferentiated swimmer cell under conditions that should elicit swarmer cell differentiation. This finding suggests that *flaA* is required for the induction of swarmer cell differentiation because mutants defective in FlaA do not differentiate. This situation is contrary to what is observed during swarmer cell differentiation of *V. parahaemolyticus* (McCarter et al., 1988; Sar et al., 1990). In this case, defects causing loss of the polar flagellum (the sensing flagellum) result in a constitutively synthesized swarmer cell. The difference between these two mutants implies that the regulatory mechanisms operating to control swarmer cell differentiation of *P. mirabilis* are probably different from those of *V. parahaemolyticus*, even though the initial signal (inhibition of flagellar rotation) triggering differentiation is the same for both species.

As indicated, FlaA$^-$ mutants do not synthesize any flagellin species. This observation may be used as an indication that the third copy of *P. mirabilis* flagellin-encoding genes, *flaC*, is not actively transcribed. Although the nucleotide sequence of *flaC* has not been determined nor the transcription of this gene assessed, the evidence obtained from FlaA$^-$ mutants argues that *flaC* is not transcribed in a *flaA* mutation or wild-type cell; or if it is transcribed, the protein is not a functional flagellin. Therefore both *flaB* and *flaC* are apparently silent copies of flagellin-encoding genes.

Interestingly, at a low frequency, Mot$^+$ revertants were found to emanate as flares from *flaA* colonies on semisolid Mot agar plates. Moreover, *flaA* Mot$^+$ revertants are capable of movement through media amended with anti-FlaA polyvalent antisera. This observation in and of itself suggests that the flagella synthesized by the revertants are antigenically distinct from those produced by wild-type cells. Because the antiserum used is polyvalent and produced in response to whole flagella rather than denatured flagellin (Belas et al., 1991a), it may be anticipated that it contains antibodies capable of binding to many flagellin epitopes. During such binding, bacteria may be tethered together flagellum-to-flagellum, thereby preventing them from achieving wild-type swimming

TABLE 7.2 Genetic Loci Associated with *P. mirabilis* Swarming Differentiation, Motility, and Behavior

Genetic Locus or Gene	Type of Mutation	Mutant Phenotype	Reference(s)
flaA	Allelic exchange	Swm⁻, Swr⁻, Elo⁻	Belas (1994); Belas & Flaherty (1994)
flaB	Allelic exchange	Wild type	Belas (1994); Belas & Flaherty (1994)
flaD	Allelic exchange	Swm⁻, Swr⁻, Elo⁻	Belas (1994); Belas & Flaherty (1994)
flhA	????	Swm⁻, Swr⁻, Elo⁻	Gygi et al. (1995a)
gidA	Tn5 insertion	Swm⁺, Swrᶜʳ, Elo⁻	Belas et al. (1995)
cld locus	Tn5 insertion	Swm⁺, Swrᶜʳ, Elo⁻	Belas et al. (1995)
rfaCD	Tn5 insertion	Swm⁺, Swrᶜʳ, Elo⁻	Belas et al. (1995)
fliG	Tn5 insertion	Swm⁺, Swrᶜʳ, Elo⁻	Belas et al. (1995)
fliL	Tn5 insertion	Swm⁺, Swrᶜʳ, Eloᶜ	Belas et al. (1995)
flgH	Tn5 insertion	Swm⁺, Swrᶜʳ, Eloᶜ	Belas et al. (1995)
galU	Tn5 insertion	Swm⁺, Swrᶜʳ, Elo⁻	Belas et al. (1995)
dapE	Tn5 insertion	Swm⁺, Swrᶜʳ, Elo⁻	Belas et al. (1995)
cpsF	???	Swm⁺, Swrᶜʳ	Hughes (1994)[a]
pepQ	Tn5 insertion	Swm⁺, Swrᶜʳ, Elo⁻	Belas et al. (1995)
Urease locus	Northern blot	NA	Allison et al. (1992b)
Hemolysin locus	Northern blot	NA	Allison et al. (1992b)

NA, not available.
[a] Personal communication.

or swarming motility. This situation is exactly what happens when anti-FlaA is added to cultures of wild-type *P. mirabilis*: The bacteria become tethered by flagellum-to-flagellum binding (R. Belas, unpublished data). In contrast, *flaA* Mot⁺ revertants are only loosely bound when the same antiserum is applied to cell suspensions, indicating that *flaA* Mot⁺ revertants synthesize flagella containing only a subset of possible FlaA epitopes.

Analyses of Southern blots from *flaA* Mot⁺ revertants revealed that a large deletion of variable size within the *flaAB* locus apparently caused the reversion from Mot⁻ to Mot⁺. For example, for one such Mot⁺ revertant, the deletion removes more than 50% of *flaA* and all of *flaB*. This observation suggests that the reversion process is a one-way event in which a deletion occurs permanently, removing a portion of DNA downstream from *flaA* and spanning into *flaB* and farther into genes downstream from *flaB*. Thus the mutation may result in a hybrid gene fusion where the 5′-end of *flaA* is fused to the 3′-end of *flaC*, yielding a functional flagellin as a consequence.

The evidence gathered from sodium dodecyl sulfate polyacrylamide gel electrophoresis (SDS-PAGE), immunoblotting of V8 protease-digested flagellin, N-terminal amino acid sequencing, and amino acid composition analyses strongly suggests that FlaA and the revertant flagellin have identical N-terminal ends but different C-termini. The hypothetical model used to explain the events producing the revertant flagellin predicts that any splice between the 5'-end of *flaA* and either *flaB* or *flaC* that produces a functional flagellum results in a motile revertant phenotype. Based on this model, the splice site joining FlaA with FlaC is thought to be located at or near *flaA*.

The discovery of multiple flagellin genes of *P. mirabilis* (Belas and Flaherty, 1994) has raised provocative questions concerning their regulation and function in association with swarmer cell differentiation. These issues gain further significance as the overproduction of flagella during swarmer cell differentiation is considered to be an important prerequisite for the colonization, invasion, and ultimate pathogenicity of this organism (Allison et al., 1992a,b). Many bacterial species have been observed to possess multiple copies of flagellin genes. For example, *Salmonella typhimurium fliC* and *fljB* (Zieg et al., 1977; Scott and Simon, 1982; Szekely and Simon, 1983) and *Campylobacter coli flaA* and *flaB* (Guerry et al., 1988, 1991; Alm et al., 1992) each encode two flagellin species that play a role in antigenic variation of the flagellum. Because flagella and specifically flagellin (H-antigen) are highly antigenic, changes in flagellin antigenicity may provide the bacteria with an effective means of side-stepping the immune response of the host (Brunham et al., 1993).

An important question is whether in vitro flagellin antigenic variation such as described here can occur in situ (e.g., during colonization of the urinary tract). The inoculation of Mot⁻ bacterial cells in semisolid media or on the surface of ''hard'' agar places a strong selection for active motility (Quadling and Stocker, 1957; Silverman and Simon, 1972, 1973, 1974). It is due to the reduction of nutrients around the cells combined with the accumulation of waste products at the point of inoculation. Spontaneous mutations, occurring at random points around the chromosome, could randomly cause the deletion and genetic fusion seen in the *flaA* Mot⁺ revertants within the large population of Mot⁻ cells. Such spontaneous Mot⁺ mutants would have a significant selective advantage compared to the bulk of nonmotile cells owing to their enhanced survival by being able to translocate actively to areas of greater nutrients and lower waste products. Such spontaneous Mot⁺ mutants would also be easy to detect by a human observer owing to the evident flare of motility they produce.

Such spontaneous mutations giving rise to an antigenically distinct flagellin would be equally likely to occur in vitro and in vivo and may well occur in FlaA⁺ as in Mot⁻ cells. Such antigenic changes would increase the survivability of a urinary pathogen. For example, urinary pathogens are confronted by secretory immunoglobulin A (IgA) as the bacteria attempt to colonize the bladder. Because flagellin is strongly antigenic, a major binding epitope for the immunoglobulins could be the flagellum. Tethering the bacteria via their flagellum would effectively prevent their motility and impose the same selective pressures in this case as those that occur in vitro. A spontaneous deletion that gave rise to an antigenically distinct flagellum would thus provide the bacteria with an avenue of escape from immobilization by the immunoglobulins. It would be a successful survival strategy, prolonging the chances of colonization by the bacteria.

Other Flagellar Genes

Inhibition of flagellar rotation is known to cause induction of swarmer cell differentiation (Belas et al., 1986; McCarter et al., 1988). It is therefore not surprising to find that mutations in flagellar genes (other than *flaA*) result in abnormal swarmer cell differentiation. For example, Belas et al. (1995) characterized the Tn5 insertion sites in *P. mirabilis* mutants defective in swarmer cell elongation (Belas et al., 1995) (Table 7.2). Most of the mutants characterized in this study were in genes associated with flagellar biosynthesis. For example, Tn5 insertions were identified in (1) *fliL* (the homolog of which in *Caulobacter crescentus* is required for flagellar gene expression and normal cell division) (Stephens and Shapiro, 1993); (2) *fliG* (a component of the flagellar switch); and (3) *flgH* (encoding the basal-body L ring); and they have been shown to play a significant role in swarmer cell elongation (Belas et al., 1995). This finding suggests that subtle effects to the flagellar motor may cause major perturbations in swarming behavior.

Similarly, Gygi et al. (1995a) analyzed *P. mirabilis* mutants defective in *flhA* (Table 7.2). *flhA*, a gene necessary for flagellar biosynthesis (Macnab, 1992), is a member of a family of putative signal-transducing receptors that have been implicated in diverse cellular processes (Carpenter and Ordal, 1993). Other members of this family include LcrD of *Yersinia pestis* and *Yersinia enterocolitica*, FlbF of *Campylobacter crescentus*, FlhA of *Bacillus subtilis* and *E. coli*, MxiA and VirH of *Shigella flexneri*, InvA of *Salmonella typhimurium*, HrpC2 of *Xanthomonas campestris*, and Hrp of phytopathogenic bacteria (Vogler et al., 1991; Galan et al. 1992; Carpenter and Ordal, 1993; Dreyfus et al., 1993; Gough et al., 1993; Wei and Beer, 1993; Lidell and Hutcheson, 1994; Miller et al., 1994).

As with *flaA*, *flaB*, and *flaD*, *P. mirabilis flhA* has a flagellar-gene-specific σ^{28} promoter (Gygi et al., 1995a). Mutations affecting the function of FlhA result in a lack of flagellar synthesis and the abnormal expression of swarmer cell differentiation (Gygi et al., 1995a). Thus in general mutations that prevent the normal function or regulation of flagellar expression in *P. mirabilis* result in the abnormal expression of swarmer cells and point to the central role of flagellar function and filament rotation in maintaining proper sensing of environmental stimuli during swarmer cell differentiation and motile behavior.

Lipopolysaccharide and Capsular Polysaccharide Genes

Another class of genes found to be important in swarmer cell differentiation and swarming behavior are involved in the synthesis of lipopolysaccharide (LPS; O antigen) and capsular polysaccharide. Belas et al. (1995) found that the transposon insertion point of some Tn5-generated mutants defective in wild-type swarmer cell elongation is located in an ORF homologous to the hypothetical 43.3-kilodalton (kDa) protein in the *E. coli* locus responsible for the LPS chain-length determinant (Belas et al., 1995). LPS chain length has been postulated to be important in swarmer cell function (Armitage et al., 1975,

1979; Armitage, 1982). The *cld* locus confers a modal distribution of chain length on the O-antigen component of LPS (Bastin et al., 1993). This protein is thought to have dehydrogenase activity, as indicated by the homology to uridine diphosphate (UDP)-glucose dehydrogenase from *Streptococcus pyogenes* and *Streptococcus pneumoniae*, which was confirmed in the *P. mirabilis* homolog by amino acid sequence comparison of the deduced amino acid sequence (Belas et al., 1995). Because *cld* comprises several ORFs, it is possible that such Tn5 insertion mutants of *P. mirabilis* are defective in maintaining the preferred O-antigen chain length. Alternatively, the mutation may be in the *rfb* gene cluster (O-antigen), which is closely linked to *cld* in *E. coli* and *S. typhimurium* (Bastin et al., 1993).

In a second *P. mirabilis* mutant, also phenotypically Elo⁻, it was determined that a mutation had occurred in *rfaD*, encoding adenosine diphosphate (ADP)-L-glycero-D-mannoheptose-6-epimerase (Belas et al., 1995). In *E. coli* and *S. typhimurium* this mutation results in altered heptose (L-glycero-D-mannoheptose) and LPS biosynthesis, and it increased outer membrane permeability (Pegues et al., 1990). The nucleotide sequence at the distal end of the cloned insert was determined to be highly homologous to *rfaC*, which encodes heptosyltransferase I (Sirisena et al., 1992). In *S. typhimurium* this mutation leads to heptose-less LPS and rough colony phenotype (Sirisena et al., 1992; Chen and Coleman, 1993). Such rough mutants have, in other bacteria, been associated with defects in motility, which may explain the Mot⁻ phenotype of this particular mutant. *rfaC* is genetically linked to *rfaD*, supporting the finding that the mutation is in *rfaD* and suggesting that the defect affects LPS structure and outer membrane permeability.

Interestingly, all of these mutations appear to affect swarmer cell elongation by preventing normal synthesis and rotation of the flagella. However, although swarming motility is affected by these mutations, swimming is not. Moreover, impairment of the flagella should result in the induction of transcription of the swarmer cell regulon, although each of these mutants is Elo⁻. Hence there may be a close connection between LPS synthesis and regulation and signal transduction and swarmer cell regulation in *P. mirabilis*, as has been shown by others (Armitage et al., 1975, 1979; Armitage, 1982).

Bacteria often produce extracellular polysaccharides (capsules and "slime") that aid cells in their adhesion to substrates. Many swarming colonies produce a clear extracellular fluid and often have a glistening, mucoid appearance. In *P. mirabilis* this material has been shown to consist of a polysaccharide matrix (Allison and Hughes, 1991a), but its composition in other bacteria has not been determined, nor have defects in swarming been correlated with specific defects in slime production. For example, *S. marcescens* excretes a surfactant (Matsuyama et al., 1992) that is not essential for swarming (Harshey, 1994; Harshey and Matsuyama, 1994).

Gygi et al. (1995b) cloned and determined the nucleotide and deduced amino acid sequence for a genetic locus of *P. mirabilis* responsible for production of capsular polysaccharide that is thought to be important to swarming motility (Gygi et al., 1995b). This region has strong homology to the *Strep-*

tococcus pneumoniae type 19F capsular polysaccharide biosynthesis genes, particularly *cpsF*. The relevance and relation of capsular polysaccharide biosynthesis and swarming migration may now be assessed as *cpsF* mutations are constructed in *P. mirabilis* and the swarming characteristics of the mutants assessed. (Some preliminary results are described below; see Multicellular Migration.)

Peptidoglycan and Cell Division Genes

As part of an analysis of the defects responsible for abnormal swarmer cell elongation, Belas et al. (1995) identified several mutations in genes required for cell wall peptidoglycan biosynthesis and in cell division and chromosome replication. Two of the mutations identified were in *galU*, affecting glucose-1-phosphate uridylyltransferase and cell wall synthesis (Jiang et al., 1991; Liu et al., 1993; Varon et al., 1993; Morona et al., 1994), and in a gene homologous to *dapE* (Bouvier et al., 1992). The latter gene encodes *N*-succinyl-L-diaminopimelic acid desuccinylase, an enzyme that catalyzes the synthesis of LL-diaminopimelic acid, one of the last steps in the diaminopimelic acid–lysine pathway leading to the development of peptidoglycan (Bouvier et al., 1992). That mutations in cell wall biosynthesis genes produce abnormal swarmer cell elongation is not unusual, as defects in such genes can produce overall effects on the ability of the cell to function.

One of the more interesting mutations that has been found to produce the Elo$^-$ Mot$^-$ Swr^{cr} phenotype is in *gidA* (glucose inhibited division), a nonessential gene in *E. coli* near *E. coli oriC* (von Meyenburg et al., 1982; Ogawa and Okazaki, 1991; Belas et al., 1995). The homology between *P. mirabilis* GidA and its *E. coli* homolog is one of the strongest observed (Belas et al., 1995). In *E. coli*, *gidA* mutations are silent on complex media, but when grown on glucose-containing media *E. coli gidA* strains produce long filamentous cells (von Meyenburg and Hansen, 1980, 1987; von Meyenburg et al., 1982). Furthermore, there is evidence that *gidA* transcription is regulated by ppGpp and is involved in the initiation of chromosomal replication (Asai et al., 1990; Ogawa and Okazaki, 1991). Thus *gidA* may function to connect glucose metabolism, ribosome function, chromosome replication, and cell division (J. Shapiro, personal communication). The function of *gidA* in swarmer cell differentiation and elongation is obscure but evidently essential for swarmer cell elongation and wild-type swarming behavior.

Sigma Factors and Other Upstream Regulatory Sequences

Nucleotide sequence analysis of cloned swarmer cell differentiation genes from *P. mirabilis* (and other swarming bacteria) has been ongoing for several years. As part of these analyses, regulatory regions of each gene have been scrutinized for evidence suggestive of unique regulatory mechanisms controlling swarmer cell differentiation and behavior. It is still too early to draw conclusions from these analyses, although two aspects of the upstream regulatory regions of swarmer cell differentiation genes associated with flagellar synthesis have been found.

The upstream regulatory regions of many flagellar genes from *Bacillus subtilis* (Mirel and Chamberlin, 1989), *E. coli* (Bartlett et al., 1988) and *S. typhimurium* (Helmann, 1991) have a unique promoter region for a flagellar-gene-specific RNA polymerase. An alternate σ subunit of RNA polymerase, referred to as σ^{28} (σ^F) in *E. coli* and σ^D in *B. subtilis*, is responsible for this specificity (Helmann, 1991). Thus it is perhaps not unexpected that the upstream region in front of the start codon to the *P. mirabilis flaA*, *flaB*, and *flaD* genes (Belas and Flaherty, 1994) and the *flhA* gene (Gygi et al., 1995a) also contains a nucleotide sequence that is similar (if not identical) to the consensus σ^{28} promoter 5'-TAAA-N$_{15}$-GCCGATAA-3' of *E. coli*. The homology between the *P. mirabilis* sequences and the consensus is good, with the σ^{28} promoter for *flaA* a direct match to the *E. coli* consensus sequence. The σ^{28} promoter of *flaD* has a single mismatch compared to the consensus, and the *flaB* promoter has two mismatches, suggesting that these genes (and probably most of the flagellar genes of *P. mirabilis*) are regulated in a manner similar to that demonstrated in *E. coli*.

What then is different about the *P. mirabilis* flagellar gene regulation that might help explain surface-inducible flagellin synthesis in the swarmer cell? The single most notable difference between the *E. coli* *fliC* regulatory region and those of *flaA* and *flaB* is the presence of a dual, direct tandem repeat (DTR) sequence 5' to the *P. mirabilis* σ^{28} promoters (Belas, 1994; Belas and Flaherty, 1994). This sequence ATAAAAA, repeated twice, is about 90 basepairs (bp) upstream from the midpoint of the -35 region of each σ^{28} promoter. The *flaA* DTR sequence is 5'-ATAAAAATAATATAAAAAAATAA-3', with two overlapping direct repeats, ATAAAAA and ATAAAAATAA. The *flaB* DTR sequence is on the opposite strand to the *flaA* DTR and has a nucleotide sequence of 3'-ATAAAAAAAGAGAGGTAGATAAAAACAAAAAAGA-5'. The *flaB* DTR has two direct repeats as well. The first of these repeats is identical to *flaA* (ATAAAAA), but the second (AAAAAAGA) does not share sequence identity with the *flaA* DTR. Searches of GenBank bacterial nucleotide sequences failed to find similar sequences in front of any other flagellar gene, but a homologous sequence is found in the upstream regulatory region of *ureR*, the regulatory gene for urease expression (Nicholson et al., 1993). Urease and flagellin are coordinately regulated as part of swarmer cell differentiation (Allison et al., 1992b). Although the function of the DTR sequences is unknown, it is tempting to speculate, because of sequence conservation and placement, that they may have a function in swarmer cell-specific gene expression, perhaps as a surface-induced enhancer of transcription (Kustu et al., 1991; Gober and Shapiro, 1992).

Multicellular Migration

Control of Swarmer Cell Behavior

Migration Cycle: Model of Extracellular Signals

In contrast to the advances at the genetic level characterizing the genes required during swarmer cell differentiation, little is known or understood about the molecular mechanisms controlling swarmer cell motile behavior and, more importantly, the cyclic events of differentiation, migration, dedifferentiation, and consolidation. Mutant phenotypes produced by transposon-insertion mutageneses have implicated a series of multicellular signaling events that presumably regulate the cyclic nature of consolidation in these bacteria, although only limited characterization of these loci has thus far been accomplished. Although the role of signals (if they exist at all) is yet to be proved, I provide here a brief synopsis of the data and a possible model to explain the observations in the following section. It must be stressed that such a model is merely speculative but hopefully points toward experiments designed to answer the questions regarding multicellular swarming migration.

One of the most astonishing observations to come from the efforts to understand the regulation of swarmer cell differentiation and multicellular swarming behavior of *P. mirabilis* is that many of the mutants defective in wild-type swarming are not null mutants (i.e., mutants completely lacking the ability to swarm on an agar surface) (Fig. 7.5A). Rather, many of the mutants are wild-type for the ability to differentiate into swarmer cells and rotate the flagellar filaments in a normal manner (Allison and Hughes, 1991a,b; Belas et al., 1991a). What makes these strains interesting seem to be defects in one or more of the genes that regulate the multicellular interactions associated with the formation of the characteristic bulls-eye colony (Fig. 7.5B–D). Such mutants, which have been referred to in my laboratory as swarming crippled mutants (Swr^{cr}), have also been constructed in other laboratories (Allison and Hughes, 1991a,b). Swr^{cr} mutants do indeed have the ability to differentiate and swarm on the appropriate surfaces, yet they are defective in orchestrating the proper series of steps required to form distinct, periodic consolidation zones (Fig. 7.5). The category of crippled mutants is, by its nature, broadly defined and somewhat pleiotropic. Individual strains within the Swr^{cr} group have a wide variety of mutant phenotypes, with the central theme being a defect in the ability to form evenly spaced or well defined consolidation zones. As shown in Figure 7.5, when compared to a swarming null mutant (Fig. 7.5A), Swr^{cr} strains fall into phenotypic classes that produce (1) nonuniform consolidation zones (Fig. 7.5B); (2) spatially narrow or broad zones (Fig. 7.5C); or (3) indeterminate consolidation zones (Fig. 7.5D—a personal favorite of the author). The first two phenotypic classes (nonuniform and narrow- or broad-spacing strains) appear to have a defect in the spatial or temporal control of the multicellular signaling presumed to control the sequence of events of dedifferentiation that give rise to the consolidation phase of swarming. It may be that in such mutants the putative signals either are generated more rapidly than in the wild-type, giving rise to faster consolidation and narrower zones, or are synthesized at a slow rate, resulting in broad zones due to a slower consolidation process. The last class of Swr^{cr} mutants appears to be completely lacking in the signals or, more likely, in the receptors for the signals controlling consolidation.

Armed with these observations, a possible hypothetical scenario has been described (Belas, 1992) that may be used to help us understand the unusual Swr^{cr} mutants that produce uneven spacing in consolidation zones; and in turn

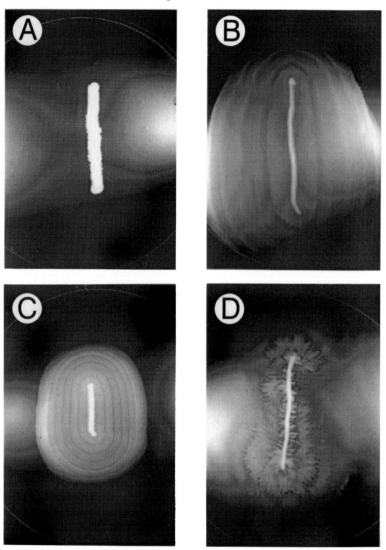

FIGURE 7.5 *P. mirabilis* mutants with defects in the spatial and temporal control of swarming behavior. Bacteria were inoculated as a 3 cm line at the center of fresh L-agar medium and incubated for 48 hours at 30°C. (**A**) Swarming null mutant (Swr⁻ Swm⁻ Elo⁻). (**B**) Swarming mutant with infrequent and variable consolidation (Swrᶜʳ Swm⁺ Elo⁺). (**C**) Swarming mutant with increased consolidation frequency (Swrᶜʳ Swm⁺ Elo⁺). (**D**) Swarming mutant lacking clearly defined consolidation, referred to as an indeterminate swarming mutant (Swrᶜʳ Swm⁺ Elo⁺).

we may be able to extrapolate to the nature of the genetic defects in indeterminate mutant strains. Such a model assumes the presence of a set of extracellular signals (in this particular case three signals, though the number was arbitrarily set). Each signal is assumed to be generated by the swarming colony, increases to a peak level, and then diminishes perhaps owing to breakdown by enzymes or other catalytic processes. As developed, the model requires that each of the three signals is produced, crests, and then diminishes simultaneously with the other signals; hence they are synchronous. Furthermore, according to the model, the cells, having the ability to sense the amplitude of signals, would regulate the differentiation cycle as a response to the sum of all of the incoming signals. These cycles of differentiation and dedifferentiation would therefore be regulated in reference to some maximum and minimum value of the sum of the signals—differentiation commencing at high signal intensity and dedifferentiation occurring at a low signal intensity. Such changes theoretically could be produced by regulatory mutations that affect the rate of production or destruction of one or more of the signal molecules. The threshold signal level for consolidation would thus become arrhythmic. The net outcome would produce a combination of narrow and broad zones, as seen in Figure 7.5B. Using the predictions of this model, it is feasible to consider the indeterminate consolidation class of mutants (Fig. 7.6A,B) as lacking all signals or possessing a central defect, such as a mutations in the receptor for these signal molecules.

It is evident that the models do simulate some of the swarming patterns observed in *P. mirabilis*. It goes without saying that the use of such models is limited, and interpretation of such depictions of signal intensity may not mirror the real world. There are many possible explanations that can be used to fit the data presented. Nonetheless, modeling does serve as a way to focus ideas for developing experimental approaches to and understanding of the nature of the defect in the Swr^{cr} mutants.

Genes and Mutant Phenotypes

Several of the genes described in the first section of this chapter are needed for wild-type multicellular migration; that is, defects have affected flagellar biosynthesis and result in failure of the cell to synthesize flagella, so the cells are nonmotile and cannot participate in migration events. However, some of the genes associated with cellular differentiation play slightly different roles in migration. Those differences, along with descriptions of some genes thought to be solely involved in multicellular migration, are examined in the following section.

Capsular Polysaccharide and Surface Migration

Proteus mirabilis swarmer cells secrete substantial amounts of a viscous biofilm that defines the leading edge of the migrating population. This substance consists of a so far uncharacterized carbohydrate that is assumed to assist migration, possibly acting as a surfactant (Stahl et al., 1983). A mutation that causes frequent consolidation has been characterized and the defect located in a gene

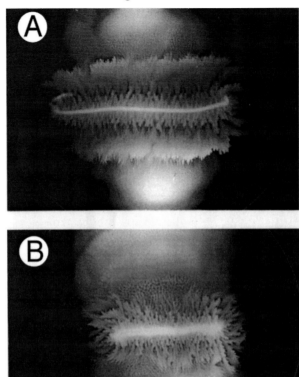

FIGURE 7.6   Variability of indeterminate swarming mutants of *P. mirabilis*. (**A**) Indeterminate consolidation mutant that produces distinct but irregular consolidation zones. (**B**) Another indeterminate consolidation mutant that produces less distinct consolidation zones and exhibits a variable consolidation behavior characterized by initial dense colonies, which give rise to low density swarms after several consolidation events.

(*cpsF*) involved in synthesis of a major, new capsular polysaccharide, referred to as CPS I (C. Hughes, personal communication). The mutant fails to polymerize CPS I, and precursors accumulate intracellularly. A second, minor capsular polysaccharide (CPS II) is still produced, though, and forms a thin capsule on the cell surface. Detailed analysis of the swarming cycle shows that the CPS I⁻ cells differentiate faster than wild-type cells. CPS I⁻ mutants do not have changes in the temporal control of consolidation pattern formation but, rather, are affected by a reduction in the motility of the cells; that is, spatial placement of the rings is shortened. These data suggest that CPS I is important in reducing friction and acts as a surfactant, perhaps performing a function similar to that of serrawettin, a surfactant of *S. marcescens* (Matsuyama et al., 1992).

Glutamine Signal: Possible Roles for a Metalloprotease and a
Proline Peptidase

Proteus mirabilis strains of diverse type produce an extracellular, EDTA-
sensitive metalloprotease that is able to cleave serum and secretory IgA1 and
IgA2, IgG, and a number of nonimmunological proteins such as gelatin and
casein (Senior et al., 1987, 1991; Loomes et al., 1990, 1992, 1993). The protease activity on IgA is such that the heavy chain of the immunoglobulin is
cleaved, which is different from classic microbial IgA proteases (Plaut, 1983).
An enzyme with the ability to cleave immunoglobulin molecules has a potential
to be a virulence factor and indeed is frequently found in the urine of patients
with *P. mirabilis* urinary infections (Senior et al., 1991).

Researchers in my laboratory have recently cloned a *P. mirabilis* genetic
locus into *E. coli* that confers to the recombinant bacteria all of the properties
associated with the metalloprotease described in the wild-type *Proteus* cells
(Wassif et al., 1995). Genetic analysis using Tn5 mutagenesis of the *P. mirabilis*
DNA has indicated that a region of approximately 4.5 kb is required for expression of the metalloprotease in *E. coli*. Using nucleotide primers to the
IS50L and IS50R regions of the transposon, the nucleotide and deduced amino
acid sequences are in the process of being analyzed. So far this analysis has
revealed the presence of an operon with three ORFs within the 4.5 kb region.
The genes associated with the ORFs correspond to a zinc metalloprotease and
two adenosine triphosphate (ATP)-dependent membrane transport proteins required for secretion of the metalloprotease to the external mileau.

It is likely that this enzyme functions as a virulence factor during urinary
tract infection. Allison et al. (1992a,b 1994) have demonstrated that the activity
of the metalloprotease is coordinately expressed with the cycles of swarmer
cell differentiation. Does this metalloprotease have any other function? A possible role is generation of the glutamine signal shown to be required for
swarmer cell differentiation and migration (Allison et al., 1993). If this is the
case, it represents a significant advance in our understanding of multicellular
migration and swarming behavior. Experiments are in progress to measure the
transcriptional control of this gene and assess its role in migration.

One of the Elo⁻ strains my laboratory recently examined (Belas et al.,
1995) is the result of a mutation in the *P. mirabilis* homolog to *pepQ*. This
gene encodes X-proline dipeptidase (Nakahigashi and Inokuchi, 1990) and
shows strong identity to other proline dipeptidases. What makes this PepQ⁻
mutant interesting is that, in addition to being defective in swarmer cell elongation (Elo⁻), the mutant produces abnormal consolidation patterns, similar to
that shown in Figure 7.6B. A possible interpretation of these data is that the
mutation in *pepQ* may function to generate the signal by cleaving larger polypeptides into transportable peptides. PepQ⁻ defects may prevent normal uptake
of these important signal molecules used by *P. mirabilis* to control swarming
behavior and consolidation (Belas, 1992). The interaction (if any) between the
metalloprotease and PepQ may be significant in the multicellular signaling that
occurs during the cyclic events of consolidation in *P. mirabilis* swarming
colonies.

Role of Amino Acid Chemotaxis

Bacteria swim by rotating rigid flagellar filaments (Silverman and Simon, 1977; Silverman, 1980). The direction of rotation is reversible and is coordinated by the components associated with chemotactic behavior (Macnab, 1987). Microorganisms such as *E. coli* swim in stretches of smooth runs interrupted by intervals of chaotic motion called tumbling. Smooth swimming and tumbling behaviors correspond to opposite directions of flagellar rotation: counterclockwise (CCW) filament rotation produces swimming, and clockwise (CW) rotation results in tumbling (Silverman and Simon, 1973). When bacteria swim up an attractant concentration gradient, the period of smooth swimming is extended, thereby allowing the cell to make progress in the direction of a higher concentration of attractant. Conversely, movement away from an attractant source or toward increasing concentration of a repellant produces more frequent tumbling response (Adler, 1983).

Historically, the role of chemotactic behavior in swarmer cell differentiation and swarming motility has been disputed. Over the last several years, however, an overwhelming amount of evidence gathered from *V. parahaemolyticus* (Sar et al., 1990), *P. mirabilis* (Belas et al., 1991a), and *S. marcescens* (O'Rear et al., 1992) has clearly identified chemotactic behavior as a major force in controlling certain aspects of differentiation and multicellular migration. Moreover, in research demonstrating swarmer cell differentiation and migration of *E. coli* and *S. typhimurium*, Harshey has unequivocally demonstrated through the use of well characterized mutants in chemotaxis that chemotactic behavior is critical for swarming of these bacteria (Harshey and Matsuyama, 1994).

There are several aspects of chemotactic behavior as it functions in swarming that require further elaboration. One aspect is analysis of the chemotactic genes of *V. parahaemolyticus* (Sar et al., 1990). Examination of the chemotaxis system of *V. parahaemolyticus* is complicated by the existence of two distinct types of motility. Analysis of the two systems was simplified by using mutants capable of only one mode of motility: Fla$^+$ Laf$^-$ (swimming only) and Fla$^-$-Laf$^+$ (swarming only) strains (Sar et al., 1990). The two systems could be evaluated separately. Chemotaxis of swimming bacteria was analyzed using the capillary system of Adler (1973). To measure chemotaxis of swarming bacteria, the viscosity of the assay medium was increased with 5% polyvinylpyrrolidone 360. The flagellar systems responded similarly with respect to attractant compounds and the concentrations that elicited the chemotactic responses. *V. parahaemolyticus* responds to serine and other attractants that in *E. coli* are recognized by the Tsr receptor (Macnab, 1992). A locus required for chemotaxis in a swimmer-only strain of *V. parahaemolyticus* was cloned and mutated with Tn5 in *E. coli*. The transposon-generated defects were transferred to Fla$^+$-Laf$^-$ and Fla$^-$-Laf$^+$ strains. Introduction of *che* mutations prevents chemotaxis into capillary tubes and greatly diminishes multicellular swarming and unicellular swimming through semisolid swimming media. Thus the two flagellar systems of *V. parahaemolyticus*, which consist of distinct motor-propeller organelles, are directed by a common chemosensory control system.

The role of glutamine chemotaxis in *P. mirabilis* swarmer cell differentiation was described in an earlier section. *P. mirabilis* grown on a basal medium does not swarm, suggesting that a chemical inducer is missing. Allison et al. (1993) showed that the amino acid glutamine, when added to the basal medium, induced swarmer cell differentiation and multicellular migration. Glutamine is sensed through a specific transduction mechanism that is independent of the cellular (nutritional) amino acid uptake system. The sensing of glutamine acts in conjunction with the monitoring of flagellar rotation by the cell, and the two signals are processed together so swarmer gene-specific expression can be induced. Thus, *P. mirabilis* is different from *V. parahaemolyticus* in that it has a specific requirement for one amino acid, glutamine.

Dienes Phenomenon

An interesting aspect of swarming behavior peculiar to *P. mirabilis* and not to any other swarming bacteria is the Dienes phenomenon (Dienes, 1946, 1947). It refers to the narrow zone of demarcation that arises between the approaching swarms of two strains of *P. mirabilis* that have been inoculated on the surface of an agar plate. Such strains are said to display mutual inhibition, to be incompatible in the Dienes test, to belong to different Dienes types, or to be Dienes test-positive (de Louvois, 1969; Skirrow, 1969; Senior, 1977). In contrast, if the approaching swarms of two strains merge imperceptibly, the strains are said to be compatible, to belong to the same Dienes type, and to be Dienes test-negative; in practice they are considered identical.

Some early descriptions of the phenomenon suggested that the approaching swarms may stop short of each other and that the zone of demarcation is sterile (Dienes, 1946, 1947; Sourek, 1968). However, closer examination of the development of the zones revealed that the approaching swarms do make an initial contact after 4–6 hours of incubation. On further incubation one or both swarms develop a thickened border of growth somewhat back from the meeting point to form the characteristic zone (Sourek, 1968), which is usually 0.5–2.0 mm wide (Sourek, 1968; Skirrow, 1969; Senior, 1977; Sohnius and Lenk, 1977). This zone first becomes apparent after 9–10 hours of incubation, is well developed by 15–18 hours (Sourek, 1968), and persists for several days (Dienes, 1946, 1947).

The Dienes phenomenon may not be attributable to a single mechanism. Several substances known or postulated to exist have been or could be considered the agent of the Dienes phenomenon. They include a nonvolatile agent proposed to induce premature induction of long forms in mixed cultures (Hughes, 1957), a postulated negative chemotaxis agent (Naylor, 1964), proticines (Sourek, 1968; Skirrow, 1969), a *Proteus* substance that is both auto-inhibitory and inhibitory to other swarming *Proteus* strains as disclosed by Grabow (1972), and the volatile amines produced by several *Proteus* species during growth on media containing adequate amino acids (Proom and Woiwod, 1951). Particulate bacteriocins abound in strains of *P. mirabilis* but are apparently not the cause of incompatibility (Sourek, 1968). Two observations are consistent with failure to identify antibiotic-like or bacteriocin-like substances: (1) isolated bacteria of one strain do not form bacteriocinogenic-type plaques in the lawn of a different strain, and (2) no boundary is formed when two strains come together on either side of a Millipore membrane filter inserted into the substrate, even though the width of the filter is much smaller than the size of the Dienes boundary. Thus it appears that direct swarm-to-swarm contact is needed for boundary formation, which is consistent with microscopic observation of the process: The two swarms interpenetrate, and boundary formation starts after only a couple of hours of contact.

The role of the Dienes phenomenon in multicellular swarming behavior is still a mystery. This phenomenon is a vital aspect of the behavior of *P. mirabilis* and is not simply a degenerative process (Dienes, 1946, 1947). A possible role for the Dienes phenomenon may be in maintenance of colonized ''territory'' by a specific strain of *Proteus*, which would prevent other strains of the same species from being able to take over metabolic resources through Dienes interaction. The answers to these questions are awaiting discovery.

Swarming of Other Bacteria

Swarming was first described for the genus *Proteus* more than a century ago and has been demonstrated in both gram-negative and gram-positive genera (Fig. 7.7), including *Vibrio* (Ulitzer, 1975a; Belas et al., 1986; McCarter et al., 1988; McCarter and Silverman, 1990); *Serratia* (Alberti and Harshey, 1990); *E. coli*, *S. typhimurium*, *Yersinia*, *Aeromonas*, and *Rhodospirillum* (Harshey, 1994; Harshey and Matsuyama, 1994); and *Bacillus* and *Clostridium* (Henrichsen, 1972; Allison and Hughes, 1991a; Harshey, 1994; Harshey and Matsuyama, 1994). These organisms can swarm on solidified agar media, usually 1–2% agar for *V. parahaemolyticus* and *P. mirabilis* and lower concentrations for *S. marcescens*, *E. coli*, and *S. typhimurium* (Alberti and Harshey, 1990; Harshey and Matsuyama, 1994). The ubiquitous occurrence of swarming among eubacteria suggests that this mode of surface translocation must play an important role in the colonization of natural environments by microorganisms.

Of the swarming genera, the swarming of *P. mirabilis* has been the most intensively investigated, with studies of *V. parahaemolyticus* and *S. marcescens* swarming following closely thereafter (Belas et al., 1986; McCarter et al., 1988; McCarter and Silverman, 1989, 1990; Alberti and Harshey, 1990; Sar et al., 1990; McCarter and Wright, 1993). Sufficient information is now available to compare the swarming of *Proteus* species, *V. parahaemolyticus*, and *S. marcescens* 274 (Table 7.3). Like *Proteus* swarmer cells, *Vibrio* and *Serratia* swarmers are elongated, nonseptate, and highly flagellated; they occur peripherally but not centrally in swarming colonies and are reversibly differentiated from the short, nonswarmer cells on transfer between liquid and solid media (Belas et al., 1986; Alberti and Harshey, 1990). In *P. mirabilis*, *V. parahaemolyticus*, and *S. marcescens* the introduction of mutations affecting chemo-

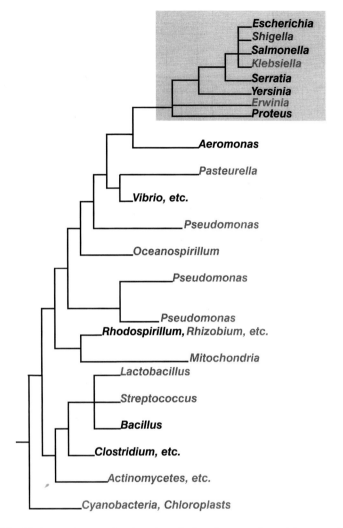

FIGURE 7.7 Phylogenetic tree showing the relations between bacterial genera in which swarming motility has been demonstrated. Those genera with at least one swarming species are indicated by black lettering. (Adapted from Ochman and Wilson, 1987, with permission).

taxis abolishes or impairs normal swarming behavior. As described earlier, all three genera monitor flagellar filament rotation to determine external conditions and, when the filament rotation becomes impaired, induce swarmer cell differentiation. Beyond these similarities there are many differences between the three genera.

Wild-type *V. parahaemolyticus* growing in a liquid medium has one sheathed polar flagellum composed of four flagellin subunits coded for by a set of *fla* (flagellin) genes (McCarter and Silverman, 1990). This flagellum is

TABLE 7.3 Comparison of Swarmer Cell Differentiation and Motile Behavior of *P. mirabilis, V. parahaemolyticus,* and *S. marcescens* 274

Characteristic	*P. mirabilis*	*V. parahaemolyticus*	*S. marcescens*
Swarmer cell length (μm)	10 to > 80	30–40	5–30
Nonseptate swarmer cells	Yes	Yes	Yes
Flagella			
Per swarmer cell	10^3–10^4	10^2–10^3 (including one polar)	10^2–10^3
Per swimmer cell	4–10	One polar	1–2
Flagellation	Uniform	Mixed	Uniform
Flagellin gene(s)	*flaA-C*, only one expressed at a time	*flaA-D* for polar flagellin; *lafA* for lateral flagellin	*hag*
Evident consolidation zones	Yes	No	No
Dedifferentiation during swarming?	Yes	Yes	Yes
Physical stimulus for differentiation	Inhibition of flagellar rotation	Inhibition of flagellar rotation	Inhibition of flagellar rotation
Chemical stimulus for differentiation/ motility	Glutamine, iron (?), zinc, (?), chemotaxis sensing machinery	Iron limitation, chemotaxis sensing machinery	Chemotaxis sensing machinery
Agar concentration (%) required to induce swarming	1.5	1.5	0.35–0.75
Extracellular slime/ surfactant	Yes	No	Yes

responsible for swimming in liquid media. In addition to this constitutively produced polar flagellum, *Vibrio* swarmer cells exhibit multiple, unsheathed peritrichously arranged flagella, referred to as lateral flagella, which are responsible for the swarming phenotype exhibited on the surface of solid media. The lateral flagella filament is coded for by the *laf* set of genes, one of which (*lafA*) codes for the single flagellin structural gene. Thus lateral flagella have a subunit composition different from that of polar flagella, and both flagella have independent motor structures (Sar et al., 1990) and are energized through different mechanisms (Atsumi et al., 1992). Consequently, *V. parahaemolyticus* is said to have "mixed flagellation." At least two conditions are required to induce transcription of the genes coding for lateral flagella and the other phenotypic changes associated with swarmer cell differentiation: physical restriction of the rotation of the polar flagellum (by solid surfaces, liquids of high viscosity, or anti-cell-surface antibodies) and iron-limiting growth conditions (McCarter et al., 1988; McCarter and Silverman, 1989, 1990).

This regulation at the level of transcription of *laf* genes in *V. parahaemolyticus* has been deduced from elegant genetic work using the *lux* genes from the luminescent bacterium *Vibrio fischeri*. These genes encode the catalytic components of the bioluminescence system, and promoterless versions have been inserted into a transposon, mini-Mu *lux* (Tet^R), which in turn was transduced by bacteriophage P1 into *V. parahaemolyticus*. Such mini-Mu *lux* (Tet^R) insertions can produce transcriptional fusions between *laf* genes and the *lux* genes carried on the transposon, the result of which is the construction of *V. parahaemolyticus* strains where the *lux* genes are under the control of the *laf* gene promoter. The resulting luminescence facilitates measurement and is non-invasive, permitting the cells to continue their processes while the level of *laf* activity is measured.

In contrast to *V. parahaemolyticus*, swarming *Proteus* strains do not have a typical pattern of mixed flagellation, where two distinct flagellar types are produced simultaneously. In wild-type *P. mirabilis* cells, the flagella of both swimmer and swarmer cells is composed of the same flagellin protein encoded by the *flaA* gene (Belas and Flaherty, 1994). Interestingly, *P. mirabilis*, like *V. parahaemolyticus*, has multiple copies of flagellin-encoding genes (referred to as *flaA*, *flaB*, and *flaC*) (Belas and Flaherty, 1994). Normally, *flaA* is expressed, and *flaB* and *flaC* are silent. However, if *flaA* is mutated, spontaneous deletions fusing the promoter region of *flaA* to either *flaB* or *flaC* may occur. This gene conversion produces a hybrid flagellin protein where FlaA sequences comprise the N-terminal end of the protein and either FlaB or FlaC sequences comprise the C-terminal portion. So, although *P. mirabilis* produces only a single flagellin species at any time, the type of flagellin produced may undergo antigenic variation.

Other differences between *P. mirabilis* and *V. parahaemolyticus* are evident. For example, glutamine is necessary for swarmer cell differentiation in *P. mirabilis*, whereas in *V. parahaemolyticus* iron limitation promotes differentiation. *P. mirabilis* can differentiate and consolidate in repetitive cycles to form concentric swarming zones, whereas *V. parahaemolyticus* and *S. marcescens* only rarely exhibit this cyclic event. Therefore it has been suggested that the swarming of *V. parahaemolyticus* and *Proteus* species is an example of convergent, rather than divergent, evolution.

In the case of *S. marcescens* 274 and also for *E. coli* and *S. typhimurium*, swarmer cell differentiation occurs only on the surface of media solidified by 0.75–0.85% agar. Although this requirement for a low agar concentration differs from *Proteus* and *Vibrio*, it is not unique among swarming bacteria. *Azospirillum brasilense* (which has mixed flagellation) swarms optimally on 0.75% agar and not at all on 1.5% agar (Hall and Krieg, 1983). *S. marcescens* 274 does not have mixed flagellation; instead, swimmer cells growing in liquid media have one or two flagella composed of the same flagellin protein found in the flagella of swarmer cells. A single gene, *hag*, codes for this flagellin protein. *Serratia* swarming is associated with the production of a slime-like material. None of a wide range of chemicals tested have been implicated in the differentiation of *S. marcescens* 274 (Alberti and Harshey, 1990) or for *V. parahaemolyticus* (Belas et al., 1986).

Conclusions and Future Directions

For most of the era during which swarming motility has been observed and studied, this form of multicellular prokaryotic behavior has been considered a curious oddity, usually more of a nuisance than a wonder and certainly not a phenomenon with any connection to other prokaryotic behaviors. Happily, those days seem to have passed. What is now becoming apparent is the idea that swarmer cell differentiation and behavior are aspects of a complex signal transduction and genetic regulatory pathway, the mechanisms of which almost certainly have implications for other eubacteria, most notably *E. coli* and *S. typhimurium.*

Although a model to explain swarmer cell differentiation and behavior is emerging, a number of important questions remain unanswered. We know that inhibition of flagellar rotation is essential for sensing the environment and inducing expression of the swarmer-cell-specific genes, yet almost nothing is known about how this signal is transduced into the cell or how it is converted to a regulator of transcription. The findings by McCarter (1994) regarding the MotX protein and its function in *V. parahaemolyticus* polar flagellum rotation and the importance of FliG in *P. mirabilis* swarmer cell differentiation (Belas et al., 1995) may be leading us to answers to some aspects of this question. In addition, although we have some limited information regarding the expression and transcriptional control of swarmer cell-specific genes (e.g., *flaA*) (Belas, 1994), most of these genes remain uncharacterized and their regulation unknown. It is intuitive that a complex series of events must occur for the periodic cycles of differentiation and dedifferentiation. Such processes must somehow link all of the cells into a single global network of communication, so each cell starts and stops the differentiation cycle at the same time. Extracellular communication via signal molecules is a possible means by which such control could be achieved, but what is the mechanism? Are the metalloprotease and proline peptidase important in swarming migration, and what is the molecular nature of the signals and the regulatory mechanisms employed in controlling the cyclic events of consolidation? We do not have clear answers to any of these questions at present.

Several research groups are actively pursuing such questions using detailed molecular analyses of the genes and mutants affecting multicellular swarming behavior. It is safe to say that many of the answers to the questions raised in this chapter will be found during the next several years. What new questions those data will bring forth, and what new mechanisms of signal transduction and genetic regulation will be unveiled can only be surmised for now. Whatever answers are found, however, they are likely to have major, profound implications on the fields of research focused on bacteria as multicellular organisms.

Acknowledgments

This work was supported by grants from the National Institutes of Health (AI27107 and DK49720) and the National Science Foundation (MCB-9206127). It is contribution 1021 from the Center of Marine Biotechnology.

References

Adler, J. (1973) A method for measuring chemotaxis and use of the method to determine optimum conditions for chemotaxis by *Escherichia coli*. *J. Gen. Microbiol.* **74**: 77–91.

Adler, J. (1983) Bacterial chemotaxis and molecular neurobiology. *Cold Spring Harbor Symp. on Quant. Biol.* **2**:803–804.

Alberti, L., and Harshey, R.M. (1990) Differentiation of *Serratia marcescens* 274 into swimmer and swarmer cells. *J. Bacteriol.* **172**:4322–4328.

Allison, C., and Hughes, C. (1991a) Bacterial swarming: an example of prokaryotic differentiation and multicellular behaviour. *Sci. Prog.* **75**:403–422.

Allison, C., and Hughes, C. (1991b) Closely linked genetic loci required for swarm cell differentiation and multicellular migration by *Proteus mirabilis. Mol. Microbiol.* **5**:1975–1982.

Allison, C., Coleman, N., Jones, P.L., and Hughes, C. (1992a) Ability of *Proteus mirabilis* to invade human urothelial cells is coupled to motility and swarming differentiation. *Infect. Immun.* **60**:4740–4746.

Allison, C., Emody, L., Coleman, N., and Hughes, C. (1994) The role of swarm cell differentiation and multicellular migration in the uropathogenicity of *Proteus mirabilis. J. Infect. Dis.* **169**:1155–1158.

Allison, C., Lai, H.C., Gygi, D., and Hughes, C. (1993) Cell differentiation of *Proteus mirabilis* is initiated by glutamine, a specific chemoattractant for swarming cells. *Mol. Microbiol.* **8**:53–60.

Allison, C., Lai, H.C., and Hughes, C. (1992b) Co-ordinate expression of virulence genes during swarm-cell differentiation and population migration of *Proteus mirabilis. Mol. Microbiol.* **6**:1583–1591.

Alm, R.A., Guerry, P., Power, M.E., and Trust, T.J. (1992) Variation in antigenicity and molecular weight of *Campylobacter coli* VC167 flagellin in different genetic backgrounds. *J. Bacteriol.* **174**:4230–4238.

Armitage, J.P. (1981) Changes in metabolic activity of *Proteus mirabilis* during swarming. *J. Gen. Microbiol.* **125**:445–450.

Armitage, J.P. (1982) Changes in the organisation of the outer membrane of *Proteus mirabilis* during swarming: freeze-fracture structure and membrane fluidity analysis. *J. Bacteriol.* **150**:900–904.

Armitage, J.P., and Smith, D.G. (1978) Flagella development during swarmer differentiation in *Proteus mirabilis. FEMS Microbiol. Lett.* **4**:163–165.

Armitage, J.P., Rowbury, R.J., and Smith, D.G. (1975) Indirect evidence for cell wall and membrane differences between filamentous swarming cells and short non-swarming cells of *Proteus mirabilis. J. Gen. Microbiol.* **89**:199–202.

Armitage, J.P., Smith, D.G., and Rowbury, R.J. (1979) Alterations in the cell envelope composition of *Proteus mirabilis* during the development of swarmer cells. *Biochim. Biophys. Acta* **584**:389–397.

Asai, T., Takanmi, M., and Imai, M. (1990) The AT richness and *gid* transcription determine the left border of the replication origin of the *E. coli* chromosome. *EMBO J.* **9**:4065–4072.

Atsumi, T., McCarter, L., and Imae, Y. (1992) Polar and lateral flagellar motors of marine *Vibrio* are driven by different ion-motive forces. *Nature* **355**:182–184.

Bahrani, F.K., Johnson, D.E., Robbins, D., and Mobley, H.L. (1991) *Proteus mirabilis* flagella and MR/P fimbriae: isolation, purification, N-terminal analysis, and serum antibody response following experimental urinary tract infection. *Infect. Immun.* **59**:3574–3580.

Bartlett, D.H., Frantz, B.B., and Matsumura, P. (1988) Flagellar transcriptional activators FlbB and FlaI: gene sequences and 5' consensus sequences of operons under FlbB and FlaI control. *J. Bacteriol.* **170**:1575–1581.

Bastin, D., Stevenson, G., Brown, P., Haase, A., and Reeves, P. (1993) Repeat unit polysaccharides of bacteria: a model for polymerization resembling that of ribosomes and fatty acid synthetase, with a novel mechanism for determining chain length. *Mol. Microbiol.* **7**:725–734.

Belas, R. (1992) The swarming phenomenon of *Proteus mirabilis. Am. Soc. Microbiol. News* **58**:15–22.

Belas, R. (1994) Expression of multiple flagellin-encoding genes of *Proteus mirabilis. J. Bacteriol.* **176**:7169–7181.

Belas, R., and Flaherty, D. (1994) Sequence and genetic analysis of multiple flagellin-encoding genes from *Proteus mirabilis. Gene* **148**:33–41.

Belas, R., Erskine, D., and Flaherty, D. (1991a) *Proteus mirabilis* mutants defective in swarmer cell differentiation and multicellular behavior. *J. Bacteriol.* **173**:6279–6288.

Belas, R., Erskine, D., and Flaherty, D. (1991b) Transposon mutagenesis in *Proteus mirabilis. J. Bacteriol.* **173**:6289–6293.

Belas, R., Goldman, M., and Ashliman, K. (1995) Genetic analysis of *Proteus mirabilis* mutants defective in swarmer cell elongation. *J. Bacteriol.* **177**:823–828.

Belas, R., Simon, M., and Silverman, M. (1986) Regulation of lateral flagella gene transcription in *Vibrio parahaemolyticus. J. Bacteriol.* **167**:210–218.

Bisset, K.A. (1973a) The motion of the swarm in *Proteus mirabilis. J. Med. Microbiol.* **6**:33–35.

Bisset, K.A. (1973b) The zonation phenomenon and structure of the swarm colony in *Proteus mirabilis. J. Med. Microbiol.* **6**:429–433.

Bisset, K.A., and Douglas, C.W. (1976) A continuous study of morphological phase in the swarm of *Proteus. J. Med. Microbiol.* **9**:229–231.

Bouvier, J., Richaud, C., Higgins, W., Bogler, O., and Stragier, P. (1992) Cloning, characterization, and expression of the *dapE* gene of *Escherichia coli. J. Bacteriol.* **174**:5265–5271.

Brogan, T.D., Nettleton, J., and Reid, C. (1971) The swarming of *Proteus* on semisynthetic media. *J. Med. Microbiol.* **4**:1–11.

Brunham, R.C., Plummer, F.A., and Stephens, R.S. (1993) Bacterial antigenic variation, host immune response, and pathogen-host coevolution. *Infect. Immun.* **61**:2273–2276.

Carpenter, P.B., and Ordal, G.W. (1993) Bacillus subtilis FlhA: a flagellar protein related to a new family of signal-transducing receptors. *Mol. Microbiol.* **7**:735–743.

Chen, L., and Coleman, W. (1993) Cloning and characterization of the *Escherichia coli* K-12 *rfa-2* (*rfaC*) gene, a gene required for lipopolysaccharide inner core synthesis. *J. Bacteriol.* **175**:2534–2540.

De Boer, S.E., Golten, C., and Scheffers, W.A. (1975) Effects of some chemical factors on flagellation and swarming of *Vibrio alginolyticus*. *Antonie van Leeuwenhoek* **41**:385–403.

De Lorenzo, M., Herrero, M., Jakubzik, U., and Timmis, K.N. (1990) Mini-Tn5 transposon derivatives for insertion mutagenesis, promoter probing, and chromosomal insertion of cloned DNA in gram-negative eubacteria. *J. Bacteriol.* **172**:6568–6572.

De Louvois, J. (1969) Serotyping and the Dienes reaction on *Proteus mirabilis* from hospital infections. *J. Clin. Pathol.* **22**:263–268.

Dick, H., Murray, R.G., and Walmsley, S. (1985) Swarmer cell differentiation of *Proteus mirabilis* in fluid media. *Can. J. Microbiol.* **31**:1041–1050.

Dienes, L. (1946) Reproductive processes in *Proteus* cultures. *Proc. Soc. Exp. Biol. Med.* **63**:265–270.

Dienes, L. (1947) Further observations on the reproduction of bacilli from large bodies in *Proteus* cultures. *Proc. Soc. Exp. Biol. Med.* **66**:97–98.

Douglas, C.W. (1979) Measurement of *Proteus* cell motility during swarming. *J. Med. Microbiol.* **12**:195–199.

Douglas, C.W., and Bisset, K.A. (1976) Development of concentric zones in the *Proteus* swarm colony. *J. Med. Microbiol.* **9**:497–500.

Dreyfus, G., Williams, A.W., Kawagishi, I., and Macnab, R.M. (1993) Genetic and biochemical analysis of *Salmonella typhimurium* FliI, a flagellar protein related to the catalytic subunit of the F0F1 ATPase and to virulence proteins of mammalian and plant pathogens. *J. Bacteriol.* **175**:3131–3138.

Falkinham, J.O.I., and Hoffman, P.S. (1984) Unique developmental characteristics of the swarm and short cells of *Proteus vulgaris* and *Proteus mirabilis*. *J. Bacteriol.* **158**:1037–1040.

Galan, J.E., Ginocchio, C., and Costeas, P. (1992) Molecular and functional characterization of the *Salmonella* invasion gene *invA*: homology of InvA to members of a new protein family. *J. Bacteriol.* **174**:4338–4349.

Gmeiner, J., Sarnow, E., and Milde, K. (1985) Cell cycle parameters of *Proteus mirabilis*: interdependence of the biosynthetic cell cycle and the interdivision cycle. *J. Bacteriol.* **164**:741–748.

Gober, J.W., and Shapiro, L. (1992) A developmentally regulated *Caulobacter* flagellar promoter is activated by 3' enhancer and IHF binding elements. *Mol. Biol. Cell* **3**: 913–926.

Gough, C.L., Genin, S., Lopes, V., and Boucher, C.A. (1993) Homology between the HrpO protein of *Pseudomonas solanacearum* and bacterial proteins implicated in a signal peptide-independent secretion mechanism. *Mol. Gen. Genet.* **239**:378–392.

Grabow, W.O.K. (1972) Growth-inhibiting metabolites of *Proteus mirabilis*. *J. Med. Microbiol.* **5**:191–204.

Guerry, P., Alm, R.A., Power, M.E., Logan, S.M., and Trust, T.J. (1991) Role of two flagellin genes in *Campylobacter* motility. *J. Bacteriol.* **173**:4757–4764.

Guerry, P., Logan, S.M., and Trust, T.J. (1988) Genomic rearrangement associated with antigenic variation in *Campylobacter coli*. *J. Bacteriol.* **170**:316–319.

Gygi, D., Bailey, M., Allison, C., and Hughes, C. (1995a) Requirement for FlhA in flagella assembly and swarm-cell differentiation by *Proteus mirabilis*. *Mol. Microbiol.* **15**:761–769.

Gygi, D., Rahman, M., Lai, H.-C., Carlson, R., Guard-Petter, J., and Hughes, C. (1995b) A cell-surface polysaccharide that facilitates rapid population migration by differentiated swarm cells of *Proteus mirabilis*. *Mol. Microbiol.* **17**:1167–1175.

Hall, P.G., and Krieg, N.R. (1983) Swarming of *Azospirillum brasilense* on solid media. *Can. J. Microbiol.* **129**:1592–1594.

Harshey, R. (1994) Bees aren't the only ones: swarming in gram-negative bacteria. *Mol. Microbiol.* **13**:389–394.

Harshey, R., and Matsuyama, T. (1994) Dimorphic transition in *E. coli* and *S. typhimurium*: surface-induced differentiation into hyperflagellate swarmer cells. *Proc. Natl. Acad. Sci. USA* **91**:8631–8635.

Helmann, J.D. (1991) Alternative sigma factors and the regulation of flagellar gene expression. *Mol. Microbiol.* **5**:2875–2882.

Henrichsen, J. (1972) Bacterial surface translocation: a survey and a classification. *Bacteriol. Rev.* **36**:478–503.

Hoeniger, J.F.M. (1964) Cellular changes accompanying the swarming of *Proteus mirabilis*. I. Observations on living cultures. *Can. J. Microbiol.* **10**:1–9.

Hoeniger, J.F.M. (1965) Development of flagella by *Proteus mirabilis*. *J. Gen. Microbiol.* **40**:29–42.

Hoeniger, J.F.M. (1966) Cellular changes accompanying the swarming of *Proteus mirabilis*. II. Observations of stained organisms. *Can. J. Microbiol.* **12**:113–122.

Hoeniger, J.F.M., and Cinits, E.A. (1969) Cell wall growth during differentiation of *Proteus* swarmers. *J. Bacteriol.* **148**:736–738.

Hoffman, P., and Falkinham, J.I. (1981) Induction of tryptophanase in short cells and swarm cells of *Proteus vulgaris*. *J. Bacteriol.* **148**:736–738.

Houwink, A.L., and van Iterson, W. (1950) Electron microscopical observations on bacterial cytology. II. A study of flagellation. *Biochim. Biophys. Acta* **5**:10–16.

Hughes, W.H. (1956) The structure and development of the induced long forms of bacteria. *Symp. Soc. Gen. Microbiol.* **6**:341–360.

Hughes, W.H. (1957) A reconsideration of the swarming of *Proteus vulgaris*. *J. Gen. Microbiol.* **17**:49–58.

Jeffries, C.D., and Rogers, H.E. (1968) Enhancing effect of agar on swarming by *Proteus*. *J. Bacteriol.* **95**:732–733.

Jiang, X., Neal, B., Santiago, F., Lee, S., Romana, L., and Reeves, P. (1991) Structure and sequence of the *rfb* (O antigen) gene cluster of *Salmonella typhimurium* (strain LT2) *Mol. Microbiol.* **5**:695–713.

Jin, T., and Murray, R.G.E. (1988) Further studies of swarmer cell differentiation of *Proteus mirabilis* PM23: a requirement for iron and zinc. *Can. J. Microbiol.* **34**: 588–593.

Kaneko, T., and Colwell, R.R. (1973) Ecology of *Vibrio parahaemolyticus* in Chesapeake Bay. *J. Bacteriol.* **113**:24–32.

Kustu, S., North, A.K., and Weiss, D.S. (1991) Prokaryotic transcriptional enhancers and enhancer-binding proteins. *Trends Biochem. Sci.* **16**:397–402.

Lidell, M.C., and Hutcheson, S.W. (1994) Characterization of the *hrpJ* and *hrpU* operons of *Pseudomonas syringae* pv. *syringae* Pss61: similarity with components of enteric bacteria involved in flagellar biogenesis and demonstration of their role in HarpinPss secretion. *Mol. Plant Microbe Interact.* **7**:488–497.

Liu, D., Haase, A., Lindqvist, L., Lindberg, A., and Reeves, P. (1993) Glycosyl transferases of O-antigen biosynthesis in *Salmonella enterica*: identification and characterization of transferase genes of groups B, C2 and E1. *J. Bacteriol.* **175**:3408–3413.

Loomes, L.M., Kerr, M.A., and Senior, B.W. (1993) The cleavage of immunoglobulin G in vitro and in vivo by a proteinase secreted by the urinary tract pathogen *Proteus mirabilis*. *J. Med. Microbiol.* **39**:225–232.

Loomes, L.M., Senior, B.W., and Kerr, M.A. (1990) A proteolytic enzyme secreted by *Proteus mirabilis* degrades immunoglobulins of the immunoglobulin A1 (IgA1), IgA2, and IgG isotypes. *Infect. Immun.* **58**:1979–1985.

Loomes, L.M., Senior, B.W., and Kerr, M.A. (1992) Proteinases of *Proteus* spp.: purification, properties, and detection in urine of infected patients. *Infect. Immun.* **60**:2267–2273.

Macnab, R.M. (1987) Flagella. In *Escherichia coli* and *Salmonella typhimurium: Molecular and Cellular Biology.* F.C. Neidhardt, J.L. Ingraham, K.B. Low, et al. (eds.), pp. 70–83. American Society for Microbiology, Washington, DC.

Macnab, R.M. (1992) Genetics and biogenesis of bacterial flagella. *Annu. Rev. Genet.* **26**:131–158.

Matsuyama, T., Kaneda, K., Nakagawa, Y., Isa, K., Hara-Hotta, H., and Isuya, Y. (1992) A novel extracellular cyclic lipopeptide which promotes flagellum-dependent and -independent spreading growth of *Serratia marcescens*. *J. Bacteriol.* **174**:1769–1776.

McCarter, L. (1994) MotX, the channel component of the sodium-type flagellar motor. *J. Bacteriol.* **176**:5988–5998.

McCarter, L., and Silverman, M. (1989) Iron regulation of swarmer cell differentiation of *Vibrio parahaemolyticus*. *J. Bacteriol.* **171**:731–736.

McCarter, L., and Silverman, M. (1990) Surface-induced swarmer cell differentiation of *Vibrio parahaemolyticus*. *Mol. Microbiol.* **4**:1057–1062.

McCarter, L., Hilmen, M., and Silverman, M. (1988) Flagellar dynamometer controls swarmer cell differentiation of *V. parahaemolyticus*. *Cell* **54**:345–351.

McCarter, L.L., and Wright, M.E. (1993) Identification of genes encoding components of the swarmer cell flagellar motor and propeller and a sigma factor controlling differentiation of *Vibrio parahaemolyticus*. *J. Bacteriol.* **175**:3361–3371.

Miller, S., Pesci, E.C., and Pickett, C.L. (1994) Genetic organization of the region upstream from the *Campylobacter jejuni* flagellar gene *flhA*. *Gene* **146**:31–38.

Mirel, D.B., and Chamberlin, M.J. (1989) The *Bacillus subtilis* flagellin gene (*hag*) is transcribed by the σ^{28} form of RNA polymerase. *J. Bacteriol.* **171**:3095–3101.

Miwatani, T., and Shinoda, S. (1971) Flagellar antigen of *Vibrio alginolyticus*. *Biken J.* **14**:389–394.

Mobley, H.L.T., and Warren, J.W. (1987) Urease-positive bacteriuria and obstruction of long-term urinary catheters. *J. Clin. Microbiol.* **25**:2216–2217.

Morona, R., Mavris, M., Fallarino, A., and Manning, P. (1994) Characterization of the *rfc* region of *Shigella flexneri*. *J. Bacteriol.* **176**:733–747.

Nakahigashi, K., and Inokuchi, H. (1990) Nucleotide sequence between *fadB* and *rrnA* from *Escherichia coli*. *Nucleic Acids Res.* **18**:6439–6439.

Naylor, P. (1964) The effect of electrolytes or carbohydrates in a sodium chloride deficient medium on the formation of discrete colonies of *Proteus* and the influence of these substances on growth in liquid culture. *J. Appl. Bacteriol.* **27**:422–431.

Nicholson, E.B., Concaugh, E.A., Foxall, P.A., Island, M.D., and Mobley, H.L. (1993) *Proteus mirabilis* urease: transcriptional regulation by UreR. *J. Bacteriol.* **175**:465–473.

Ochman, H., and Wilson, A. (1987) Evolutionary history of enteric bacteria. In *Escherichia coli and Salmonella typhimurium: Cellular and Molecular Biology*, J. Ingraham, B.B. Low, M. Schaechter, and H. Umbarger (eds.), pp. 1649–1654. American Society for Microbiology, Washington, DC.

Ogawa, T., and Okazaki, T. (1991) Concurrent transcription from the *gid* and *mioC* promoters activates replication of an *Escherichia coli* minichromosome. *Mol. Gen. Genet.* **230**:193–200.

O'Rear, J., Alberti, L., and Harshey, R.M. (1992) Mutations that impair swarming motility in *Serratia marcescens* 274 include but are not limited to those affecting chemotaxis or flagellar function. *J. Bacteriol.* **174**:6125–6137.

Pegues, J., Chen, L., Gordon, A., Ding, L., and Coleman, W., Jr. (1990) Cloning, expression, and characterization of the *Escherichia coli* K-12 *rfaD* gene. *J. Bacteriol.* **172**:4652–4660.

Plaut, A. (1983) The IgA1 proteases of pathogenic bacteria. *Annu. Rev. Microbiol.* **37**: 603–622.

Proom, H., and Woiwod, A. (1951) Amine production in the genus *Proteus*. *J. Gen. Microbiol.* **5**:930–938.

Quadling, C., and Stocker, B.A.D. (1957) The occurrence of rare motile bacteria in some non-motile *Salmonella* strains. *J. Gen. Microbiol.* **17**:424–436.

Sar, N., McCarter, L., Simon, M., and Silverman, M. (1990) Chemotactic control of the two flagellar systems of *Vibrio parahaemolyticus*. *J. Bacteriol.* **172**:334–341.

Scott, T.N., and Simon, M.I. (1982) Genetic analysis of the mechanism of the *Salmonella* phase variation site specific recombination system. *Mol. Gen. Genet.* **188**: 313–321.

Senior, B., Loomes, L., and Kerr, M. (1991) The production and activity in vivo of *Proteus mirabilis* IgA protease in infections of the urinary tract. *J. Med. Microbiol.* **35**:203–207.

Senior, B.W. (1977) The Dienes phenomenon: identification of the determinants of compatibility. *J. Gen. Microbiol.* **102**:235–244.

Senior, B.W. (1983) *Proteus morgani* is less frequently associated with urinary tract infections than *Proteus mirabilis*—an explanation. *J. Med. Microbiol.* **16**:317–322.

Senior, B.W., Albrechtsen, M., and Kerr, M.A. (1987) *Proteus mirabilis* strains of diverse type have IgA protease activity. *J. Med. Microbiol.* **24**:175–180.

Shinoda, S., Miwatani, T., and Fujino, T. (1970) Antigens of *Vibrio parahaemolyticus*. II. Existence of two different subunits in the flagella of *Vibrio parahaemolyticus* and their characterization. *Biken J.* **13**:241–247.

Silverman, M. (1980) Building bacterial flagella. *Q. Rev. of Bio.* **55**:395–408.

Silverman, M.R., and Simon, M.I. (1972) Flagellar assembly mutants in *Escherichia coli*. *J. Bacteriol.* **112**:986–993.

Silverman, M., and Simon, M. (1973) Genetic analysis of flagellar mutants in *Escherichia coli*. *J. Bacteriol.* **113**:105–113.

Silverman, M., and Simon, M. (1974) Assembly of hybrid flagellar filaments. *J. Bacteriol.* **118**:750–752.

Silverman, M., and Simon, M.I. (1977) Bacterial flagella. *Annu. Rev. Microbiol.* **31**: 397–419.

Sirisena, D., Brozek, K., MacLachlan, P., Sanderson, K., and Raetz, C. (1992) The *rfaC* gene of *Salmonella typhimurium*: cloning, sequencing, and enzymatic function in heptose transfer to lipopolysaccharide. *J. Biol. Chem.* **267**:18874–18884.

Skirrow, M.B. (1969) The Dienes (mutual inhibition) test in the investigation of *Proteus* infections. *J. Med. Microbiol.* **2**:471–477.

Sohnius, I., and Lenk, V. (1977) Investigations made to test the demarcation line method (DLM) after Dienes for its suitability in the epidemiology of *Proteus mirabilis*. *Zentralbl. Bakteriol.* [*Orig.*] **164**:384–389.

Sourek, J. (1968) On some findings concerning Dienes's phenomenon in swarming *Proteus* strains. *Zentralbl. Bakteriol.* [*Orig.*] **208**:419–427.

Stahl, S.J., Stewart, K.R., and Williams, F.D. (1983) Extracellular slime associated with *Proteus mirabilis* during swarming. *J. Bacteriol.* **154**:930–937.

Stephens, C., and Shapiro, L. (1993) An unusual promoter controls cell-cycle regulation and dependence on DNA replication of the *Caulobacter fliLM* early flagellar operon. *Mol. Microbiol.* **9**:1169–1179.

Story, P. (1954) *Proteus* infections in hospitals. *J. Pathol. Bacteriol.* **68**:55–62.

Szekely, E., and Simon, M. (1983) DNA sequence adjacent to flagellar genes and evolution of flagellar-phase variation. *J. Bacteriol.* **155**:74–81.

Ulitzur, S. (1975a) Effect of temperature, salts, pH, and other factors on the development of peritrichous flagella in *Vibrio alginolyticus. Arch. Microbiol.* **104**:285–288.

Ulitzer, S. (1975b) The mechanism of swarming of *Vibrio alginolyticus. Arch. Microbiol.* **104**:67–71.

Varon, D., Boylan, S., Okamoto, K., and Price, C. (1993) *Bacillus subtilis gtaB* encodes UDP-glucose pyrophosphorylase and is controlled by stationary phase transcription factor sigma-B. *J. Bacteriol.* **175**:3964–3971.

Vogler, A.P., Homma, M., Irikura, V.M., and Macnab, R.M. (1991) *Salmonella typhimurium* mutants defective in flagellar filament regrowth and sequence similarity of FliI to F0F1, vacuolar, and archaebacterial ATPase subunits. *J. Bacteriol.* **173**:3564–3572.

Von Meyenburg, K., and Hansen, F. (1980) The origin of replication, *oriC*, of the *Escherichia coli* chromosome: genes near *oriC* and construction of *oriC* deletion mutants. *ICN UCLA Symp.* **19**:137–159.

Von Meyenburg, K., and Hansen, F. (1987) Regulation of chromosome replication. In *Escherichia coli and Salmonella typhimurium: Molecular and Cellular Biology*, F. Neidhardt, J. Ingraham, K. Low, et al. (eds.), pp. 1555–1577. American Society for Microbiology, Washington, DC.

Von Meyenburg, K., Jorgensen, B., Neilsen, J., and Hansen, F. (1982) Promoters of the *atp* operon coding for the membrane-bound ATP synthase of *Escherichia coli* mapped by *Tn10* insertion mutations. *Mol. Gen. Genet.* **188**:240–248.

Wassif, C., Cheek, D., and Belas, R. (1995) Molecular analysis of a metalloprotease from *Proteus mirabilis. J. Bacteriol.* **177**:5790–5798.

Wei, Z.M., and Beer, S.V. (1993) HrpI of *Erwinia amylovora* functions in secretion of harpin and is a member of a new protein family. *J. Bacteriol.* **175**:7958–7967.

Williams, F.D., and Schwarzhoff, R.H. (1978) Nature of the swarming phenomenon in *Proteus. Annu. Rev. Microbiol.* **32**:101–122.

Zieg, J., Silverman, M., Hilmen, M., and Simon, M. (1977) The metabolism of phase variation. In *Molecular Approaches to Eucaryotic Genetic Systems*, G. Wilcox (eds.), pp. 25–35. Academic Press, Orlando, FL.

8

Myxobacterial Multicellularity

LAWRENCE J. SHIMKETS & MARTIN DWORKIN

Bacteria are usually thought of as unicellular creatures. This book and the two American Society for Microbiology (ASM) conferences that preceded it emphasize that the view of bacteria as solitary organisms is more a reflection of the techniques microbiologists have used to study microorganisms than an accurate view of the microbial world. The myxobacteria, with their dramatic multicellularity and cell–cell interactions emphatically belie the strict unicellular paradigm.

Microbial development, especially among eukaryotic microbes such as the cellular slime molds, is often viewed as an alternation between a unicellular and a multicellular stage. Although one might be tempted to view myxobacterial development in a similar fashion, a closer look leads one instead to the view that the entire life cycle is pervaded by social behavior, including cooperative feeding, group motility, aggregation and cohesion, and rippling and fruiting body formation. The goal of this chapter is to describe and emphasize these myxobacterial cell–cell interactions that may have been nature's earliest experiments on its way to functional multicellularity. Detailed descriptions of myxobacterial biology have been offered by Reichenbach and Dworkin (1992), Shimkets (1990), Dworkin (1996), and Dworkin and Kaiser (1993).

Most of the recent research on myxobacterial development and behavior has been done on *Myxococcus xanthus*, which has served as a model myxobacterium. We are well aware of the danger of extrapolating to generalizations about all myxobacteria from studies on a single experimentally tractable species.

Myxobacterial Life Cycle

The following description of the life cycle of *M. xanthus* is intended to set the stage for subsequent, more detailed discussions of the individual social phenomena. The life cycle of *M. xanthus*, illustrated in Figure 8.1, can be seen as two separate but interlocking cycles of growth and development. Under the proper conditions, *M. xanthus* can be grown exponentially in liquid culture consisting of a complex source of nitrogen, with a minimum generation time of about 4 hours.

If three conditions are satisfied, *M. xanthus* enters the developmental phase of its life cycle: (1) perception of a nutritional shift-down (Dworkin, 1963; Manoil and Kaiser, 1980); (2) presence on a solid surface so gliding motility is permitted; and (3) presence at a sufficiently high cell density (Wireman and Dworkin, 1975; Shimkets and Dworkin, 1981). Under these conditions, the

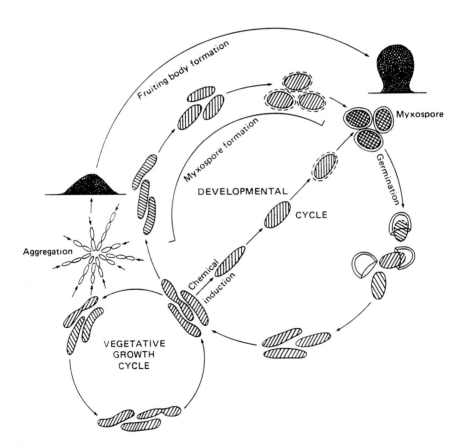

FIGURE 8.1   *M. xanthus* life cycle. The fruiting body is not drawn to scale but is a few hundredths of a millimeter in diameter, in contrast to the vegetative cells, which are about 5 × 0.7 to 7 × 0.7 μm. (Reprinted from Dworkin, 1985, with permission)

cells glide into aggregation centers, following tactic clues that have not yet been defined. In addition to the visible manifestations of the initiation of development, a series of signaling events is set in motion that persist throughout development (Kaiser and Kroos, 1993). These events are accompanied by an ordered program of gene expression measured by the activation of a set of Tn5 *lac* insertions in developmentally regulated genes (Kroos and Kaiser, 1987).

It is almost diagnostic of the behavior of certain genera of myxobacteria that developmental aggregation is accompanied by the behavioral phenomenon known as ''rippling'' (Fig. 8.2). Neither the function nor the mechanism of rippling is understood; the phenomenon is discussed in greater detail later in the chapter.

During the early stages of aggregation, a portion of the population begins to lyse. The lysis reaches a peak during aggregation and can result in death of as many as 80–90% of the cells. The surviving cells convert to resistant resting cells, the myxospores. There is no evidence that the process is programmed in a fashion analogous to apoptosis in vertebrate tissues (Cohen et al., 1992); it may be purely stochastic. Nevertheless, it is another interesting example of widely different groups of organisms manifesting a similar population behavior.

The morphological culmination of aggregation is the formation of the fruiting body. The genus *Myxococcus* forms one of the simplest myxobacterial fruiting bodies (Fig. 8.3). Biologists have been intrigued by the problem of the generation of form. More often than not it has resisted analysis; and morpho-

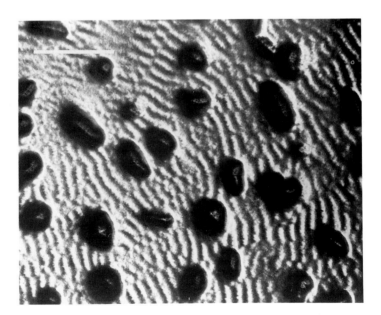

FIGURE 8.2   Induction of *M. xanthus* ripples by peptidoglycan components. Culture with 2.5 mM each *N*-acetylglucosamine, D-alanine, and diaminopimelate. Bar = 400 μm. (Reprinted from Shimkets and Kaiser, 1982, with permission)

FIGURE 8.3   *Myxococcus fulvus* fruiting bodies. (Reprinted from Dworkin and Kaiser, 1993, with permission)

genesis, in the literal sense, remains the aspect of developmental biology that has been most difficult to penetrate. Apropos, it is exciting that work on fruiting body morphogenesis in *M. xanthus* begins to offer some promise. This work is discussed in more detail under Directed Movement During Fruiting Body Formation (p. 237).

The final stage in the formation of the mature fruiting body is the cellular morphogenesis of the rod-shaped vegetative cells to round, optically refractile,

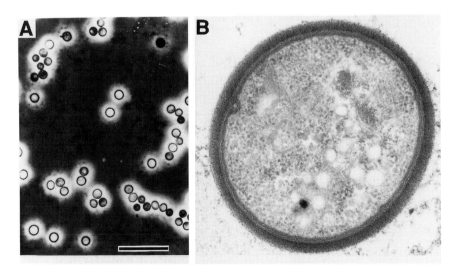

FIGURE 8.4   Myxospores of *M. xanthus*. **(A)** Phase contrast micrograph. Bar = 10 μm. **(B)** Electron micrograph of a thin section. (×75,000) (**A**, reprinted from Rosenberg, 1984; **B**, reprinted from Dworkin, 1979; with permission)

resistant, metabolically quiescent myxospores (Fig. 8.4). Spore formation has not been examined in the same detail as the earlier stages of development.

Genetic Nomenclature

Genetic analysis of *M. xanthus* social behavior has involved systematic isolation of mutants with altered vegetative and developmental properties. Characterization of these mutants is in progress and should reveal the regulatory networks that coordinate social behavior. Genetic approaches for analysis of myxobacterial behavior were reviewed by Gill and Shimkets (1993). A comprehensive list of the known genes, their mnenomics, their map locations, and their phenotypic traits can be found in Shimkets (1993). Table 8.1 is a summary of the loci described in this chapter.

Myxobacterial Social Behavior

Cooperative Growth

It has been proposed that the raison d'être for the social quality of the myxobacterial life cycle is the optimization of feeding (Dworkin, 1972). The tendency of the myxobacteria to feed collectively as a swarm on macromolecules was demonstrated by the density-dependent growth of *M. xanthus* on protein (Rosenberg et al., 1977). They showed that the growth rate of *M. xanthus* on unhydrolyzed casein was a direct function of the cell density and, in turn, that the cell density was directly proportional to the amount of excreted protease and finally to the accumulation of hydrolyzed casein in the medium. It thus seems clear that while the myxobacteria are in their motile phase the social quality of their behavior optimizes the quality of their feeding. This phenomenon has been referred to as the myxobacterial wolf-pack effect (Dworkin, 1973). While in the sessile stage of its life cycle, the fruiting body maintains the potential for a high density population upon germination and release of the cells from their quiescent state.

One cannot help but wonder about the function of the more elaborate fruiting structures produced by myxobacteria, such as *Stigmatella* and *Chondromyces* (Fig. 8.5). Note that the fruiting bodies contain multiple sacs (sporangioles), each of which is filled with resting cells. Germination in these myxobacteria consists of a sporangiole opening and releasing a swarm of cells. It is feasible that the explanation for the morphologic complexity of these fruiting bodies lies with the process of optimal packaging of myxospores. The cells in each of the sporangioles may represent an optimal swarm size, and this process may increase the efficiency of dispersal of the clone.

TABLE 8.1 Some genes of *Myxococcus xanthus* controlling cell behavior

Gene Symbol	Role	Gene Product
aglB	Adventurous gliding	Unknown
asgA	Production	Histidine protein kinase/response regulation
asgB	Production	DNA binding protein
asgC	Production	Signal factor
bsgA	Production	ATP-dependent protease
cglB1	Contact adventurous gliding	Unknown
cglF1; *cglF2*	Contact adventurous gliding	Unknown
csgA	C-signal production	Short chain alcohol dehydrogenase
dsgA	D-signal production	Translation IF-3
dsp	Dispersed growth	Unknown
frzA	Frizzy	CheW-like
frzB	Frizzy	Motility regulation
frzCD	Frizzy	Methyl-accepting chemotaxis protein (MCP)
frzE	Frizzy	Phosphokinase
frzF	Frizzy	MCP methyltransferase
frzG	Frizzy	Demethylation of MCP
mglA	Mutual gliding	G protein
mglB	Mutual gliding	Stimulates mglA
sgl-3119	Social gliding	Unknown
sglA1	Social gliding	Unknown
stk-1907	Sticky	Unknown
tglA	Transient social gliding	Unknown

Cell–Cell Attachments

A tendency to maintain close proximity with adjacent cells is manifested in the cells' behavior during feeding (Rosenberg et al., 1977), fibril formation (Behmlander and Dworkin, 1991), rippling (Shimkets and Kaiser, 1982), aggregation (Shimkets and Dworkin, 1981), signaling (Kaiser and Kroos, 1993), fruiting body formation (Sager and Kaiser, 1993a,b), and social motility (Hodgkin and Kaiser, 1977). The two systems that have been examined most carefully in the context of cell–cell proximity are social motility (McBride et al., 1993) and C-signaling (Kim and Kaiser, 1990a,c).

Two extracellular organelles have been variously implicated in the social behavior of the cells: pili and fibrils. Pili, first demonstrated in myxobacteria by MacRae and McCurdy (1975), have been shown to be regulated by the social motility of *M. xanthus*. Hodgkin and Kaiser (1977) found that certain pairs of nonmotile mutants temporarily reacquired motility when placed in close proximity to each other. Subsequently, Kaiser (1979) showed that the loss of social motility was correlated with the loss of pili; and, most interestingly, when *tgl* and *sgl* social motility mutants were placed in close proximity to each other the *tgl* mutants acquired both social motility and pili. The mechanism of this stimulation is thought to be due to transfer of the Tgl gene product from *tgl*⁺ to *tgl*⁻ cells to temporarily restore an intact social motility system (Hodg-

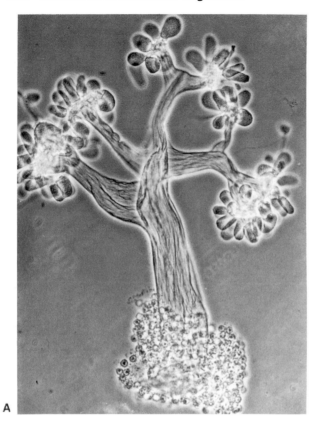

A   B

FIGURE 8.5   (**A**) *Chondromyces crocatus* fruiting body. Bar = 100 μm. (**B**) *Stigmatella aurantiaca* fruiting body. Bar = 50 μm. (**A, B**, reprinted from Reichenbach, 1993, with permission)

kin and Kaiser, 1977). Kaiser (1979) concluded that "it appears that pili permit cells that are close to one another to move." This conclusion was strengthened by the experiments of Rosenbluh and Eisenbach (1992), who showed that the mechanical depilation of *M. xanthus* resulted in the loss of social motility. The genetics of social motility is discussed under Social Motility, (p. 229).

Another organelle that may play a role in social behavior emerged from the discovery by Arnold and Shimkets (1988a,b) that cohesion and social behavior of *M. xanthus* was correlated with the presence of extracellular organelles called fibrils. Fibrils are long appendages about 30 nm in diameter arranged peritrichously around the cell (Fig. 8.6) and composed of approximately equal amounts of polysaccharide and protein (Behmlander and Dworkin, 1994). Arnold and Shimkets (1988a) showed that if they prevented fibril formation by treating the cells with the diazo dye congo red the cells were unable to undergo cell–cell cohesion, were blocked at the earliest stages of development, and showed no signs of social behavior. An A^+S^- (adventurous-positive, social-negative; see below) mutant, *dsp*, that lacked fibrils displayed the same spec-

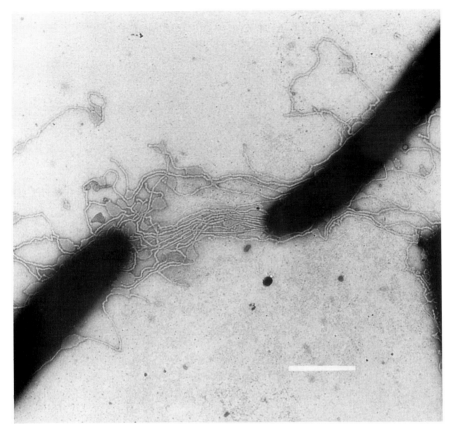

FIGURE 8.6   Transmission electron micrograph of negatively stained wild-type cells of *M. xanthus* showing the fibrils. Bar = 1 μm. (Courtesy of J.W. Arnold and L.J. Shimkets)

trum of properties. On the basis of these observations, Arnold and Shimkets (1988a) suggested that the fibrils play a role in the social behavior of the cells. Behmlander and Dworkin (1991) subsequently characterized the fibrils morphologically (Fig. 8.7) and chemically and showed that they were formed only under conditions of high cell density. Convincing evidence for the role of fibrils in the social behavior of the cells was obtained when it was shown that purified wild-type fibrils restored the ability to cohere and to form aggregates, fruiting bodies, and myxospores to *dsp* mutants (Chang and Dworkin, 1994).

Rosenbluh and Eisenbach (1992) speculated that pili may act as tactile sensors that monitor cell–cell proximity and at a properly high cell density may signal the cell to produce extracellular fibrils, which then mediate the cell–cell attachments characteristic of social behavior. This theory fails to explain the fact that the *tgl* social motility mutants that lack pili nevertheless regain motility and pili when placed close to *tgl*[+] cells. Alternatively, there

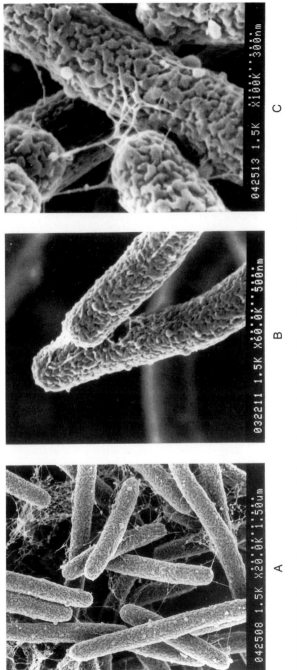

FIGURE 8.7 Low voltage scanning electron micrographs of vegetative cells of *M. xanthus*. (**A**) Cluster of cells with extensive fibrillar connections. (**B**) Cells from submerged culture anchored to the substratum at one end. (**C**) High magnification of the cell surface and fibril-anchoring point. The corrugated surface probably indicates shrinkage of the cells even though they have been fixed and critical-point dried. (Reprinted from Behmlander and Dworkin, 1991, with permission)

may be another signal, possibly *Tgl* protein, the exchange of which may precede a pilus-mediated signal.

The other system for which the role of cell–cell proximity has been examined is transmission of the C-signal between cells (see Developmental Signal Exchange, below). Nonmotile mutants cease expression of developmentally regulated genes at approximately the same point in the developmental program as *csgA* (*C-si*gnaling) mutants (Kroos et al., 1988). Kim and Kaiser (1990b) found that the nonmotile mutant *mglA* was unable to rescue development of the *csgA* mutant, despite the fact that *mglA* was able to produce normal amounts of the C-signal. To rationalize this apparent contradiction, they proposed that motility was required for effective exchange of the C-signal between cells. They suggested that the motility was necessary to align the cells closely enough to each other so that effective signal exchange could occur. In an elegantly simple experiment, Kim and Kaiser (1990a) supported this proposal by showing that nonmotile cells placed in narrow grooves etched in an agar surface, were sufficiently aligned to express a C-signal-dependent developmental gene. These results suggest that mechanical alignment improves transmission of the C-signal.

The mechanism(s) by which these proximity/contact-requiring interactions occur is unknown. One can envisage two possible roles for extracellular appendages such as pili and fibrils: They may be involved in sensing the proximity of adjacent cells, and they may play a role in maintenance of the physical interactions between cells during social motility, aggregation, and fruiting body formation. At this time, neither the precise roles played by these two classes of organelles nor the interactions between them is clear.

Social Motility

The movement and behavior of gliding myxobacterial cells is controlled by two multigene systems known as A (adventurous) and S (social), which determine whether cells move as individuals or groups, respectively (Hodgkin and Kaiser, 1979a,b). Mutations in the A system genes eliminate movement of isolated cells, and movement is possible only if two cells are within one cell length (Hodgkin and Kaiser, 1979a; Kaiser and Crosby, 1983) (Fig. 8.8, A^-S^+). The S system can most easily be thought of as contact-dependent motility. Mutations in the S system genes reduce group movement, but cells are still motile via the A system (Hodgkin and Kaiser, 1979b). For example, cells with a mutation in a *dsp* gene move primarily as individuals (Shimkets, 1986a) (Fig. 8.8, A^+S^-). Loss of motility occurs only when both systems are disrupted by individual mutations in each system or by a single mutation in the *mglA* gene, the only gene known to be shared by the two systems (Hodgkin and Kaiser, 1979a,b) (Fig. 8.8, A^-S^-). A host of additional genes that belong to neither system (A or S) modify the behavior of gliding cells such as the *frz* genes, which are similar to the enteric chemotaxis genes in terms of structure and function (McBride et al., 1989, 1992; McCleary et al., 1990; McCleary and Zusman, 1990a,b; McBride and Zusman, 1993).

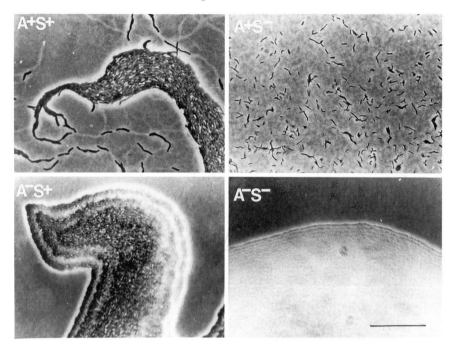

FIGURE 8.8   Effect of A (adventurous) and S (social) motility mutations on *M. xanthus* cell behavior. Bar = 50 μm. (Reprinted from Shimkets, 1986a, with permission)

The S system contains at least 11 loci, *sglA* to *sglH*, *sgl-3119*, *tgl* and *dsp*, none of which has been characterized at the level of DNA sequence (Hodgkin and Kaiser, 1979a,b; Shimkets, 1986a,b; Dana and Shimkets, 1993). The discovery that S motile cells require close proximity for motility argues that some of these genes encode a specific cell–cell recognition system. As has been pointed out, the cell surface of *M. xanthus* contains two types of appendage— pili and fibrils—that are capable of spanning short distances to interact with nearby cells and are likely to be involved in different aspects of the cell–cell recognition process. All S⁻ mutants with the exception of *dsp* are unable to produce pili, which are typically located on the cell poles (Kaiser, 1979; Arnold and Shimkets, 1988a,b). Although the mechanical removal of pili appears to eliminate S motility (Rosenbluh and Eisenbach, 1992), attempts to directly demonstrate a function for the pili in social motility have been limited by an inability to purify pili and then replace them exogenously. Representatives of all known S system mutant groups have reduced ability to produce fibrils (Arnold and Shimkets, 1988a,b; Dana and Shimkets, 1993). The fibrils are most likely involved in cell–cell cohesion as well as attachment to solid surfaces.

The utility of the two motility systems is partially mechanical and partially behavioral. The A and S systems enable cells to respond differently to a variety of surfaces. A motile cells (A⁺S⁻) move more rapidly on solid surfaces, whereas S motile cells (A^-S^+) move more rapidly on semisolid surfaces (Shi and Zusman, 1993). This observation may be a clue that different mechanisms of movement are employed by each of the systems. The A and S systems also allow the cells to establish a balance between dispersal and contiguity. Foraging for new food sources may provide the selective advantage for adventurous motility. Conversely, the advantages of being in a swarm are an accelerated growth rate during cooperative feeding (Rosenberg et al., 1977) and the opportunity to produce spores during fruiting body development.

Dispersal and contiguity also provide the extremes from which flow many of the classes of animal behavior; and the A and S systems may provide the basis for analogous classes of bacterial behavior. Animal behavior can be divided into several categories based on the overall objective or consequence of the behavior (Wallace, 1979). Examples of animal behavioral strategies include aggression, defense of individual or group territories, competition, affiliation with other individuals of the same species, cooperation to cope with some problem, and coordination or reciprocal regulation of individual actions throughout the colony. These behaviors also appear to be manifested in microorganisms (Oleskin, 1993). Indeed, many of these behaviors have been documented by time-lapse microscopy of the myxobacteria (Reichenbach et al., 1965a,b) and flow naturally from the dual motility systems. As analysis of microbial behavior progresses and the functions of the A and S genes become known, it may be possible to explain the molecular basis of these behaviors with myxobacteria.

Rippling

Myxococcus xanthus displays a form of multicellular movement referred to as rippling because of the resemblance of this behavior to ripples on the surface of water (Fig. 8.2) (Reichenbach, 1965; Reichenbach et al., 1965a,b). Rippling cells form a series of equidistant ridges that are parallel to each other and move in a processive manner so that a wave is propagated outward from a point in the interior of the colony. Rippling is induced by peptidoglycan and its components N-acetylglucosamine, N-acetylmuramic acid, diaminopimelate, and D-alanine (Shimkets and Kaiser, 1982). It is usually observed under conditions where there is substantial hydrolysis of eubacterial cell walls, for example when myxobacteria prey on *Micrococcus luteus* cells (Reichenbach et al., 1965a,b; Shimkets and Kaiser, 1982).

A curious feature about rippling is that the entire colony goes through 5-hour cycles of vigorous rippling followed by a quiescent period (Reichenbach, 1965). This sequence indicates the involvement of an extensive cell–cell signaling network that can propagate a signal throughout the colony. Despite extensive rippling over extended periods, there is little net cell accumulation (Shimkets and Kaiser, 1982), suggesting that individual cells move in cyclic paths. Fluorescently labeled cells were tracked with confocal laser microscopy and were found to move in the direction of wave propagation, reversing direction after about one wavelength (approximately 10 cell lengths) (Sager and

Kaiser, 1995). This finding suggests a model for rippling in which end-to-end collisions between waves of cells increases the probability that individual cells reverse their direction of movement (Sager and Kaiser, 1995). Cells are thus locked into repetitive cycles of equidistant backward and forward motion, and there is little net movement despite the fact that traveling waves are realized.

Rippling requires a cell–cell signaling system known as C, which is also essential for fruiting body development. All the known C-signal mutations map to a gene known as *csgA* (Shimkets et al., 1983; Shimkets and Asher, 1988), which encodes cell surface-associated protein (Hagen and Shimkets, 1990; Kim and Kaiser, 1990c,d; Shimkets and Rafiee, 1990). Although the details of ripple induction remain to be worked out, it is likely that induction begins with hydrolysis of peptidoglycan into small, soluble fragments. Myxobacteria produce a variety of extracellular peptidoglycan hydrolytic enzymes (Haska, 1972, 1974; Sudo and Dworkin, 1972). Peptidoglycan components, in turn, act as transcriptional inducers of the CsgA gene (Li et al., 1992).

It has been proposed that oriented collisions between cells initiate C-signaling, which in turn causes individual cells to reverse their direction of movement (Sager and Kaiser, 1995). Addition of CsgA to cells increases the mean frequency of cell reversal threefold. Furthermore, dilution of C-signaling cells with CsgA-deficient cells increases the wavelength of the ripple as the probability of a productive collision decreases. These results argue that CsgA regulates basic aspects of wave propagation. This system, in its elegance and simplicity, resembles the patterns of multicellular movement preceding cellular differentiation in developing embryos (Sager and Kaiser, 1995).

Fruiting Body Development

Fruiting bodies are the end-product of a complex developmental program that begins with a nutritional shift-down and progresses through programmed changes in gene expression and cell behavior. The behavioral aspect of the program begins with rippling, proceeds through aggregation and fruiting body formation, and culminates in sporulation (Shimkets, 1990; Dworkin and Kaiser, 1993). Fruiting body development of *Myxococcus* involves approximately 10^5 cells, requires intricate social interactions, and is dependent on directed movement.

Developmental Signal Exchange

Fruiting body development is regulated by a series of intercellular signals. The primary evidence for these signals and the strategy for identifying them involves mixing nondeveloping mutants and looking for extracellular complementation of development. Certain pairwise mixtures of nondeveloping mutants form wild-type levels of fruiting bodies and spores, suggesting that one mutant is able to produce an extracellular molecule that the other mutant is incapable of making or secreting (McVittie et al., 1962; Hagen et al., 1978). Altogether five complementation groups have been identified (A–E) (Hagen et al., 1978;

Downard et al., 1993). These results could be interpreted to mean that there is one signal requiring five synthetic steps or five different signals. The developmental phenotypes of the mutants are sufficiently similar that they do not distinguish between these possibilities; all of the mutants fail to form both fruiting bodies and spores. Kroos and Kaiser (1987) observed that the various mutant groups exhibited different patterns of developmental gene expression, arguing that the mutants blocked development at slightly different points in the developmental program. These results suggest that the mutant classes represent five different signals. Remarkably, all five mutant groups block developmental gene expression within the first 6 hours of development, prior to any morphological changes (Fig. 8.9). Failure to supply the requisite signal results in premature termination of morphological development and cessation of developmental gene expression at specific points in the developmental program (Kroos and Kaiser, 1987).

Mutations in any of the three *asg* (A *s*ignal) loci cause the developmental program to arrest just 2 hours after initiation (Kuspa et al., 1986). All A-signaling mutants are deficient in the production of extracellular proteases and as a consequence have reduced levels of A-factor, a mixture of amino acids and peptides (Kuspa et al., 1986; Kuspa and Kaiser, 1989). Development may be restored by adding low concentrations of amino acids or peptides that contain these amino acids (Kuspa et al., 1992b; Plamann et al., 1992). A factor is found in amounts roughly proportional to the cell density, and A-signaling is likely used to determine if the cell density is sufficiently high for the initiation of development (Kuspa et al., 1992a). Dilution of wild-type cells to densities lower than what can support production of a threshold level of A-factor inhibits development, and development can be restored by adding either A-factor or a mixture of proteases (Kuspa et al., 1992a). Furthermore, plating an *asgB* mutant, which produces 5–10% of the wild-type level of A-factor, at abnormally high ($10\times$) cell densities generates sufficient A-factor to restore development. The *asgA*, *asgB*, and *asgC* mutants may decrease protease export in different ways (Kuspa and Kaiser, 1989). The *asgB* gene appears to encode a DNA-binding protein with a helix-turn-helix motif near the C-terminus (Plamann et al., 1994) and may be a transcriptional activator of protease genes. The *asgA* gene encodes a signal transduction protein with properties of a histidine protein kinase and homology with response regulators (Plamann et al., 1995; Li and Plamann, 1996). The *asgA* gene encodes the major sigma factor of RNA polymerase holoenzyme (Davis et al., 1995).

The B mutations map to a gene known as *bsgA* (Gill and Cull, 1986; Gill and Bornemann, 1988; Gill et al., 1988), which encodes an ATP-dependent protease related to protease La (Lon) of *Escherichia coli* (Gill et al., 1993). The BsgA gene product is clearly intracellular (Gill and Bornemann, 1988), and the manner by which this protein regulates production of an extracellular signal is not clear.

All mutations that render a cell unable to produce the C-signal map to a gene known as *csgA* (Shimkets et al., 1983; Shimkets and Asher, 1988). Several lines of evidence argue that CsgA is associated with the cell surface. First,

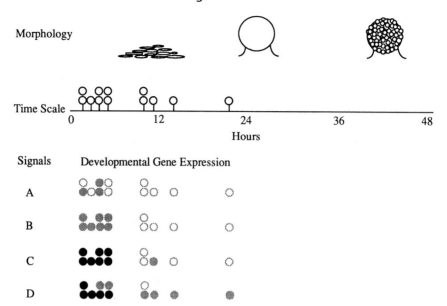

FIGURE 8.9 Effect of A, B, C, and D mutations on developmental gene expression. The developmental markers are derived from a set of Tn5 *lac* insertions in genes normally expressed during wild-type development. The insertions themselves have no obvious adverse effects on development even though they disrupt developmentally regulated genes. The time of expression of each insertion, denoted by a circle above the time scale, is indicated by the position of the insertion above the time scale. The insertions depicted are ΩDK4408, 2 hours; ΩDK4494, 2 hours; ΩDK4521, 3 hours; ΩDK4455, 4 hours; ΩDK4469, 4 hours; ΩDK4273, 5 hours; ΩDK4411, 5 hours; ΩDK4406, 10 hours; ΩDK4506, 10 hours; ΩDK4473/DK4414, 11 hours; ΩDK4500/DK4403, 13 hours; and ΩDK4435, 22 hours. The time of expression of each insertion is the arithmetic mean of those reported by Cheng and Kaiser (1989a), Kroos and Kaiser (1987), and Li and Shimkets (1993). The morphology of wild-type cells at various developmental times is depicted above the time scale. Aggregation begins 8–12 hours after starvation induction. Fruiting bodies without spores are formed by 24 hours, and sporulation is complete by about 48 hours. The effect of A (Kroos and Kaiser, 1987; Kuspa et al., 1986), B (Kroos and Kaiser, 1987), C (Kroos and Kaiser, 1987; Li and Shimkets, 1993), or D (Cheng and Kaiser, 1989a) mutations on the expression of this set of insertions is depicted by the coloration of the circles. Black indicates wild-type levels of expression, gray indicates partial expression, and white indicates no expression. Each mutant group blocks the developmental program of gene expression at a different place. The results suggest that each mutant group represents a different signal rather than different steps in the production of a single signal.

polyclonal antibodies raised against a LacZ–CsgA fusion protein in *E. coli* inhibit fruiting body morphogenesis and sporulation, recreating the phenotype of *csgA* mutants (Shimkets and Rafiee, 1990). Second, colloidal gold-labeled anti-CsgA antibodies bind to the cell surface and extracellular matrix (Shimkets and Rafiee, 1990). Third, CsgA is found in the cell pellet following high speed centrifugation of disrupted cells and requires detergent for solubilization (Kim and Kaiser, 1990c,d). Finally, separation of *csgA*[+] and *csgA*[−] cells with a 0.45 μm pore size membrane prevented extracellular complementation (Kim and Kaiser, 1990c). Together these results present a strong argument that surface-associated CsgA mediates the signaling process. Collisions between cell surfaces appears to be important in C-signal transmission, as discussed under Cell–Cell Attachments and Rippling (see above). Specifically, end-to-end collisions between cells appear to mediate C-signal exchange and at the same time control cell reversal frequency (Sager and Kaiser, 1995). Recently Sogaard-Anderson and Kaiser (1996) have shown that exchange of the C-signal decreases the rate of motility reversal when it occurs in the context of developmental aggregation. They have proposed a model in which the C-signal stimulates the Frz cascade and decreases cell reversal frequency in a way that preferentially leads cells into an aggregate. The chemical nature of the C-signal is unknown. CsgA could be the signal and serve as a paracrine, or short-range, hormone (Kim and Kaiser, 1990a,b,c,d). Recently a class of csgA suppressors has led to the alternative hypothesis that CsgA produces the C-signal through an enzymatic activity. Two transposon insertions, socC559 and soc-560 efficiently restored C-signaling and development to csgA null mutants (Lee and Shimkets, 1994; 1996). These mutations are located in the socABC operon which contains: socA, a member of the short chain alcohol dehydrogenase (SCAD) family, like csgA; socB, a membrane anchoring protein; and socC, a negative autoregulator of socABC expression. The two insertions inactivate socC leading to a 30 to 100-fold increase in socA transcription (Lee and Shimkets, 1996). SocA and CsgA could be folded into similar 3-dimensional structures, using the $3\alpha/20\beta$-hydroxysteroid dehydrogenase x-ray crystal structure as a model, in which the coenzyme binding site and catalytic site were preserved and juxtaposed (Lee et al., 1995). This prompted a closer examination of the putative CsgA active site by site-directed mutagenesis of codons conserved in SCAD members. A T6A amino acid substitution in the coenzyme binding pocket eliminated NAD^+ binding in vitro and resulted in the production of biologically inactive CsgA in vivo. Similarly, S135T and K155R substitutions in the catalytic site resulted in biologically inactive CsgA. Together these results present a strong argument for the role of CsgA as an enzyme. Unfortunately, the substrate and product have not been identified. Collectively the SCAD family has an enormous range of substrates including carbohydrates, steroids, isoprenoids, phenolic compounds and hydroxylated acyl compounds. Thus determining the substrate is not likely to be an easy task.

The D mutations map to a gene known as dsgA, which is essential for both growth and development (Cheng and Kaiser, 1989a,b). DsgA has 50% amino acid identity to translation initiation factor 3 (IF3; infC) from E. coli, which helps the ribosome select the initiation codon on the mRNA (Kalman et al., 1994). DsgA reacts with anti-IF3 serum and is present at relatively constant levels during M. xanthus vegetative growth and development. Furthermore, the dsgA$^+$ allele from M. xanthus restores normal function to an E. coli infC362 missense mutant. Therefore there is little doubt that DsgA serves as the translation initiation factor in M. xanthus. Neither the manner by which dsgA functions in the production of the D-signal nor the chemical structure of the D-signal are known. In E. coli AUU is the start codon for infC translation, and this atypical start codon is important for the posttranslational regulation of infC expression (Gold, 1988). When IF3 levels are low, the *infC* message is translated because IF3 shows selectivity for the typical initiation codons. Translation of *infC* from AUU continues until the levels of IF3 are high enough to restrict translation from an AUU initiation codon, thus regulating the cellular level of IF3. Translation of *dsgA* begins with another unusual start codon, AAC (Cheng et al., 1994). These results suggest that production of the D-signal may involve translation of mRNA with an unusual start codon.

The E-mutations (Downard et al., 1993) map to genes that encode proteins E1α and E1β, which constitute the E1 component of a branched-chain ketoacid dehydrogenase (Downard and Toal, 1995; Toal et al., 1995). These dehydrogenases are part of a pathway that converts branched-chain ketoacids to coenzyme A (CoA) derivatives of the short branched-chain fatty acids isovalerate, methylbutyrate, and isobutyrate. The *esg* mutants have greatly reduced levels of long branched-chain fatty acids, which are normally synthesized from the short branched-chain fatty acid precursors. Furthermore, exogenous isovalerate restores branched chain fatty acid biosynthesis to an *esg* mutant and corrects the developmental defect. The current model for E-signaling is that shortly after initiation of development phospholipase activity liberates branched-chain fatty acids, which are passed between cells to function as short-range signals (Downard and Toal, 1995). It was earlier proposed that phospholipase plays a role in *M. xanthus* development by liberating autocides, a complex of fatty acids that are responsible for developmental autolysis (Mueller and Dworkin, 1991). Thus the developmental release of fatty acids from cell surface phospholipids may play a role as a developmental signaling event as well as in developmental autolysis.

The utility of multiple developmental signals is not fully appreciated, as we lack an understanding of the precise functions of each of the five signals. The first signaling step using the A-signal seems to serve primarily as an indicator of cell density (Fig. 8.9). The production of A-factor is constitutive and proportional to the cell density (Kuspa et al., 1992a). Production of the threshold level of A-factor serves to indicate that there are enough cells to form a fruiting body, and development progresses to the next step. Somewhat later in development is the requirement for the E-signal, which also signals cell proximity by fatty acid transfer. The last known signal is the C-signal, which plays a central role in development by activating expression of more than half of the developmentally regulated genes (Kroos and Kaiser, 1987; Li and Shimkets, 1993). CsgA transcription, measured with a *lacZ* reporter assay, appears to increase steadily throughout development (Hagen and Shimkets, 1990; Li et al., 1992). CsgA may act as an extracellular developmental timer, inducing successive developmental stages at higher threshold concentrations. Addition of low levels of CsgA induces early developmental gene expression and aggregation, but somewhat higher levels are required to induce late gene expression and sporulation (Kim and Kaiser, 1991). Similarly, reduction of *csgA* expression by mutation of the regulatory region results in developmental progress that is proportional to the level of expression observed with the mutant (Li et al., 1992). Therefore CsgA seems to serve as a sort of developmental timer whose extracellular concentration determines the pace of development.

Directed Movement During Fruiting Body Formation

Scanning electron microscopy of developing *Stigmatella aurantiaca* cells suggest a series of morphological changes that accompany selection of an aggregation center (Vasquez et al., 1985). Initially cells glide as groups in circles or spirals around the perimeter of an aggregation focus. Several streams of cells then join the cells encircling the aggregation center. Cells appear to enter the aggregate from the surrounding ring of cells by changing the pitch of their movement to create an ascending spiral. The encircling ring of cells is still apparent late during aggregation when the aggregate extends well off the agar surface. During stalk formation the pattern of movement changes and the cells immediately next to the aggregate on the agar surface become oriented vertical to the agar surface and appear to hoist the aggregate off the agar surface.

Fruiting body formation of *M. xanthus* is somewhat simpler than that of *S. aurantiaca* because there is no stalk on the fruiting body. Scanning electron micrographs of *M. xanthus* fruiting bodies did not reveal the same type of cellular organization. Most notably, the ring of cells encircling the aggregation center was not apparent. Nevertheless, the cells within the fruiting body appear to be organized in spiral patterns, some parallel to the agar surface and others perpendicular to it (O'Connor and Zusman, 1989). Confocal laser microscopy was used to examine *M. xanthus* fruiting body formation. Because confocal microscopy is not disruptive the same fruiting body can be examined throughout development. Cells within the nascent aggregate appear to be arranged in horizontal spirals, and the fruiting body seems to consist of two distinct domains (Sager and Kaiser, 1993b). The outer domain comprises motile, densely packed, rod-shaped cells that travel in a circular trajectory inside the fruiting body. The inner domain comprises spherical, sporulating cells in a less ordered, threefold less dense pattern. Cells in the outer domain are optimized for C-signal exchange, being both motile and aligned end to end (Kim and Kaiser, 1990a,b; Sager and Kaiser, 1993b). In contrast, cells in the inner domain would not be expected to perform extensive C-signaling, as they are neither aligned or motile. This spatial separation of the two cell types provides the framework for a homogeneous population to begin the process of separating into spatial domains with different functions. Sager and Kaiser (1993a) proposed that as cells in the horizontal spiral begin to convert to spores, and consequently become nonmotile, the movement of cells in the spiral sweeps them toward the center of the fruiting body where they enter the spore domain.

To determine if motile cells can move spores, Sager and Kaiser (1993a) spread myxospores from a kanamycin-sensitive strain on agar that promotes development and then applied cells from a kanamycin-resistant strain. This strain was then allowed to develop for 22 hours—long enough to form fruiting bodies but not long enough to sporulate. The preexisting spores were swept away from the fruiting body, suggesting that moving cells have sufficient motive force to move spores. It remains to be determined if this movement can be spatially and temporally coordinated to corral spores into the inner domain of a fruiting body. Because the confocal technique lacks the resolution to follow the movement of individual cells or spores, direct proof of this model must be obtained by other means.

The mechanism by which *M. xanthus* cells are attracted to an aggregation center and form a fruiting body is virtually unknown. It is possible that a chemoattractant is involved, as the methylation state of the methyl-accepting protein FrzCD increases at the time of mound formation (McBride and Zusman, 1993). However, no chemoattractant has yet been identified.

Myxobacterial Territoriality

Specific cell−cell recognition is an integral feature of multicellularity. Invariably, recognition is based on specific interactions between cell surface macromolecules. The molecular basis of this recognition has been studied in simple systems such as sponges (Humphreys et al., 1977) and more recently by studying cadherins and integrins in cultured mammalian cells (Hynes, 1992a,b). The reluctance of certain genera of the cellular slime molds to form chimeric fruiting bodies has also been shown to reflect the interactions of specific cell surface adhesion sites (Gerisch et al., 1980).

The ability of the myxobacteria to move as a social group, to aggregate, and to build multicellular fruiting bodies generates the question: Will the swarm of one strain or species of myxobacteria merge with the swarm of another and build a mixed fruiting body? In other words, can the myxobacteria also distinguish self from nonself? Previous observations in one of our laboratories (M.D.) had shown that when swarms of *Stigmatella aurantiaca* and *M. xanthus* encountered each other, they initially joined but at some subsequent time sorted themselves out and eventually formed separate fruiting bodies (unpublished observations).

This question was addressed quantitatively and in more detail by Smith and Dworkin (1994), who showed that two closely related species of *Myxococcus* (*M. xanthus* and *M. virescens*) when mixed together likewise formed their fruiting bodies only with cells of their own species. Furthermore, under some conditions fruiting body formation by one of the two species was completely inhibited. This sorting out seemed less a matter of recognition of self based on characteristic cell surface molecules than a result of the production by each species of an extracellular inhibitor that, like a bacteriocin, killed only the closely related species. The species whose fruiting bodies ultimately dominated the territory appeared to be a function of the relative potency of the putative bacteriocins and the ratio of the numbers of the two species.

Conclusion

As should be obvious from this review, a great deal of progress has been made in understanding the physiology, biochemistry, and genetics of myxobacteria. There have also been exciting advances in our understanding of some of the signals exchanged by them. The myxobacteria manifest a variety of social behaviors (dramatically illustrated by the time-lapse motion picture films of Hans Reichenbach (Reichenbach et al., 1965a,b), which continue to puzzle and intrigue investigators in the field.

1. What are the forces that hold cells together as they swarm over an agar surface?
2. What is the relation between the peculiar circling movement of these swarms and the formation of fruiting body aggregates?
3. How do the cells construct their elaborate fruiting bodies?
4. What is the mechanism of their gliding motility?
5. What is the function and mechanism of rippling?
6. How do the swarms perceive the presence of objects at a distance?

It is not unreasonable to hope that the fundamental information that has been gathered will eventually lead to an understanding of the biology of the myxobacteria at the next higher level of organization and behavior.

Acknowledgments

L.J.S. was supported by National Science Foundation grant MCB9304083. M.D. was supported by U.S. Public Health Service grant GMS 19957 from the National Institutes of Health.

References

Arnold, J.W., and Shimkets, L.J. (1988a) Cell surface properties correlated with cohesion in *Myxococcus xanthus*. *J. Bacteriol.* **170**:5771–5777.

Arnold, J.W., and Shimkets, L.J. (1988b) Inhibition of cell-cell interactions in *Myxococcus xanthus* by Congo red. *J. Bacteriol.* **170**:5765–5770.

Behmlander, R.M., and Dworkin, M. (1991) Extracellular fibrils and contact mediated cell interactions in *Myxococcus xanthus*. *J. Bacteriol.* **173**:7810–7821.

Behmlander, R.M., and Dworkin, M. (1994) Biochemical and structural analysis of the extracellular matrix fibrils of *Myxococcus xanthus*. *J. Bacteriol.* **176**:6295–6303.

Chang, B.-Y., and Dworkin, M. (1994) Isolated fibrils rescue cohesion and development in the Dsp mutant of *Myxococcus xanthus*. *J. Bacteriol.* **176**:7190–7196.

Cheng, Y., and Kaiser, D. (1989a) *dsg*, a gene required for cell-cell interaction early in *Myxococcus* development. *J. Bacteriol.* **171**:3719–3726.

Cheng, Y., and Kaiser, D. (1989b) *dsg*, a gene required for *Myxococcus* development, is necessary for cell viability. *J. Bacteriol.* **171**:3727–3731.

Cheng, Y., Kalman, L.V., and Kaiser, D. (1994) The *dsg* gene of *Myxococcus xanthus* encodes a protein similar to translation initiation factor IF3. *J. Bacteriol.* **176**: 1427–1433.

Cohen, J.J., Duke, R.C., Fadok, V.A., and Sellins, K.S. (1992) Apoptosis and programmed cell death in immunity. *Annu. Rev. Immunol.* **10**:267–293.

Dana, J.R., and Shimkets, L.J. (1993) Regulation of cohesion-dependent cell interactions in *Myxococcus xanthus. J. Bacteriol.* **175**:3636–3647.

Davis, J.M., Mayor, J., and Plamann, L. (1995) A missense mutation in *rpoD* results in a A-signalling defect in *Myxococcus xanthus. Molec. Microbiol.* **18**:943–952.

Downard, J., and Toal, D. (1995) Branched-chain fatty acids: the case for a novel form of cell-cell signaling during *Myxococcus xanthus* development. *Mol. Microbiol.* **16**: 171–175.

Downard, J., Ramaswamy, S.V., and Kil, K.-S. (1993) Identification of *esg*, a genetic locus involved in cell-cell signaling during *Myxococcus xanthus* development. *J. Bacteriol.* **175**:7762–7770.

Dworkin, M. (1963) Nutritional regulation of morphogenesis in *Myxococcus. J. Bacteriol.* **86**:67–72.

Dworkin, M. (1972) The myxobacteria: new directions in studies of procaryotic development. *C.R.C. Crit. Rev. Microbiol.* **1**:435–452.

Dworkin, M. (1973) Cell-cell interactions in the myxobacteria. In *Microbial Differentiation*, J.M. Ashworth and J.E. Smith (eds.), pp. 123–142. Cambridge University Press, Cambridge.

Dworkin, M. (1979) Spores, cysts and stalks. In *The Bacteria*, Vol. VII, J.R. Sokatch and L.N. Ornsten (eds.), pp. 1–84. Academic Press, Orlando, FL.

Dworkin, M. (1985) *Developmental Biology of the Bacteria.* Benjamin/Cummings, Menlo Park, CA.

Dworkin, M. (1996) Recent advances in the social and developmental biology of the myxobacteria. *Microbial. Rev.* **60**:70–102.

Dworkin, M., and Kaiser, D. (1993) *Myxobacteria II.* American Society for Microbiology, Washington, DC.

Gerisch, G., Krelle, H. Bozzaro, S., Eitle, E., and Guggenheim, R. (1980) Analysis of cell adhesion in *Dictyostelium* and *Polysphondylium* by the use of Fab. In *Cell Adhesion and Motility*, A.S.G. Curtis and J.D. Pitts (eds.), pp. 293–307. Cambridge University Press, Cambridge.

Gill, R.E., and Bornemann, M.C. (1988) Identification and characterization of the *Myxococcus xanthus bsgA* gene product. *J. Bacteriol.* **170**:5289–5297.

Gill, R.E., and Cull, M.G. (1986) Control of developmental gene expression by cell-to-cell interactions in *Myxococcus xanthus. J. Bacteriol.* **168**:341–347.

Gill, R.E., and Shimkets, L.J. (1993) Genetic approaches for the analysis of myxobacterial behavior. In *Myxobacteria II*, M. Dworkin and D. Kaiser (eds.), pp. 129–155. American Society for Microbiology, Washington, DC.

Gill, R.E., Cull, M., and Fly, S. (1988) Genetic identification and cloning of a gene required for developmental cell interactions in *Myxococcus xanthus. J. Bacteriol.* **170**:5279–5288.

Gill, R.E., Karlok, M., and Benton, D. (1993) *Myxococcus xanthus* encodes an ATP-dependent protease which is required for developmental gene transcription and intercellular signaling. *J. Bacteriol.* **175**:4538–4544.

Gold, L. (1988) Post transcription mechanisms in *Escherichia coli. Annu. Rev. Biochem.* **57**:199–233.

Hagen, D.C., Bretscher, A.P., and Kaiser, D. (1978) Synergism between morphogenetic mutants of *Myxococcus xanthus. Dev. Biol.* **64**:284–296.

Hagen, T.J., and Shimkets, L.J. (1990) Nucleotide sequence and transcriptional products of the *csg* locus of *Myxococcus xanthus. J. Bacteriol.* **172**:15–23.

Haska, G. (1972) Extracellular lytic enzymes of *Myxococcus virescens*. III. Characterization of two endo-β-*N*-acetylglucosaminidases. *Physiol. Plant* **27**:139–142.

Haska, G. (1974) Extracellular lytic enzymes of *Myxococcus virescens*. IV. Purification and characterization of a D-alanyl-E-*N*-lysine endopeptidase. *Physiol. Plant* **31**: 252–256.

Hodgkin, J., and Kaiser, D. (1977) Cell-to-cell stimulation of movement in nonmotile mutants of *Myxococcus xanthus*. *Proc. Natl. Acad. Sci. U.S.A.* **74**:2938–2942.

Hodgkin, J., and Kaiser, D. (1979a) Genetics of gliding motility in *Myxococcus xanthus* (Myxobacterales): genes controlling movement of single cells. *Mol. Gen. Genet.* **171**:167–176.

Hodgkin, J., and Kaiser, D. (1979b) Genetics of gliding motility in *Myxococcus xanthus* (Myxobacterales): two gene systems control movement. *Mol. Gen. Genet.* **171**: 177–191.

Humphreys, S., Humphreys, T., and Sano, J. (1977) Organization and polysaccharides of sponge aggregation factor. *J. Supramol. Struct.* **7**:339–351.

Hynes, R.O. (1992a) Integrins: versatility, modulation and signalling in cell adhesion. *Cell* **69**:11–25.

Hynes, R.O. (1992b) Specificity of cell adhesion in development: the cadherin superfamily. *Curr. Opin. Genet. Dev.* **2**:621–624.

Kaiser, D. (1979) Social gliding is correlated with the presence of pili in *Myxococcus xanthus*. *Proc. Natl. Acad. Sci. U.S.A.* **76**:5952–5956.

Kaiser, D., and Crosby, C. (1983) Cell movement and its coordination in swarms of *Myxococcus xanthus*. *Cell Motil.* **3**:227–245.

Kaiser, D., and Kroos, L. (1993) Intercellular signaling. In *Myxobacteria II*, M. Dworkin and D. Kaiser (eds.), pp. 257–283. American Society for Microbiology, Washington, DC.

Kalman, L.V., Cheng, Y.L., and Kaiser, D. (1994) The *Myxococcus xanthus dsg* gene product performs functions of translation initiation factor IF3 in vivo. *J. Bacteriol.* **176**:1434–1442.

Kim, S.K., and Kaiser, D. (1990a) Cell alignment required in differentiation of *M. xanthus*. *Science* **249**:926–928.

Kim, S.K., and Kaiser, D. (1990b) Cell motility is required for the transmission of C-factor, an intercellular signal that coordinates fruiting body morphogenesis of *Myxococcus xanthus*. *Genes Dev.* **4**:896–905.

Kim, S.K., and Kaiser, D. (1990c) C-factor: a cell-cell signaling protein required for fruiting body morphogenesis of *M. xanthus*. *Cell* **61**:19–26.

Kim, S.K., and Kaiser, D. (1990d) Purification and properties of *Myxococcus xanthus* C-factor, an intercellular signalling protein. *Proc. Natl. Acad. Sci. U.S.A.* **87**:3635–3639.

Kim, S., and Kaiser, D. (1991) C-factor has distinct aggregation and sporulation thresholds during *Myxococcus* development. *J. Bacteriol.* **173**:1722–1728.

Kroos, L., and Kaiser, D. (1987) Expression of many developmentally regulated genes in *Myxococcus* depends on a sequence of cell interactions. *Genes Dev.* **1**:840–854.

Kroos, L., Hartzell, P., Stephens, K., and Kaiser, D. (1988) A link between cell movement and gene expression argues that motility is required for cell-cell signaling during fruiting body development. *Genes Devel.* **2**:1677–1685.

Kuspa, A., and Kaiser, D. (1989) Genes required for developmental signalling in *Myxococcus xanthus*: three *asg* loci. *J. Bacteriol.* **171**:2762–2772.

Kuspa, A., Kroos, L., and Kaiser, D. (1986) Intercellular signaling is required for developmental gene expression in *Myxococcus xanthus*. *Dev. Biol.* **117**:267–276.

Kuspa, A., Plamann, L., and Kaiser, D. (1992a) A-signalling and the cell density requirement for *Myxococcus xanthus* development. *J. Bacteriol.* **174**:7360–7369.

Kuspa, A., Plamann, L., and Kaiser, D. (1992b) Identification of heat-stable A-factor from *Myxococcus xanthus*. *J. Bacteriol.* **174**:3319–3326.

Lee, B.-U., Lee, K., Mendez, J., and Shimkets, L.J. (1995) A tactile sensory system of *Myxococcus xanthus* involves an extracellular NAD(P)$^+$-containing protein. *Genes Dev.* **9**:2964–2973.

Lee, K., and Shimkets, L.J. (1994) Cloning and characterization of the socA locus which restores development to *Myxococcus xanthus* C-signaling mutants. *J. Bacteriol.* **176**:2200–2209.

Lee, K., and Shimkets, L.J. (1996) Suppression of a signaling defect during *Myxococcus xanthus* development. *J. Bacteriol.* **178**:977–984.

Li, S., Lee, B., and Shimkets, L.J. (1992) *csgA* expression entrains *Myxococcus xanthus* development. *Genes Dev.* **6**:401–410.

Li, S.F., and Shimkets, L.J. (1993) Effect of *dsp* mutations on the cell-to-cell transmission of CsgA in *Myxococcus xanthus*. *J. Bacteriol.* **175**:3648–3652.

Li, Y., and Plamann, L. (1996) Purification and in vitro phosphorylation of *Myxococcus xanthus AsgA* protein. *J. Bacteriol.* **178**:289–292.

MacRae, T.H., and McCurdy, H.D. (1975) Ultrastructural studies of *Chondromyces crocatus* vegetative cells. *Can. J. Microbiol.* **21**:1815–1826.

Manoil, C., and Kaiser, D. (1980) Accumulation of guanosine tetraphosphate and guanosine pentaphosphate in *Myxococcus xanthus* during starvation and myxospore formation. *J. Bacteriol.* **141**:297–304.

McBride, M.J., and Zusman, D.R. (1993) FrzCD, a methyl-accepting taxis protein from *Myxococcus xanthus*, shows modulated methylation during fruiting body formation. *J. Bacteriol.* **175**:4936–4940.

McBride, M.J., Hartzell, P., and Zusman, D.R. (1993) Motility and tactic behavior of *Myxococcus xanthus*. In *Myxobacteria II*, M. Dworkin and D. Kaiser (eds.), pp. 285–305. American Society for Microbiology, Washington, DC.

McBride, M.J., Kohler, T., and Zusman, D.R. (1992) Methylation of FrzCD, a methyl-accepting taxis protein of *Myxococcus xanthus*, is correlated with factors affecting cell behavior. *J. Bacteriol.* **174**:4246–4257.

McBride, M.J., Weinberg, R.A., and Zusman, D.R. (1989) Frizzy aggregation genes of the gliding bacterium *Myxococcus xanthus* show sequence similarities to the chemotaxis genes of enteric bacteria. *Proc. Natl. Acad. Sci. U.S.A.* **86**:424–428.

McCleary, W., and Zusman, D. (1990a) FrzE of *M. xanthus* is homologous to both CheA and CheY of *Salmonella typhimurium*. *Proc. Natl. Acad. Sci. U.S.A.* **87**:5898–5902.

McCleary, W., and Zusman, D. (1990b) Purification and characterization of the *M. xanthus* FrzE protein shows that it has autophosphorylation activity. *J. Bacteriol.* **172**:6661–6668.

McCleary, W., McBride, M., and Zusman, D. (1990) Developmental sensory transduction in *M. xanthus* involves methylation and demethylation of FrzCD. *J. Bacteriol.* **172**:4877–4887.

McVittie, A., Messik, F., and Zahler, S.A. (1962) Developmental biology of *Myxococcus*. *J. Bacteriol.* **84**:546–551.

Mueller, C., and Dworkin, M. (1991) Effects of glucosamine on lysis, glycerol formation, and sporulation in *Myxococcus xanthus*. *J. Bacteriol.* **173**:7164–7175.

O'Connor, K.A., and Zusman, D.R. (1989) Patterns of cellular interactions during fruiting body formation in *Myxococcus xanthus*. *J. Bacteriol.* **171**:6013–6024.

Oleskin, A.V. (1993) Supramolecular-level interactions in the populations of microorganisms. *Microbiology* **62**:243–252.

Plamann, L., Davis, J.M., Cantwell, B., and Mayor, J. (1994) Evidence that *asgB* encodes a DNA-binding protein essential for growth and development of *Myxococcus xanthus*. *J. Bacteriol.* **176**:2013–2020.

Plamann, L., Li, Y., Cantwell, B., and Mayor, J. (1995) The *Myxococcus xanthus asgA* gene encodes a novel signal protein required for multicellular development. *J. Bacteriol.* **177**:2014–2020.

Plamann, L., Kuspa, A., and Kaiser, D. (1992) Proteins that rescue A-signal-defective mutants of *Myxococcus xanthus*. *J. Bacteriol.* **174**:3311–3318.

Reichenbach, H. (1965) Rhythmische Vorgange bei der Schwarmentfaltung von Myxobakterien. *Ber. Dtsch. Bot. Ges.* **78**:102–105.

Reichenbach, H. (1993) Biology of the myxobacteria: ecology and taxonomy. In *Myxobacteria II*, M. Dworkin and D. Kaiser (eds.), pp. 13–62. American Society for Microbiology, Washington, DC.

Reichenbach, H., and Dworkin, M. (1992) The myxobacteria. In *The Prokaryotes*, 2nd ed., A. Balows, H.G. Truper, M. Dworkin, W. Harder, and K.-H. Schleifer (eds.), pp. 3416–3487. Springer-Verlag, New York.

Reichenbach, H., Heunert, H., and Kuczka, H. (1965a) *Myxococcus* spp. (Myxobacterales)-Schwarmentwicklung und bildung von protocysten. Film E779. Institute of Wissen., Göttingen.

Reichenbach, H., Huenert, H.H., and Kuczka, H. (1965b) Schwarmentwicklung und Morphogenese bei Myxobakterien—*Archangium, Myxococcus, Chondrococcus, Chondromyces*. Film C893. Institute of Wissen. Film. Göttingen.

Rosenberg, E. (ed.) (1984) *Myxobacteria: Development and Cell Interactions*. Springer-Verlag, New York.

Rosenberg, E., Keller, K.H., and Dworkin, M. (1977) Cell density-dependent growth of *Myxococcus xanthus* on casein. *J. Bacteriol.* **129**:770–777.

Rosenbluh, A., and Eisenbach, M. (1992) Effect of mechanical removal of pili on gliding motility of *Myxococcus xanthus*. *J. Bacteriol.* **174**:5406–5413.

Sager, B., and Kaiser, D. (1993a) Spatial restriction of cellular differentiation. *Genes Dev.* **7**:1645–1653.

Sager, B., and Kaiser, D. (1993b) Two cell-density domains within the *Myxococcus xanthus* fruiting body. *Proc. Natl. Acad. Sci. U.S.A.* **90**:3690–3694.

Sager, B., and Kaiser, D. (1995) Intercellular C-signaling and the traveling waves of *Myxococcus. Genes Dev.* **8**:2793–2804.

Shi, W., and Zusman, D.R. (1993) The two motility systems of *Myxococcus xanthus* show different selective advantages on various surfaces. *Proc. Natl. Acad. Sci. U.S.A.* **90**:3378–3382.

Shimkets, L.J. (1986a) Correlation of energy-dependent cell cohesion with social motility in *Myxococcus xanthus*. *J. Bacteriol.* **166**:837–841.

Shimkets, L.J. (1986b) Role of cell cohesion in *Myxococcus xanthus* fruiting body formation. *J. Bacteriol.* **166**:842–848.

Shimkets, L. (1990) Social and developmental biology of the myxobacteria. *Microbiol. Rev.* **54**:473–501.

Shimkets, L.J. (1993) The myxobacterial genome. In *Myxobacteria II*, M. Dworkin and D. Kaiser (eds.), pp. 85–107. American Society for Microbiology, Washington, DC.

Shimkets, L.J., and Asher, S.J. (1988) Use of recombination techniques to examine the structure of the *csg* locus of *Myxococcus xanthus*. *Mol. Gen. Genet.* **211**:63–71.

Shimkets, L.J., and Dworkin, M. (1981) Excreted adenosine is a cell density signal for the initiation of fruiting body formation in *Myxococcus xanthus*. *Dev. Biol.* **84**:51–60.

Shimkets, L.J., and Kaiser, D. (1982) Induction of coordinated cell movement in *Myxococcus xanthus*. *J. Bacteriol.* **152**:451–461.

Shimkets, L., and Rafiee, H. (1990) CsgA, an extracellular protein essential for *M. xanthus* development. *J. Bacteriol.* **172**:5299–5306.

Shimkets, L.J., Gill, R.E., and Kaiser, D. (1983) Developmental cell interactions in *Myxococcus xanthus* and the *spoC* locus. *Proc. Natl. Acad. Sci. U.S.A.* **80**:1406–1410.

Smith, D., and Dworkin, M. (1994) Territorial interactions between two *Myxococcus* species. *J. Bacteriol.* **176**:1201–1205.

Sogaard-Anderson, L., and Kaiser, D. (1996) C factor, a cell-surface-associated intercellular signaling protein stimulates the cytoplasmic Frz signal transduction system in *Myxococcus xanthus*. *Proc. Natl. Acad. Sci. U.S.A.* **93**:2675–2679.

Sudo, S., and Dworkin, M. (1972) Bacteriolytic enzymes produced by *Myxococcus xanthus*. *J. Bacteriol.* **110**:236–245.

Toal, D., Clifton, S.W., Roe, B.A., and Downard, J. (1995) The *esg* locus of *Myxococcus xanthus* encodes the E1α and E1β subunits of a branched-chain keto acid dehydrogenase. *Mol. Microbiol.* **16**:177–189.

Vasquez, G.M., Qualls, F., and White, D. (1985) Morphogenesis of *Stigmatella aurantiaca* fruiting bodies. *J. Bacteriol.* **163**:515–521.

Wallace, R.A. (1979) *Animal Behavior: Its Development, Ecology, and Evolution.* Goodyear Publishing, Santa Monica, CA.

Wireman, J.W., and Dworkin, M. (1975) Morphogenesis and developmental interactions in *Myxococcus xanthus*. *Science* **189**:516–522.

9

Oral Microbiology and Coaggregation

PAUL E. KOLENBRANDER

Human oral bacteria survive in a lotic, or flowing, environment. Normal salivary flow makes the process of adherence to a surface of central importance and makes the genes encoding adherence-relevant information essential for microbial colonization. The human oral cavity is colonized by more than 500 different kinds of bacteria (for review: Moore and Moore, 1994); 300 species have been named. Such bacterial diversity and convenient access to their habitat provides a scientific paradise for oral microbial ecologists.

Consistent with other ecosystems, species diversity of the colonized oral surfaces increases temporally. Certain species of *Streptococcus* and *Actinomyces* predominate as the primary colonizers of a freshly cleaned tooth surface during the first 4 hours after cleaning (Nyvad and Kilian, 1987). Other genera then adhere to the host surface or accrete onto the bacteria that already occupy sites on the surface. Accretion can occur one cell at a time or with a previously formed mixed cell-type coaggregate. Coaggregates are defined as a mixture of two or more genetically distinct cell types interacting by specific cell–cell recognition to form a multicellular network. Coaggregations are highly specific in that a strain has a defined set of partner cell types. In addition to adherence, the accreting cells must be metabolically compatible with the biofilm to succeed in oral colonization. Accretion and development of biofilms involve many adherence and metabolic events and are beyond the scope of this chapter. Here the focus is on aspects of one adherence-relevant property—oral bacterial coaggregation—and its potential role in colonization of the human oral cavity.

Coaggregation Assays

Coaggregation is mediated by the recognition and binding of complementary surface molecules on the partner cell surfaces. It was first reported in 1970 (Gibbons and Nygaard, 1970) and is easily measured by mixing a dense suspension of one cell type with a dense suspension of its partner cell type (Fig. 9.1). The cell types are homogeneous unclumped suspensions of about 1×10^9 cells/ml (Fig. 9.1, tubes 1 and 2) which after mixing immediately form coaggregates (Fig. 9.1, tube 3) that may settle within 30 seconds to the bottom of the tube (Fig. 9.1, tube 4). Many of the partnerships are dissociated by lactose (McIntire et al., 1978) to an evenly turbid appearance (Fig. 9.1, tube 5). More than 1000 strains have been tested with this simple assay for their ability to coaggregate. Surprisingly, coaggregation partnerships are not random.

Intergeneric coaggregations between pairs of bacteria are shown in Fig. 9.2. By microscopic examination coaggregates appear variously as clumps of just a few cells (Fig. 9.2D,G) to three-dimensional opaque clumps (Fig. 9.2A, top; 9.2I,J). They can exhibit a "corncob" morphology (Fig. 9.2H) or a subtle alignment of the small cells on the larger, longer partner cells (Fig. 9.2F, arrows). Cell–cell adherence is shown between two actinomyces cells interacting with the terminal cell of a chain of four streptococci (Fig. 9.A, arrow) and between a pair of streptococci and surrounding veillonellae (Fig. 9.2I, arrows). Coaggregation can be reversed (Fig. 9.2A,D) by adding lactose (Fig. 9.2C,E). Spontaneously occurring coaggregation-defective actinomyces mutants, when

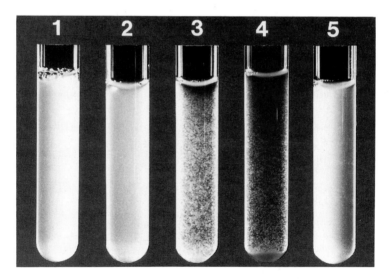

FIGURE 9.1 Visual assay of coaggregation between genetically distinct cell types. A portion of the suspension of one cell type (tube 1) is mixed with a portion of the suspension of a second cell type (tube 2) and observed immediately (tube 3) or after 30 seconds (tube 4). With some aggregated partners the addition of 60 mM lactose completely reverses the coaggregation, resulting in a homogeneous mixed cell type suspension (tube 5).

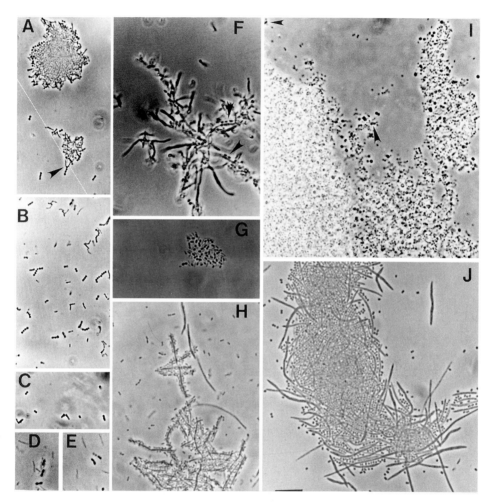

FIGURE 9.2 Phase-contrast micrographs of intergeneric coaggregations between (**A**) *Actinomyces naeslundii* T14V and *Streptococcus oralis* 34; (**D**) *Capnocytophaga ochracea* ATCC33596 and *Streptococcus oralis* ATCC 55229 (formerly *Streptococcus sanguis* H1); (**F**) *Actinomyces naeslundii* ATCC 51655 (formerly strain PK606) and *Streptococcus gordonii* ATCC 51656 (formerly strain PK488); (**G**) *Prevotella loescheii* PK1295 and *Streptococcus oralis* 34; (**H**) *Fusobacterium nucleatum* PK1594 and *Actinobacillus actinomycetemcomitans* Y4: (**I**) *Veillonella atypica* PK1910 and *Streptococcus oralis* 34; and (**J**) *Fusobacterium nucleatum* PK1594 and *Veillonella atypica* PK1910. (**B**) Mixed cell suspension of *Streptococcus oralis* 34 and the coaggregation-defective mutant of *Actinomyces naeslundii* T14V. The appearance after adding 60 mM to a mixed cell suspension of (**C**) *Actinomyces naeslundii* T14V and *Streptococcus oralis* 34 or (**E**) *Capnocytophaga ochracea* ATCC 33596 and *Streptococcus oralis* ATCC 55229. (**A, F, I**) Arrows indicate cell–cell interactions in coaggregates. (**I**) Arrow in the middle indicates a rosette, (**H**) Corncob arrangement is clearly discerned. Bar = 10 μm.

mixed with the streptococcal partner cells of the parent actinomyces strain (Fig. 9.2B), are unable to coaggregate with them (compare Fig. 9.2A and 9.2B).

Coaggregation can be quantitatively determined using a radioactive cell type mixed with unlabeled partner cells (Fig. 9.3). After mixing the cell types the coaggregates are sedimented by low speed centrifugation, and the radioactivity in the coaggregates is calculated as the percent of the input radioactivity. This method is particularly useful with multigeneric coaggregates because the behavior of the radioactive cell can be measured in the presence of many other cell types. Accordingly, each cell type in the multigeneric coaggregates can be labeled and its behavior determined (Kolenbrander and Andersen, 1986).

The coaggregating partners used for the experimental results presented in Fig. 9.3 are *Prevotella loescheii* PK1295 and *Streptococcus oralis* C104. To determine the behavior or reactivity of a cell in a coaggregation partnership it is important to establish a saturation curve, which can be accomplished by holding constant the radioactive cell concentration and increasing the number of unlabeled partner cells. A typical saturation curve is shown with the open

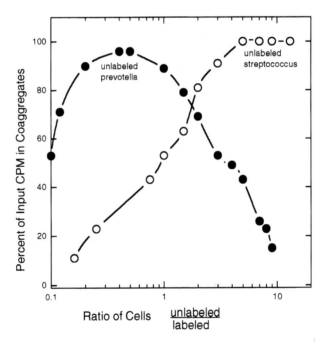

FIGURE 9.3   Percent of input radioactivity in coaggregates formed between unlabeled *Streptococcus oralis* C104 and radioactively labeled *Prevotella loescheii* PK1295 (open circles) or between unlabeled *Prevotella loescheii* PK1295 and labeled *Streptococcus oralis* C104 (closed circles). The number of labeled cells was kept constant while increasing the number of unlabeled partner cells. The logarithmic plot is used to expand the region of cell ratios between 0.5 and 4.0, where the greatest changes in coaggregation occur. See text for details. (Redrawn from data of Kolenbrander and Andersen, 1988)

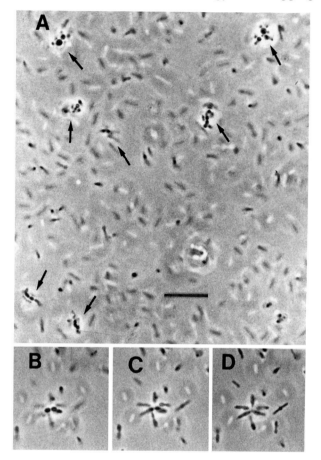

FIGURE 9.4 (A) Population of rosettes formed by mixing a 1:50 concentration of *Strepto-coccus oralis* C104/*Prevotella loescheii* PK1295 (Kolenbrander and Andersen, 1988). After mixing, the suspension was centrifuged at 2000 rpm for 2 minutes. The supernatant is examined by phase contrast microscopy. Arrows indicate rosettes consisting of one or a few streptococci surrounded by prevotellae in a mileau of unbound prevotellae. (**B–D**) Photomicrographs at three focal planes of the same rosette between *Streptococcus oralis* C104 and *Prevotella loescheii* PK1295 at a cell ratio of 1:10 (Kolenbrander and Andersen, 1988). Streptococcal cells are prominent in the center of the rosette (**B**), and the splayed prevotellae cells become more discernible (**C**) and prominent (**D**) as they surround the center cells. Bar = 10 μm.

circles in Figure 9.3, where the addition of more unlabeled partner results in sedimentable coaggregates. A linear increase in the percent of radioactive prevotellae in coaggregates is seen with increasing numbers of unlabeled partner streptococci (Fig. 9.3, open circles). All of the radioactivity is in the coaggregates at about a fivefold excess of partner cells.

An atypical saturation curve, shown with the closed circles in Figure 9.3, is the result of an unusual property of the prevotellae. The prevotellae are difficult to sediment in pure culture in that their sedimentation requires longer

times and higher speeds of centrifugation than those needed to sediment streptococci and most other bacteria. Thus the coaggregation saturation curve looks
quite different when the streptococci are radioactively labeled (Fig. 9.3, closed
circles). Maximum numbers of streptococci are coaggregated when they are in
two- to threefold excess of the unlabeled partner prevotellae. As the numbers
of prevotellae are increased to five to tenfold excess, there is a dramatic decrease in radioactivity in the sediment after standard low speed centrifugation.
The reason for this phenomenon is the formation of rosettes (Fig. 9.4A, arrows)
which consists of streptococci surrounded by prevotellae (Fig. 9.4B–D—three
views of the same rosette). The unlabeled cells buoy up the radioactive streptococci and make it appear as if no coaggregation has occurred, when in fact
all of the streptococci are coaggregated with surrounding prevotellae.

Rosettes such as those seen in Figure 9.4 and corncobs (Fig. 9.2H) (the
rosette equivalent when the central cell is a long rod) occur when the cell ratio
is ≥10:1, and they are commonly found at the periphery of dental plaque
(Listgarten et al., 1975). Such coaggregates may enhance the probability for
accretion of the participating cells. Each of the cells at the periphery of the
rosette is capable of coaggregation; and so if any one of the cells contacts an
already bound partner, the entire rosette can become part of the biofilm. The
likelihood of this event occurring has been demonstrated experimentally by
adding a third cell type as partner of the peripheral cells of the rosette: The
entire rosette becomes a sedimentable multigeneric coaggregate (Kolenbrander
and Andersen, 1988). The central cell of the rosette is sequestered and unable
to coaggregate with its partners, but the peripheral cells behave independently
and coaggregate with all partners tested (Kolenbrander and Andersen, 1988).
More complex multigeneric coaggregates comprised of at least nine distinct
cell types have been studied (Kolenbrander and Andersen, 1986). Such
mixtures, resembling the composition of plaque, were shown by the radioac-
tivity-based assay to be a network of coaggregating pairs that interact independently (Kolenbrander and Andersen, 1986).

Genera That Coaggregate

Most if not all human oral bacteria coaggregate (Kolenbrander, 1989, 1991;
Kolenbrander and London, 1993). Pairwise combinations of seven genera in
seven partnerships are shown in Fig. 9.2. Pairwise tests have been conducted
on strains from most of the frequently isolated oral genera (Table 9.1). The list
in Table 9.1 is a compilation from studies conducted in several laboratories
and is based primarily on the data summarized by Moore and Moore (1994.
Of the 20 genera most frequently isolated, collectively, from plaque samples
taken from healthy and diseased sites, 17 have been tested for coaggregation;
nearly all of the strains from all 17 genera tested positive. Although frequently
isolated, *Bacteroides*, *Campylobacter*, and *Gemella* have not been tested for
coaggregation. Two other genera, *Corynebacterium* and *Rothia*, exhibit coaggregation but are not frequently isolated.

TABLE 9.1 Relation of Genera Most Frequently Isolated from Human Dental Plaque and Other Oral Sites with Their Ability to Coaggregate

Human Oral Bacteria Most Frequently Isolated in Health and Disease	Coaggregation Observed
Actinobacillus	Yes
Actinomyces	Yes
Bacteroides	[a]
Campylobacter	[a]
Capnocytophaga	Yes
Eikenella	Yes
Eubacterium	Yes
Fusobacterium	Yes
Gemella	[a]
Haemophilus	Yes
Lactobacillus	Yes
Peptostreptococcus	Yes
Porphyromonas	Yes
Prevotella	Yes
Propionibacterium	Yes
Selenomonas	Yes
Streptococcus	Yes
Treponema	Yes
Veillonella	Yes
Wolinella	Yes

[a] Strains from these genera have not been tested for coaggregation.

Before taxonomic reclassification, members of the genera *Bacteroides* and *Gemella* showed coaggregation with other bacteria (Kolenbrander and Andersen, 1984; Kolenbrander et al., 1989). The bacterial strains that had been tested are now part of the genera *Prevotella*, *Porphyromonas*, and *Streptococcus*. Two species of oral *bacteroides*, *B. forsythus* and *B. gracilis*, are frequently isolated but neither has been tested as a coaggregation partner. It is likely that *B. forsythus* (closely related to species of *Porphyromonas*) (Tanner et al., 1994) and *Gemella morbillorum* (related to species of *Streptococcus*) (Tanner et al., 1994) also coaggregate. *B. gracilis* is closely related to species of *Campylobacter* and may be a nonmotile *Campylobacter* species (Tanner et al., 1994); both are related to the coaggregating *Wolinella*. Hence it is likely that these untested species are partners of some of the genera listed in Table 9.1.

Specificity of Coaggregation

Intergeneric partnerships are not formed randomly. Where strains within a given genus coaggregate with strains of another genus, the reactions represent a defined set of partnerships. When two or more strains exhibit the same kind of coaggregations with the same set of partners, they comprise a coaggregation group. For example, six coaggregation groups of streptococci and six coaggregation groups of actinomyces have been delineated (Kolenbrander, 1988, 1989) and account for more than 90% of the more than 200 tested strains of *Actinomyces naeslundii* and the streptococci belonging to *S. gordonii*, *S. oralis*, and *S. sanguis*. The veillonellae were grouped on the basis of their coaggregations with the streptococci and actinomyces and comprise four groups (Hughes et al. 1988).

The criteria for establishing coaggregation groups are (1) coaggregation with a set of partner strains, (2) inhibition of coaggregation by lactose or other sugars, (3) protease (or heat) sensitivity of the coaggregation ability of either or both cell types, and (4) simultaneous loss of a cluster of coaggregation partnerships by coaggregation-defective (Cog⁻) mutants. So far, all coaggregations reported have been pairs in which one cell type bears a surface molecule that is inactivated by heat or protease treatment and the other cell type bears a complementary molecule that is insensitive to these treatments (Kolenbrander, 1989).

Four kinds of coaggregation are depicted in Figure 9.5 to illustrate these criteria for coaggregation grouping. The figure does not show all known kinds of coaggregation and is not intended to be more than an example. The names of the cell types could be changed to other genera, but coaggregation has been studied most extensively with the oral streptococci and actinomyces. The four actinomyces cells represent four distinct partner cell types. The center streptococcus is a composite of four distinct cell types, each of which is represented as the portion (the 90 degree arc) of the streptococcal surface that faces each of the respective actinomyces cells. Each coaggregation is mediated by a complementary set of symbols. The symbol with a stem represents a heat- and protease-sensitive coaggregation mediator and is called an adhesin. The symbol without a stem, the receptor, is insensitive to heat or protease and may contain carbohydrate, protein, or both. Three complementary sets (adhesin–receptor pairs) are depicted. The triangle-shaped receptor on the actinomyces (Fig. 9.5, upper right) and its complementary adhesin on the streptococcus represent a unimodal coaggregation. Heat treatment of only one of the cell types (streptococcus) is sufficient to prevent coaggregation. The triangle-shaped symbols represent an interaction with no known inhibitors.

The rectangle-shaped receptor on the center streptococcal cell represents a carbohydrate-containing molecule, and the coaggregations mediated by that molecule are sensitive to addition of 60 mM lactose (McIntire et al., 1988). Sugar inhibition characterizes the interaction as a lectin/carbohydrate-mediated coaggregation, where the lectin (adhesin) is borne on the actinomyces (Fig. 9.5, lower right) and the carbohydrate is expressed by the streptococcus. The interaction is unimodal, as heat treatment of the actinomyces is sufficient to prevent coaggregation.

The cell pair depicted on the bottom left is a bimodal coaggregation because both cell types must be heated to prevent coaggregation. Lactose does

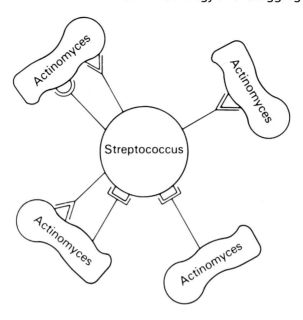

Complementary Surface Components		Coaggregation Inhibited by 60mM Lactose
on Streptococcus	on Actinomyces	
Heat Inactivated (85°C/30')	Heat Stable:	
		No
		No
Heat Stable:	Heat Inactivated (85°C/30')	
		Yes

FIGURE 9.5   Interactions between streptococci and actinomyces. See text for details.

not inhibit this coaggregation with untreated cell types. Although lactose inhibits the coaggregation mediators represented by the rectangle-shaped symbols, the coaggregation still occurs by the molecules represented by the triangle-shaped symbols. It is an example of the independent nature of each complementary set of coaggregation mediators. Heating only the streptococcus inactivates the adhesin depicted as a triangle and renders the coaggregation with unheated actinomyces inhibitable by 60 mM lactose. Heating only the actinomyces makes the bimodal coaggregation indistinguishable from the unimodal coaggregation between the streptococcus and the actinomyces in the upper right of the figure.

The fourth kind of coaggregation (Fig. 9.5, upper left) is unimodal because heating only the streptococcus is sufficient to prevent coaggregation. However, two adhesin−receptor pairs are depicted. Their separate activities cannot be detected because each masks the other's function. Unmasking can be accomplished by isolating Cog⁻ mutants that have an altered activity in one of the adhesins. Such Cog⁻ mutants are easily isolated as spontaneous mutants from the population by mixing the parent strain with nonviable partner cells. If the coaggregation is strong, such as that shown in Figure 9.1, tube 4 (or after low speed centrifugation), the supernatant fluid contains the putative Cog⁻ mutants. The cycle of addition of partner cells to the supernatant, mixing, and low speed centrifugation is repeated five or six times, and the final supernatant is plated on a suitable agar medium. Colonies are picked into microtiter wells, and the colony mass is suspended in buffer. An aliquot is transferred to a replica microtiter plate, and a suspension of partner cells is added and mixed to identify colonies expressing the Cog⁻ phenotype.

For example, assume that the center streptococcus represents an oral streptococcal strain, and that it expresses on its surface two adhesins (represented by the triangle and semicircle symbols with stems) and the receptor for the lactose-inhibitable coaggregation (rectangle-shaped symbol without a stem). If a Cog⁻ mutant of the streptococcus that was unable to coaggregate with the actinomyces in the upper right was desired, the parent streptococcus would be mixed with the actinomyces and the Cog⁻ mutant isolated by the procedure described above. Further testing of the mutant would reveal that it is unaltered during coaggregation (remains lactose-inhibitable) with the actinomyces in the lower right; it now exhibits lactose-inhibitable coaggregation with the actinomyces in the lower left; and the semicircle-mediated coaggregation is unmasked in the coaggregation depicted with the actinomyces in the upper left. These four kinds of coaggregation illustrate the basic principles of delineating coaggregation groups. Each interaction functions specifically and independently of the others.

Of the 17 genera examined (Table 9.1), strains of *Fusobacterium nucleatum* exhibit the broadest range of partners (Kolenbrander et al., 1989). Collectively, the strains of *F. nucleatum* coaggregate with all of the other 16 genera. Individually, each strain of *F. nucleatum* exhibits a specific set of partners that includes members of many but not all genera. Many coaggregations between *F. nucleatum* and its gram-negative partners are inhibited by lactose (Kolenbrander and Andersen, 1989, Kolenbrander et al., 1989). The mediator of these coaggregations on the fusobacterium surface seems to be the same, as Cog⁻ mutants of *F. nucleatum* that were isolated by their failure to coaggregate with one gram-negative partner have simultaneously lost their ability to coaggregate with all of the lactose-inhibitable coaggregation partners but retained coaggregation with lactose-noninhibitable partnerships. This mediator has been reported to be a 42 kDa major outer membrane protein (Kinder and Holt, 1993). Because of this unusual universal coaggregation trait and the fact that they are some of the most numerous bacteria in dental plaque (Moore and Moore 1994), the fusobacteria may be the coaggregation hubs around which other bacteria attach either before or after accretion. Indeed, corncob formations with fuso-bacteria in the center (Fig. 9.2H) are morphologically the same as those that are common at the plaque periphery (Listgarten et al., 1975).

Only a few of the surface adhesins and receptors that mediate specific coaggregation reactions have been characterized (for reviews: Kolenbrander, 1991; Kolenbrander and London, 1992, 1993). Two adhesins are of particular interest. The 75 kDa adhesin on *Prevotella loescheii* PK1295 (London and Allen, 1990) (Fig. 9.2G) is encoded by the *plaA* gene. The *plaA* gene product is a lectin that recognizes galactoside moieties on both prokaryotic and eu-karyotic cells (Weiss et al., 1989). Expression is mediated by a programmed frameshifting hop (Manch-Citron and London, 1994), a rare occurrence among prokaryotes. During translation of the RNA message, 28 nucleotides appear not to be read. This segment of DNA is flanked by two ochre codons, and ribosomal transit across the 3′ stop codon is followed immediately by a +1 frameshift (Manch-Citron and London, 1994). The role of the frameshifting is not yet clear, although in viruses and some bacteria frameshifting itself serves to en-code more than one protein within a gene or to fuse two or more proteins into a polyprotein complex (Atkins et al., 1990). Whether more than one protein is encoded by the *plaA* gene is not known.

The second adhesin, ScaA, is a streptococcal coaggregation adhesin borne on *Streptococcus gordonii* PK488; it mediates coaggregation with *Actinomyces naeslundii* PK606 (Fig. 9.2F). ScaA is 34.8 kDa, and homologs of nearly the same size are expressed on most human oral streptococci that coaggregate with *A. naeslundii* PK606 as well as some that do not coaggregate (Andersen et al., 1993). The *scaA* gene encoding ScaA has been cloned and is part of a putative ATP-binding cassette (ABC) operon that encodes for three functional kinds of protein (Kolenbrander et al., 1994). The ABC proteins have been characterized in a variety of gram-negative organisms and include one or two ATPases, one or two hydrophobic membrane proteins, and a substrate-binding protein that initiates transport of the substrate into the cell at the expense of an ATP mol-ecule (for reviews: Ames, 1992; Nikaido and Saier, 1992; Fath and Kolter, 1993). In gram-positive organisms such as the streptococci, the substrate-bind-ing protein is a lipoprotein, presumably because it must anchor itself to the cytoplasmic membrane. Substrates such as peptides (Alloing et al., 1994) and sugars (Russell et al., 1992) are known to be transported by the ABC proteins. ScaA of *S. gordonii* PK488 contains the signal peptidase II cleavage site for lipoproteins and is 91% identical to the lipoprotein SsaB (Ganeshkumar et al., 1993), which mediates adherence of another oral streptococcus, *S. sanguis* 12, to a salivary receptor (Ganeshkumar et al., 1988, 1991). Both *S. gordonii* PK488 and the oral *S. parasanguis* FW213, which expresses the ScaA homolog FimA (Fenno et al., 1993) also bind to salivary receptors. The relation of a lipoprotein surface adhesin that mediates (1) oral streptococcal coaggregation with actinomyces, (2) binding to salivary receptors, and probably (3) transport of small molecules for energy requirements for cellular growth is an area of

active research interest. Such dual/multiple functions of these surface lipoprotein adhesins presents ideas on potentially useful methods of isolating adher-
ence-altered mutants and on the roles played by these proteins in recognizing
small molecules and communicating with other bacteria within the biofilm.

Available Surfaces

The human oral cavity contains two distinctive surfaces: mucosal and tooth.
One basic difference in the two surfaces is that the soft mucosal tissue sloughs
its external layer every few days, whereas the tooth is retained. Accordingly,
the attached bacteria are sloughed with the mucosal cells. In contrast, the tooth
remains colonized until some form of treatment, mechanical or chemical, removes the adherent biofilm.

The mucosal surfaces include the tongue, palate, cheek, and gingival tissue
surrounding the teeth (Fig. 9.6). The bacterial species inhabiting the mucosal
tissue throughout the oral cavity are also found in saliva, which is considered
to be the medium by which the bacteria move from place to place. The hard
enamel surface of the tooth, although in contact with the tongue, is colonized
by distinct bacteria (Gibbons and van Houte, 1971; Frandsen et al., 1991). For
example, whereas *S. gordonii* is isolated infrequently from mucosal surfaces,
it is one of the more common isolates from mature dental plaque. The tongue
dorsum contains high numbers of *S. salivarius*, but this bacterial species is

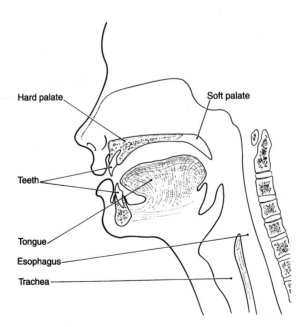

FIGURE 9.6   Sagittal section of human head and neck.

present in low numbers and is often not isolated in samples of dental plaque. *S. sanguis* is found in high numbers on teeth and on the buccal mucosa but is not found on the tongue dorsum (Frandsen et al., 1991). Actinomyces are found on both mucosal and tooth surfaces, as are the veillonellae. *Veillonella parvula* predominates in subgingival dental plaque, and *Veillonella atypica* and *Veillonella dispar* occupy the tongue (Hughes et al., 1988).

The ecological significance of coaggregation on both surfaces was shown by the observation that veillonellae isolated from human tongue coaggregate with *S. salivarius*, and veillonellae isolated from subgingival sites coaggregate with *S. sanguis* and other plaque bacteria (Hughes et al., 1988).

Temporal Changes in Microbial Population

During the first few days after birth, the oral cavity of the human infant becomes colonized almost exclusively by streptococci (Smith et al., 1993; Cole et al., 1994). Although the streptococci constitute approximately 95% of the culturable bacteria at 101 days, other bacteria such as *A. naeslundii* and veillonellae also appear (McCarthy et al., 1965). The predominant streptococcal species in predentate infants are *S. salivarius* (Carlsson et al., 1970a), *S. oralis*, and *S. mitis* biovar 1 (Smith et al., 1993; Cole et al., 1994). *S. sanguis* appears after tooth eruption (Carlsson et al., 1970b), as do fusobacteria and veillonellae. The mutans streptococci, which are recognized as the etiologic agent of caries, are acquired at the median age of 26 months during a discrete period called the "window of infectivity" (Caufield et al., 1993).

The development of adult dentition is accompanied by a varied population of bacteria composed of at least 500 taxa (Moore and Moore, 1994). The most numerous species from subgingival plaque samples from healthy sites are gram-positive (Table 9.2, top group). Of the 10 bacterial species representing the highest percentage of randomly obtained isolates, the only gram-negative one is *F. nucleatum*. The other nine represent nearly 50% of the isolates identified. Three of the nine are actinomyces and four are streptococci, which indicates the predominance of these bacteria in plaque from healthy sites. Gingivitis, an inflammatory condition of the tissue surrounding the teeth, is accompanied by a shift to high numbers of gram-negative bacteria (Table 9.2, middle group). The fusobacteria now become the predominant population, increasing 10,000-fold. Of the six species with the highest numbers, four are gram-negative and constitute 21% of the isolates. The gram-positive flora decreased from 49% to 22% in samples obtained from healthy sites and gingivitis-affected sites, respectively.

Table 9.2 is arranged so the ranking of a species can be determined in health, with gingivitis, and with the types of periodontitis called moderate, adult, juvenile, and severe. Each is defined clinically by inflammation of the gingiva, degrees of loss of alveolar bone, and number of sites of inflammation and bone loss. The species ranking of *F. nucleatum* and *A. naeslundii* remain high at all stages of health and disease. Whereas *F. nucleatum* is a common

TABLE 9.2 Most Numerous Species from Subgingival Crevices of Subjects with Various Periodontal Health Conditions[a]

Species Rank[b]	No., by Periodontal Condition					
	Health	Gingivitis	Moderate	Adult	Juvenile	Severe
Health						
A. naeslundii[c]	1	2	1	1	2	4
F. nucleatum	**2**	**1**	**2**	**2**	**3**	**1**
S. sanguis	3	14	32	3	23	41+
S. oralis	4	24	17	8	8	41
S. intermedius	5	19	14	11	36	9
P. micros	6	7	3	4	10	6
A. meyeri	7	43+	30	31	20	31
S. gordonii	8	40	35	47+	42+	41+
G. morbillorum	9	31	44+	27	37	41+
A. odontolyticus	10	18	22	12	18	41+
Gingivitis						
F. nucleatum	**2**	**1**	**2**	**2**	**3**	**1**
A. naeslundii	1	2	1	1	2	4
L. uli	38+	3	9	9	7	5
C. concisus	**30**	**4**	**44+**	**6**	**42+**	**41+**
S. sputigena	**36**	**5**	**15**	**10**	**39**	**23**
B. gracilis	**17**	**6**	**21**	**5**	**25**	**25**
P. micros	6	7	3	4	10	6
A. anaerobius ID	38+	8	20	23	16	29
A. israelii	21	9	38	22	34	21
L. rimae	11	10	29	13	13	12
Severe periodontitis						
F. nucleatum	**2**	**1**	**2**	**2**	**3**	**1**
E. nodatum	38+	13	11	14	4	2
E. timidum	33	12	10	7	9	3
A. naeslundii	1	2	1	1	2	4
L. uli	38+	3	9	9	7	5
P. micros	6	7	3	4	10	6
Eubacterium D06	38+	43	13	47+	1	7
P. intermedia	**38+**	**33**	**7**	**46**	**22**	**8**
S. intermedius	5	19	14	11	36	9
Streptococcus D39	38+	43+	44+	47+	42+	10

[a]Species rank is based on data obtained from Moore and Moore (1994).
[b]Listed in order of percent of isolates obtained at random. Boldface type indicates gram-negative bacteria.
[c]*Actinomyces israelii, Actinomyces meyeri, Actinomyces naeslundii, Actinomyces odontolyticus; Bacteroides gracilis; Campylobacter concisus; Eubacterium nodatum, Eubacterium timidum; Fusobacterium nucleatum; Gemella morbillorum; Lactobacillus rimae, Lactobacillus uli; Peptostreptococcus anaerobius, Peptostreptococcus micros; Prevotella intermedia; Selenomonas sputigena; Streptococcus gordonii, Streptococcus intermedius, Streptococcus oralis, Streptococcus sanguis.*

pathogen of other body sites, *A. naeslundii* is not (Moore and Moore, 1994). Because there is a significant rise in the relative concentration of *F. nucleatum* with the appearance of gingivitis, it seems probable that it is associated with gingivitis. Likewise, the relative concentration of *A. naeslundii* tends to decrease, suggesting that it is more relevant to maintaining a healthy oral microbial community.

The population of bacteria from sites showing severe periodontitis commonly contain three species of *Eubacterium* (Table 9.2, bottom group). These eubacteria and *Prevotella intermedia* are considered potential periodontal pathogens because they are frequently found in high numbers in samples taken from diseased sites but are found in low numbers in samples from healthy sites. It is evident from the data in Table 9.2 that no single bacterial species is solely responsible for causing periodontal disease. Other species, such as *Actinobacillus actinomycetemcomitans*, *Porphyromonas gingivalis*, *Bacteroides forsythus*, *Campylobacter rectus*, and *Treponema* spp. are suspected periodontal pathogens. They are frequently isolated from diseased sites and are usually absent from healthy sites. Moreover, their numbers decrease as the clinical condition of the site improves. Thus periodontal disease seems to be an ecological disease in that several species or combinations of species contribute to its development.

Potential Role of Coaggregation in Oral Colonization

Acquired Pellicle

The eruption of teeth introduces a dramatic change in the oral cavity. Bacteria that were unable to attach to the mucosal surfaces or to compete for colonization of the mucosal surface may now have the opportunity to colonize (Fig. 9.7). The first coating on a professionally cleaned tooth surface is the acquired pellicle, which is shown as an amorphous layer along the length of the tooth surface (Fig. 9.7A). This layer is composed of host-derived molecules such as glycoproteins, mucins, proline-rich proteins, phosphate-containing proteins, and enzymes such as α-amylase, as well as bacterial extracellular glucans and pieces of bacterial debris. Each of these molecules is a potential receptor for bacterial adherence, and presumably there are others that await discovery. None of these recognition and binding processes requires energy, which may be advantageous to human oral bacteria and allows them to accrete in the absence of food. Thus bacterial plaque maintains its integrity and does not fall apart under conditions of normal diurnal feeding habits of the host.

Primary Colonization

Primary colonizers must adhere to components of the acquired pellicle (Fig. 9.7B). The best studied host-derived receptor is a group of acidic proline-rich proteins (Gibbons et al., 1988, 1991). One binding site is the Pro-Gln dipeptide at the carboxy-terminus, which appears to be the smallest recognized receptor for *Streptococcus gordonii* (Gibbons et al., 1991). A different region is recognized by some actinomyces that express type 1 fimbriae, which enable them to bind to proline-rich protein-coated hydroxyapatite beads (Gibbons et al.,

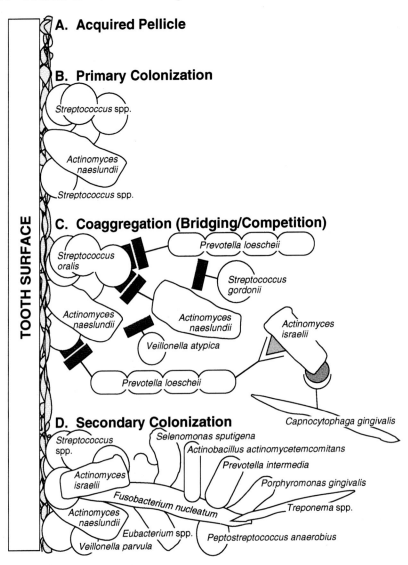

FIGURE 9.7 (**A, B**) Temporal relation of bacterial colonization of the acquired pellicle coating on the surface of a freshly cleaned tooth. (**C**) Some of the mechanisms of coaggregation. (**D**) Proposed involvement of competitive events (*top*) and bridging (*bottom*) when establishing the microbial community.

1988). *Actinomyces naeslundii* and *Fusobacterium nucleatum* bind to the phosphoprotein statherin, and *S. gordonii* binds to α-amylase (Scannapieco et al., 1989) but not statherin. Several other bacteria, such as the capnocytophagae, haemophili, prevotellae, and porphyromonads, bind to unknown receptors on saliva-coated hydroxyapatite, a model surface for enamel. Bacterial debris, including cell wall fragments, is an excellent source of receptors in the acquired pellicle. They in turn can be bound to other fragments or host-derived molecules. Salivary lysozyme is able to produce a constant supply of potential receptor fragments that can bind and be integrated into the pellicle or can coat bacteria suspended in saliva. Therefore the nascent surface of an acquired pellicle contains an immense array of receptors available for accretion by the primary colonizing bacteria.

Although the variety of available receptors may be great, the bacteria isolated during the first 4 hours after cleaning is a highly selected group. Streptococci constitute 47–85% of the cultivatable bacteria, and actinomyces comprise a large part of the remainder (Nyvad and Kilian, 1987). The most common streptococci are *S. sanguis*, *S. oralis*, and *S. mitis* biovar 1, all of which produce immunoglobulin A1 (IgA1) protease (Nyvad and Kilian, 1987). *S. gordonii*, which does not produce IgA1 protease but does bind to host-derived proline-rich proteins and α-amylase, varies between 0 and 22% (Nyvad and Kilian, 1990). Because of the preponderance of IgA1 protease-producing streptococci among the early colonizers, compared to only 17% in mature dental plaque, it was postulated that this property may confer an ecological advantage on the primary colonizing streptococci (Frandsen et al., 1991). The ability to evade the host secretory IgA defense against mucosal pathogens by disabling the IgA could be advantageous to the protease-producing bacteria. However, a comparison of 4-hour plaque of IgA-deficient and normal individuals revealed no difference in the numbers of IgA1 protease-producing streptococci in the early plaque of these individuals (Reinholdt et al., 1993). These results suggest that oral bacteria engage more than one ecological advantage for establishing colonies on host surfaces.

Coaggregation (Bridging/Competition)

One of these ecological advantages may be an organism's ability to coaggregate with other colonizing bacteria (Fig. 9.7B,C). Coaggregations can be intergeneric, intrageneric, or multigeneric. Intrageneric coaggregation seems to be frequent only among the streptococci (Kolenbrander et al., 1990). Among 122 strains tested from 11 genera for the property of intrageneric coaggregation, only the streptococci and a few actinomyces showed this ability (Kolenbrander et al., 1990). All of the streptococcal coaggregations were inhibited by lactose or *N*-acetyl-D-galactosamine. In a study of 71 strains of viridans streptococci, most of the 11 *S. oralis* strains were coaggregation partners of the 17 *S. sanguis* strains and 10 *S. gordonii* strains tested (Hsu et al., 1994). Only a few examples of coaggregation were observed between *S. gordonii* and *S. sanguis* strains. Coaggregations involving *S. mitis* strains and *S. anginosus* strains with any of the viridans streptococci were even more rare (Hsu et al., 1994). These data support the proposal that the cell surface of an adherent *S. oralis* may be the principal target of coaggregation competition between and among colonizing streptococci. Considering that streptococci and actinomyces exhibit intrageneric coaggregation and comprise most of the primary colonizers, they may have gained this dominance, at least partially, through their capability of intrageneric coaggregation.

The tooth surface coated with an acquired pellicle favors colonization by bacteria that both recognize exposed receptors and grow after adhering. Facultative anaerobes such as the streptococci and actinomyces seem best adapted to this environment. Actinomyces adhere during the first 4 hours (Nyvad and Kilian, 1987) but are much slower-growing organisms than the streptococci. Given an equal probability of binding and number of initial cells bound, the streptococci would quickly outnumber the actinomyces and any of the other bacteria that are capable of adhering to the acquired pellicle. Moreover, as each bacterial cell binds to the pellicle, its surface becomes available for adherence. The intrageneric coaggregation ability of streptococci and actinomyces would therefore be a significant advantage, as it would provide an additional mechanism for evading clearance from the oral cavity by salivary flow. It is curious that the fusobacteria predominate in mature dental plaque samples from both healthy and diseased sites (Table 9.2), but they are not commonly found during the first few hours after the tooth surface is cleaned. Two factors probably contribute to these observations. First, they require an anaerobic atmosphere for growth and the increasingly more diverse early colonizers prepare the environment by making the microniche progressively more anaerobic. Second, the surviving fusobacterial cells may be immediately coated with 10 to 50 coaggregating partner cells and thus the relative members of fusobacteria are decreased dramatically. Their coaggregation ability to form corncobs where the surrounding cells maintain an anaerobic microniche for the fusobacterium may protect them from death. Under ensuing favorable conditions, accretion of the corncob may take place, as may growth of both constituent cell types.

As in any ecosystem, it is expected that oral colonization is also subject to competitive events. Many coaggregations, both intrageneric and intergeneric, are inhibited by lactose (Kolenbrander, 1989; Kolenbrander and Andersen, 1989; Kolenbrander et al., 1989, 1990; Kolenbrander and London, 1992). A cell that can express a receptor recognized by partner cell types of several genera could thus present a surface for which each of the cell's partners would compete. An example is depicted in Figure 9.7C, where *S. oralis* expresses a receptor (stippled rectangle) that is recognized by adhesins (complementary stippled rectangles with stems) on *P. loescheii* and *A. naeslundii*. Two other cell types, *S. gordonii* and *V. atypica*, express adhesins that are capable of recognizing the same receptor on *S. oralis* and can actively compete with *P. loescheii* for binding (Kolenbrander and Andersen, 1985).

Cog⁻ mutants are useful for understanding the relation of surface molecules on cells that exhibit identical coaggregation properties. Cog⁻ mutants of *S. oralis* have been selected for failure to coaggregate with *P. loescheii*. They simultaneously lost the ability to coaggregate with all four partners of the parent strain (Fig. 9.7C). Likewise, Cog⁻ mutants of *S. oralis* unable to coaggregate with *S. gordonii* simultaneously lost their coaggregation with all four partners depicted. Cog⁻ mutants that are unable to coaggregate with *S. oralis* have been isolated for each of the four partners depicted (*P. loescheii, S. gordonii, A.*

naeslundii, and *V. atypica*). Cog⁻ mutants of *P. loescheii* do not compete with their parent for coaggregation with *S. oralis*; Cog⁻ mutants of *A. naeslundii* do not compete with *P. loescheii* for coaggregation with *S. oralis*, whereas the parent strain of *A. naeslundii* does compete with *P. loescheii* (Kolenbrander and Andersen, 1985; Kolenbrander et al., 1985); the Cog⁻ mutants of the other two strains have not been tested in the competition assay.

The structural and functional relation among the four adhesins are being actively researched. Only the 75 kDa adhesin from *P. loescheii* PK1295 has been purified and shown to mimic the functional properties of the cell that bears it (London and Allen, 1990). An isofunctional adhesin has been identified as a 45 kDa protein of *V. atypica* PK1910 (Hughes et al. 1992). The adhesin was absent in Cog⁻ mutants of *V. atypica*, and the mutant-absorbed antiserum against the parent strain blocked coaggregation between the parent and *S. oralis* 34. The adhesins from *S. gordonii* DL1 and *Actinomyces* serovar WVA963 strain PK1259, which exhibit lactose-inhibitable coaggregation with *S. oralis* 34, are being studied by using Cog⁻ mutants to identify potential adhesins.

In addition to competition among cohabitants of an econiche, there is co-operation. With regard to adherence, cooperation may take the form of bridging between two cell types that are unable to coaggregate (Fig. 9.7C). For example, *P. loescheii*, the partner in common, can bridge a coaggregation between the *Streptococcus oralis* adherent to the acquired pellicle and *Actinomyces israelii* (Kolenbrander et al., 1985). In turn, *A. israelii* can bridge between *Capnocytophaga gingivalis* and *P. loescheii*. Each coaggregation in such a sequential accretion behaves independently of the others (Kolenbrander and Andersen, 1986). Bridging is distinct from competition because the coaggregation mediators (represented as a rectangle, triangle, and semicircle in Figure 9.7C) on the respective partners are different for each portion of the bridging event, whereas competition is active when the same kind of receptor (or the adhesin) is the coaggregation mediator (represented as the rectangle in Fig. 7C).

Secondary Colonization

The numbers and diversity of bacteria increase in dental plaque in the absence of oral hygiene. The primary colonizers during the first few hours after the teeth are cleaned are generally the same as the species that are found in plaque samples obtained from normal healthy sites (Table 9.2, top group). Fusobacteria are present in high numbers in plaque from healthy sites, and fusobacteria bind to statherin, a component of the acquired pellicle. Gingivitis-affected sites are occupied by an increasing number of gram-negative bacteria, and the clearly prominent one is *Fusobacterium nucleatum*. Periodontally diseased sites contain *F. nucleatum* in high numbers, but most of the remainder are various species of gram-positive bacteria (Table 9.2, bottom group).

Secondary colonization is considered here to be the process that occurs when the acquired pellicle is already occupied with primary colonizers, and so direct attachment to the acquired pellicle is minimized but not necessarily absent. Accretion could occur by cell surface recognition of an adherent cell by a suspended cell. Considering that streptococci, actinomyces, and fusobacteria are the predominant primary colonizers of the acquired pellicle, secondary colonizers would be capable of recognizing their surfaces. Results from numerous studies of coaggregation potential of more than 1000 strains of oral bacteria, representing the genera shown in Fig. 9.7, revealed a distinct temporal selection of coaggregation partners. Early colonizers, such as the streptococci and actinomyces, coaggregate with many other genera that are also found in healthy plaque but in lower numbers (Kolenbrander, 1989, 1991; Kolenbrander et al., 1989, 1990), some of which are the prevotellae, veillonellae, capnocytophagae, and propionibacteria. All of these bacteria coaggregate with fusobacteria. Bacteria such as porphyromonads, treponemes, actinobacilli, selenomonads, and eubacteria are found in low numbers in healthy sites, but they increase in number temporally in the absence of oral hygiene. Only a few porphyromonads of the latter group of secondary colonizers coaggregate with streptococci (Stinson et al., 1992) and actinomyces (Li et al., 1991), and only one treponeme coaggregates with a porphyromonad (Grenier, 1992); in contrast, members of all of the species coaggregate with fusobacteria.

Many of the interactions between *F. nucleatum* and gram-negative partners are inhibited by lactose. Cog⁻ fusobacterium mutants that fail to coaggregate with *P. gingivalis*, a receptor-bearing partner of a lactose-inhibitable coaggregation, have been isolated. Most of these mutants simultaneously lost the ability to coaggregate with all other lactose-inhibitable coaggregation partners, (e.g., actinobacilli, capnocytophagae, prevotellae, selenomonads, and veillonellae) while retaining lactose-noninhibitable coaggregations with actinomyces, peptostreptococci, propionibacteria, and streptococci. These results illustrate two important aspects of coaggregation that are seen in most coaggregations among oral bacteria. First, a coaggregation-mediating surface molecule recognizes its complementary molecule on the surfaces of strains from several genera. Second, most strains have more than one kind of coaggregation mediator, and these mediators may bear little or no functional resemblance to each other. At least five have been identified on *A. naeslundii* PK606 (Kolenbrander and Andersen, 1990). This capability ensures that the cell has adherence potential in the presence of inhibitors of one kind of coaggregation but not the other(s). Thus to have an inhibitory effect on the integrity of the plaque biofilm it seems necessary to treat bacterial plaque with an inhibitor or set of inhibitors that could interfere with adherence in a broader sense than the specific coaggregations outlined in this chapter.

Climax Community

The breadth of coaggregation partnerships among the fusobacteria and the predominance of fusobacteria in dental plaque in health and diseased states directs our attention to the potential significance of their coaggregations during colonization of themselves as well as their partners. Clearly, fusobacterial corncob formations with partners have been observed frequently at the periphery of plaque biofilm (Listgarten et al., 1975). An important aspect in the development of climax communities is the temporal relationship of the inhabitants. In the oral cavity, coaggregation partners illustrate a strong temporal relationship. Primary colonizers coaggregate extensively among each other and with fusobacteria; they bridge and promote accretion of some of the early wave of secondary colonizers, which also coaggregate with fusobacteria. In contrast, the final group of secondary colonizers (e.g., the selenomonads, actinobacilli, eubacteria, and treponemes) coaggregate almost exclusively with fusobacteria. As had been noted with other coaggregations, these are also highly specific. This coaggregation pattern of limited partnerships of the late colonizers suggests a role for the fusobacteria as assistants or coaggregation bridges in the colonization of the plaque biofilm. Thus it is proposed that the major adherence-relevant factors in the formation of the climax community are the fusobacterium bridges between the final group of secondary colonizers and the plaque biofilm. It follows that interdicting the colonization of fusobacteria selectively should significantly decrease the number of secondary colonizers, many of which are considered potential periodontal pathogens. Similarly, preserving a quorum of early colonizers should be advantageous to the host and prevent opportunistic pathogens from occupying sites on the tooth surface.

Plaque as a Multicellular Organism

The interacting network of coaggregating bacteria comprising dental plaque can be considered collectively as a multicellular organism. Microscopic viewing techniques reveal that morphologically distinct bacteria of dental plaque appear to be attached to each other in a variety of morphological shapes. Words such as corncobs, rosettes, and test tube brushes are used to describe some of these multicellular arrangements, which contribute to the fabric of the plaque biofilm. Presumably, these arrangements occur because the cells benefit by the interaction. One likely advantage would accrue from metabolic communication. The coaggregation partners could pass end-products of metabolism among the cells, leading to communication for energy requirements or for removal of molecules from the econiche. The result would be growth of the cells, species diversity of the community, and a thicker, denser biofilm. Attendant to species diversity is the production of antimicrobial compounds by accreting members of the biofilm. Some oral bacteria, such as *Streptococcus mutans*, produce bacteriocins that inhibit members of their own species; they are called mutacins (Novak et al., 1994). One of these compounds is a 3245 dalton lantibiotic, so called because it contains lanthionine (Novak et al., 1994). Obviously not all bacteria are affected by these compounds, so there is an active mechanism in the microbial community to inactivate them. The ability of one organism to inactivate such molecules may be its sole cooperative contribution to its microniche, and it uses other molecules excreted by its near neighbors.

Metabolic communication and species diversity are two critical attributes of a plaque biofilm. Only a few examples of metabolic exchange have been studied in any detail (for review: Kolenbrander, 1991). They include the utilization by veillonellae of lactate produced by streptococci from glucose, stimulation of growth of wolinellae by formate (excreted by prevotellae), or satis-fying a porphyromonad's requirement for heme by wolinellae-excreted protoheme. Further study in this area of cross-feeding would be productive, as the end-products and other metabolites present in spent culture media after pure culture growth of most of the oral bacteria are known. This information can be used to design synthetic consortia and test the potential for predicted food chains and cross-feedings to maintain the consortia. Biofilms can be observed in situ by confocal laser microscopy (Wolfaardt et al., 1994), so the relative positions of a each member of the biofilm can be determined.

The principal inhabitants of the biofilm are those listed in Table 9.1 and 9.2 and are the ones exchanging metabolic end-products among the other ex-creted and exchanged molecules. They also are the ones that exhibit coaggre-gation and must be responsive to competition and cooperation both metaboli-cally and in cell-surface interactions. Rosettes have been shown in vitro to sequester the cell-surface interactive ability of the central cell (Kolenbrander and Andersen, 1988). Likewise, it is conceivable that the exterior cells can protect the central cell from harmful molecules by disarming them. The exterior cells could also process unusable molecules into energy-deriving substrates for the central cell; this proposal is equally tenable for the central cell of rosettes or corncobs. Corncob arrangements consisting of a fusobacterial cell in the center and coaggregation partners surrounding it are easily made in vitro simply by mixing fusobacteria with a 10-fold excess of partner cells (Kolenbrander, 1991). Considering the diversity of species that form partnerships in corncobs and rosettes, such multicellular formations may be central to the development of plaque biofilm. Although some advances have been made in understanding the adherence-relevant events in biofilms, little is known about the metabolic interplay among the composite cells leading to the growth of biofilms. Coor-dination of cellular adherence and cellular growth is the essence of the biodi-versity of biofilms and is an area in need of investigation. Results obtained from such studies will improve our understanding of the multicellular nature of coaggregates composed of genetically distinct individual bacterial cells.

References

Alloing, G., dePhilip, P., and Claverys, J.-P. (1994) Three highly homologous mem-brane-bound lipoproteins participate in oligopeptide transport by the Ami system of the gram-positive *Streptococcus pneumoniae*. *J. Mol. Biol.* **241**:44–58.

Ames, G.F.-L. (1992) Bacterial periplasmic permeases as model systems for the super-family of traffic ATPases, including the multidrug resistance protein and the cystic fibrosis transmembrane conductance regulator. *Int. Rev. Cytol.* **137A**:1–35.

Andersen, R.N., Ganeshkumar, N., and Kolenbrander, P.E. (1993) Cloning of the *Strep-tococcus gordonii* PK488 gene, encoding an adhesin which mediates coaggregation with *Actinomyces naeslundii* PK606. *Infect. Immun.* **61**:981–987.

Atkins, J. F., Weiss, R. B., and Gesteland, R. F. (1990) Ribosome gymnastics—degree of difficulty 9.5, style 10.0. *Cell* **62**:413–423.

Carlsson, J., Grahnén, H., Jonsson, G., and Wikner, S. (1970a). Early establishment of *Streptococcus salivarius* in the mouth of infants. *J. Dent. Res.* **49**:415–419.

Carlsson, J., Grahnén, H., Jonsson, G., and Wikner, S. (1970b). Establishment of *Streptococcus sanguis* in the mouths of infants. *Arch. Oral Biol* **15**:1143–1148.

Caufield, P.W., Cutter, G.R., and Dasanayake, A.P. (1993) Initial acquisition of mutans streptococci by infants: evidence for a discrete window of infectivity. *J. Dent. Res.* **72**:37–45.

Cole, M.F., Evans, M., Fitzsimmons, S., et al. (1994) Pioneer oral streptococci produce immunoglobulin A1 protease. *Infect. Immun.* **62**:2165–2168.

Fath, M.J., and Kolter, R. (1993) ABC transporters: bacterial exporters. *Microbiol. Rev.* **57**:995–1017.

Fenno, J. C., Shaikh, A., and Fives-Taylor, P. (1993) Characterization of allelic replacement in *Streptococcus parasanguis*: transformation and homologous recombination in a "nontransformable" streptococcus. *Gene* **130**:81–90.

Frandsen, E.V.G., Pedrazzoli, V., and Kilian, M. (1991) Ecology of viridans streptococci in the oral cavity and pharynx. *Oral Microbiol. Immunol.* **6**:129–133.

Ganeshkumar, N., Arora, N., and Kolenbrander, P.E. (1993) Saliva-binding protein (SsaB) from *Streptococcus sanguis* 12 is a lipoprotein. *J. Bacteriol.* **175**:572–574.

Ganeshkumar, N., Hannam, P.M., Kolenbrander, P.E., and McBride, B.C. (1991) Nucleotide sequence of a gene coding for a saliva-binding protein (SsaB) from *Streptococcus sanguis* 12 and possible role of the protein in coaggregation with actinomyces. *Infect. Immun.* **59**:1093–1099.

Ganeshkumar, N., Song, M., and McBride, B.C. (1988) Cloning of a *Streptococcus sanguis* adhesin which mediates binding to saliva-coated hydroxyapatite. *Infect. Immun.* **56**:1150–1157.

Gibbons, R.J., and Nygaard, M. (1970) Interbacterial aggregation of plaque bacteria. *Arch. Oral Biol.* **15**:1397–1400.

Gibbons, R.J., and van Houte, J. (1971) Selective bacterial adherence to oral epithelial surfaces and its role as an ecological determinant. *Infect. Immun.* **3**:567–573.

Gibbons, R.J., Hay, D.I., and Schlesinger, D.H. (1991) Delineation of a segment of adsorbed salivary acidic proline-rich proteins which promotes adhesion of *Streptococcus gordonii* to apatitic surfaces. *Infect. Immun.* **59**:2948–2954.

Gibbons, R.J., Hay, D.I., Cisar, J.O., and Clark, W.B. (1988) Absorbed salivary proline-rich protein 1 and statherin: receptors for type 1 fimbriae of *Actinomyces viscosus* T14V-J1 on apatitic surfaces. *Infect. Immun.* **56**:2990–2993.

Grenier, D. (1992) Demonstration of a bimodal coaggregation reaction between *Porphyromonas gingivalis* and *Treponema denticola*. Oral Microbiol. Immunol. **7**: 280–284.

Hsu, S.D., Cisar, J.O., Sandberg, A.L., and Kilian, M. (1994) Adhesive properties of viridans streptococcal species. *Microb. Ecol. Health Dis.* **7**:125–137.

Hughes, C.V., Andersen, R.N., and Kolenbrander, P.E. (1992) Characterization of *Veillonella atypica* PK1910 adhesin-mediated coaggregation with oral *Streptococcus* spp. *Infect. Immun.* **60**:1178–1186.

Hughes, C.V., Kolenbrander, P.E., Andersen, R.N., and Moore, L.V.H. (1988) Coaggregation properties of human oral *Veillonella* spp.: relationship to colonization site and oral ecology. *Appl. Environ. Microbiol.* **54**:1957–1963.

Kinder, S.A., and Holt, S.C. (1993) Localization of the *Fusobacterium nucleatum* T18 adhesin activity mediating coaggregation with *Porphyromonas gingivalis* T22. *J. Bacteriol.* **175**:840–850.

Kolenbrander, P.E. (1988) Intergeneric coaggregation among human oral bacteria and ecology of dental plaque. *Annu. Rev. Microbiol.* **42**:627–656.

Kolenbrander, P.E. (1989) Surface recognition among oral bacteria: multigeneric coaggregations and their mediators. *Crit. Rev. Microbiol.* **17**:137–159.

Kolenbrander, P.E. (1991) Coaggregation: adherence in the human oral microbial ecosystem. In *Microbial Cell–Cell Interactions*, M. Dworkin (ed.), pp. 303–329. American Society for Microbiology, Washington, DC.

Kolenbrander, P.E., and Andersen, R.N. (1984) Cell-to-cell interactions of *Capnocytophaga* and *Bacteroides* species with other oral bacteria and their potential role in development of plaque. *J. Periodont. Res.* **19**:564–569.

Kolenbrander, P.E., and Andersen, R.N. (1985) Use of coaggregation-defective mutants to study the relationships of cell-to-cell interactions and oral microbial ecology. In *Molecular Basis of Oral Microbial Adhesion*, S.E. Mergenhagen, and B. Rosan (eds.), pp. 164–171. American Society for Microbiology, Washington, DC.

Kolenbrander, P.E., and Andersen, R.N. (1986) Multigeneric aggregations among oral bacteria: a network of independent cell-to-cell interactions. *J. Bacteriol.* **168**:851–859.

Kolenbrander, P.E., and Andersen, R.N. (1988) Intergeneric rosettes: sequestered surface recognition among human periodontal bacteria. *Appl. Environ. Microbiol.* **54**:1046–1050.

Kolenbrander, P.E., and Andersen, R.N. (1989) Inhibition of coaggregation between *Fusobacterium nucleatum* and *Porphyromonas (Bacteroides) gingivalis* by lactose and related sugars. *Infect. Immun.* **57**:3204–3209.

Kolenbrander, P.E., and Andersen, R.N. (1990) Characterization of *Streptococcus gordonii (S. sanguis)* PK488 adhesin-mediated coaggregation with *Actinomyces naeslundii* PK606. *Infect. Immun.* **58**:3064–3072.

Kolenbrander, P.E., and London, J. (1992) Ecological significance of coaggregation among oral bacteria. *Adv. Microb. Ecol.* **12**:183–217.

Kolenbrander, P.E., and London, J. (1993) Adhere today, here tomorrow: oral bacterial adherence. *J. Bacteriol.* **175**:3247–3252.

Kolenbrander, P.E., Andersen, R.N., and Ganeshkumar, N. (1994) Nucleotide sequence of the *Streptococcus gordonii* PK488 coaggregation adhesin gene, *scaA*, and ATP-binding cassette. *Infect. Immun.* **62**:4469–4480.

Kolenbrander, P.E., Andersen, R.N., and Holdeman, L.V. (1985) Coaggregation of oral *Bacteroides* species with other bacteria: central role in coaggregation bridges and competitions. *Infect. Immun.* **48**:741–746.

Kolenbrander, P.E., Andersen, R.N., and Moore, L.V.H. (1989) Coaggregation of *Fusobacterium nucleatum, Selenomonas flueggei, Selenomonas infelix, Selenomonas noxia*, and *Selenomonas sputigena* with strains from 11 genera of oral bacteria. *Infect. Immun.* **57**:3194–3203.

Kolenbrander, P.E., Andersen, R.N., and Moore, L.V.H. (1990) Intrageneric coaggregation among strains of human oral bacteria: potential role in primary colonization of the tooth surface. *Appl. Environ. Microbiol.* **56**:3890–3894.

Li, J., Ellen, R.P., Hoover, C.I., and Felton, J.R. (1991) Association of proteases of *Porphyromonas (Bacteroides) gingivalis* with its adhesion to *Actinomyces viscosus*. *J. Dent. Res.* **70**:82–86.

Listgarten, M.A., Mayo, H. E., and Tremblay, R. (1975) Development of dental plaque on epoxy resin crowns in man. *J. Periodontol.* **46**:10–26.

London, J., and Allen, J. (1990) Purification and characterization of a *Bacteroides loeschei* adhesin that interacts with procaryotic and eucaryotic cells. *J. Bacteriol.* **172**:2527–2534.

Manch-Citron, J.N., and London, J. (1994) Expression of the *Prevotella* adhesin gene (*plaA*) is mediated by a programmed frameshifting hop. *J. Bacteriol.* **176**:1944–1948.

McCarthy, C., Snyder, M.L., and Parker, R.B. (1965) The indigenous oral biota of man. I. The newborn to the 1-year-old infant. *Arch. Oral Biol.* **10**:61–70.

McIntire, F.C., Crosby, L.K., Vatter, A.E., et al. (1988) A polysaccharide from *Streptococcus sanguis* 34 that inhibits coaggregation of *S. sanguis* 34 with *Actinomyces viscosus* T14V. *J. Bacteriol.* **170**:2229–2235.

McIntire, F.C., Vatter, A.E., Baros, J., and Arnold, J. (1978) Mechanism of coaggregation between *Actinomyces viscosus* T14V and *Streptococcus sanguis* 34. *Infect. Immun.* **21**:978–988.

Moore, W.E.C., and Moore, L.V.H. (1994) The bacteria of periodontal diseases. *Periodontology 2000* **5**:66–77.

Nikaido, H., and Saier Jr., M.H. (1992) Transport proteins in bacteria; common themes in their design. *Science* **258**:936–942.

Novak, J., Caufield, P.W., and Miller, E.J. (1994) Isolation and biochemical characterization of a novel lantibiotic mutacin from *Streptococcus mutans*. *J. Bacteriol.* **176**:4316–4320.

Nyvad, B., and Kilian, M. (1987). Microbiology of the early colonization of human enamel and root surfaces in vivo. *Scand. J. Dent. Res.* **95**:369–380.

Nyvad, B., and Kilian, M. (1990) Comparison of the initial streptococcal microflora on dental enamel in caries-active and in caries-inactive individuals. *Caries Res.* **24**:267–272.

Reinholdt, J., Friman, V., and Kilian, M. (1993) Similar proportions of immunoglobulin A1 (IgA1) protease-producing streptococci in initial dental plaque of selectively IgA-deficient and normal individuals. *Infect. Immun.* **61**:3998–4000.

Russell, R.R.B., Aduse-Opoku, J., Sutcliffe, I.C., Tao, L., and Ferretti, J.J. (1992) A binding protein-dependent transport system in *Streptococcus mutans* responsible for multiple sugar metabolism. *J. Biol. Chem.* **267**:4631–4637.

Scannapieco, F.A., Bergey, E.J., Reddy, M.S., and Levine, M.J. (1989) Characterization of salivary α-amylase binding to *Streptococcus sanguis*. *Infect. Immun.* **57**:2853–2863.

Smith, D.J., Anderson, J.M., King, W.F., van Houte, J., and Taubman, M.A. (1993) Oral streptococcal colonization of infants. *Oral Microbiol. Immunol.* **8**:1–4.

Stinson, M.W., Haraszthy, G.G., Zhang, S.L., and Levine, M.J. (1992) Inhibition of *Porphyromonas gingivalis* adhesion to *Streptococcus gordonii* by human submandibular-sublingual saliva. *Infect. Immun.* **60**:2598–2604.

Tanner, A., Maiden, M.F.J., Paster, B.J., and Dewhirst, F.E. (1994) The impact of 16S ribosomal RNA-based phylogeny on the taxonomy of oral bacteria. *Periodontology 2000* **5**:26–51.

Weiss, E.I., London, J., Kolenbrander, P.E., and Andersen, R.N. (1989) Fimbria-associated adhesin of *Bacteroides loeschei* that recognizes receptors on procaryotic and eucaryotic cells. *Infect. Immun.* **57**:2912–2913.

Wolfaardt, G.M., Lawrence, J.R., Robarts, R.D., Caldwell, S.J., and Caldwell, D.E. (1994) Multicellular organization in a degradative biofilm community. *Appl. Environ. Microbiol.* **60**:434–446.

IV.

Examining Multicellular Populations

10

Flow Cytometry

Useful Tool for Analyzing Bacterial Populations Cell by Cell

BERNHARD HAUER & HEINZ EIPEL

Flow cytometry is an extension of the methodological repertoire in the hands of microbiologists wherein structural and functional parameters can be measured on a cell-by-cell basis. Not only can bacterial populations be monitored and subpopulations identified, this instrument can also sort individual cells. The importance of flow cytometry for microbiology has been stressed by Lloyd (1993) and others (Hutter, 1978; Shapiro, 1990; Tanke and van der Keur, 1993).

In this chapter, we introduce flow cytometry and discuss its applications to the study of biological functions, heterogeneity during bacterial starvation and survival, and microbial ecology. Flow cytometry has already demonstrated that microbial populations, even as pure cultures, are more heterogeneous than one would expect. We are confident that flow cytometry can lead to more unexpected discoveries in microbiology.

Principles of Flow Cytometry

Flow cytometry should not be considered an alternative to microscopy and image processing. It is, rather, an indispensable supplementary method if subpopulations at low concentrations within populations must be identified and counted at high statistical accuracy. If flow cytometry is used in combination with cell sorting, the growth of new colonies seeded with single cells selected from subpopulations of the original population can be studied. Because flow cytometry has become a useful tool for the microbiologist, the method is briefly described. Some instrumental details are of special interest for the analysis of small objects such as bacteria.

As the name implies, flow cytometry allows measurement of physical parameters of cells carried by a flow through a sensing volume. The first flow cytometer was designed to count blood cells in a flow chamber using optical detection by darkfield illumination (Crosland-Taylor, 1953). It is remarkable that this early device made use of hydrodynamic focusing to keep the cells within a well defined observation volume and away from the optical windows.

The well known Coulter Counter was the first commercially available flow cytometer. Only one parameter (cell volume) could originally be assayed with this instrument, although sorting capabilities were added later to this basic volume-sensing machine (Fulwyler, 1965). Over the following years this early cell sorter was upgraded by introducing laser-excited fluorescence (Bonner et al., 1972).

Another instrument for ultrarapid cell analysis, again based on optical spectroscopy (Kamentsky et al., 1965), could measure light absorption at 253.5 nm and scattered light at 410 nm. A later version of this instrument used laser light to obtain useful optical signals at longer wavelengths.

An instrument to measure cell fluorescence at high precision was next designed (Dittrich and Göhde, 1969). Its prime application was to determine the DNA content of cells, aiming mainly at oncologic applications. Independently, a similar flow cytometer was developed, also based on epifluorescence microscopy (Sprenger et al., 1971).

So far, these flow cytometers had been constructed mainly for medical applications and, consequently, work with mammalian cells. For microbiologic applications the instruments had to be two to three orders of magnitude more sensitive to cope with the much smaller cell volumes. As it turned out, the sensitivity could easily be increased using the current basic instruments, but increasing the signal-to-background noise ratio called for some special modifications of the optical and hydrodynamic design. An interesting solution to these problems was a setup based on a hydrodynamically focused sample stream within a sheath current delivered by a nozzle as a tiny jet to hit a coverslip in a fluorescence microscope (Lindmo and Steen, 1979; Steen and Boye, 1980). The oil immersion objective of the microscope looks through from the opposite side of the coverslip and forms a confocal image of the sample stream on a small aperture in front of a photomultiplier. This process allows illumination of a well defined small volume and effectively reduces the pickup of background stray light. Excellent analytical work on bacteria has been done using this setup. Sorting, however, does not seem possible with this instrument.

The arrangement shown in Figure 10.1 may be used for analyzing and sorting bacterial populations. A laser beam is focused into an elliptical spot 50 μm wide and 3 μm high within a quartz cuvette enclosing a square channel 250 × 250 μm wide. The sample stream carrying the cells is hydrodynamically focused within a sheath current down to a small diameter of the same magnitude as the size of the cells in the sample. The sheath current carries the cells through the laser beam spot at a speed of 2 m/s. A point-like object travel-

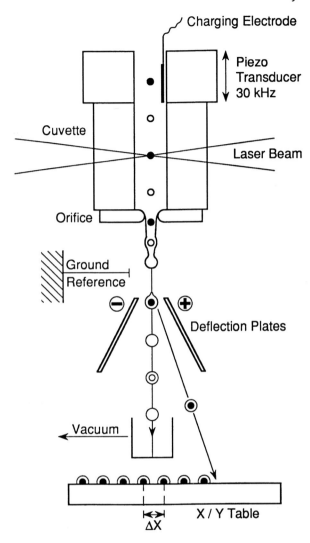

FIGURE 10.1 Cell sorter.

ing through the 3 μm laser focus thus stays for only 3/2 = 1.5 μs within the beam.

A lens behind the flow cuvette picks up the "forward light scatter" signal of the objects traversing the focus in the direction of the laser beam (Figure 10.2). A black stripe just in front of this lens prevents the direct laser beam to enter the forward light scatter path. The pickup lens is focused to produce a sharp image onto an aperture designed to observe only the tiny volume of the hydrodynamically focused sample stream. By this means, almost all of the stray light from the cuvette and lens surfaces is prevented to reach the detector.

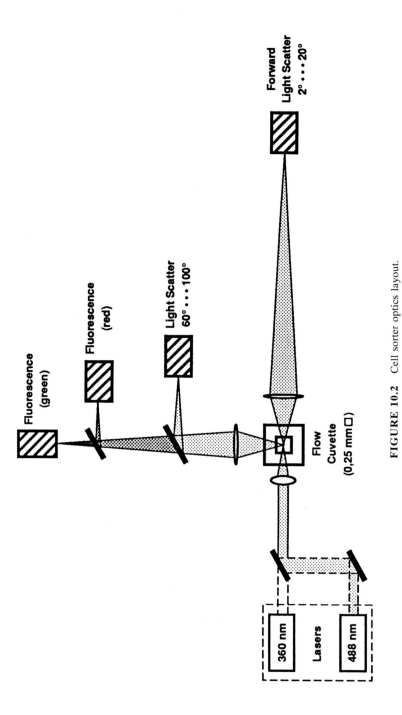

FIGURE 10.2   Cell sorter optics layout.

276

Orthogonally to the direction of the sample current and the laser beam another pickup lens is focused to the sample stream. Again, a suitably dimensioned aperture hole is used to stop light from other areas than from the illuminated cells. After passing through this aperture, the light is split up according to its wavelengths by dichroic mirrors and interference filters. The first mirror deflects the light at the laser wavelength to the "right angle light scatter" detector. Light from sample fluorescence at wavelengths longer than the laser light is passed straight on through this mirror. The next mirrors split up the fluorescent light at short and at long wavelength ("green fluorescence" and "red fluorescence," in case the most favored wavelength 488 nm of an argon ion laser is used for excitation).

The electrical signals from the four detectors ("forward" and "right angle" scatter, "green" and "red" fluorescence) are processed in parallel according to pulse height and pulse integral and are subsequently digitized for further processing and storage in a fast computer.

Any two of these eight measured parameters may be plotted against each other to obtain a "cytogram" showing one dot on an x/y plane for each measured particle. The software allows us to encircle any desired regions within these cytograms in order to analyze and sort subpopulations separately. Regions within different cytograms may be combined logically. For example, in a cytogram "forward versus right angle light scatter" a subpopulation may be selected to allow only these signals to appear in a second cytogram "red versus green fluorescence." In the latter cytogram a region may be placed to sort out cells from a population of interest.

Sorting the cells is physically accomplished by placing a small nozzle (usually of 50–100 μm diameter) at the end of the measurement cuvette to obtain a small, well defined jet of sheath current carrying the sample in its center. This jet is broken up into a pearl string of equally sized droplets by moving the cuvette with its nozzle up and down by the force of a piezoelectric transducer. The optimal droplet frequency depends mainly on the flow speed within the cuvette and the size of the nozzle, usually lying in the range of 15–50 kHz. The sample dilution and the ratio of sample to sheath flow rate allow us to set up a practical operation mode wherein, on average, 1 in 10–20 droplets contains a cell. To separate cells from wanted populations, certain single droplets containing a desired cell must be deflected out of the path of the many unwanted droplets. This task is accomplished by applying an electrical charge of approximately 100 volts to the desired droplets just at the right moment, when they begin to separate from the continuous water jet. A few centimeters farther down, the file of droplets passes through an intense static transverse electrical field in the order of 5 kV/cm. The uncharged droplets continue straight on, whereas the charged droplets are deflected to land on a Petri dish or microtiter plate, which is advanced into the next position automatically after each sort event.

For optimal results from the analysis and sorting of microbial populations, a few technical details must be given special attention: The signal-to-background noise of a flow cytometer depends on a sharp optical focus in the sample volume as well as on a sharp, stable hydrodynamic focus of the sample stream. Most commercially available cell sorters illuminate the sample in a free jet in air to avoid expensive cuvette constructions and to speed up sorting. This method works well with large cells. It also works satisfactorily to some extent when analyzing bacteria. However, when the piezoelectric transducer is energized to split up the stream into droplets for sorting, the outer shape of the water jet changes periodically at the droplet frequency, making detection and quantitation of small light scatter signals difficult, especially if the laser beam must be focused to a small area the size of the microbes. ''Forward light scatter'' is a valuable parameter in microbiological work, as a clean scatter signal allows us to exactly switch on and off the integration period for weak fluorescence signals and thus helps lower the detection limit of the fluorescence parameters. Illumination in a cuvette is desirable to obtain clean light scatter signals from particles as small as bacteria under sorting conditions.

To reduce background fluorescence noise from unbound dye within the sample stream, it is of utmost importance to optically focus the laser beam down vertically to as small a size as possible. The second and third dimension of the illuminated sample stream is determined by a sharp, stable hydrodynamic focus, which can best be accomplished within a cuvette.

It should also be kept in mind that a small optical focus allows us to sort bacteria according to their length/diameter ratio by obtaining a cytogram of the integrated pulse versus pulse height signals, as elongated objects orient their long axis in parallel to the flow. This cell length/diameter ratio has turned out to be an interesting additional parameter for studying microbial subpopulations.

Following the above-mentioned general design rules, even the smallest bacteria can be detected, and the signal-to-noise ratio obtained from them is sufficient for high yield sorting. Some typical applications are outlined here to illustrate the practical value of flow cytometry in the field of microbiology.

Analysis of Microbial Functions

Flow cytometry allows rapid characterization of a bacterial population on an individual cell basis. In microbiology most data about chemical and morphological changes represent mean values of a population and do not provide information about individual cells within the population. Characterization of individual cells in a flow cytometer has revealed that cellular heterogeneity is far greater than is normally assumed and a hitherto unrecognized aspect of the physiology of microorganisms (Kell et al., 1991). In the quantitative analysis of such heterogeneity, flow cytometry is an important technique.

Flow cytometry can be used to count bacteria; and by measuring light scatter signals at various angles their size can be estimated and some information on internal structures obtained. Even minute bacteria (0.2 μm) have been studied in marine picoplankton samples (Legendre and Yentsch, 1989). One major achievement of flow cytometry has been the discovery of the prochlorophytes (Chisholm et al., 1988). These organisms are abundant in the water column in low light environments. The cells are so sensitive to light they lyse under the continuous illumination of an epifluorescence microscope. The laser illumination in a flow cytometer is short (1.5 μs) and thus allows detection of these light-sensitive organisms.

Subpopulations can be isolated and examined further with a sorting device. They not only can be collected as a pool, but individual cells may be seeded onto an agar plate in a regular manner and scored for their colony-forming ability (Figure 10.3).

With various dyes, fluorescent substrates, or probes, DNA or specific cellular constituents can be quantified spectrophotometrically. Even simultaneous measurements of different cellular parameters are possible. Some cellular compounds emit autofluorescence signals that can be used to identify and sort out specific cells.

Flow cytometry is close to being an established method for studying the microbial cell cycle. The DNA content of bacteria can be measured with high affinity dyes for DNA, such as DAPI or propidium iodide (Hutter, 1978a,b, 1981b). By using a combination of dyes, the guanine and cytosine (G + C) content can be measured (Sanders et al., 1990). The cell cycle of a number of organisms (e.g., *Escherichia coli, Bacillus subtilis, Caulobacter*, cyanobacteria and prochlorophytes) has been studied with this elegant, rapid method. Various cell cycle parameters (e.g., the replication period, postreplication period, chromatin organization, and ploidy levels) have been studied and mathematical mod-

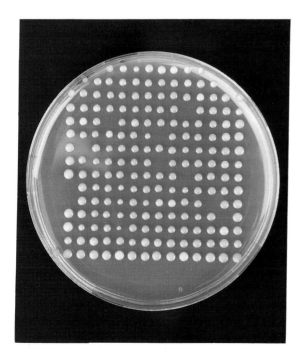

FIGURE 10.3 Cells of *Arthrobacter* sp. Lu 3184 were sorted on agar-plates in regular order, and colonies formed upon incubation.

els established (Steen and Boye, 1980; Skarstad et al., 1985, 1986; Bernander and Nordstrom, 1990; Binder and Chisholm, 1990; Allman et al., 1991; Boye and Lobner-Olesen, 1991; von Freiesleben and Rasmussen, 1992). The cell cycle has also been studied in natural populations. For example, most prochlorophytes are found in G1 in nitrogen-depleted waters and respond quickly to nitrogen pulses (Vaulot and Partensky, 1992). Propagation, effects of copy number, segregation, and instability of plasmids have also been analyzed (Seo et al., 1985; Seo and Bailey, 1987; Wittrup et al., 1990). Flow cytometry is a useful tool for analyzing the influence of certain mutations in the cell cycle and in DNA replication. The *gyrB*, *mioC*, *dnaA*, *recA*, *tus*, *dam*, *min*, and *ori* mutants have been analyzed (Kogoma et al., 1985; Skarstad et al., 1986, 1989; Skarstad and Boye, 1988; Lobner-Olesen et al., 1989; Dasgupta et al., 1991; von Freiesleben and Rasmussen, 1991; Lobner-Olesen and Boyd, 1992).

Biological activities that have been measured by flow cytometry include respiratory activities (Kaprelyants and Kell, 1993b), transport systems (Molenaar et al., 1992), enzyme activities such as that of β-galactosidase (Alvarez et al., 1993), and viability (Diaper et al., 1992; Kaprelyants, 1992b; Nebe von Caron and Badley, 1995). With the continuing development of new dyes, fluorescent substrates, and oligonucleotides, the list of possibly measurable parameters is rapidly expanding.

Flow cytometry is a promising tool for biotechnology. The rapid analysis of DNA and protein content, size distribution, and viability on a cell-by-cell basis is helpful for fermentation monitoring and control (Scheper et al., 1987; Mueller and Schmidt, 1989; Mueller et al., 1992). Flow cytometry was demonstrated to be useful in strain improvement for the production of gramicidin S (Azuma et al., 1992), astaxanthin (An et al., 1991), lipase (Betz et al., 1984), and poly-3-hydroxybutyrate (Mueller et al., 1995). It was also helpful in isolating *Escherichia coli* cells, which express the cloned surface antigen of the oil-degrading *Acinetobacter calcoaceticus* RAG 1 (Minas et al., 1988). Contaminants in food can be detected by flow cytometry (Donelly and Baigent, 1986; Patchett et al., 1991). Even mycoplasmas can be detected with a benzoxazinone derivate (Monsigny et al., 1991).

Another interesting application is its use as a tool when screening for new antimicrobial activities. It has been demonstrated that flow cytometry is sufficiently sensitive to measure changes in bacterial morphology, inner architecture, and DNA content after exposure to antibiotics (O'Gorman and Hopfer, 1991; Gant et al., 1993). Toxicological studies are also possible by flow cytometry. The effects of heavy metal intoxication (Hutter et al., 1981a), surfactants, or herbicides (Gala and Giesy, 1990) on microorganisms have been monitored.

Analysis of Microbial Survival Strategies

There is a good correlation between viable counts (colony-forming units) and direct counts (based on acridine orange-stained particles) in exponentially growing cultures. However, in the late stationary phase or under conditions of changing nutrient availability or exposure to various forms of physical stress or in natural environments, the correlation between total cell count and the number of colony-forming units differs significantly. The usual argument given is that the difference in numbers would be due to dead cells. More recently it has been suggested that cells unable to form a colony might be in a different physiological state and therefore would not be able to do so. Flow cytometry allows precise analysis of several parameters and even the sorting of individual cells from a subpopulation on agar plates. Especially cell size, morphology, DNA content, number of genomes, membrane potential, and the ability of individual cells to form a colony are studied.

Subpopulations of nonculturable but viable (NCBV) cells were detected in stationary cultures of *Aeromonas salmonicida* (Morgan et al., 1991). This gram-negative bacterium is a fish pathogen and causes furunculosis in salmonids. The behavior of the microorganism was studied in sterile lake water over 21 days. The number of colony-forming units (cfu) declined from 1×10^5 cfu/ml at the beginning to a constant value of 9×10^1 cfu/ml. There was some variability in these numbers, probably due to the nutrient level of the water. The difference in numbers could be explained in two ways: (1) The cells die off but remain intact; or (2) there is a subpopulation of NCBV cells. The cells are morphologically intact (but smaller) and contain DNA, RNA, and bound fatty acids. The reason they are termed viable is because they maintain a membrane potential. A rhodamine 123-binding subpopulation is seen in the flow cytometer. This dye is taken up and maintained by cells with a membrane potential. Incubation with gramicidin S removes this potential, and the fluorescence disappears.

So far, no conditions have been found that revive NCBV cells so they can regain their colony-forming ability. In mixed samples the viability of nonculturable cells could not be determined because it was impossible to discriminate between cells of *A. salmonicida* and other bacteria. However, with dual labeling of cells using rhodamine and a fluorescence-labeled antibody, it should be possible.

The behavior of dormancy has been described for *Micrococcus luteus* (Kaprelyants and Kell, 1993a). Dormancy, or the inability to form a colony, occurs during an extended stationary phase. Such a phase following batch culturing has been shown to result in a decrease of colony-forming units. After 60–80 days the percent of colony-forming units reached an apparently steady value of 0.05%. At the same time the number of total counts dropped to about 60–70% of the level seen at the beginning of the stationary phase. Flow cytometry showed that, in contrast to the stable DNA content, the protein content of the cells decreased significantly. The scatter signal indicated that such cells are more homogeneous and smaller than growing cells. In contrast to *A. salmonicida*, a membrane potential with rhodamine 123 could be shown. The surprising observation was that upon resuscitation in fresh medium the status of dormancy was reversible. There was no increase in total counts, but the small dormant cells were first converted to large, active cells that were now fluorescent with rhodamine 123. Only later were they able to develop colony-forming units again.

The extent to which individual cells accumulate rhodamine 123 can be rapidly assessed by flow cytometry and reflects the three distinguishable physiological states exhibited by the culture: nonviable, viable, and nonviable but resuscitable. After 2.5 months in the stationary phase 50% of the dormant cells are still able to revert. After 7 months 25% can be resuscitated to the active state, but these cells do not form colonies on agar. In chemostat cultures at low dilutions, up to 40% of these cells persist as small, dormant forms.

Another fish pathogenic bacterium whose behavior upon starvation was studied with flow cytometry is *Yersinia ruckeri* (Thorsen et al., 1992). The strain is the etiological agent of enteric red mouth disease of salmonids in fresh water. The strain was isolated for the first time in 1985 during an outbreak in sea water.

In unsupplemented water and up to 20% salinity no detectable changes in colony-forming units were found over at least 4 months. At 35% salinity the survival potential was markedly reduced. The number of colony-forming units dropped below the detection limit of 3 cfu/ml after 32 days. Cell size and DNA content were studied with flow cytometry. Survival was examined employing the direct viable counts (DVC) method. Starved cells were inoculated in tryptone soya broth containing nalidixic acid 1 µg/ml. This concentration was inhibiting cell division and thereby producing elongated cells. After preservation in 70% ethanol and staining with ethidium bromide, cells were analyzed by flow cytometry. The viable but nonculturable state was not found for *Y. ruckeri*. The DVC counts were in good agreement with the colony-forming units, but morphological changes took place during starvation. Cells tended to become more coccoid, which resulted in a reduction of the light scatter signal.

The DNA measurements indicated that after starvation the ongoing round of replication was completed, and after 24 hours of starvation discrete DNA peaks appeared in the flow cytometer. Varying fractions of cells have three or more genomes (up to 20%). Whether these multigenomic cells are a result of starvation is an open question. After continued starvation, DNA peaks tended to become broader, probably the result of DNA degradation.

These results suggest a high survival capacity. *Y. ruckeri* may survive for long periods in fresh water and brackish environments after an outbreak of enteric red mouth disease. There might, however, be differentiation among cells in a population undergoing starvation in order to adjust to environmental changes.

The result of carbon starvation in *Pseudomonas putida* is an active non-dormant but stress resistant state. The cell size, shape, and DNA content were analyzed with flow cytometry (Givskov et al., 1994). Cells became smaller, and their shape changed to round or coccoid forms; they eventually became exceedingly small. The DNA content was also diminished upon starvation. After 4 hours of starvation, 90% of the cells contained only one chromosome. After 24 hours the distribution of cells carrying one and two chromosomes stabilized at 95% and 5%, respectively. Upon starvation, the cells acquired a pattern of cross protection and were able to survive more extreme conditions, such as high or low temperature, elevated osmolarity, and the presence of solvents or oxidative agents. The cells did not survive well under conditions of phosphate or nitrogen starvation. Upon proper conditions of starvation, *Pseudomonas putida* develops into forms that exhibit a higher resistance than their growing counterparts, and growth is reinitiated as soon as the necessary nutrients are supplied.

The Tn5 bleomycin resistance gene confers improved survival and growth advantage on *E. coli* (Blot et al., 1994). The *ble* gene is known to decrease the death rate of *E. coli* during the stationary phase. Bleomycin resistance is produced by improved DNA repair, which requires coding of the host genes *aidC* and *polA* for an alkylation-inducible protein and DNA polymerase I, respectively. The fraction of viable *E. coli* cells was estimated by the ability of a single cell to form a colony. Using a flow cytometer cells were sorted onto plates in regular order, and colony-forming units were scored after overnight incubation. Strains carrying both the *aidC* and *ble* genes showed increased survival relative to all other strains, but the $aidC^+ ble^+$ strain was also capable of improved survival during the stationary phase. If the viability measurements are presumed to indicate that the bacteria will contribute to the next generation, the $aidC^+ ble^+$ strain is likely to display a better growth rate during continuous cultures.

Arthrobacter sp. Lu 3184 is a bacterium used in our laboratory for specific hydroxylation of aromatic compounds. With flow cytometry we analyzed how many cells in a resuspended colony are able to form a colony. We found that the ability of the cells to form a colony depends on the history (age, growth medium) and the medium on which the cells are supposed to form a colony (Figure 10.4).

In aged cultures we could identify three cell types.

Type I: cells that form no colony at all
Type II: cells that form a colony on minimal medium and on nutrient broth
Type III: cells that form a colony on minimal medium but not on nutrient broth

Flow cytometric and microscopic analysis of the bacteria gives no indication that there are morphological changes over time. When colonies grow older the number of type I cells increases and the number of type II cells decreases. Type III is an interesting population. A spontaneous ribose-utilizing mutant of *Arthobacter* sp. Lu 3184 growing in nutrient broth consists almost exclusively of type III cells. Overnight cultures in nutrient broth gave no colony-forming units on nutrient broth agar, although 90% of the cells placed on minimal medium agar did form colonies. The cells are not dead or killed immediately on nutrient broth-agar, as demonstrated by placing cells of *Arthrobacter* sp. Lu 3184 Rib on isopore membranes. These filters were laid on nutrient broth or minimal medium and incubated. After a certain time had passed the filters were transferred to the opposite medium (Figure 10.5). These transfer experiments

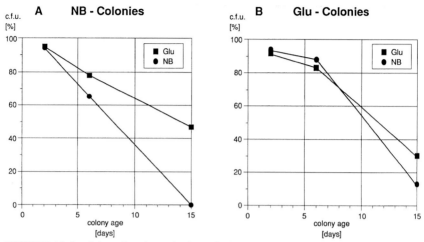

FIGURE 10.4 Colony-forming units in *Arthrobacter* sp. Lu 3184 colonies. At a given time colonies from nutrient broth (**A**) and glucose minimal medium (**B**) were resuspended. With the cell sorter, 100 single cells were placed on nutrient broth agar and glucose minimal agar. After incubation the colonies were counted.

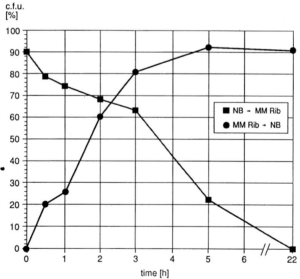

FIGURE 10.5 Transfer experiment with *Arthrobacter* sp. Lu 3184 Rib. With the cell sorter, cells from an overnight nutrient broth culture were put on filter membranes. After a certain time passed, the filters were transferred to another plate. Colony-forming units were scored after a 24-hour incubation.

showed that if the first cell division occurs on minimal medium, colonies can again be formed on nutrient broth. After one cell division on nutrient broth, however, the ability of the cells to form a colony on minimal medium drops. These results indicate that bacteria of a similar morphology might be different with regard to their response in forming a colony.

Microbial Ecology Studies

When studying mixed microbial populations one must identify, monitor, and isolate individual bacteria. Rapid characterization of bacterial populations can be accomplished by flow cytometry. Light scatter and DNA measurements, immuno- and oligonucleotide-based fluorescent probes or other functional probes are the basis of microbial ecology studies. Therefore the apparent intentions are to count the number of bacterial cells and to distinguish organisms by measuring morphological and physiological characteristics. A cytometer with a sorting device is helpful in this regard because it allows us to sort and collect individual cells from a subpopulation.

Clinically relevant microorganisms have been discriminated on the basis of forward and wide-angle light scatter signals (Boye et al., 1983, Allman et al., 1992). With DNA content as an additional parameter, different species can be discriminated even better.

The variability of scatter signals depends on growth conditions, as seen in studies with soil bacteria (Christensen et al., 1993). On the basis of DNA and its G + C content, these investigators differentially quantified mixtures of bacteria. Dual-laser flow cytometry was used by Sanders et al. (1990) to analyze different species of bacteria according to the molar percentage of G + C without the need for DNA extraction. Two cell populations were separated in mixtures containing two species that differed in base composition by as little as 4% G + C. Bacteriophage T4 was detected and analyzed inside *E. coli* using the same technique (Sanders et al., 1991). At 35 minutes after T4 infection, T4 DNA-containing cells were distinguishable from *E. coli* cells, which contain little or no T4 DNA.

Using species-specific fluorescent probes it is possible to quantify species in defined mixtures of bacteria. For example, combinations of 16 S rRNA targeted oligonucleotid probes were used to identify *Desulfovibrio gigas*, *Desulfobacter hydroglenophilus*, and *E. coli*. As few as 3% of the total were clearly visible as a well separated population (Amann et al., 1990; Wallner et al., 1993).

Labeling bacteria with specific antibodies is a convenient method for distinguishing species. *Staphylococcus aureus* was labeled with fluorescein isothiocyanote (FITC)-IgG. Protein A of *S. aureus* binds to immunoglobulin-binding protein. A high degree of specificity was shown in eutrophic lake water containing 10^6-10^7 bacteria per milliliter. The lower limit of detection in lake water was around 10^3 *S. aureus* cells per milliliter. In pure cultures the limit was 10^1-10^2 cells/ml (Edwards et al., 1993). Flow cytometry has significantly improved our understanding of aquatic bacteria (Legendre, 1989; Frankel et al., 1990; Andreoli et al., 1993), and procedures for direct analysis of water samples have been devised (Robertson and Button, 1989). Flow cytometry resolves most or all of the biomass and probes populations on a cell-by-cell basis. Significant differences in cell characteristics during growth, observed by flow cytometry, suggested substantial population diversity. Subtle environmental effects can be quantified (Button et al., 1993).

Aquatic bacteria are ubiquitous and comprise a major component of the aquatic biomass. Most are small and resist growth to high densities in the laboratory. They are present at about 10^6 cells/ml, which is ideal for flow cytometry.

Aquatic bacteria 0.05 μm^3 in volume were clearly resolved according to DNA content by staining with DAPI. Direct analysis of the bacteria in marine and freshwater samples without interference from algae or sediments is possible. Cytograms indicated one or more clearly resolved subpopulations of bacteria that were substantially smaller in size and DNA content than the typically classified laboratory organisms.

The interactions between marine bacteria and dissolved-phase and beached hydrocarbons were studied after the Exxon Valdez oil spill (Button et al., 1992). Changes in the characteristics of the indigenous bacteria and associated rates of hydrocarbon oxidation were found in response to the oil spill. Samples were analyzed by flow cytometry. Cell size was determined by forward light scatter, and DNA content was measured after staining with DAPI. Biomass was computed from cell counts as well as the mean cell volume, assuming a cell density of 1.04. A bacterial biomass of 2–14 mg/kg appeared in apparent response to the new carbon as the energy source. These sediment organisms were small at $\approx$ 0.06 μm^3 in volume, low in DNA at 4.1 fg/cell, and unlike the aquatic bacteria obtained by enrichment culture but similar to the oligobacteria in the water column. This biomass is thought to be composed mostly of hydrocarbon-oxidizing bacteria. Biomass washed from oiled gravel and measured directly by flow cytometry was nearly 10 times that washed from the control. The naturally developed population should consume at a specific growth rate of 0.2 h^{-1} and a constant population all the oil on the gravel within 5 days; however, substantial quantities remained after 2 years. It has been suggested that a large population consists of carbon- and energy-starved induced hydrocarbon oxidizers with metabolisms limited by the physical and molecular recalcitrance of the heavier components. A small rod-shaped bacterium (0.05–0.06 μm^3, 1.0–1.5 fg DNA/cell) was isolated from marine environments, and flow cytometry was used to characterize the isolates (Shut et al., 1993). Isolates obtained on relatively high-nutrient agar plates with respect to cell volume and apparent DNA content are identical to the cells initially obtained in the obligately oligotrophic bacterial dilution culture.

Flow cytometry is also used as a rapid, reliable method to test the efficiency of strains producing antimicrobial products. The interaction between co-cultured *Listeria monocytogenes* and an antagonistic *Leuconostoc* strain producing an anti-*Listeria* bacteriocin was studied (Hechard et al., 1992). Light scattering

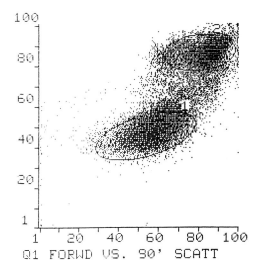

100

80

60

40

20

1

1   20   40   60   80   100

Q1   FORWD VS. 90' SCATT

FIGURE 10.6   Forward versus 90-degree light scatter signal cytogram was used to identify two strains (1:2nII2, 2: 2nIII) in a mixed population.

at different angles allowed visualization of individual populations of bacteria within a mixed culture in real time. It was concluded that *L. mesenteroides* exerted a bactericidal effect against *L. monocytogenes* that led to cell lysis.

Flow cytometric separation was used to study the cyanobacterial symbiont biosynthesis of chlorinated metabolites by the tropical marine sponge *Dysidea herbacea* (Unson and Falkner, 1993). Phycoerythrin fluorescence was chosen as the sorting parameter for separation of the cyanobacterium *Oscillatoria spongeliae* from sponge cells. Upon chemical analysis it was demonstrated that the group of polychlorinated compounds is limited to the cyanobacterium, whereas sesquiterpenoids are found only in sponge cells.

We used flow cytometry and cell sorting to characterize a mixed population growing on a column for the denitrification of drinking water (Shloe et al., 1994). A single strain was isolated by classical methods that could denitrify the water and metabolize a copolymer of polyhydroxybutyrate and hydroxyvalerate. The strain could be identified as a subpopulation on the basis of forward and 90-degree scatter signals (Figure 10.6). The purity of the strain was verified by sorting on agar plates and scoring colonies that grow from single cells.

References

Allmann, R., Schjerren, T., and Boye, E. (1991) Cell cycle parameters of *Escherichia coli* K-12. *J. Bacteriol.* **173**:7970–7974.

Allmann, R., Hann, A.C., Manchee, R., and Lloyd, D. (1992) Characterization of bacteria by multiparameter flow cytometry. *J. Appl. Bacteriol.* **73**:438–444.

Alvarez, A.M., Ibáñez, M., and Rotger, R. (1993) β-Galactosidase activity in bacteria measured by flow cytometry. *Biotechniques* **15**:974–976.

Amann, R.I., Binder, B.J., Olson, R.J., Chisholm, S.W., Devereux, R., and Stahl, D.A. (1990) Combination of 16 S ribosomal RNA-targeted oligonucleotide probes with flow cytometry for analyzing mixed microbial populations. *Appl. Environ. Microbiol.* **56**:1919–1925.

An, G.H., Bielich J., Auerbach, R., and Johnson, E.A. (1991) Isolation and characterization of carotenoid hyperproducing mutants of yeast by flow cytometry and cell sorting. *Biotechnology* **9**:70–73.

Andreoli, C., Scarabel, L.R., and Tolomio, C. (1993) The distribution of photoantotrophic picoplankton in the Terra Nova Ross Sea Antarctica during astral summer 1989–1990. *Arch. Hydrobiol. Suppl.* **96**:123–132.

Azuma, T., Harrison, G.I., and Demain, A.L. (1992) Isolation of a gramicidin S hyperproducing strain of *Bacillus brevis* by use of a fluorescence activated cell sorting system. *Appl. Microbiol. Biotechnol.* **38**:173–178.

Bernander, R., and Nordstrom, K. (1990) Chromosome replication does not trigger cell division in *Escherichia coli*. *Cell* **60**:365–374.

Betz, J.W., Aretz, W., and Härtel, W. (1984) Use of flow cytometry in industrial microbiology for strain improvement programs. *Cytometry* **5**:145–150.

Binder, B., and Chisholm, S.W. (1990) Relationship between DNA cycle and growth rate in *Synechococcus* sp. strain PCC 6301. *J. Bacteriol.* **172**:2313–2319.

Blot, M., Hauer, B., and Monnet, G. (1994) The Tn5 bleomycin resistance gene confers improved survival and growth advantage on *Escherichia coli*. *Mol. Gen. Genet.* **242**:595–601.

Bonner, W.A., Hulett, H.R., Sweet, R.G., and Herzenberg, L.A. (1972) Fluorescence activated cell sorting. *Rev. Sci. Instrum.* **43**:404–409.

Boye, E., and Lobner-Olesen, A. (1991) Bacterial growth control studied by flow cytometry. *Res. Microbiol.* **142**:131–135.

Boye, E., Steen, H.B., and Skarstad, K. (1983) Flow cytometry of bacteria: a promising tool in experimental and clinical microbiology. *J. Gen. Microbiol.* **129**:973–980.

Button, D.K., Robertson, B.R., McIntosh, D., and Juttner, F. (1992) Interactions between marine bacteria and dissolved-phase and beached hydrocarbons after the Exxon Valdez oil spill. *Appl. Environ. Microbiol.* **58**:243–251.

Button, D.K., Shut, F., Quang, P., Martin, R., and Robertson, B.R. (1993) Viability and isolation of marine bacteria by dilution culture theory procedures and initial results. *Appl. Envirion. Microbiol.* **59**:881–891.

Chisholm, S.W., Olsen, R.J., Zettler, E.R., Waterbury J., Goericke, R., and Welschmeyer N. (1988) A novel free-living prochlorophyte occurs at high cell concentrations in the oceanic euphotic zone. *Nature* **334**:340–343.

Christensen, H., Bakken, L.R., and Olsen, R.A. (1993) Soil bacterial DNA and biovolume profiles measures by flow cytometry. *FEMS Microbiol. Ecol.* **102**:129–140.

Crosland-Taylor, P.J. (1953) A device for counting small particles suspended in a fluid through a tube. *Nature* **171**:37–38.

Dasgupta, S., Bernander, R., and Nordstrom, K. (1991) In-vivo effect of the *tus* mutation on cell division in an *Escherichia coli* strain where chromosome replication is under the control of plasmid R1. *Res. Microbiol.* **142**:177–180.

Diaper, J.P., Tither, K., and Edwards, C. (1992) Rapid assessment of bacterial viability by flow cytometry. *Appl. Microbiol. Biotechnol.* **38**:268–272.

Dittrich, W., and Göhde, W. (1969) Impulsfluorometrie bei Einzelzellen in Suspensionen. *Z. Naturforsch.* **24b**:360–361.

Donelly, C.W., and Baigent, G.J. (1986) Method for flow cytometric detection of *Listeria monocytogenes* in milk. *Appl. Environ. Microbiol.* **52**:689–695.

Edwards, C., Diaper, J.P., Porter, J., and Pickup, R. (1993) Applications of flow cytometry in bacterial ecology. In *Flow Cytometry in Microbiology*, D. Lloyd (ed.), pp. 121–129, Springer Verlag, Berlin.

Frankel, S.L., Binder, B.J., Chisholm, S.W., and Shapiro, H.M. (1990), A high-sensitivity flow cytometer for studying picoplankton. *Limnol. Oceanogr.* **35**:1164–1169.

Fulwyer, M.J. (1965) An electronic particle separator with potential biological application. *Science* **150**:371–372.

Gala, W.R., and Giesy, J.P. (1990) Flow cytometric techniques to assess toxicity to algae. *ASTM Spec. Tech. Publ.* **1096**:237–246.

Gant, V.A., Warnes, G., Philipps, I., and Savidge, G.F. (1993) The application of flow cytometry to the study of bacterial responses to antibiotics. *J. Med. Microbiol.* **39**: 147–154.

Givskov, M., Eberl, L., Moller, S., Poulsen, L.K., and Molin, S. (1994) Responses to nutrient starvation in *Pseudomonas putida* KT2442: analysis of general cross-protection, cell shape and macromolecular content. *J. Bacteriol.* **176**:7–14.

Hechard, Y., Jayat, C., Lettellier, F., Julien, R., Cenatiempo, Y., and Ratinaud, M.H. (1992) On-line visualization of the competitive behavior of antagonistic bacteria. *Appl. Environ. Microbiol.* **58**:3784–3786.

Hutter, K.J., and Eipel, H. (1978a) DNA determination of yeast by flow cytometry. *FEMS Microbiol. Lett.* **3**:35–38.

Hutter, K.J., and Eipel, H. (1978b) Flow cytometric determinations of cellular substances in algae, bacteria, molds and yeasts. *Antonie van Leeuwenhock* **44**:269–282.

Hutter, K.J., Eipel, H.E., and Stoehr, M. (1981a) Flow cytometric analysis of microbial cell constituents after heavy metal intoxication. *Acta Pathol. Microbiol. Scand. Suppl.* **274**:317–322.

Hutter, K.J., Stoehr, M., and Eipel, H. (1981b) Simultaneous DNA and protein measurements of microorganisms. *Acta Pathol. Microbiol. Scand. Suppl.* **274**:100–102.

Kamentsky, L.A., Melamed, M.R., and Derman, H. (1965) Spectrophotometer: new instrument for ultrarapid cell analysis. *Science* **150**:630–631.

Kaprelyants, A.S., and Kell, D.B. (1992) Rapid assessment of bacterial viability and vitality by rhodamine 123 and flow cytometry. *J. Appl. Bacteriol.* **72**:410–422.

Kaprelyants, A.S., and Kell, D.B. (1993a) Dormancy in stationary-phase cultures of *Micrococcus luteus*. *Appl. Environ. Microbiol.* **59**:3187–3196.

Kaprelyants, A.S., and Kell, D.B. (1993b) The use of 5-cyano-2,3-ditolyltetrazolium chloride and flow cytometry for the visualization of respiratory activity in individual cells of *Micrococcus luteus*. *J. Microbiol. Methods* **17**:115–122.

Kell, D.B., Ryder, H.M., Kaprelyants, A.S., and Westerhoff, H.V. (1991) Quantifying heterogeneity: flow cytometry of bacterial cultures. *Antonie von Leeuwenhoek* **60**: 145–158.

Kogoma, T., Skarstad, K., Boye, E., von Meyenburg, K., and Steen, H.B. (1985) Rec A protein acts at the initiation of stable DNA replication in *rnh* mutants of *Escherichia coli* K-12. *J. Bacteriol.* **163**:439–444.

Legendre, L., and Yentsch, C.M. (1989) Overview of flow cytometry and image analysis in biological oceanography and limnology. *Cytometry* **10**:501–510.

Lindmo, T., and Steen, H.B. (1979) Characteristics of a simple, high-resolution flow cytometer based on a new flow configuration. *Biophys. J.* **28**:33–44.

Lloyd, D. (ed.) (1993) *Flow Cytometry in Microbiology*. Springer-Verlag, Berlin.

Lobner-Olesen, A., and Boye, E. (1992) Different effects of mioC transcription on initiation of chromosomal and minichromosomal replication in *Escherichia coli*. *Nucleic Acids Res.* **20**:3029–3036.

Lobner-Olesen, A., Skarstad, K., Hansen, F.G., von Meyenburg, K., and Boye E. (1989) The DNA-A protein determines the initiation mass of *Escherichia coli* K-12. *Cell* **57**:881–889.

Minas, W., Sahar, E., and Gutnick, D. (1988) Flow cytometric screening and isolation of *Escherichia coli* clones with express surface antigens of the oil degrading microorganism *Acinetobacter calcoaceticus* RAG-1. *Arch. Microbiol.* **105**:432–437.

Molenaar, D., Bolhuis, H., Abee, T., Poolmann, B., and Konings, W.N. (1992) The efflux of a fluorescent probe is catalyzed by an ATP-driven extrusion system in *Lactococcus lactis*. *J. Bacteriol.* **174**:3118–3124.

Monsigny, M., Midoux, P., Depierreux, C., Bebear, C., LeBris, M.-T., and Valeur, B. (1991) Benzoxazinone kanamycin A conjugate: a new fluorescent probe suitable to detect mycoplasmas in cell cultures. *Biol. Cell* **70**:101–106.

Morgan, J.A.W., Cranwell, P.A., and Pickup, R.W. (1991) Survival of *Aeromonas salmonicida* in lake water. *Appl. Environ. Microbiol.* **57**:1777–1782.

Mueller, S., and Schmidt, A. (1989) Flow cytometric determination of yeast sterol content. *Acta Biotechnol.* **9**:89–93.

Mueller, S., Loesche, A., and Bley, T. (1992) Flow cytometric investigation of sterol content and proliferation activity of yeast. *Acta Biotechnol.* **12**:365–375.

Mueller, S., Loesche, A., Bley, T., and Scheper, T. (1995) A flow cytometric approach for characterization and differentiation of bacterial during microbial processes. *Appl. Microbiol. Biotechnol.* **43**:93–101.

Nebe-von Caron, G., and Badley, R.A. (1995) Viability assessment of bacteria in mixed populations using flow cytometry. *J. Microscopy* **179**:55–66.

O'Gorman, M.R.G., and Hopfer, R.L. (1991) Amphotericin B susceptibility testing of *Candida* species by flow cytometry. *Cytometry* **12**:743–747.

Patchett, R.A., Back, J.P., Pinder, A.C., and Kroll, R.G. (1991) Enumeration of bacteria in pure cultures and in foods using a commercial flow cytometer. *Food Microbiol.* **8**:119–126.

Robertson, B.R., and Button, D.K. (1989) Characterizing aquatic bacteria according to population cell size and apparent DNA content by flow cytometry. *Cytometry* **10**: 70–76.

Sanders, C.A., Yajko, D.M., Hyun, W., et al. (1990) Determination of guanine-plus-cytosine content of bacterial DNA by dual-laser flow cytometry. *J. Gen. Microbiol.* **136**:359–365.

Sanders, C.A., Yajko, D.M., Nassos, P.S., Hyun, W.C., Fulwyer, M.J., and Hadley, W. (1991) Detection and analysis by dual-laser flow cytometry of bacteriophage T4 DNA inside *Escherichia coli*. *Cytometry* **12**:167–171.

Scheper, T., Hitzmann, B., Rinas, U., and Schügerl, K. (1987) Flow cytometry of *Escherichia coli* for process monitoring. *J. Biotechnol.* **5**:139–148.

Schloe, K., Hauer, B., Biedermann, J., Wais, S., Staniszewski, M., and Süßmuth, R. (1995) A pure strain or a mixed population as a starter culture for the denitrification of drinking water? (Submitted)

Seo, J.-H., and Bailey, J.E. (1987) Cell cycle analysis of plasmid-containing *Escherichia coli* HB101 populations with flow cytometry. *Biotechnol. Bioeng.* **30**:297–305.

Seo, J.-H., Srienc, F., and Bailey, J.E. (1985) Flow cytometry analysis of plasmid amplification in *Escherichia coli*. *Biotechnol. Prog.* **4**:181–188.

Shapiro, H.M. (1990) Flow cytometry in laboratory microbiology: new directions. *ASM News* **11**:584–588.

Shut, F., de Vries, E.J., Gottschal, J.C., et al. (1993) Isolation of typical marine bacteria by dilution culture growth maintenance and characteristics of isolates under laboratory conditions. *Appl. Environ. Microbiol.* **59**:2150–2160.

Skarstad, K., and Boye, E. (1988) Perturbed chromosomal replication in *rec A* mutants of *Escherichia coli. J. Bacteriol.* **170**:2549–2554.

Skarstad, K., Boye, E., and Steen, H.B. (1986) Timing of initiation of chromosome replication in individual *Escherichia coli* cells. *EMBO J.* **5**:1711–1718.

Skarstad, K., Lobner-Olesen, A., Atlung, T., von Meyenburg, K., and Boye, E. (1989) Initiation of DNA replication in *Escherichia coli* after overproduction of the DNA A protein. *Mol. Gen. Genet.* **218**:50–56.

Skarstad, K., Steen, H.B., and Boye, E. (1985) *Escherichia coli* DNA distributions measured by flow cytometry and compared with theoretical computer simulations. *J. Bacteriol.* **163**:661–668.

Sprenger, E., Böhm, N., and Sandritter, W. (1971) Flow-through fluorescence cytophotometry for ultra-rapid DNA measurement in large cell populations. *Histochemie* **26**:238–257.

Steen, H.B., and Boye, E. (1980) Bacterial growth studied by flow cytometry. *Cytometry* **1**:32–36.

Tanke, H.J., and van der Keur, M. (1993) Selection of defined cell types by flow-cytometric cell sorting. *TIBTECH* **11**:55–62.

Thorsen, B.K., Enger, O., Norland, S., and Hoff, K.A. (1992) Long-term starvation survival of *Yersinia ruckeri* at different salinities studied by microscopical and flow cytometric methods. *Appl. Environ. Microbiol.* **58**:1624–1628.

Unson, M.D., and Falkner, D.J. (1993) Cyanobacterial symbiont biosynthesis of chlorinated metabolites from *Dysidea herbacea* porifera. *Experientia* **49**:349–353.

Vaulot, D., and Partensky, F. (1992) Cell cycle distributions of prochlorophytes in the north western Mediterranean Sea. *Deep Sea Res. [A] Oceanogr. Res. Pap.* **39**: 727–742.

Von Freiesleben, U., and Rasmussen, K.V. (1992) The level of supercoiling affects the regulation of DNA replication in *Escherichia coli. Res. Microbiol.* **143**:655–663.

Von Freiesleben, U., and Rasmussen, K.V. (1991) DNA-replication in *Escherichia coli gyrB ts* mutants analyzed by flow cytometry. *Res. Microbiol.* **142**:223–228.

Wallner, G., Amann, R., and Beisker, W. (1993) Optimizing fluorescent in situ hybridization with rRNA-targeted oligonucleotide probes for flow cytometric identification of microorganisms. *Cytometry* **14**:136–143.

Wittrup, K.D., Bailey, J.E., Ratzkin, B., and Patel, A. (1990) Propagation of an amplifiable recombinant plasmid in *Saccharomyces cerevisiae*: flow cytometry studies and segregated modeling. *Biotechnol. Bioeng.* **35**:565–577.

11

In Situ Analyses of Microbial Populations with Molecular Probes

Phylogenetic Dimension

NORMAN K. FRY, LUTGARDE RASKIN,

RICHARD SHARP, ELIZABETH W. ALM,

BRUCE K. MOBARRY, & DAVID A. STAHL

It is within the context of the early diversification of microbial life on earth that the first examples of apparent multicellularity are found, most conspicuously in the fossil remains of microbial mat communities (Cohen, 1989). The study of the interdependence of microbial populations, which may display aspects of multicellularity, is contained within the field of microbial ecology. It is from this vantage that we approach multicellular behavior among microorganisms. For the purposes of this chapter, we consider multicellularity to be manifested by specific mechanisms of inter- and intrapopulation communication. Such communication may be mediated by products of catabolism, as well as by more specific biochemical mediators. Although "metabolic signals" seem to lack specificity, we suggest that the specificity of any mode of communication can be fully evaluated only within the context of community. Mechanisms of communication evolved within communities and are unlikely to be interpretable outside that context. We consider microbial ecology to be inseparable from microbial evolution. This point is perhaps self-evident; if interpopulation communication defines the structure of contemporary microbial communities and microorganisms evolved as parts of communities, full appreciation of microbial multicellularity must consider possible co-evolution of interacting populations.

Implicit in the evolution of communities is the selective advantage conferred on populations that function as communities. Microbiologists have frequently asked why there are no "supermicrobes," organisms of extraordinary catabolic range. The absence of such organisms is most evident in anaerobic communities, where populations of relatively restricted metabolic capabilities cooperate in the complete mineralization of organic material. For example, methanogens and most sulfate-reducing bacteria are incapable of fermenting carbohydrates. Answers to why this is so will likely come only from complete knowledge of community structure, resource allocation, and the capacity of communities to respond to changing environmental conditions as intact communities. The latter includes consideration of homeostatic behavior conferred by multicellularity. To move beyond speculation, the essential first step is defining the individual populations that make up a community. It is an almost forgotten embarrassment that to date no microbial community has been described at the level of all contributing populations.

Ecologists have recorded the history of plants and animals for centuries, and it has served as an essential foundation for the incorporation of new observations and hypotheses in macroecology. In contrast, there is virtually no description of the natural history of microorganisms available to microbial ecologists (Stahl, 1993). Estimates of the fraction of microorganisms in the biosphere that awaits isolation vary from 80%–95% (Wayne et al., 1987; Bull et al., 1992). Studies suggest that even these figures are underestimates, and it is increasingly evident that traditional methods of isolation and classification are insufficient to define community structure. New methods are now available. In particular, developments in molecular biology have provided the tools for direct studies of population structure. Initial molecular surveys have revealed a startling discrepancy between organisms isolated from a natural habitat (and thought to be representative of the habitat) and the molecular determination of their abundance. Most ``molecular isolates''—microorganisms discovered and identified by direct analysis of environmental nucleic acids [e.g., ribosomal RNA (rRNA) or ribosomal DNA (rDNA) sequence analysis]—are not represented in pure culture collections. Nearly complete reliance on pure culture phenotype has also placed microbiologists in the awkward position of having to infer the community behavior of an organism from observations of its behavior in isolation. The connection between the pure culture phenotype and environmental activities must be treated as tenuous; and although the benefits of the pure culture description have been profound, microbiology has only a superficial understanding of diversity and community.

Foundations of Molecular Ecology: Path to a Comprehensive Description of Communities

The ease by which nucleic acid sequence can now be determined has resulted in the routine use of molecular data in phylogenetic studies (Zuckerkandl and Pauling, 1965; Woese, 1987). Protein gene sequences employed in phylogenetic studies to date include ATPases, protein elongation factors, and cytochromes. But, the comparative sequencing of the rRNAs has had the most far-reaching influence on microbiology. The establishment of a robust phylogeny based on comparative sequence analyses of the rRNAs provided the first consistent taxo-

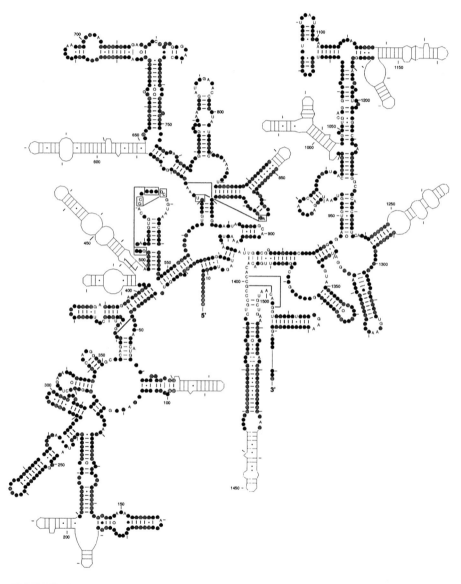

FIGURE 11.1 Phylogenetic walk, three domains, 51 representative sequences from (eu)Bacteria, Archaea, and Eukarya. The consensus of the defined alignment is superimposed over the *Escherichia coli* 16S rRNA secondary structure. Lines designate positions with size variation (or complete deletion) within that data set. Positions present in all sequences in the data set are denoted with its nucleotide if it is invariant or a bullet if it is variable in composition. The density of the bullet is proportional to its conservation. (Reprinted from Gutell, 1992, with permission).

294

nomic description of microorganisms. This event was an essential prelude to the unambiguous description of their communities.

Ribosomal RNAs: The Ultimate Molecular Chronometers?

Although all of the molecules listed above generally fulfill the criteria for phylogenetic inference, the rRNAs are the most widely used (Woese, 1987). Certain attributes of the rRNAs favor their use as molecular chronometers. The genes that code for rRNAs (5S-, 16S-, and 23S-like) are among the most highly conserved known (Brenner, 1986; Woese, 1987). The 16S rRNA molecule is a linear mosaic of differing sequence conservation (Figure 11.1). Regions that vary sufficiently slowly allow inference of relationships between members of the three domains (Bacteria, Eukarya, Archaea) (Woese et al., 1990) (Figure 11.2), whereas the more variable regions provide for discrimination among organisms of approximate genus and species rank differences. Regional differences in sequence conservation also provide the basis for designing nucleic acid probes of varying specificity (see below). The 5S rRNA molecule provides relatively little sequence (approximately 120 nucleotides) and therefore has been most useful for studying distant phylogenetic relationships. The large 16S molecule (approximately 1500 nucleotides) provides much more information and is the most widely used for phylogenetic inference. The 23S rRNA mol-

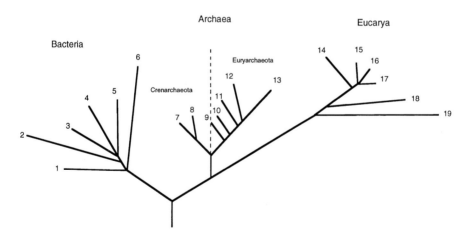

FIGURE 11.2   Universal phylogenetic tree in rooted form, showing the three domains Bacteria, Archaea, and Eukarya. Branching order and branch lengths are based on rRNA sequence comparisons. The position of the root was determined by comparing (the few known) sequences of pairs of paralogous genes that diverged from each other before the three primary lineages emerged from their common ancestral condition. The numbers on the branch tips correspond to the following groups of organisms. **Bacteria**: 1, Thermotogales; 2, flavobacteria and relatives; 3, cyanobacteria; 4, purple bacteria; 5, gram-positive bacteria; 6, green nonsulfur bacteria. **Archaea**: kingdom Crenarchaeota: 7, genus *Pyrodictium*; 8, genus *Thermoproteus*; the kingdom Euryarchaeota: 9, Thermococcales; 10, Methanococcales; 11, Methanobacteriales; 12, Methanomicrobiales; 13, extreme halophiles. **Eukarya**: 14, animals; 15, ciliates; 16, green plants; 17, fungi; 18, flagellates; 19, microsporidia. (Adapted from Woese, 1987, with permission).

ecule (approximately 3000 nucleotides) contains about twice as much information as the 16S rRNA and has been used to supplement the 16S rRNA data for resolving closely spaced evolutionary branching (Olsen, 1988)

State of the Art

The identification of contributing populations is an essential prelude to community characterization, but this information alone is insufficient. Communities are dynamic entities, and cellular activity and gene expression vary within and among populations in response to a changing local environment. This local environment may be determined primarily by surrounding populations. Although the influence of surrounding organisms is unarguable, the specificity of these interactions is unknown. Two general categories of community can be imagined. Communities may be composed of a somewhat arbitrary assemblage of populations, each contributing to a general class of biochemical transformation(s) required for community processes. If so, communities developing in separate locales but of similar environment and activity could be composed of a vast variety of populations. Interacting populations would be co-adapted rather than co-evolved. Alternatively, communities may be highly integrated units displaying specifically evolved homeostatic mechanisms. If this condition is more representative, the variety of possible population arrangements must be more restricted.

Connecting populations to their interactive roles in community processes is not easy using available techniques. Traditional approaches have been primarily correlative, for example relating the consumption of different electron acceptors or donors in the presence or absence of specific metabolic inhibitors (e.g., of sulfate reduction, methanogenesis, and nitrification). Therefore the study of interactions (multicellularity) almost certainly depends on the continued development of molecular techniques to interrogate the activities of individual cells and populations within a community context. The potential for monitoring microorganisms in time and space using molecular techniques together with microscopy and digital image analysis is an exciting area of rapid techniques development. The traditional methods of enrichment and isolation remain fundamental to characterization of microorganisms. Contemporary studies of microbial community generally combine traditional microbiology with other analytical techniques, including those of molecular biology. While immunochemical techniques also remain an important resource (Currin et al., 1990; Visser et al., 1991; Ahring et al., 1993), the application of molecular genetics has provided the first unified conceptual and technical framework for the explicit characterization of natural systems.

We begin this chapter with a conceptual overview and brief discussion of the technical state of the art. The basic analytical approaches are depicted in Figure 11.3. Some of them are described in greater detail in appropriate sections of this chapter. For the remainder, in order to provide a connection between technology and ecology, we discuss three communities of varying physical structure and cooperative behavior. The community examples are taken mostly

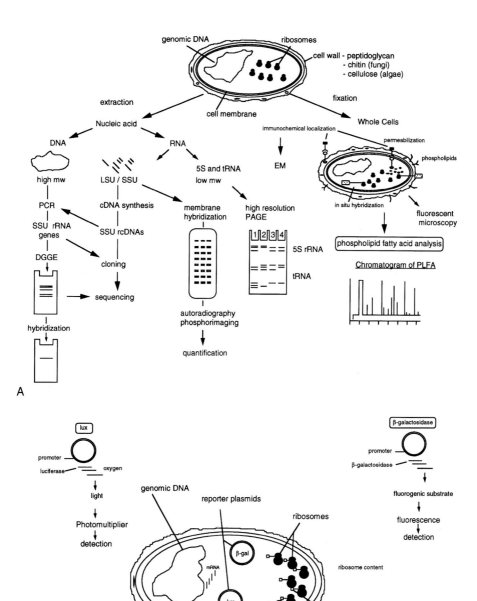

FIGURE 11.3 Molecular and biochemical approaches to characterizing microbial community structure and activity. (**A**) Methods based on nucleic acid, cell wall, protein, and fatty acid signatures. (**B**) Reporter systems for measuring single cell activity. See text for detailed discussion of the various approaches. DGGE, denaturing gradient gel electrophoresis; rcDNA, recombinant DNA.

from our own research experience. The common analytical framework for each community is phylogenetic. Although they provide ample examples of multi-cellular behavior, they do not represent comprehensive coverage of well studied environmental systems, and we apologize for any omissions in this regard. We briefly introduce the three communities here.

1. *Biofilms:* Surface-attached microbial consortia are ubiquitous and often dominate the microbial presence in the open environment. We have used laboratory reactor systems to study the complex spatial relationships of single cells and population interactions in relation to system processes.

2. *Microbial mats*: These microbial systems may represent the most an-cient of communities on earth. However, contemporary communities develop only in environments that experience limited grazing by higher life forms, for example in thermal environments or those of high salin-ity. These communities display a remarkable spatial and functional or-ganization of interacting microbial populations.

3. *Rumen*: The rumen is a gut environment in which a diverse collection of anaerobic microbial populations cooperate in accomplishing the bio-degradation of plant material.

Phylogenetic Dimension

Every organism embodies its ancestry and is not intelligible apart from that ancestry. This sentiment is not new, but it has yet to be fully incorporated into microbiology. A major goal when writing this chapter was to develop the con-nection between the evolution of communities and interacting associated mi-crobial populations. Microbes have generally been examined in isolation, even though they mostly evolved in the context of communities and so cannot be fully understood apart from community. Because the cause of evolutionary change must in large part derive from interactions with other community mem-bers, there is an abiding relation between microbial evolution and microbial ecology. Moreover, because community is dependent on the cooperative be-havior of its many components (microbial populations), the community may be the unit of selection (Wolfaardt et al., 1994). Demonstration of this concept is not yet possible with our current understanding of community structure. Indeed, the appropriate questions may not yet be formulated because of our lack of information on what might be referred to as ''natural history'' (Stahl, 1993). How tightly integrated are microbial communities? What are the hall-marks of a stable community as defined by population structure? What are the limits defined by physicochemical surroundings within which a community remains functionally and structurally intact?

Ecology developed with little consideration of microorganisms due mostly to methodological limitations, although microbial communities have existed for more than 3.5 billion years (Cohen, 1989). The oldest known microfossils have been found in stromatolites (the fossilized remains of ancient microbial mats) and are of an antiquity unmatched by any community at a higher trophic level. If the microbial mat systems of today are the direct descendants of those captured in the fossil record, the evolution of mat communities spans nearly the same time period as the biosphere itself. Thus mat communities may provide one of the clearest records of the relation between the evolution of microorganisms and the communities they constitute. If the evolution of major phylogenetic groups has been shaped by cooperative behavior, their phylogeny should be reflected in community structure and function. There is considerable confusion, however, concerning the ability to infer physiology from phylogenetic affiliation. There are many examples of close phylogenetic relationships between organisms that have remarkably different physiological attributes, for example close relationships between phototrophs and autotrophs and between autotrophs and heterotrophs (for examples: Lane et al., 1992). Conversely, there are examples of phylogenetically defined groups that are remarkably coherent in terms of their physiological characteristics. The *desulfovibrio* (Devereux et al., 1989, 1992) and methanogenic groups (Rouvière et al., 1992; Raskin et al., 1994b) are notable examples of phylogenetic and physiological coherence. Other examples include ammonia-oxidizing (Head et al., 1993; Teske et al., 1994) and methane-oxidizing (Alvarez-Cohen et al., 1992) bacteria. We suggest that part of this confusion results from failure to recognize ecologically significant features, another consequence of using the pure culture phenotype to infer environmental activity. We believe that as we learn more about the activities of various microbial populations by directly observing them in their natural environment, we will find that the phylogenetically cohesive characters are closely aligned with their roles in microbial communities. It is this hypothesis that structures the analytical foundation of many of the studies of multicellular systems discussed in this chapter.

Technical Overview

Only within a phylogenetic framework has the staggering diversity of environmental populations become fully accessible to explicit study. The sequencing of biopolymers (most commonly the rRNAs) recovered directly from the environment is now the preferred method of phylogenetic characterization and is the emphasis of this chapter. Several methods have been applied to the recovery of rRNA (gene) sequences directly from the environment: direct sequencing of 5S rRNA (Stahl et al., 1985), cDNA cloning (reverse transcriptase) (Ward et al., 1990, 1992; Weller et al., 1991), ''shotgun'' cloning (Schmidt et al., 1991), and polymerase chain reaction (PCR) amplification with specific primers using rRNA or DNA as template (Giovannoni et al., 1990; Liesack et al., 1991; Fuhrman et al., 1992; Reysenbach et al., 1992). There are many technical caveats associated with these techniques, and the reader is referred to specific references for more detailed considerations of each method. The derived sequences serve to explicitly describe community population structure, but this phase is only the first step. Community structure must be connected to community processes. Techniques for directly measuring specific cellular processes in environmental populations, such as transcription, are less well developed. Although we discuss some aspects of these evolving techniques, our intent is to provide more general connections between process and structure. We restrict detailed discussion to the use of the rRNAs as hybridization targets.

DNA Probes

Nucleic Acid Sequence Databases

The availability of sequence information, supporting software, and the ability to synthesize DNA oligonucleotides has revolutionized the design and use of DNA hybridization probes and PCR primers. The principal sources of nucleic acid sequence data are (1) the European Bioinformatics Institute (EBI) databases (Emmert et al., 1994), which in collaboration with the DNA Database of Japan (Mishima) and GenBank, National Center for Biotechnology Information (NCBI), Bethesda, USA (Benson et al., 1994), maintains and distributes the EMBL Nucleotide Sequence Database (Rice et al., 1993); (2) the Ribosomal Database Project (RDP) (Maidak et al., 1994) and (3) the compilation of small ribosomal subunit RNA structures (Neefs et al., 1993). In addition to sequence data, some of these databases provide supportive software (e.g., alignment editors and analytical functions). Although an immense resource, the current data structure is not suited to the study of environmental systems, the reason being that the existing sequence collections provide a limited collection of associated data relevant to the environmental range and activity of the organisms associated with the deposited sequences. There is a clear need for more comprehensive information content. For example, although the rRNAs are becoming the standard for describing environmental populations, there is virtually no formal link between sequence, physiology, and habitat. Development of these links is essential to connect the sequence information to environmental processes.

Ribosomal RNAs: Phylogenetic Nesting of Probes

The application of increasingly specific phylogenetic probes to samples derived from the same environment has been termed *phylogenetic nesting*. Design of nested probes is achieved by selecting probe target sites in relation to the phylogeny, such that phylogenetic groups are encompassed by individual probes. The phylogenetic groups, as shown in Figure 11.2, have a natural hierarchy that is matched to the probe target groups. Those sequences (organisms) identified by a specific probe also hybridize to probes designed for the larger phylogenetic group. The nested probe format provides for systematic characterization of environmental diversity. If our understanding of the environmental diversity for a phylogenetic group is complete, quantification should be additive. For example, quantification using species-specific probes should add up to the value obtained using the corresponding genus-specific probe. Lower values for summed species quantification could imply unrecognized diversity with the genus. This strategy was used to identify novel fiber-digesting populations (Lin and Stahl, 1995). At the highest taxonomic level, the proportion of Archaea, Bacteria, and Eukarya (expressed as a per cent of a ''universal'' probe) should total 100%. We have designed and characterized phylogenetically nested probes for fiber-digesting bacteria of the genus *Fibrobacter*, methanogens, ammonia-oxidizing and sulfate-reducing bacteria. The application of these probes to natural system studies is discussed in appropriate sections (see below).

The reader is referred to other sources for more detailed discussion of probe design and application (Stahl and Amann, 1991; Amann et al., 1995; Raskin et al., 1995a). We stress one general caution regarding the use of hybridization probes to characterize natural diversity: Probes should be considered tools to be refined through application (as discussed below) and the availability of additional sequence information from culture and environmental sources.

Hybridization Formats

Generally, one of two basic nucleic acid hybridization formats have been used to quantify environmental populations: whole cell hybridization and hybridization following nucleic acid extraction. Whole cell hybridization (often referred to as in situ hybridization) is used in conjunction with flow cytometry or microscopic analyses. Microscopy is often combined with image analysis. Hybridization of extracted nucleic acid (DNA or RNA) typically involves immobilization of the extracted nucleic acid on nylon membranes prior to hybridization and quantification. Extraction of nucleic acid from environmental samples is technically difficult. Because of the wide variety of environments colonized by microorganisms and their differing resistance to breakage (e.g., Gram-negative, Gram-positive, acid-fast bacteria, eukaryotes) there is no universal extraction procedure suitable for efficient, unbiased recovery of DNA or RNA. For recovery of intact high-molecular-weight rRNA, we have found a mechanical lysis protocol to be the best available (Stahl et al., 1988; Lin and Stahl, 1995). Glass or zirconium/silica beads are used with a low pH buffer-equilibrated phenol (with or without guanidine) and sodium dodecyl sulfate (SDS) to recover intact microbial rRNA from a wide range environments (Figure 11.4). This technique is less well suited to DNA isolation because extended mechanical breakage results in the shearing of DNA. Although low-molecular-weight DNA may serve as an acceptable hybridization target, it is not acceptable for PCR-based techniques. For example, low-molecular-weight DNA contributes to the generation of chimeric amplification products derived from fragmented rRNA genes from different organisms (Liesack et al., 1991).

In addition to methods for the direct recovery or identification of nucleic acid sequences in the environment, other methods are available that provide a molecular overview of population diversity, including low-molecular-weight RNA profiling (Höfle, 1992) and denaturing gradient gel electrophoresis (Muyzer et al., 1993; Muyzer and de Waal, 1994). These techniques provide an

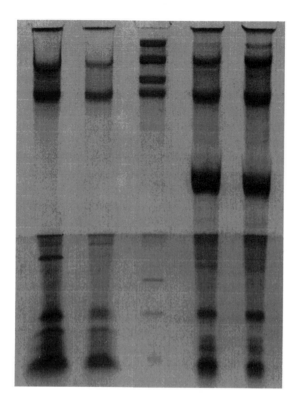

FIGURE 11.4   Polyacrylamide gel of nucleic acids extracted from freshwater sediments and microbial mats. Lane 1: from Lake Michigan sediment (0.–0.5 cm deep), from Fox Point, Michigan. Lane 2: from Lake Michigan sediment (0.5–1.0 cm deep), Fox Point, Michigan. Lanes 4 and 5: two different microbial mat samples from Bird Shoal, North Carolina. Lane 3: 125 ng of mixed standards consisting of equal concentrations of RNA isolated from pure cultures of *Escherichia coli* and *Zea mays*.

assessment of changing community structure through time and space. The reader is referred to appropriate references for discussion of their methods.

Image Analysis of Microbial Communities

Although microscopy has often hinted at structure in fixed or aggregated microbial communities, it was only with the advent of molecular techniques in combination with image analysis that full reconstruction of three-dimensional structure of microbial communities became possible. Advances in light microscopy, including *f*luorescent *i*n *s*itu *h*ybridization (FISH), confocal microscopy, and digital image analysis are changing the way microbiologists observe natural microbial systems. This spatial dimension, with an emphasis on tracking individual microorganisms in both time and space, will likely provide entirely new insights on cooperativity and competition among microorganisms within communities.

Whole Cell Hybridization: Observation and Quantification of Single Cells

Single cells can be identified following hybridization of fixed specimens to DNA probes complementary to the rRNAs. The microbial cell wall, with the exception of some thick-walled microorganisms, is made permeable to short DNA probes during standard fixation. Specific populations can then be quantified using microscopy or flow cytometry (Amann et al., 1990a). Microscopic visualization of single cells following hybridization is label-dependent; radioisotopic and nonradioisotopic reporter systems have been used. These reporters are in general use in molecular and cellular biology and have been the subject of a number of reviews (Stahl and Amann, 1991; Amann et al., 1995). Because most of our experience has been with fluorescent dye-labeled probes (FISH), we restrict our discussion to this format.

Three-Dimensional Optical Techniques for In Situ Hybridization

Individual cells of various populations can be resolved by hybridization with probes labeled with different fluorochromes. The technique is relatively straightforward. In contrast, study of their three-dimensional relation relies on sophisticated image acquisition and analysis techniques. These methods are only briefly considered here, and the reader is referred to the literature for more complete descriptions. First, a series of two-dimensional sections must be produced, which can be done physically or optically (Brakenhoff et al., 1989). The two-dimensional sections are then reassembled to form a three-dimensional image of the sample (Figure 11.5).

Physical Sectioning

Samples are prepared for microscopy by first embedding in paraffin or plastic for sectioning using a microtome. Our direct experience is restricted to thin sectioning of paraffin-embedded specimens. The main advantage of thin sections (1.5–3.0 μm) is the small cross section the light traverses, which reduces the amount of light scattering (haze) from out-of-focus images originating from sample material above and below the plane of focus (Agard et al., 1989). One disadvantage of sectioning is that the three-dimensional reconstruction requires the individual slices to be brought into register. This problem has been addressed in part with image analysis software, most recently the ROSS program (Dr. Muriel Ross, Biocomputation Center, MS239-11, NASA Ames Research Center, Moffett Field, CA). Another strategy is to cut thicker sections. They have more mechanical integrity, which is important for three-dimensional reconstruction; but it contributes more out-of-focus haze. These technical prob-

Physical sectioning of biofilm samples
for 3-dimensional optical imaging

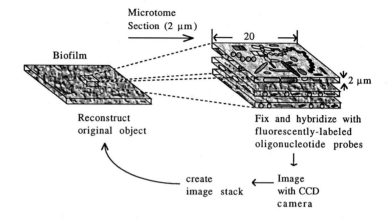

Optical sectioning with Confocal
Laser Scanning Microscopy

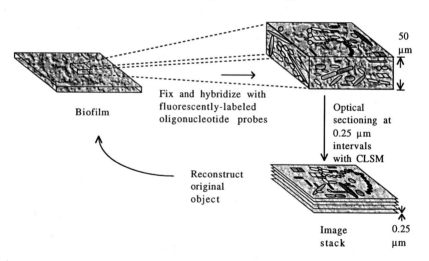

FIGURE 11.5 Three-dimensional reconstruction of a natural microbial community. Samples are fixed (see text) and then sliced physically or optically. Fixed samples are embedded in a paraffin block for sectioning with a microtome. The practical limit for physical sectioning of a biofilm for fluorescent microscopy is 2 μm. Sections are mounted on slides, and fluorescently labeled probes are hybridized to the RNA in the sample. Images of the sections are registered and stacked to achieve a three-dimensional reconstruction of the original sample. The confocal laser scanning microscope (CLSM) can optically section a hybridized sample as thick as 50 μm in 0.25 μm slices due to rejection of out-of-focus light and an intense light source, making physical sectioning unnecessary. The resultant images are stacked and reassembled into a three-dimensional reconstruction of the original sample.

lems have been greatly reduced by the introduction of optical sectioning techniques, discussed in the following sections.

Optical Sectioning

Two methods are available for removing out-of-focus light: confocal microscopy and deconvolution software. Confocal microscopy is a well established technique for producing high resolution three-dimensional images (Wilson, 1989; Brakenhoff et al., 1990; Richardson, 1990; Lawrence et al., 1991; Caldwell et al., 1992). Lasers are the best light sources, but arc lamps have also been used, particularly if a broad selection of wavelengths is required. A major advantage of confocal microscopy is that thick specimens (as thick as 100 μm) can be imaged. Laser confocal microscopes, because of the intense light source needed to illuminate the sample through the pinhole, bleach the fluorescent dye as the specimen is scanned (Wells et al., 1989). This problem has been partly alleviated with introduction of the tandem confocal microscope (Wright et al., 1989).

Deconvolution algorithms have also been developed to deblur images (Moss, 1992). These algorithms remove out-of-focus light, reducing the haze and forming a sharp image. With deconvolution software, as with confocal microscopes, a narrow bandwidth (preferably monochromatic) light source gives the best results. Analysis of thick specimens (approximately 25 μm) is also possible. Other advantages include the ability to use any contrast technique (e.g., differential interference contrast microscopy) or phase contrast, and less photobleaching than confocal microscopy (particularly important for samples containing little fluorescent label). The most significant disadvantage is the large amount of computing time required. The most important trade-off is speed of image reconstitution versus accuracy of reconstitution (especially photon recovery).

Multifilter Analysis

Fluorescent in situ hybridization (FISH) allows the use of multiple probes, each with a different fluorescent label. Labels are available from the ultraviolet to the near infrared, and so more than one biological target can be labeled at a time. In addition, more targets can be visualized than the number of available fluorescent dyes by using a combination of dyes on individual probes. Image processing is used to create pseudocolor representation of targets fluorescing at multiple wavelengths.

Image Analysis Software

Submicron image sections must be assembled to reconstruct the biological structure. Such reconstruction is referred to as *volume rendering*. Volume rendering converts a serial collection of image sections to a three-dimensional representation called the image stack. Image stacks can be manipulated (e.g., rotated and sliced) in order to study the structure of the biological sample. A large number of these stacks produce an enormous file. For instance, a $512 \times 512 \times 64$ section stack at 8 bit resolution requires 32 megabytes of memory. In order to manipulate a stack of this size, 64 or more megabytes of free RAM memory is required. Image analysis software has been written for Macintosh and MS-DOS computers. Software has also been developed for UNIX work stations and Silicon Graphics computers (SGI). A list of image analysis software available from private vendors or for the public domain is available from NIH Image (National Institutes of Health, Bethesda, MD).

Molecular Measures of In Situ Processes and Activity

Detection of Proteins

The identification of specific microbial populations using immunodetection methods has a long history in environmental microbiology (Schmidt, 1973; Macario and Conway de Macario, 1992; Conway de Macario et al., 1993). The combination of cell-specific immunochemical staining with microscopy provided the first general format for in situ analyses of single cells. More recently this format has been used to determine the microscopic distribution of specific gene products. For example, nitrogenase distribution in marine cyanobacterial cells (*Trichodesmium*) has been assessed using fluorescence microscopy in combination with fluorescent antibody for nitrogenase and transmission electron microscopy following postsection immunocolloidal gold staining (Paerl et al., 1989). A well recognized limitation of this method is the need to verify the specificity of the immunochemical reagent. Many questions of specificity, however, could be more fully evaluated using a combination of techniques. For example, the combined application of nucleic acid and antibody stains to the study of complex environmental systems serves to specifically link protein distribution with individual cells of a population and gene expression.

Detection of Functional Genes

Methods to study community dynamics must include techniques to assess distribution and expression of genes within single cells of individual populations. These methods are only now being developed. There are two basic formats: detection of mRNA and detection of reporter-gene products derived from reporter genes fused to the promoter of the gene of question. The latter technique is not directly amenable to nondestructive in situ analysis, as the genetic construct must be characterized in the laboratory and reintroduced into the natural setting. Use of reporter-gene products has been primarily restricted to laboratory systems or as biological sensors to measure the presence of specific environmental solutes (Burlage and Kuo, 1994).

Detection of mRNA Transcripts

The in situ detection of natural populations of mRNA by hybridization to whole cells (Hahn et al., 1993; Madan and Nierzwicki-Bauer, 1993) or hybridization to nucleic acid immobilized on membrane supports (Pichard and Paul, 1991) has been reported only during the last few years and so far has been restricted to high copy number messages (e.g., from plasmid-encoded genes). General application requires substantial improvement in sensitivity and specificity. In particular, the development of a robust method for single cell studies is essential for community analyses.

Gene Fusion Reporter Systems

The detection of gene expression using reporter-gene fusions is a standard technique in molecular genetics. Application of this technique to complex microbial systems has been primarily limited to the monitoring of target compounds or intermediates during biodegradation (Heitzer et al., 1994). The target-gene promoter is fused to a promoterless reporter gene, (e.g., *lacZ* or *lux*CDABE). The fusion construct can be either plasmid-encoded or introduced into the host genome via a variety of genetic techniques, including transposon delivery (Menn et al., 1993). The lux system offers a format immediately amenable to single-cell microscopic detection but requires the use of highly sensitive image acquisition equipment [e.g., cooled charge-coupled device (CCD) cameras]. An important development has therefore been the application of fluorogenic substrates for detection of β-galactosidase activity in single cells using standard fluorescence microscopic techniques and flow cytometry (Plovins et al., 1994).

Use of Cellular Ribosome Content as a General Activity Measure for Single Cells

The correlation between cellular ribosome content and growth rate was one of the earliest, most fundamental observations in microbial physiology (Schaecter et al., 1958). In principle, the correlation can be used to gauge the activity of environmental populations. For example, one study used RNA/DNA ratios to evaluate the growth rate and biomass of marine microorganisms (Dortch et al., 1983). Ideally, this determination is made at the level of the individual cells comprising different environmental populations, as is now possible using rRNA-targeted probes for whole cell hybridization. Probe-conferred fluorescence corresponds to individual cell ribosome content and, by inference, to single-cell activity.

Application of this technique has been limited. Some unresolved issues concern biases introduced by variable probe penetration and possible growth rate-associated change in cell wall permeability. Growth rate related change in ribosome abundance has been examined for relatively few environmentally relevant microorganisms. Other factors, such as the influence of transient physical and biological changes (e.g., temperature and starvation) on this ratio, are mostly unstudied. The few studies published to date have invested most effort in defining the growth rate-associated changes in ribosome content of the study organism(s) (Kemp et al., 1993; Kerkhof and Ward, 1993; Poulsen et al., 1993). We anticipate that the more immediate use of this technique will be restricted to qualitative studies, comparing the ''activities'' of individual populations in relation to a changing environment or community processes. It could also offer a useful complement to measures of gene expression as revealed by the detection of specific proteins or mRNA transcripts.

Studies of Natural Systems

The remainder of this chapter considers several complex microbial communities. They were introduced earlier, and we restrict this more detailed discussion to systems with which we have had direct experimental involvement: microbial mats, biofilms, and the rumen habitat. Although research goals are system-specific, they share a common feature: Each community mediates one (or more) relatively specific chemical transformations that can be related to microbial community structure. All of these systems are anaerobic.

General Themes of Interpopulation Associations in Anaerobic Habitats

This introduction is intended to provide only a glimpse of the complexity of anaerobic communities. The anaerobic decomposition of organic material is carried out through the cooperation of several general physiological categories of microorganisms (Schink, 1988). In the absence of sulfate, four microbial groups are generally recognized to participate in the anaerobic degradation of organic compounds.

1. Hydrolytic and fermentative bacteria, which degrade complex biopolymers (e.g., plant cell wall components), monomers, and oligomers into acetate, hydrogen, carbon dioxide, and a mixture of short-chain fatty acids, alcohols, succinate, and lactate
2. Proton-reducing acetogenic bacteria, which convert fatty acids, alcohols, succinate, and lactate to acetate, hydrogen, and carbon dioxide
3. Hydrogen-oxidizing methanogens, which convert hydrogen and carbon dioxide (as well as other quantitatively less important compounds) to methane and water
4. Acetoclastic methanogens, which convert acetate to methane and carbon dioxide

Other microbial groups are present in anaerobic communities, but their importance in natural ecosystems is not yet completely understood. For example, microorganisms have been identified that can produce butyrate and propionate from acetate and ethanol (Morotomi et al., 1988). Homoacetogens and methanogens compete for the same substrates. Methanogenesis is thermodynamically more favorable, but homoacetogens appear to dominate in some anaerobic environments (Dolfing, 1988; Wolin and Miller, 1993). In the presence of sulfate, sulfate-reducing bacteria can compete with methanogens and proton-reducing acetogenic bacteria for their common substrates. These general functional groups, each almost certainly comprised of multiple populations, must function in concert. In this regard, natural anaerobic microbial communities embody characteristics of multicellularity.

Biofilms

Background Information

Microbial growth on surfaces, or *biofilm*, is increasingly recognized to be a ubiquitous, often dominant form of microbial life in the environment (Costerton et al., 1987, 1994). There is also an indication that surface-associated microorganisms differ in many ways from planktonic cells growing in batch culture. For example, expression of genes involved in exopolysaccharide synthesis is modulated by association with substratum (Davies et al., 1993). The most profound difference between planktonic and attached modes of growth is the variety of microniches created by the biofilm matrix-associated populations. The environment experienced by individual organisms within the community is largely determined by neighboring populations and the local structure of the biofilm. Each cell lives in a unique microniche that is sustained by, and in turn sustains, nearby cells. Some of these local structural features have now been brought into clear view using molecular biology, microscopy, and image analysis (Costerton et al., 1994). To the extent that we understand biofilm communities, their physical and metabolic structure is not unlike that of a multicellular organism. An immediate question raised by this view of biofilm structure is the specificity of cooperation. Are the associations determined by rather general metabolic themes? Alternatively, do specific, co-evolved associations exist?

Molecular Measures of Biofilm Community Structure

A long-standing project in our laboratory is the study of anaerobic biofilm communities in reactor systems (Amann et al., 1992b; Raskin, 1993; Raskin et al., 1995a). Notably, these communities contain representatives of most of the major phylogenetic groups of sulfate-reducing and methanogenic microorganisms. Thus although the systems are in a sense ''artificial'' because they are not open environmental systems, they seem to demonstrate comparable phylogenetic group representation.

Structural and Functional Analysis of Reactor Communities

We used two sets of oligonucleotide hybridization probes to follow the microbial structural changes in four anaerobic biofilm reactors in response to perturbation of the influent medium composition. The first set of probes (Table

TABLE 11.1 Oligonucleotide Hybridization Probes and PCR Primer Pairs

Probe Name[a]	Target Group[b]	Probe or Primer Sequence 5' to 3'	Original Reference	Old Probe/ Primer Name
S-Univ1392aR-15	All organisms	ACGGGCGGTGTGTRC	Pace et al. (1986)	UNIV1400
Probes for domains				
S-D-Arch344aR-20	Archaea	TCGCGCCTGCTGCTCCCCGT	Raskin et al. (1994b)	ARC344
S-D-Arch915aR-20	Archaea	GTGCTCCCCCGCCAATTCCT	Amann et al. (1990b)	ARC915
S-D-Bact338aR-18	Bacteria	GCTGCCTCCCGTAGGAGT	Amann et al. (1990a)	EUB338
S-D-Euca502aR-16	Eucarya	ACCAGACTTGCCCTCC	Amann et al. (1990a)	EUK516
Probes for methanogens				
S-O-Mcoc1109aR-20	Methanococcales	GCAACATAGGGCACGGGTCT	Raskin et al. (1994b)	MC1109
S-F-Mbac310aR-22	Methanobacteriaceae	CTTGTCTCAGGTTCCATCTCCG	Raskin et al. (1994b)	MB310
S-F-Mbac1174aR-22	Methanobacteriaceae	TACCGTCGTCCACTCCTTCCTC	Raskin et al. (1994b)	MB1174
S-O(-1)-Mmic1200aR-21	Methanomicrobiales except Methanosarcinaceae	CGGATAATTCGGGGCATGCTG	Raskin et al. (1994b)	MB1200
S-F-Msar860aR-21	Methanosarcinaceae	GGCTCGCTTCACGGCTTCCCT	Raskin et al. (1994b)	MSMX860
S-F(-1)-Msar1414aR-21	Methanosarcinaceae except Methanosaeta	CTCACCCATACCTCACTCGGG	Raskin et al. (1994b)	MS1414
S-G-Msar821aR-24	Methanosarcina	CGCCATGCCTGACACCTAGCGAGC	Raskin et al. (1994b)	MS821
S-G-Msae825aR-23	Methanosaeta	TCGCACCGTGGCCGACACCTAGC	Raskin et al. (1994b)	MX825
Probes for fiber-digesting bacteria				
S-G-Fibro225aR-21	Fibrobacter	AATCGGACGCAAGCTCATCCC	Stahl et al. (1988)	Fibro
S-S-F.suc649aR-20	Fibrobacter succinogenes	TGCCCCTGAACTATCCAAGA	Amann et al. (1990b)	Succ

Probe	Target organism	Sequence	Reference	Designation
S-Ss-F.s.suc207aR-21	*Fibrobacter succinogenes* subspecies *succinogenes* (Group 1)	CCATACCGATAAATCTCTAGT	Stahl et al. (1988)	*B. succinogenes* S85 and A3c
S-Ss-F.s.suc628aR-22	*Fibrobacter succinogenes* subspecies *succinogenes* (Group 1)	GATCCAGTTCGGACTGCAGAGC	Lin et al. (1994)	Sub1
S-Ss-F.suc(2)628aR-22	*Fibrobacter succinogenes* Group 2 (formerly subspecies *elongatus*)	AACCCAGTTCGGACTGCAGGTC	Lin et al. (1994)	Sub2
S-Ss-F.suc(3)628aR-22	*Fibrobacter succinogenes* Group 3 (formerly subspecies *elongatus*)	GGTGCAGTCCGAACTGCAGGCC	Lin et al. (1994)	Sub3
S-S-F.int136aR-20	*Fibrobacter intestinalis*	CGGTTGTTCCGGAATGCGGG	Lin et al. (1994)	Int
S-Ss-F.int207aR-21	*Fibrobacter intestinalis* (strains SR7 and NR9)	CCGCATCGATGAATCTTTCGT	Stahl et al. (1988)	*B. succinogenes* DR7 and NR9
S-Ss-F.int1241aR-24	*Fibrobacter intestinalis* (strains DR7 and NR9)	GCCCCGCTGCCCATTGTACCGCCC	Amann et al. (1990b)	*F. intestinales* DR7 and NR9
S-S-L.multi207aR-24	*Lachnospira multiparus*	CTTATACCACCGGAGTTTTTCA	Stahl et al. (1988)	*L. multiparus*
Probes for sulfate-reducing bacteria				
S-F-Dsv687aR-16	Desulfovibrionaceae	TACGGATTTCACTCCT	Devereux et al. (1992)	SRB687R
S-Dsb804aR-18	Desulfobacter group	CAACGTTTACTGCGTGGA	Devereux et al. (1992)	SRB804R
S-G-Dsbm221aR-20	*Desulfobacterium*	TGCGCGGACTCATCTTCAAA	Devereux et al. (1992)	SRB221R
S-G-Dsb129aR-18	*Desulfobacter*	CAGGCTTGAAGGCAGATT	Devereux et al. (1992)	SRB129R
S-G-Dsb220aR-20	*Desulfobacter*	TMCGCARACTCATCCCCAAA	Amann et al. (1990a)	DSBACT

(continued)

TABLE 11.1 *Continued*

Probe Name[a]	Target Group[b]	Probe or Primer Sequence 5′ to 3′	Original Reference	Old Probe/ Primer Name
S-Dcoc814aR-18	Desulfococcus group (D. multivorans, Desulfobotulus sapovorans, Desulfosarcina variabilis)	ACCTAGTGATCAACGTTT	Devereux et al. (1992)	SRB814R
S-G-Dsbb660aR-20	Desulfobulbus	GAATTCCACTTTCCCCTCTG	Devereux et al. (1992)	SRB660R
S-S-Dsm.sp647aR-19	Desulfuromonas species-like organisms	TCTCCCGTATTCAAGTCTG	Amann et al. (1992b)	Population type 1
S-S-Dsv.sp647aR-19	Desulfovibrio species-like organisms (strain PT-2)	TCTCCCGAACTCAAGTCCA	Amann et al. (1992b)	Population type 2
PCR primer pairs S-Srb402aF-18 S-Univ907aR-22	Selective recovery of sulfate-reducing bacteria	cccgggatCCTGACGCAGCIACGCCG[c] tctagaagcttCCCGTCAATTCCTTTGAGTTT	Amann et al. (1992b)	SRB385F UNIV907R

[a]Probe and primer names have been standardized as follows: S or L for small or large subunit rRNA as the target; letter(s) designating the taxonomic level targeted; letters designating the target group of the oligonucleotide probe or primer; nucleotide position (E. coli numbering) in the target where the 3′ end of the probe or primer binds; letter designating the version of the probe or primer (e.g., a for version 1, b for version 2); F or R for forward or reverse direction; number indicating the length in nucleotides of the probe or primer.

[b]Probe specificity was checked with CHECK-PROBE on the RDP, November 1994.

[c]Lower-case letters indicate restriction enzyme recognition sites.

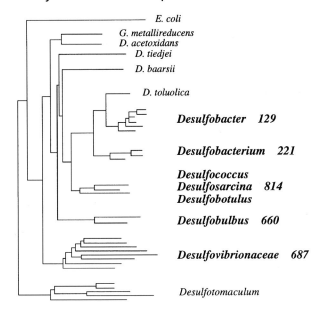

E. coli
G. metallireducens
D. acetoxidans
D. tiedjei
D. baarsii
D. toluolica

Desulfobacter 129

Desulfobacterium 221

Desulfococcus
Desulfosarcina 814
Desulfobotulus

Desulfobulbus 660

Desulfovibrionaceae 687

Desulfotomaculum

FIGURE 11.6 Probe target groups in relation to the phylogeny of SRB. The five probes shown here target 16S-like rRNA sequences unique to different phylogenetic groups of SRB. The probe sequences and demonstration of target group specificity has been reported previously (Devereux et al., 1992). Sulfate-reducing bacteria shown in this figure not targeted by the available probes include *Desulfotomaculum* species, *Desulfomonile tiedjei*, *Desulfobacula toluolica*, and *Desulfoarculus baarsii*. Genera not shown or of uncertain affiliation, in addition to those indicated, include: *Archaeoglobus*, *Desulfonema*, and *Thermosulfobacterium* (Widdel and Bak, 1992). Non-sulfate-reducing species indicated are *Escherichia coli*, *Geobacter metallireducens*, and *Desulfuromonas acetoxidans*. Numbers refer to the corresponding probe target position on the *E. coli* 16S rRNA. (From Devereux et al., 1992, with permission.)

11.1, Figure 11.6) was designed to quantify phylogenetic groups of Gram-negative mesophilic sulfate-reducing bacteria (SRB). The second collection of probes was specific for groups of methanogens (Table 11.1, Figure 11.7). In addition, universal and domain-specific (Archaea, Bacteria, Eukarya) probes were used for more general assessments of community structure. This approach gives a phylogenetic overview to community structure. Extensive discussions, including details on hybridizations and chemical analyses employed, may be found elsewhere (Amann et al., 1992b; Raskin, 1993; Raskin et al., 1995a,b). Some of the conclusions drawn from this study, in the context of multicellular behavior, are summarized below.

1. Methanogens and SRB coexisted, regardless of sulfate availability, as electron acceptors. This concept contrasts with the general view that SRB have a competitive advantage over methanogens in the presence of sulfate, a perception supported by culture-based studies of many environments (e.g., Ward and Winfrey, 1985; Isa et al., 1986; Gibson et al., 1988); and by kinetic and thermodynamic considerations (Ward and Winfrey, 1985; Widdel, 1988). There

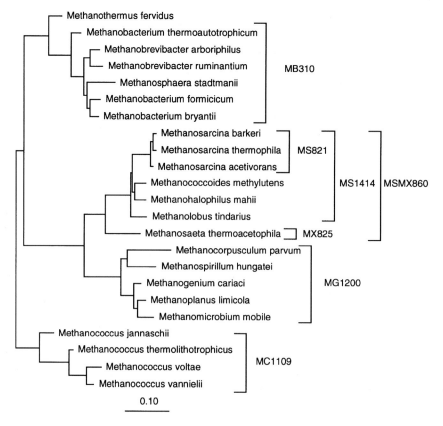

FIGURE 11.7   Probe target groups in relation to the phylogeny of methanogens (Vogels et al., 1980). The seven probes shown here target 16S-like rRNA sequences unique to different phylogenetic groups of methanogens. The probe sequences and demonstration of target group specificity has been reported previously (Raskin et al., 1994a). Methanogens shown in this figure not targeted by the available probes include *Methanothermus fervidus*. Numbers refer to the corresponding probe target position on the *E. coli* 16S rRNA. The maximum likelihood tree was adapted from the larger tree provided by the Ribosomal RNA Database Project (Maidak et al., 1994).

have also been reports of coincident activities (Isa et al., 1986; Nielsen, 1987; Barlaz et al., 1989; Parkin et al., 1990; Pochart et al., 1992; Yoda et al., 1992). Our observations using direct molecular quantification suggest that factors other than kinetics and thermodynamics are important for explaining interactions between methanogens and SRB. Such factors may include syntrophy, mass-transfer limitations, microbial colonization, and adhesion.

2. Several groups of SRB were present at significant levels in the presence and absence of sulfate. A *desulfovibrio*-like population was the dominant population in methanogenic biofilms. We suspect that this population was stabilized by interspecies hydrogen transfer in the absence of sulfate. Hydrogen-consuming methanogens (or acetogens) can substitute for sulfate in the general role of electron acceptor. This type of association has been demonstrated by recon-

struction of syntrophic associations from pure cultures (Bryant et al., 1977; Traore et al., 1983; Tasaki et al., 1993), but this study clearly documented its community significance. The suggested interaction between SRB and hydrogen-consuming populations now raises the question of the specificity of this association. Do specific populations of methanogens associate with specific populations of sulfidogens?

3. *Methanosarcina* and *Methanosaeta* (formerly *Methanothrix*) species coexisted under all conditions tested. The utilization of acetate as a substrate for methanogenesis among recognized methanogens is confined to species of *Methanosarcina* and *Methanosaeta*. Because of their lower (K_s) values and minimum substrate thresholds, *Methanosaeta* are capable of using acetate at concentrations below the minimum threshold value of *Methanosarcina* (Raskin, 1993), which should give *Methanosaeta* a competitive advantage over *Methanosarcina* under such conditions. The bulk liquid and effluent acetate concentrations in our biofilm reactors were below the minimum threshold values reported for most *Methanosarcina* species, suggesting that *Methanosaeta* species would be the only acetate-utilizing methanogens present. However, we observed comparable abundance of both. Their coexistence suggested that the minimum thresholds for the *Methanosarcina* species in our biofilm reactors were different from those reported in the literature, or bulk liquid acetate concentrations were significantly different from acetate concentrations in the biofilm. This finding raises the specter of an inadequate representation of environmental populations by the pure culture collection and prediction of environmental distribution or activity based on pure culture studies.

Microscopic Studies of Distribution and Activity of Specific Biofilm Population Members

Analysis of 16S rRNA sequence information recovered directly from biofilm samples was used to provide a higher-resolution analysis of population architecture. The PCR analysis was used to selectively amplify 16S rRNA gene sequences from organisms affiliated with the delta subdivision of the Proteobacteria (Table 11.1) (Amann et al., 1992b). Most gram-negative mesophilic SRB are affiliated with this subdivision. Two molecular isolates related to the described SRB were recovered (Amann et al., 1992b). We restrict this discussion to the subsequent study of one of these molecular isolates, identified by direct sequence characterization as a close relative of *Desulfovibrio vulgaris* (Amann et al., 1992b; Kane et al., 1993). More detailed studies of this organism's contribution to the biofilm community was initiated by designing an oligonucleotide probe specific for the recovered sequence (probe population type 2/S-s-Dsv.sp.647aR-19) (Table 11.1) (Amann et al., 1992b). Whole-cell hybridization using a fluorescent dye-labeled version of the probe was used to study its spatial distribution within the biofilm (Figure 11.8). Subsequently, this probe was used to monitor the enrichment and isolation of this organism (strain PT-2) from biofilm samples (Figure 11.8) (Kane et al., 1993). Using enrichment

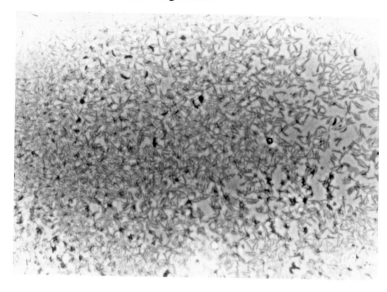

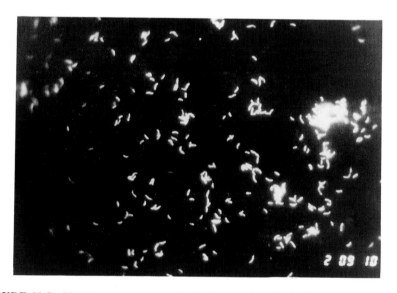

FIGURE 11.8 Photomicrographs of reactor biofilm whole-cell hybridizations with a fluorescent-dye-labeled PT-2 probe. The phase-contrast (**top**) and fluorescence (**bottom**) photomicrographs do not correspond to the same field (Amann et al., 1992b).

procedures for *D. vulgaris*, the closest identified relative, strain PT-2, was isolated. After isolation, strain PT-2 was grown in five different media (five different electron donors) in order to establish a relation between cellular composition and specific growth rate (Figure 11.9) (Poulsen et al., 1993). The rRNA content of single cells was estimated by whole-cell hybridization with a fluo-

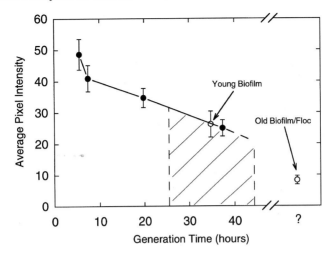

FIGURE 11.9   Growth rates of sulfate-reducing bacteria biofilm populations inferred through the use of in situ hybridization with fluorescence-labeled probes in conjunction with digital microscopy. The fluorescent signals (mean pixel intensity from images of hybridized cells) of cells from young versus old biofilm (○) were compared to signals from cells grown in pure culture at different growth rates (●). The cross hatching shows the range of growth rates spanned by the standard deviation derived from 50 cells in the young biofilm. The standard deviation for the old biofilm is based on nine cells, and the standard deviations for the pure culture cells are each based on 20–25 cells. (Data from Poulsen et al., 1993).

rescent version of the probe (Poulsen et al., 1993) in combination with image analysis. These analyses served as the basis for direct assessment of the growth rate (activity) of the PT-2 population in biofilms (Poulsen et al., 1993).

Initial studies have served primarily to explore some methodological avenues to the study of biofilm communities. Developing biofilms were obtained by placing glass coverslips in the reactor for 7 days. Established biofilm was taken from glass beads that had been in the reactor since its start-up. Figure 11.9 shows the mean pixel intensities for PT-2 cells in the two biofilm samples in relation to the standard curve derived from pure culture studies. Cells of strain PT-2 in the new biofilm were apparently growing with an average generation time of approximately 35 hours (specific growth rate of 0.029 h^{-1}), whereas the PT-2 cells in the established biofilm appeared to grow much more slowly. Generation times and specific growth rates within the established biofilm could not be estimated because the mean pixel intensity did not fall within the range of mean pixel intensities determined using pure cultures. We anticipate that this system will provide a useful model for the study of interpopulation associations. As a more complete picture of population makeup is developed, hypothesized interactions can be tested directly. For example, the proposed interaction between hydrogen-producing and hydrogen-consuming populations may have a spatial basis.

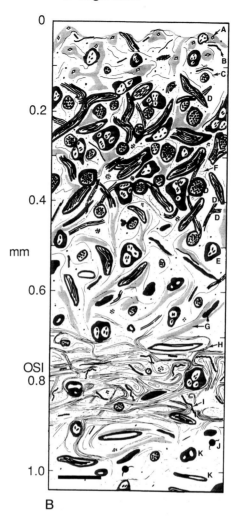

B

FIGURE 11.10 (**A**) Vertical section of the Octopus Spring, Yellowstone National Park, cyanobacterial mat (ca. $50°-55°C$, pH 8). The top green layer is the interval during which cyanobacteria (*Synechococcus* species) occur, embedded in a fabric of filamentous bacteria, some of which are green nonsulfur bacteria of the genus *Chloroflexus*. The deeper red to orange layers include *Chloroflexus* to a few millimeters below the surface, after which they are merely layers in which recalcitrant caretenoid pigments accumulate. The bioactive zone is only within the top 5 mm, below which decomposition is slow. (**B**) Daytime depth distribution of bacterial populations in a cyanobacterial mat from the Exportadora del Sal. Although the oxygen−sulfide interface (OSI) is shown at 0.8 mm, it may occur between 0.7 and 4.0 mm depending on the time of the year and the salinity of the pond; A, diatoms; B, *Spirulina* sp. (cyanobacteria); C, *Oscillatoria* spp. (cyanobacteria); D, *Microcoleus chthonoplastes* (cyanobacteria); E, nonphotosynthetic bacteria; F, unicellular cyanobacteria; G, fragments of bacterial mucilage; H, *Chloroflexus* spp. (green bacteria capable of anoxygenic photosynthesis if light penetrates to the OSI); I, *Beggiatoa* spp. (nonphotosynthetic sulfide-oxidizing bacteria); J, unidentified grazer; K, abandoned cyanobacterial sheaths. (**A**: reprinted from Ward; **B**: reprinted from DesMarais).

FIGURE 11.10 (A) Vertical section of the Octopus Spring, Yellowstone National Park, cyanobacterial mat (ca. 50°–55°C, pH 8). The top green layer is the interval during which cyanobacteria (*Synechococcus* species) occur, embedded in a fabric of filamentous bacteria, some of which are green nonsulfur bacteria of the genus *Chloroflexus*. The deeper red to orange layers include *Chloroflexus* to a few millimeters below the surface, after which they are merely layers in which recalcitrant caretenoid pigments accumulate. The bioactive zone is only within the top 5 mm, below which decomposition is slow.

Microbial Mats

Two microbial mat systems have so far been characterized by molecular phylogenetic techniques: a hot spring cyanobacterial mat community located at Octopus Spring, Yellowstone National Park (Ward et al., 1989, 1992; Weller et al., 1991) and a saline microbial mat located at a solar evaporate at Guerrero Negro, Mexico (D'Amelio et al., 1989; Risatti et al., 1994). Figure 11.10 shows an example of an Octopus Spring Mat and a saline mat. The studies of the Yellowstone hot spring and associated mat system using rRNA-based comparative analyses were among the first to use molecular phylogenetic techniques to explicitly assess natural microbial diversity (Stahl et al., 1985; Giovannoni et al., 1990; Ward et al., 1990). These and subsequent molecular surveys have confirmed the initial indication of the limited extent to which the pure culture collection describes the microbial diversity of these habitats. These studies have now more directly integrated the molecular comparisons with the functional and physiological dimensions of the mat communities.

Octopus Hot Spring Microbial Mat

The areal extent of the Octopus Spring Mat (and thermal mat systems in general) is restricted to a relatively well defined temperature interval within the thermal gradient originating from the hot spring source. The upper temperature limit (approximately 72°C) is defined by the loss of community integrity, and below about 42°–50°C metazoan grazers degrade the community. Studies by Ward and associates have revealed that *Synechococcus* ''species'' are restricted to relatively narrow temperature intervals within and between mat systems (Ruff-Roberts et al., 1994). This distribution generally corresponds to the optimum growth temperatures for the individual populations. A straightforward hypothesis is that individual *Synechococcus* populations perform the same general community function (oxygenic photosynthesis) but are further specialized within that role, in this case differing in their optimum temperature for growth. This theory is consistent with the earlier suggested relation between phylogeny and ecology. Members of specific phylogenetic lineages are predicted to serve related roles (within communities), and diversification within lineages reflects further specialization within that role. A linkage between ecology and phylogeny is more apparent in the results of our characterization of a saline microbial mat, discussed below.

Saline Mat

The microbial mat examined in our study was from a saline evaporation pond of the Exportadora de Sal (Baja, California) in which dissimilatory sulfate reduction is the major electron-accepting pathway for electrons provided by organic carbon mineralization (Canfield and DesMarais, 1993), as is the case for many marine systems (Skyring et al., 1983). It has been shown that the highest rates of sulfate reduction in these mats coincide with the photo-oxic

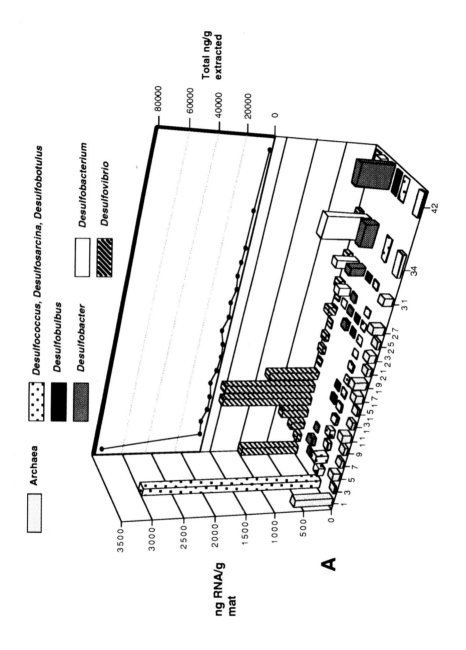

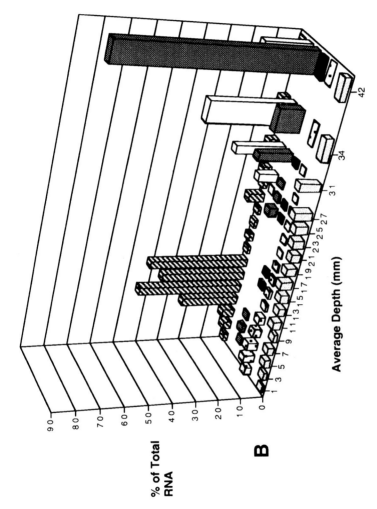

FIGURE 11.11 Relative and absolute abundance of archaeal and different SRB target group rRNA in nucleic acid extracted from mat samples obtained from different depths. (**A**) Measure of absolute population size (nanograms of target organism rRNA per gram of mat). (**Inset**) Total nanograms of 16S rRNA recovered per gram of mat as measured by the universal probe. (**B**) Fractional representation of each target group (nanograms of rRNA detected by group-specific probes presented as a percentage of nanograms of RNA) detected by the universal probe (Stahl et al., 1988). Indicated depths are the midpoint of the depth range for a given mat section. Values from replicate samples have been averaged.

zone in the upper 3–4 mm of the mat (Canfield and DesMarais, 1991). This finding seems to contradict the conventional view that bacterial sulfate reduction is an obligatory anaerobic process.

The group-specific probes for SRB described in the previous section were used to explore the phylogenetic dimension of microbial community structure and niche in more detail (Risatti et al., 1994). The results of this study are depicted in Figure 11.11. The abundance of each target group is displayed in two ways: (1) as the amount of each target group rRNA per gram of mat material within a depth interval (Figure 11.11A); and (2) as the fraction of total rRNA recovered from each interval (Figure 11.11B) (Stahl et al., 1988). Notably, most of the major known groups of SRB were observed in this single community. This observation is reminiscent of the comparable diversity of phylogenetic groups observed in the biofilm systems.

The mat presents a conspicuous spatial dimension to trophic structure. It is a laminated structure in which photosynthetic and aerobic organisms are restricted to the upper regions. Anaerobes are generally thought to be restricted

FIGURE 11.12   SEM of a partly digested Whatman No. 1 filter paper fiber from 24-hour enrichment cultures inoculated with rumen contents from a steer (this page) and a water buffalo (p. 323). Bar = 5 μm. (Courtesy of K.-J. Cheng, Agriculture & Agri-Food Canada Research Station, Lethbridge, Canada) Figure continues.

to depths below the photooxic zone during the day, but some migrate to the vicinity of the surface during the night, when there is a net depletion of oxygen in the absence of photosynthesis (Figures 11.11B, 11.12). Increasing depth corresponds to increasing distance from primary photosynthetic production. The most notable result of our study was the observation that distinct populations of SRB, as defined by phylogenetic affiliation, are restricted to well defined depth intervals within the mat community. It is likely that this highly structured distribution is intimately linked to the flow of carbon and energy through the community, implying that members of each phylogenetic group perform specific, interrelated metabolic functions in the community. At this time we can offer only general inferences about the functional significance of the observed structure. For example, the recognized acetate-oxidizing SRB genera *Desulfobacter* and *Desulfobacterium* (Widdel and Pfennig, 1984) are restricted to the greater depths, consistent with their occupancy of a higher trophic level. The desulfobacter, which specialize in the nearly exclusive use of acetate for carbon and energy, dominate the deepest levels analyzed and therefore must define processes at these depths within the mat community. Although highest acetate production is expected in the suboxic zone, imme-

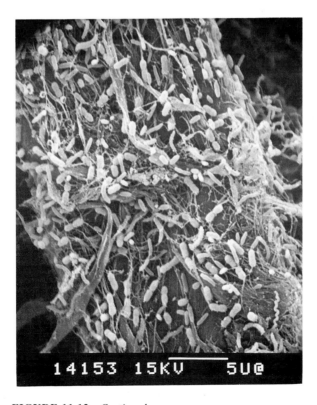

FIGURE 11.12 *Continued*

diately below the zone of photosynthesis, the molecular data suggest that the desulfobacter have a competitive advantage only in regions of anticipated lower acetate concentration deeper in the mat. This observation is consistent with our biofilm study (above). We were surprised during our biofilm studies to observe that the desulfobacter were a minor biofilm population, as they are generally assumed to be the major acetate-consuming SRB. We suspect that their distribution and inferred function in the saline mat is more representative of their community role. These results suggest that the major acetate-catabolizing SRB have yet to be identified. The dominance of the desulfovibrios below the photooxic zone is probably due to diffusion of preferred substrates (e.g., hydrogen and lactate) and nutrients from the top horizons containing the primary producing photosynthesizers and fermenting populations (Jørgensen and Cohen, 1977; Fründ and Cohen, 1992). Surprisingly, between 13 and 23 mm there is a distinct decrease in SRB. This finding is in contrast to the high sulfate-reduction rates (Canfield and DesMarais, 1993), which indicate the presence of SRB. Hence the presence of another unknown or untargeted (Figure 11.6) group of SRB is suggested that may utilize metabolites (e.g., acetate) originating from the overlying desulfovibrio.

The comparison of biofilm and mat communities suggests a clear correspondence between phylogeny and community. Both systems demonstrate a comparable complexity of phylogenetic group representation. Molecular studies of the mat population architecture revealed a remarkable correlation between spatial and phylogenetic dimensions, suggestive of tissue-like multicellular organization. Microbial mats may therefore represent one of the better examples of bacterial multicellularity.

Rumen

Microbial fermentation occurs in the gastrointestinal tracts of all mammals. In some, a portion of the gastrointestinal tract has evolved to promote significant microbial fermentation of components of the host's diet. For ruminants (e.g., cows, sheep, goats, deer), the fermentation occurs in a complex pregastric chamber, the rumen. Ruminants are distinguished from most other mammals by their ability to digest plant polysaccharides, such as cellulose, hemicellulose, and pectin. The enzymes active in digestion are not synthesized by the host but by their indigenous microbial populations. It is generally recognized that digestion of the complex array of plant cell wall polymers requires the action of a diverse assemblage of microbial populations varying in time and space.

Fermentation in the Rumen

The initial steps of microbial fermentation in the rumen are similar to those in other anaerobic habitats, except that the host animal is a major consumer of fatty acid intermediates. Ingested plant polysaccharides are converted to short-chain volatile fatty acids (VFAs) and the gases methane and carbon dioxide.

Ruminants derive most of their carbon and energy requirements from direct absorption of the VFAs across the rumen wall. Methanogens function in a capacity similar to that noted for other anaerobic habitats (above): Their consumption of hydrogen promotes substrate degradation and the formation of more oxidized organic acids. It has been suggested that hydrogen- and carbondioxide consuming acetogens could fill the hydrogen-consuming role of methanogens, as occurs in some termites (Kane et al., 1993) and in other environments previously discussed (Dolfing, 1988; Schink, 1988; Wolin and Miller, 1993). Again, however, virtually nothing is known of the interpopulation associations or nutritional factors that may serve to stabilize an acetogenic, rather than a methanogenic, community (Pochart et al., 1992). A more intimate interpopulation association with methanogens is observed for rumen ciliate protozoa, which depend on a hydrogen-evolving fermentation. Hydrogen-consuming methanogens are found to be associated with either the surface or the cytoplasm of these ciliates (Vogels et al., 1980; Finlay et al., 1994).

Within this constellation of interacting bacteria, the fiber-digesting (cellulolytic) assemblage assumes a central position at the first trophic level of the microbial food chain. The host animal and fermenting microbial groups are dependent on the cellulolytics to depolymerize plant cell wall components. The highly polymerized and cross-linked cell wall of plants presents a complex spectrum of insoluble substrates that have yet to be fully characterized (Chesson et al., 1982). This lack of information has created a special difficulty with regard to standard techniques of microbial classification. The principal substrates for fiber-digesting bacteria are relatively undefined insoluble polymers. The soluble substrate range (a standard criterion for microbial identification) does not address the niche occupied by these bacteria, nor does it provide a useful foundation for classification. As a consequence, standard microbiological techniques have failed to fully categorize all of the known fiber-digesting populations now available in pure culture (Hudman and Gregg, 1989) (see below).

We restrict any detailed discussion here to the fiber-digesting bacteria. Although few details are known, it is recognized that this assemblage works synergistically in the degradation of plant fiber (Dinsdale et al., 1978), and specific interpopulation interactions are anticipated. For example, although *Fibrobacter* species release pentoses during the degradation of fiber, they are unable to metabolize them (Collings and Yokoyama, 1980). Rather, they are restricted to the fermentation of glucose and glucose polymers, and other ruminal populations consume the pentoses (Huang and Forsberg, 1990; Avgustin et al., 1994). Although an example of an apparently loose metabolic association from which the fibrobacter derive no apparent benefit, we believe that more specific and mutualistic associations also structure this community. The plant surface promotes a spatial dimension to community structure, as displayed by attachment of the principal fiber-digesting bacteria to this substrate. We anticipate that spatial patterns of surface-attached populations may also reflect preferred or essential interactions (Figure 11.12) (Czerkawski and Cheng, 1988). Before population interactions can be identified, though, the populations must first be defined. Our principal objective here is to indicate progress toward resolving outstanding questions of population structure, which provides the background for a discussion of the initial studies.

Revised Perspectives of Ruminal Population Diversity

The full extent of ruminal genetic diversity is being revealed for many significant groups of ruminal bacteria: *Bacteroides* (Manarelli et al., 1991), *Prevotella* (*Bacteroides*) *ruminicola* (Avgustin et al., 1994), *Selenomonas ruminantium* (Ning et al., 1991), *Butyrivibrio fibrosolvens* (Manarelli, 1988; Hudman and Gregg, 1989), *Ruminococcus* (Ezaki et al., 1994), and *Fibrobacter* (Montgomery et al., 1988). This diversity has been determined by several genetic criteria, including comparative 16S rRNA sequencing, DNA/DNA similarity, restriction endonuclease digestion patterns, and restriction fragment length polymorphisms. It was the creation of the new genus *Fibrobacter* and delineation of two species for isolates previously characterized as *Bacteroides succinogenes* (Montgomery et al., 1988) that stimulated similar investigations of other ruminal and colonic bacteria.

Fibrobacter and *Ruminococcus* appear to be the predominant cellulolytic bacteria of the rumen. Strains of *F. succinogenes* form a dominant part of the fiber-digesting ruminal microbiota of cattle and sheep (Huang and Forsberg, 1990). In contrast, *Fibrobacter intestinalis* strains are more frequently isolated from the intestines and ceca of nonruminants (Amann et al., 1990b, 1992a; Lin et al., 1994). We emphasize that these initial observations of habitat range must be qualified with consideration of the relatively few survey data now available. Also, we remind the reader that there are, as yet, no defining phenotypic characteristics to differentiate among fibrobacter. The use of genetic criteria is currently the only robust method of identification (Stahl et al., 1988; Amann et al., 1992a), and we must acknowledge our near-complete ignorance of the roles these genetically distinct populations play in the ruminal ecosystem.

Use of Phylogenetically Nested Probes to Identify Novel Fibrobacter Populations

Even greater diversity among fibrobacter has been suggested by molecular analyses. Recognition of additional novel lineages resulted from the systematic application of a nested set of rRNA-targeted hybridization probes to survey the natural distribution and abundance of members of this genus. In samples of bovine ruminal fluid, total *Fibrobacter* rRNA approximately equaled the sum of *F. intestinalis* and *F. succinogenes* rRNA. In contrast, although the sum of the subspecies probes accounted for a major fraction of *F. succinogenes*, the sum was less than the amount determined using the species-specific probe, suggesting greater diversity within the species than is currently recognized in pure culture. Even greater inconsistencies were apparent in the goat rumen and equine cecum. In the goat *F. intestinalis* was undetectable, and the *F. succino-*

genes species-specific probe could account for only half of the total *Fibrobacter* quantified. In the equine cecum combined use of the six *Fibrobacter* probes revealed that *F. succinogenes* was the only quantitatively important *Fibrobacter* species present. However, the virtual absence of subspecies target groups suggested the presence of one or more novel *Fibrobacter*. This possibility was confirmed using the genus-specific oligonucleotide probe as a primer in PCR analysis for selective amplification of *Fibrobacter* 16S rDNA sequences from the equine cecum (Lin and Stahl, 1995). Sequencing the cloned amplification products revealed two unique *Fibrobacter*-like sequences. Although phylogenetically affiliated with *F. succinogenes*, they are not closely related to any described subspecies. These studies again emphasize our limited appreciation of environmental diversity, even among intensively studied groups.

Response of Ruminal Populations to System Perturbation

The first molecular insights into the community dynamics of the rumen were provided by studies in our laboratory (Stahl et al., 1988). Species-specific and group-specific oligonucleotide probes were developed to quantify 16S rRNA of *F. succinogenes* and *Lachnospira multiparus* in the bovine rumen before, during, and after perturbation of the ecosystem with the ionophore monensin. Monensin is a feed additive used to improve ruminant feed conversion. The in vivo changes associated with addition of this and other ionophores to the diet include decreased acetate and increased propionate production, increased amino acid production, and decreased methane production. These changes can in part be attributed to a semiselective inhibition of Gram-positive populations, but the relative contribution of altered fermentation by the resident community versus an alteration in community structure (and associated fermentation patterns) remains unresolved (Wallace, 1992). Our initial study was not designed to resolve these outstanding questions but, rather, to invoke a community response that could be used to evaluate the utility of molecular probes to study community dynamics and highlight possible population associations.

Consistent with expectation, relative numbers of the Gram-positive *L. multiparus* were depressed about twofold by monensin addition (Stahl et al., 1988), and a transient 5- to 10-fold increase in numbers was observed following antibiotic removal from the diet. However, the Gram-negative fibrobacter populations demonstrated a much more complex response. Three successions of different fibrobacter populations were observed over a period of several weeks after addition of the antibiotic to the diet. Long-term studies of animals on antibiotic-free diets have shown these same three populations to coexist stably (D.A. Stahl, unpublished data), with *F. succinogenes* subspecies 3 generally predominating in bovine ruminants. It is apparent that the normal population balance was transiently disrupted by the antibiotic, allowing minor populations to transiently replace the dominant population. These results suggest that both cooperative and competitive interactions influence ruminal community structure, but we have virtually no understanding of the nature of these interactions.

Stable coexistence suggests that the various *Fibrobacter* populations serve distinct community roles (niches) and therefore generally do not compete with each other directly. Distinct roles might be defined by different substrate preference or by providing similar supportive roles to different groups of interacting bacteria. The latter possibility is of particular interest with respect to specific interpopulation interactions. If such interactions have a physical dimension (e.g., physical association on the surface of plant fiber), it might be observed using the in situ hybridization techniques previously described.

The Road Ahead

By now it should be evident to the reader that microbiology is a discipline at a crossroads. The pure culture description has failed to provide a reasonable accounting of microbial populations in natural or engineered communities. For lack of tools to do so, investigators concerned with the critical questions of community structure comprise a microbiology subdiscipline (''microbial ecology'') that suffers the comparison to studies of a single population within the confines of a test tube. Yet we argue that this subdiscipline is microbiology. Microorganisms evolved in the context of communities, and so the many facets of regulatory and physiological variation can find meaning only in the context of evolution and community. We remind the reader, however, that no natural microbial community has been fully resolved at the level of contributing populations. With this handicap, current appreciation of interpopulation associations (multicellularity) can only be feeble. Our intention was to provide perspective, balancing the past with a glimpse of future studies now possible using the expanded suite of molecular techniques.

It is essential that characteristics identified and interpreted by pure culture study now be reevaluated within the framework of community. In this regard there is much that we have not been able to address. For example, it is well recognized that accessory DNA elements (e.g., plasmids and transposons) and DNA exchange contribute to the survival of individual populations. However, if we accept the premise that virtually all microbial populations evolved in the context of community and that communities are a unit of selection, so-called accessory DNA should also be treated as a feature of community. The contribution of these genetic elements to community structure (e.g., homeostasis) is unexplored. For example, community response to changing environment may be derived in substantial part from the availability of an associated pool of accessory DNA. If so, it would be highly informative to track the movement of accessory DNA at the level of single cells in multipopulation systems. The development of general methods to identify individual genes and their expression at the level of single cells would open entirely new worlds to observation. Ultimately, it is be at this level of observation that the importance of intracellular communication of multicellular behavior will be realized.

Acknowledgments

N.K.F. is the recipient of DOE grant D.A.S.–ONR, NSF, Lyonnaise des Eaux, & USDA grants; LR NSF grant number BES94–10476. L.R. is the recipient of grant number BES94–10476. D.A.S. received NSF DEB 9408243, ONR N00014–94–1–1171, USDA: 93–37206–9642.

References

Agard, D., Yasushi, H., Shaw, P., and Sedat, J. (1989) Fluorescence microscopy in three dimensions. *Methods Cell Biol.* **30**:353–377.

Ahring, B.K., Schmidt, J.E., Winther-Nielsen, M., Macario, A.J.L., and Conway de Macario, E. (1993) Effect of medium composition and sludge removal on the production, composition, and architecture of thermophilic (55°C) acetate-utilizing granules from an upflow anaerobic sludge blanket reactor. *Appl. Environ. Microbiol.* **59**:2538–2545.

Alvarez-Cohen, L., McCarty, P.L., Boulygina, E., Hanson, R.S., Brusseau, G.A., and Tsien, H.C. (1992) Characterization of a methane-utilizing bacterium from a bacterial consortium that rapidly degrades trichloroethylene and chloroform. *Appl. Environ. Microbiol.* **58**:1886–1893.

Amann, R.I., Binder, B.J., Olson, R.J., Chisholm, S.W., Devereux, R., and Stahl, D.A. (1990a) Combination of 16S rRNA-targeted oligonucleotide probes with flow cytometry for analyzing mixed microbial populations. *Appl. Environ. Microbiol.* **56**: 1919–1925.

Amann, R.I., Krumholz, L., and Stahl, D.A. (1990b) Fluorescent-oligonucleotide probing of whole cells for determinative, phylogenetic, and environmental studies in microbiology. *J. Bacteriol.* **172**:762–770.

Amann, R.I., Lin, C., Key, R., Montgomery, L., and Stahl, D.A. (1992a) Diversity among *Fibrobacter* isolates: towards a phylogenetic classification. *Syst. Appl. Microbiol.* **15**:23–31.

Amann, R.I., Ludwig, W., and Schleifer, K.-H. (1995) Phylogenetic identification and in situ detection of individual microbial cells without cultivation. *Microbiol. Rev.* **59**:143–169.

Amann, R.I., Stromley, J., Devereux, R., Key, R., and Stahl, D.A. (1992b) Molecular and microscopic identification of sulfate-reducing bacteria in multispecies biofilms. *Appl. Environ. Microbiol.* **58**:614–623.

Avgustin, G., Wright, F., and Flint, H.J. (1994) Genetic diversity and phylogenetic relationships among strains of *Prevotella (Bacteroides) ruminicola* from the rumen. *Int. J. System. Bacteriol.* **44**:246–255.

Barlaz, M.A., Schaeffer, D.M., and Ham, R.K. (1989) Bacterial population development and chemical characteristics of refuse decomposition in a simulated sanitary landfill. *Appl. Environ. Microbiol.* **55**:55–65.

Benson, D.A., Bogushi, M., Lipman, D.J., and Ostell, J. (1994) GenBank. *Nucleic Acids Res.* **22**:3441–3444.

Brakenhoff, G.J., van der Voort, H.T.M., Oud, J.L., and Mans, A. (1990) Potentialities and limitations of confocal microscopy for the study of 3-dimensional biological structures. In *Optical Microscopy for Biology*, B. Herman and K. Jacobson (eds.), pp. 19–28) Wiley-Liss, New York.

Brakenhoff, G.J., van der Voort, H.T.M., van Spronsen, E.A., and Nanninga, N. (1989) Three-dimensional imaging in fluorescence by confocal scanning microscopy. *J. Microsc.* **153**:151–159.

Brenner, D.J. (1986) Classification of Legionellaceae: current status and remaining questions. *Isr. J. Med. Sci.* **22**:620–632.

Bryant, M.P., Campbell, L.L., Reddy, C.A., and Crabill, M.R. (1977) Growth of *Desulfovibrio* in lactate or ethanol media low in sulfate in association with H_2 utilizing methanogenic bacteria. *Appl. Environ. Microbiol.* **33**:1162–1169.

Bull, A.T., Goodfellow, M., and Slater, J.H. (1992) Biodiversity as a source of innovation in biotechnology. *Annu. Rev. Microbiol.* **46**:219–252.

Burlage, R.S., and Kuo, C.-T. (1994) Living biosensors for the management and manipulation of microbial consortia. *Annu. Rev. Microbiol.* **48**:291–309.

Caldwell, D.E., Korber, D.R., and Lawrence, J.R. (1992) Confocal laser microscopy and digital image analysis in microbial ecology. *Adv. Microb. Ecol.* **12**:1–67.

Canfield, D.E., and DesMarias, D.J. (1991) Aerobic sulfate reduction in microbial mats. *Science* **251**:1471–1473.

Canfield, D.E., and DesMarais, D.J. (1993) Biogeochemical cycles of carbon, sulfur, and free oxygen in a microbial mat. *Geochim. Cosmochim. Acta* **57**:3971–3984.

Chesson, A., Stewart, C.S., and Wallace, R.J. (1982) Influence of plant phenolic acids on growth and cellulolytic activity of rumen bacteria. *Appl. Environ. Microbiol.* **44**:597–603.

Cohen, Y. (1989) In *Microbial Mats: Physiological Ecology of Benthic Microbial Communities*, Y. Cohen and E. Rosenberg (eds.), pp. xv–xvii. American Society for Microbiology, Washington, DC.

Collings, G.F., and Yokoyama, M.T. (1980) Gas-liquid chromotography for evaluating polysaccharide degradation by *Ruminococcus flavefaciens* C94 and *Bacteroides succinogenes* S85. *Appl. Environ. Microbiol.* **39**:566–571.

Conway de Macario, E., Macario, A.J.L., Mok, T., and Beveridge, T.J. (1993) Immunochemistry and localization of the enzyme disaggregatase in *Methanosarcina mazei. J. Bacteriol.* **175**:3115–3120.

Costerton, J.W., Cheng, K.-J., Geesey, G.G., et al. (1987) Bacterial biofilms in nature and disease. *Annu. Rev. Microbiol.* **41**:435–464.

Costerton, J.W., Lewandowski, Z., DeBeer, D., Caldwell, D., Korber, D., and James, G. (1994) Biofilms, the customized microniche. *J. Bacteriol.* **176**:2137–2142.

Currin, C.A., Pearl, H.W., Suba, G.K., and Alberte, R.S. (1990) Immunofluorescence detection and characterization of N_2-fixing microorganisms from aquatic environments. *Limnol. Oceanogr.* **35**:59–71.

Czerkawski, J.W., and Cheng, K.-J. (1988) The rumen microbial ecosystem. In *Compartmentation in the Rumen*, P.N. Hobson (ed.), pp. 361–385. Elsevier, New York.

D'Amelio, E.D., Cohen, Y., and DesMarais, D.J. (1989) Comparative functional ultrastructure of two hypersaline submerged cyanobacterial mats: Guerrero Negro, Baja California Sur, Mexico, and Solar Lake, Sinai, Egypt. In *Microbial Mats: Physiological Ecology of Benthic Microbial Communities*, Y. Cohen and E. Rosenberg (eds.), pp. 97–113. American Society for Microbiology, Washington, DC.

Davies, D.G., Chakrabarty, A.M., and Geesey, G.G. (1993) Exopolysaccharide production in biofilms: substratum activation of alginate gene expression by *Pseudomonas aeruginosa. Appl. Environ. Microbiol.* **59**:1181–1186.

Devereux, R., Delaney, M., Widdel, F., and Stahl, D.A. (1989) Natural relationships among sulfate-reducing eubacteria. *J. Bacteriol.* **171**:6689–6695.

Devereux, R., Kane, M.D., Winfrey, J., and Stahl, D.A. (1992) Genus- and group-specific hybridization probes for determinative and environmental studies of sulfate-reducing bacteria. *Sys. Appl. Microbiol.* **15**:601–609.

Dinsdale, D., Morris, E.J., and Bacon, J.S.D. (1978) Electron microscopy of the microbial populations present and their modes of attack on various cellulosic substrates undergoing digestion in the sheep rumen. *Appl. Environ. Microbiol.* **36**: 160–168.

Dolfing, J. (1988) Acetogenesis. In *Biology of Anaerobic Microorganisms*, A.J.B. Zehnder (ed.), pp. 417–468. Wiley, New York.

Dortch, Q., Roberts, T.L., Clayton, J.R., and Ahmed, S.I. (1983) RNA/DNA ratios and DNA concentrations as indicators of growth rate and biomass in planktonic marine organisms. *Mar. Ecol. Prog. Ser.* **13**:61–71.

Emmert, D.B., Stoehr, P.J., Stoesser, G., and Cameron, G.N. (1994) The European Bioinformatics Institute (EBI) databases. *Nucleic Acids Res.* **22**:3445–3449.

Ezaki, T., Li, N., Hashimoto, Y., Muira, H., and Yamamoto, H. (1994) 16S ribosomal DNA sequences of anaerobic cocci and proposal of *Ruminococcus hansenii* comb. nov. and *Ruminococcus productus* comb. nov. *Int. J. Syst. Microbiol.* **44**:130–136.

Finlay, B.J., Esteban, G., Clark, K.J., Williams, A.G., Embley, T.M., and Hirt, R.P. (1994) Some rumen ciliates have endosymbiotic methanogens. *FEMS Microbiol. Lett.* **117**:157–162.

Fründ, C., and Cohen, Y. (1992) Diurnal cycles of sulfate reduction under oxic conditions in cyanobacterial mats. *Appl. Environ. Microbiol.* **58**:70–77.

Fuhrman, J.A., McCallum, K., and Davis, A.A. (1992) Novel major archaebacterial group from marine plankton. *Nature* **356**:148–149.

Gibson, G.R., MacFarlane, G.T., and Cummings, J.H. (1988) Occurrence of sulphate-reducing bacteria in human faeces and the relationship of dissimilatory sulphate reduction to methanogenesis in the large gut. *J. Appl. Bacteriol.* **65**:103–111.

Giovannoni, S.J., Britschgi, T.B., Moyer, C.L., and Field, K.G. (1990) Genetic diversity in Sargasso Sea bacterioplankton. *Nature* **345**:60–63.

Gutell, R.R. (1992) Evolutionary characteristics of 16S and 23S rRNA structures. In *Proceedings of the Conference on the Origin and Evolution of Prokaryotic and Eukaryotic Cells*, H. Hartman and K. Matsuno (eds.), pp. 243–309. World Scientific, River Edge, New Jersey.

Hahn, D., Amann, R.I., and Zeyer, J. (1993) Detection of mRNA in *Streptomyces* cells by whole-cell hybridization with digoxigenin-labeled probes. *Appl. Environ. Microbiol.* **59**:2753–2757.

Head, I.M., Hiorns, W.D., Embley, T.M., McCarthy, A.J., and Saunders, J.R. (1993) The phylogeny of autotrophic ammonia-oxidizing bacteria as determined by analysis of 16S ribosomal RNA gene sequences. *J. Gen. Microbiol.* **139**:1147–1153.

Heitzer, A., Malachowsky, K., Thonnard, J.E., Bienkowski, P.R., White, D.C., and Sayler, G.S. (1994) Optical biosensor for environmental on-line monitoring of naphthalene and salicylate bioavailability with an immobilized bioluminescent catabolic reporter bacterium. *Appl. Environ. Microbiol.* **60**:1487–1494.

Höfle, M.G. (1992) Bacterioplankton community structure and dynamics after large-scale release of nonindigenous bacteria as revealed by low-molecular weight RNA analysis. *Appl. Environ. Microbiol.* **58**:3387–3394.

Huang, L., and Forsberg, C.W. (1990) Cellulose digestion and cellulase regulation and distribution in *Fibrobacter succinogenes* subsp. *succinogenes* S85. *Appl. Environ. Microbiol.* **56**:1221–1228.

Hudman, J.F. and Gregg, K. (1989) Genetic diversity among strains of bacteria from the rumen. *Curr. Microbiol.* **19**:313–318.

Isa, Z., Grusenmeyer, S., and Verstraete, W. (1986) Sulfate reduction relative to methane production in high-rate anaerobic digestion: technical aspects. *Appl. Environ. Microbiol.* **51**:572–579.

Jørgensen, B.B., and Cohen, Y. (1977) Solar Lake (Sinai). 5. The sulfur cycle of the benthic cyanobacterial mats. *Limnol. Oceanogr.* **22**:657–666.

Kane, M.D., Poulsen, L.K., and Stahl, D.A. (1993) Monitoring the enrichment and isolation of sulfate-reducing bacteria by using oligonucleotide hybridization probes designed from environmentally derived 16S rRNA sequences. *Appl. Environ. Microbiol.* **59**:682–686.

Kemp, P.F., Lee, S., and LaRoche, J. (1993) Estimating the growth rate of slowly growing marine bacteria from RNA content. *Appl. Environ. Microbiol.* **59**:2594–2601.

Kerkhof, L., and Ward, B.B. (1993) Comparison of nucleic acid hybridization and fluorometry for measurement of the relationship between RNA/DNA ratio and growth rate in a marine bacterium. *Appl. Environ. Microbiol.* **59**:1303–1309.

Lane, D.J., Harrison Jr., A.P., Stahl, D.B.P., Giovannoni, S.J., Olsen, G.J., and Pace, N.R. (1992) Evolutionary relationships among sulfur- and iron-oxidizing eubacteria. *J. Bacteriol.* **174**:269–278.

Lawrence, J.R., Korber, D.R., Hoyle, B.D., Costerton, J.W., and Caldwell, D.E. (1991) Optical sectioning of microbial biofilms. *J. Bacteriol.* **173**:6558–6567.

Liesack, W.H., Weyland, H., and Stackebrandt, E. (1991) Potential risks of gene amplification by PCR as determined by 16S rDNA analysis of a mixed culture of strict barophilic bacteria. *Microb. Ecol.* **21**:191–198.

Lin, C., and Stahl, D.A. (1995) Taxon-specific probes for the cellulolytic genus *Fibrobacter* reveal abundant and novel equine-associated populations. *Appl. Environ. Microbiol.* **61**:1348–1351.

Lin, C., Flesher, B., Capman, W.C., Amann, R.I., and Stahl, D.A. (1994) Taxon specific hybridization probes for fiber-digesting bacteria suggest novel gut-associated *Fibrobacter*. *Sys. Appl. Microbiol.* **17**:418–424.

Macario, A.J.L., and Conway de Macario, E. (1993) Antibody probes and immunotechnology for direct identification of microbes in complex ecosystems. In *Trends in Microbial Ecology: Proceedings of the Sixth International Symposium on Microbial Ecology*. Spanish Society for Microbiology, Barcelona, Spain.

Madan, A.P., and Nierzwicki-Bauer, S.A. (1993) In situ detection of transcripts for ribulose-1,5-bisphosphate carboxylase in cyanobacterial heterocysts. *J. Bacteriol.* **175**:7301–7306.

Maidak, B.L., Larsen, N., McCaughey, M.J., et al. (1994) The Ribosomal Database project. *Nucleic Acids Res.* **22**:3485–3487.

Manarelli, B.M. (1988) Deoxyribonucleic acid relatedness among strains of the species *Butyrivibrio fibrosolvens*. *Int. J. Sys. Bacteriol.* **38**:340–347.

Manarelli, B.M., Ericsson, L.D., Lee, D., and Stack, R.J. (1991) Genetic diversity and phylogenetic relationships among strains of *Prevotella (Bacteroides) ruminicola* from the rumen. *Int. J. Sys. Bacteriol.* **44**:246–255.

Menn, F.-M., Applegate, B.M., and Sayler, G.S. (1993) NAH plasmid-mediated catabolism of anthracene and phenanthrene to naphthoic acids. *Appl. Environ. Microbiol.* **59**:1938–1942.

Montgomery, L., Flesher, B., and Stahl, D.A. (1988) Transfer of *Bacteroides succinogenes* (Hungate) to *Fibrobacter* gen. nov. as *Fibrobacter succinogenes* comb. nov.

and description of *Fibrobacter intestinalis* sp. nov. *Int. J. Syst. Bacteriol.* **38**: 430–435.

Morotomi, M., Ohno, T., and Mutai, M. (1988) Rapid and correct identification of intestinal *Bacteroides* spp. with chromosomal DNA probes by whole-cell dot blot hybridization. *Appl. Environ. Microbiol.* **54**:1158–1162.

Moss, V.A., Jenkinson, D., McEwan, E., and Elder, H.Y. (1990) Automated 3-D reconstruction. *J. Microscopy* **158**:187–200.

Muyzer, G., and de Waal, E.C. (1994) Determination of the genetic diversity of microbial communities using DGGE analysis of PCR-amplified 16S rDNA. In *Microbial Mats: Structure, Development and Environmental Significance*, L.J. Stal and P. Caumette (eds.), pp. 207–214. Springer-Verlag, Berlin.

Muyzer, G., de Waal, E.C., and Uitterlinden, A.G. (1993) Profiling of complex microbial populations by denaturing gradient gel electrophoresis analysis of polymerase chain reaction-amplified genes coding for 16S rRNA. *Appl. Environ. Microbiol.* **59**:695–700.

Neefs, J.-M., Van de Peer, Y., De Rijk, P., Chapelle, S., and De Wachter, R. (1993) Compilation of small ribosomal subunit RNA structures. *Nucleic Acids Res.* **21**: 3025–3049.

Nielsen, P.H. (1987) Biofilm dynamics and kinetics during high-rate sulfate reduction under anaerobic conditions. *Appl. Environ. Microbiol.* **53**:27–32.

Ning, Z., Attwood, G.T., Lockington, R.A., and Brooker, J.D. (1991) Genetic diversity in ruminal isolates of *Selenomonas ruminantium*. *Curr. Microbiol.* **22**:279–284.

Olsen, G.J. (1988) Phylogenetic analysis using ribosomal RNA. *Methods Enzymol.* **164**: 793–812.

Pace, N.R., Stahl, D.A., Lane, D.J., and Olsen, G.J. (1986) The analysis of natural microbial populations by ribosomal RNA sequences. In *Advances in Microbial Ecology*, K.C. Marshall (ed.), pp. 1–55. Plenum Press, New York.

Paerl, H.W., Priscu, J.C., and Brawner, D.L. (1989) Immunochemical localization of nitrogenase in marine *Trichodesmium* aggregates: relationship to N_2 fixation potential. *Appl. Environ. Microbiol.* **55**:2965–2975.

Parkin, G.F., Lynch, N.A., Kuo, W.-C., Van Keuren, E.L., and Bhattacharya, S.K. (1990) Interaction between sulfate reducers and methanogens fed acetate and propionate. *Res. J. Water Pollution Control Fed.* **62**:780–788.

Pichard, S.L., and Paul, J.H. (1991) Detection of gene expression in genetically engineered microorganisms and natural phytoplankton populations in the marine environment by mRNA analysis. *Appl. Environ. Microbiol.* **57**:1721–1727.

Plovins, A., Alvarez, A.M., Ibanez, M., Molina, M., and Nombela, C. (1994) Use of fluorescein-di-β-D-galactosidase (FDG) and C_{12}-FDG as substrates for β-galactosidase detection by flow cytometry in animal, bacterial, and yeast cells. *Appl. Environ. Microbiol.* **60**:4638–4641.

Pochart, P., Doré, J., Lémann, F., Goderel, I., and Rambaud, J.-C. (1992) Interrelations between populations of methanogenic Archaea and sulfate-reducing bacteria in the human colon. *FEMS Microbiol. Lett.* **98**:225–228.

Poulsen, L.K., Ballard, G., and Stahl, D.A. (1993) Use of rRNA fluorescence in situ hybridization for measuring the activity of single cells in young and established biofilms. *Appl. Environ. Microbiol.* **59**:1354–1360.

Raskin, L. (1993) *Structural and Functional Analysis of Anaerobic Biofilm Communities: An Integrated Molecular and Modeling Approach.* Ph.D. thesis, University of Illinois at Urbana-Champaign.

Raskin, L., Amann, R.I., Poulsen, L.K., Rittmann, B.E., and Stahl, D.A. (1995b) Use of ribosomal RNA based molecular probes for characterization of complex microbial communities in anaerobic biofilms. *Water Sci. Technol.* **31**:261–272.

Raskin, L., Capman, W.C., Sharp, R., and Stahl, D.A. (1995a) Molecular ecology of gastrointestinal ecosystems. In *Gastrointestinal Microbiology and Host Interactions*, R.I. Mackie, B.A. White, and R.E. Isaacson (eds.). Chapman & Hall, London (in press).

Raskin, L., Poulsen, L.K., Noguera, D.R., Rittmann, B.E., and Stahl, D.A. (1994a) Quantification of methanogenic groups in anaerobic biological reactors by oligonucleotide probe hybridization. *Appl. Environ. Microbiol.* **60**:1241–1248.

Raskin, L., Stromley, J.M., Rittmann, B.E., and Stahl, D.A. (1994b) Group-specific 16S rRNA hybridization probes to describe natural communities of methanogens. *Appl. Environ. Microbiol.* **60**:1232–1240.

Reysenbach, A.-L., Giver, L.J., Wickham, G.S., and Pace, N.R. (1992) Differential amplification of rRNA genes by polymerase chain reaction. *Appl. Environ. Microbiol.* **58**:3417–3418.

Rice, C.M., Fuchs, R., Higgins, D.G., Stoehr, P.J., and Cameron, G.N. (1993) The EMBL data library. *Nucleic Acids Res.* **21**:2967–2971.

Richardson, M. (1990) Confocal microscopy and 3-D visualization. American Laboratory **22**:19–24.

Risatti, J.B., Capman, W.C., and Stahl, D.A. (1994) Community structure of a microbial mat: the phylogenetic dimension. *Proc. Natl. Acad. Sci. USA* **91**:10173–10177.

Rouvière, P., Mandelco, L., Winker, S., and Woese, C.R. (1992) A detailed phylogeny for the Methanobacteriales. *Sys. Appl. Microbiol.* **15**:363–371.

Ruff-Roberts, A.L., Kuenen, J.G., and Ward, D.M. (1994) Distribution of cultivated and uncultivated cyanobacteria and *Chloroflexus*-like bacteria in hot spring microbial mats. *Appl. Environ. Microbiol.* **60**:697–704.

Schaecter, M, Maaløe, O., and Kjeldgaard, N.O. (1958) Dependency on medium temperature of cell size and chemical composition during balanced growth of *Salmonella typhimurium*. *J. Gen. Microbiol.* **19**:592–606.

Schink, B. (1988) Principles and limits of anaerobic degradation environmental and technological aspects. In *Biology of Anaerobic Microorganisms*, A.J.B. Zehnder (ed.), pp. 771–846. Wiley, New York.

Schmidt, E.L. (1973) Fluorescent antibody techniques for the study of microbial ecology. *Bull. Ecol. Res. Commun. NFR* **17**:67–76.

Schmidt, T.M., DeLong, E.F., and Pace, N.R. (1991) Analysis of a marine picoplankton community by 16S rRNA gene cloning and sequencing. *J. Bacteriol.* **173**:4371–4378.

Skyring, G.W., Chambers, L.A., and Bauld, J. (1983) Sulfate reduction in sediments colonized by cyanobacteria, Spencer Gulf, South Australia. *Aust. J. Mar. Freshwater Res.* **34**:359–374.

Stahl, D.A. (1993) The natural history of microorganisms. *ASM News* pp. 609–613.

Stahl, D.A., and Amann, R. (1991) Development and application of nucleic acid probes. In *Nucleic Acid Techniques in Bacterial Systematics*, E. Stackebrandt and M. Goodfellow (eds.), pp. 205–248 Wiley, Chichester.

Stahl, D.A., Flesher, B., Mansfield, H., and Montgomery, L. (1988) Use of phylogenetically based hybridization probes for studies of ruminal microbial ecology. *Appl. Environ. Microbiol.* **54**:1079–1084.

Stahl, D.A., Lane, D.J., Olsen, G.J., and Pace, N.R. (1985) Characterization of a Yellowstone hot spring microbial community by 5S rRNA sequences. *Appl. Environ. Microbiol.* **49**:1379–1384.

Tasaki, M., Kamagata, Y., Nakamura, K., Okamura, K., and Minami, K. (1993) Acetogenesis from pyruvate by *Desulfotomaculum thermobenzoicum* and differences in pyruvate metabolism among three sulfate-reducing bacteria in the absence of sulfate. *FEMS Microbiol. Lett.* **106**:259–264.

Teske, A., Alm, E., Regan, J.M., Toze, S., Rittmann, B.E., and Stahl, D.A. (1994) Evolutionary relationships among ammonia- and nitrite-oxidizing bacteria. *J. Bacteriol.* **176**:6623–6630.

Traore, A.S., Fardeau, M.-L., Hatchikian, C.E., Le Gall, J., and Belaich, J.-P. (1983) Energetics of growth of a defined mixed culture of *Desulfovibrio vulgaris* and *Methanosarcina barkeri*: interspecies hydrogen transfer in batch and continuous cultures. *Appl. Environ. Microbiol.* **46**:1152–1156.

Visser, F.A., Van Lier, J.B., Macario, A.J.L., and Conway De Macario, E. (1991) Diversity and population dynamics of methanogenic bacteria in a granular consortium. *Appl. Environ. Microbiol.* **57**:1728–1734.

Vogels, G.D., Hoppe, W.F., and Stumm, C.K. (1980) Association of methanogenic bacteria with rumen ciliates. *Appl. Environ. Microbiol.* **40**:608–612.

Wallace, R.J. (1992) Rumen microbiology, biotechnology and ruminant nutrition: the application of research findings to a complex microbial ecosystem. *FEMS Microbiol. Lett.* **100**:529–534.

Ward, D.M., and Winfrey, M.R. (1985) Interactions between methanogenic and sulfate-reducing bacteria in sediments. In *Advances in Aquatic Microbiology*, H.W. Jannasch and P.J.L. Williams (eds.), pp. 141–179. Academic Press, Orlando, FL.

Ward, D.M., Bateson, M.M., Weller, R., and Ruff-Roberts, A.L. (1992) Ribosomal RNA analysis of microorganisms as they occur in nature. In *Advances in Microbial Ecology*, K.C. Marshall (ed.), pp. 219–286. Plenum Press, New York.

Ward, D.M., Weller, R., and Bateson, M.M. (1990) 16S rRNA sequences reveal numerous uncultured microorganisms in a natural community. *Nature* **345**:63–65.

Ward, D.M., Weller, R., Shiea, J., Castenholz, R.W., and Cohen, Y. (1989) Hot spring microbial mats: anoxygenic and oxygenic mats of possible evolutionary significance. In *Microbial Mats: Physiological Ecology of Benthic Microbial Communities*, Y. Cohen and E. Rosenberg (eds.), pp. 3–15. American Society for Microbiology, Washington, DC.

Wayne, L.G., Brenner, D.J., Colwell, R.R., et al. (1987) Report of the ad hoc committee on reconciliation of approaches to bacterial systematics. *Int. J. Sys. Bacteriol.* **37**: 463–464.

Weller, R., Weller, J.W., and Ward, D.M. (1991) 16S rRNA sequences of uncultivated hot spring cyanobacterial mat inhabitants retrieved as randomly primed cDNA. *Appl. Environ. Microbiol.* **57**:1146–1151.

Wells, K.S., Sandison, D.R., Strickler, J., and Webb, W.W. (1989) Quantitative fluorescence imaging with laser scanning confocal microscopy. In *The Handbook of Biological Confocal Microscopy*, J. Pawley (ed.), IMR Press, Madison, WI.

Widdel, F. (1988) Microbiology and ecology of sulfate- and sulfur-reducing bacteria. In *Biology of Anaerobic Microorganisms*, A.J.B. Zehnder (ed.), pp. 469–585. Wiley, New York.

Widdel, F., and Bak, F. (1992) Gram-negative mesophilic sulfate-reducing bacteria. In *The Prokaryotes*, A. Balows, H.G. Tuper, M. Dworkin, W. Harder, and K.H. Schleifer (eds.), pp. 3352–3378. Springer-Verlag, Berlin.

Widdel, F., and Pfennig, N. (1984) Dissimilatory sulfate- or sulfate-reducing bacteria. In *Bergey's Manual of Systematic Bacteriology*, N.R. Kreig and J.G. Holt (eds.), pp. 663–679. Williams & Wilkins, Baltimore.

Wilson, T. (1989) Three-dimensional imaging in confocal systems. *J. Microsc.* **153**: 161–169.

Woese, C.R. (1987) Bacterial evolution. *Microbiol. Rev.* **51**:221–271.

Woese, C.R., Kandler, O., and Wheelis, M.L. (1990) Towards a natural system of organisms: proposal for the domains Archaea, Bacteria and Eucarya. *Proc. Natl. Acad. Sci. USA* **87**:4576–4579.

Wolfaardt, G.M., Lawrence, J.R., Robarts, R.D., Calwell, S.J., and Caldwell, D.E. (1994) Multicellular organization in a degradative biofilm community. *Appl. Environ. Microbiol.* **60**:434–446.

Wolin, M.J., and Miller, T.L. (1993) Bacterial strains from human feces that reduce CO_2 to acetic acid. *Appl. Environ. Microbiol.* **59**:3551–3556.

Wright, S.J., Schatten, H., Simerly, C., and Schatten, G. (1989) Three-dimensional fluorescence imaging with the tandem scanning confocal microscope. In *Optical Microscopy for Biology*, B. Herman and K. Jacobson (eds.), pp. 29–44. Wiley-Liss, New York.

Yoda, R., Macario, A.J.L., and Conway de Macario, E. (1992) Immunochemical differences among *Methanosarcina mazei* S-6 morphologic forms. *J. Bacteriol.* **174**: 4683–4688.

Zuckerkandl, E., and Pauling, L. (1965) Molecules as documents of evolutionary history. *J. Theor. Biol.* **8**:357–366.

V.

Physical View of Bacterial Multicellularity

12

Physical and Genetic Consequences of Multicellularity in *Bacillus subtilis*

NEIL H. MENDELSON, BACHIRA SALHI, &
CHEN LI

If indeed life began as simple replicative processes that progressed to cellular entities with more complex phenotypes, achieving multicellularity must have been a key evolutionary step leading to the diversity now extant (Kaufmann, 1993). Multicellularity provides a means to scale up structures, produce more elaborate ones, make new materials, and generate greater forces than can single cells. Multicellularity fosters cell diversification, yet it implies cooperativity and the shifting of selection from operating at the level of individual cells to that of the multicellular entity. Although multicellularity is best understood in eucaryotic systems, it is also a significant way of life for procaryotes.

This chapter focuses on two aspects of multicellularity in *Bacillus subtilis*: one concerned with the self-assembly of a highly organized multicellular form known as a macrofiber (Mendelson, 1978) and the other with elaborate gene expression patterns found in conventional colonies (Salhi and Mendelson, 1993; Mendelson and Salhi, 1996). Both systems illustrate that multicellular organization causes constraints, and constraints give rise to responses that result in complexity. The outcome of bacterial multicellularity is therefore, at least in these two systems, a higher level of organization and function than can be achieved in individual cells.

Bacterial Macrofibers

Multicellular Filaments

The tendency for wild-type *Bacillus subtilis* to grow as multiseptate filaments rather than individual cells was noted long ago and was shown to be correlated with growth rate and multifork DNA replication (Paulton, 1970, 1971). At rapid growth rates cells can maintain a balance between DNA content and cell mass, but the process of cell separation cannot keep pace; consequently, filaments arise. Wild-type strains grown at low density in rich media generally behave in this manner. The multiseptate filaments in such cultures eventually break down into shorter chains and individual cells as the cell density becomes high, growth rate slows, and sufficient autolytic enzyme becomes available to cleave septa (Doyle and Koch, 1987). Mutants deficient in cell wall-hydrolyzing enzyme activity (*lyt*), in contrast, grow into long chains of cells that are stable for extended periods even when cell density becomes high (Fein and Rogers, 1976). They provide an ideal way to generate and maintain multicellular filaments. Because filament length integrity is a requisite for macrofiber stability, we have used *lyt* mutants to prolong macrofiber longevity. It is important to note, however, that the *lyt* mutation itself does not cause macrofiber production, nor does it govern macrofiber phenotype other than macrofiber longevity.

Macrofiber Structure

Bacterial macrofibers are considerably more complex structurally than their subunit cellular filaments. They consist of bundles of helically twisted filaments packed parallel to one another and twisted together into a fiber. The shaft of a macrofiber terminates in loops usually at both ends. The multicellular organization in a fiber shaft and terminal loop are illustrated in Figure 12.1. Panels 1 through 6 are through-focal planes starting on the upper surface of the fiber and ending on the lower surface. The shaft contains eight cellular filaments organized in a right-handed helix of moderate twist. The full cylindrical volume of the shaft is occupied by cells. Cell septa can be seen at various locations, giving some indication of the relative length of individual cells vis-a-vis their twist and curvature. Cells are connected to one another by septa within filaments but not in other ways. Filaments are held together only by their entwinement. They are free to slide and rearrange themselves in response to forces generated by growth (Mendelson, 1982). In the loop one can see that two filament pairs, each consisting of a double-helical structure, are themselves twisted together into a four-stranded bundle. The integrity of the four-stranded bundle can be seen where it, in turn, twists together with itself to form the fiber shaft. Clearly, macrofiber structure is based on hierarchies of helices, and the building plan must involve a folding and plying process (Mendelson, 1992).

Heritability of Twist States

If one starts with a single spore or vegetative cell and allows it to grow into a macrofiber, the mature fiber has a phenotype consisting of its helix hand and degree of twist (expressed as turns per millimeter). The single mature fiber can then be disrupted into single cells or even spheroplasts (Briehl and Mendelson, 1987), each of which is capable of growing into a new macrofiber. When the culture conditions used to produce the initial fiber and the second-generation fibers are kept constant, the phenotype of the initial and progeny fibers are identical. Helix hand and degree of twist therefore "breed true." They are heritable properties, and as expected one can isolate mutants with different helix hands, degrees of twist, or both (Favre et al., 1986). Because the phenotype of the multicellular form is governed genetically, it must reflect the growth properties of the individual cells. An individual cell must therefore have a helix hand and degree of twist even if its shape is not helical. The meaning of helix hand and degree of twist at the level of individual cells is discussed later. It pertains to the relation between cylindrical and helical shape (Mendelson and Thwaites, 1989).

The phenotype of a cell or a multicellular form such as a macrofiber is subject to influence by the environment and to control by genes. Helix hand and degree of twist are passed from parent to progeny when the environment is constant but not if the environment changes in certain parameters. For example, if cells are taken from a right-handed macrofiber of 20 turns/mm grown at 20°C and seeded into fresh medium at 48°C, the new fibers that arise are all left-handed with a twist of about 15 turns/mm. Their progeny can be either left- or right-handed depending on the temperature at which they are grown (Favre et al., 1985). The switching of one phenotype to another can be effected by a number of environmental factors other than temperature, and it can occur any time during the growth of a fiber, not just when starting anew from individual cells (Mendelson, 1988). A right-handed fiber such as that in Figure 12.1, when shifted to conditions leading to left-handedness, untwists all of its right-handedness and then twists back together in the opposite helix hand. This remarkable process has been recorded by time-lapse cinematography and studied in detail. Analysis of helix hand inversion kinetics suggests that helix hand and degree of twist become established when new cell surface is assembled. Hand and degree of twist are retained so long as the individual cell walls retain their corresponding structural organization (Favre et al., 1985).

Role of the Cell Wall

Changing the state of a macrofiber requires building sufficient new cell wall of an alternate organization to replace the old wall material corresponding to the initial state. Normal cell wall upwelling and turnover processes during growth are involved. Cross-linking of peptidoglycan is implicated: One of the most potent affectors of twist state establishment is the concentration of D-alanine in the environment (Surana et al., 1988). D-Alanine is the amino acid found in the peptide moiety of peptidoglycan in *B. subtilis* through which the glycan backbones become cross-linked. Maintaining a phenotypic state, once established, is another matter. It requires that new wall of the same state as that already present be added during growth; therefore growth conditions must remain constant. The glycan backbone of peptidoglycan must remain intact; therefore enzymes that cleave peptidoglycan must be avoided (Favre et al., 1986). Moreover, there can be no interference with the electrostatic interactions normally operating in the cell wall: The pH must be maintained and certain

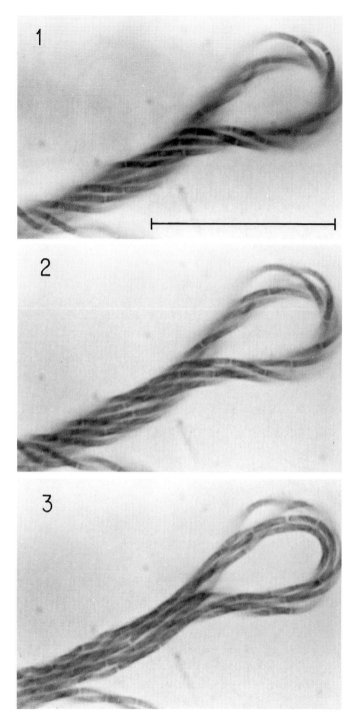

FIGURE 12.1 Phase contrast micrograph optical sections of an eight-stranded right-handed macrofiber. Six of nineteen optical sections are shown in order, from top to bottom. 1, section 5; 2, section 7; 3, section 11; 4, section 12; 5, section 16; 6, section 17. Bar = 25 μm. (Reprinted from Mendelson, 1992, with permission) Figure continues.

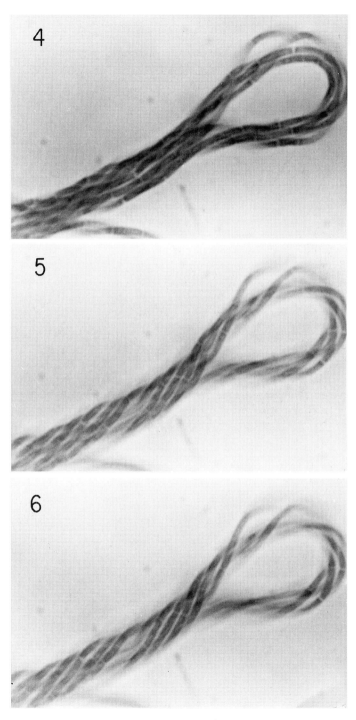

FIGURE 12.1 *Continued*

343

ions such as bromide avoided. When these conditions are satisfied, macrofibers retain their helix hand and degree of twist and faithfully reproduce their phenotype (Mendelson et al., 1985).

Blocked Rotation and the Cylinder–Helix Deformation

If one could measure the growth geometry of an individual cell from which a macrofiber arises, it would be possible to predict both the helix hand and the degree of twist of the mature macrofiber (Mendelson and Thwaites, 1989). One could do so because as the cell cylinder elongates with growth it would twist along a screw axis, the pitch and hand of which are related to the helical form eventually observed in the macrofiber. The conversion of cell growth motions of the cylindrical cell to helical shape in the multicellular form seems to be induced by an impediment to the twisting motions (Mendelson, 1982, 1992). The twisting motions can be blocked by the attachment of cell poles to the spore coat during spore outgrowth, by touching the filament to itself or something else in the environment, or by viscous drag of the cell surface as it moves in the growth medium. The folding processes brought about by blocked rotation is akin to introducing negative torsion; hence the helix hand of the cell cylinder screw axis (and the twisting motions it causes during elongation) are the same as that of the helix once formed. Similarly, the degree of twist of the helical form reflects the geometry of the screw axis growth pattern of the cell itself. The geometry of the relation has been described (Mendelson, 1976; Mendelson et al., 1984). Using theory adapted from that developed for the description of multifilament helically twisted textile yarns and assuming a constant cell diameter of 0.8 μm, we calculated that the pitch angle of the cell screw axis would have to range from 0 to 8 degrees to produce macrofiber twist over the range of 0–57 turns/mm, as was observed.

Bacterial macrofibers are themselves cylindrical because nearly all of their cells are aligned with the fiber axis. More pertinent is that the growth axis of the individual cells is also aligned with the fiber axis. Figure 12.2 illustrates this process in an idealized model. Twisting about the cell cylinder axis during growth gives rise to bending and writhing motions that distort the form of the cell filament until it touches itself, forming a loop. The loop then plies into a double helix, and the process is repeated. Each cycle gives rise to a structure with an increased number of cell filaments packed parallel to one another; and because every cycle involves folding and plying in the same helix hand as that already present, all the filaments pack uniformly into a tight bundle. The first two and in some cases three folding/loop closure cycles have been measured in a number of clones of right- and left-handed structures (Mendelson et al., 1995).

Figure 12.3 shows the filament growth kinetics as well as the lengths at which folds 1 and 2 took place in a typical FJ7 clone. The rate of turning (angular velocity) of the loops remains constant during the period of closure when the loop twists together, forming a new fiber shaft. Some differences have been found, however, between the angular velocities of the first, second,

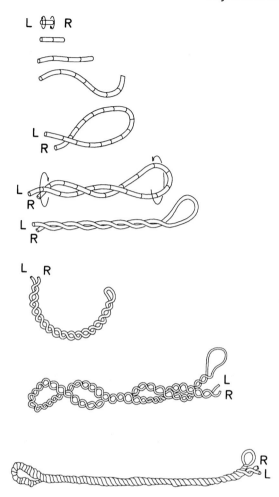

FIGURE 12.2 Idealized model of macrofiber initiation from growth of a single cell in a solution of 1 cps viscosity. Twisting (right-handed) and writhing lead to touching and plying together of a double-strand right-handed structure. The process repeats giving rise to a four-stranded right-handed fiber. L and R refer to the original cell poles. (Reprinted from Mendelson and Thwaites, 1990, with permission)

and third closure cycles in the same clone, suggesting that there may be some mechanical differences between the first and subsequent cycles. During all cycles the tangential velocity decreases as the loop gets tighter and tighter and its radius decreases. The change in tangential velocity gives the false impression when viewing time-lapse films of the process that immediately after loop formation the twisting motions are accelerated. Writhing motions cease when a loop is formed, and it appears that the dynamic energy in the structure then is utilized for twisting and loop closure. The constant rate of these motions suggests that they too are linked to individual cell growth.

Stochastic and Determinative Processes

Figure 12.4 shows that a mature macrofiber differs from an ideal structure, such as the one shown in Figure 12.2. The origin of such differences depends

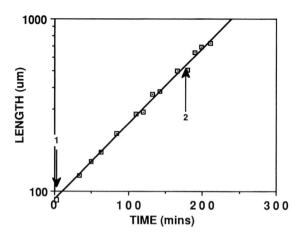

FIGURE 12.3 Growth following spore germination of strain FJ7. Clone lengths were measured with a Numonics Graphic Calculator from videographic prints taken from a time-lapse film. Arrows 1 and 2 indicate the times of the first and second foldings. (Reprinted from Mendelson, 1992, with permission)

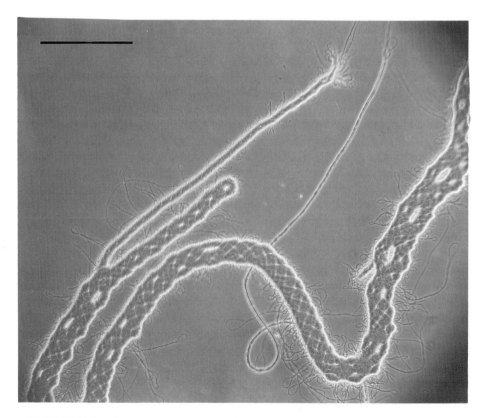

FIGURE 12.4 Phase-contrast micrograph of a mature right-handed macrofiber. Bar = 200 μm. (Reprinted from Mendelson, 1988, with permission)

on the strain and environmental conditions. Vegetative cells can initiate fiber formation in at least four ways: by buckling or by folding in U, J, or V configurations. The course of events following a U-fold illustrates that various stochastic events—such as the size of the loop formed initially at a fold, the direction of plying from the point of contact forming a loop, the geometry and location of a second fold, and the behavior of single-cell filaments with respect to the remainder of the structure—contribute to the unique detailed character of each fiber. For example, branched structures such as T- and Y- shaped fibers can arise. A route leading to eight-strandedness by the folding and plying together of three subunits (one two-strand and two three-strand units) rather than by plying of two four-strand units has been observed. Occasionally a loop formed by a single filament snaps open rather than plying together into a double-strand helix. When this event occurs, the filament continues to grow and writhe until eventually it touches itself again, forming a successful node from which plying proceeds.

Measuring Forces from Motions

The rapid motion observed when a loop snaps open provides a means to calculate forces associated with filament motions that are concerned with writhing and twisting. These forces are of interest because they originate within the developing structure and drive the dynamics of building the multicellular form. Motions play a central role in macrofiber morphogenesis but not the motions associated with motility that are so significant in other systems of bacterial multicellularity. Instead, macrofiber motions stem from the growth of individual cells within the filament, particularly from the geometry of the cell surface assembly process. Figure 12.5 depicts the method used to measure snap-opening motions. The velocity of motion and the length of the moving filament were determined from time-lapse video films. The force required to generate the snap motion was calculated from the viscous drag that had to be overcome as the cell filament swept through the growth medium using the following relation derived from the classical physical model of a rigid rod moving through a solution of known viscosity.

$$\tau = (\gamma/3)(A^3)(\partial\Psi/\partial t) \qquad (1)$$

where τ is the torque, γ is related to viscosity, A is the length of the filament that moves, and $\partial\Psi/\partial t$ is the change of angle shown in Figure 12.5 as a function of time. The torque values obtained from 16 snap-open sequences analyzed in this way were in the range of 10^{-8}–10^{-10} dyne-cm.

Although the physical model applied to analyze snap opening motion is based on determining the amount of force required to overcome the resistance to moving an object of length A through a solution, it is reasonable to assume that the production of force responsible for the motions is also related to Eq. (1). If so, the number of cells in the structure must be the most significant factor because the number of cells governs the length of the filament (A). The force that drives the snap motion can arise and operate to cause a snap only

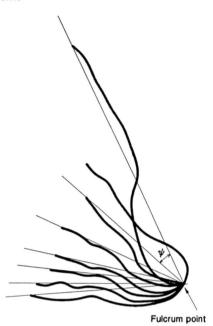

FIGURE 12.5 Snap-opening motion of a clone of strain RHX from video film frames showing the fulcrum point and angles measured. Length of the moving arm was 495 μm. (Reprinted from Mendelson et al., 1995)

Fulcrum point

when the cells are linked into a filament. The same holds for other forces that drive the self-assembly of macrofibers. Thus the organization of cells in a multicellular population is a significant factor in the mechanical behavior of the system. A cohesive structure can produce and utilize force in ways not possible when a comparable biomass is not so organized. Macrofibers are a good example of this principle. The twist and supercoiling processes on which macrofiber morphogenesis is based require physical attachment of cells to one another so a string-like entity is formed. Moreover, because the forces responsible for macrofiber twist and supercoiling are generated by cell growth, the mechanics of fiber formation can operate only when the string-like entity itself arises by cell growth. Forming a string-like entity from cells already present elsewhere does not suffice unless an alternate method is used to introduce twist into the structure after it has formed. Keeping cells together physically during their growth has special significance therefore in governing the outcome of multicellularity.

Limitations Due to Geometry and Mechanics

Even though all the cells in a given macrofiber have the same helix hand and twist state, the pitch angle of the cell filaments that lie on the surface of a fiber with respect to the fiber axis differs at different stages of fiber growth. For instance, compare the two-strand and four-strand helical forms in Figure 12.2. As fibers increase in diameter by repetitive folding and plying cycles, the fil-

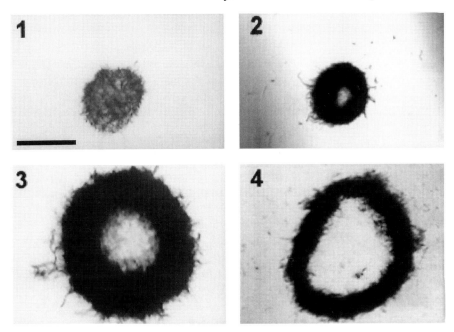

FIGURE 12.6   Macrofiber ball structures and their decay. Progressive stages of breakdown are shown in structures from various clones selected to illustrate the process. All clones were produced by strain FJ7 grown in TB medium plus magnesium at 20°C. Breakdown always began in the interior of the ball form and progressed outward, eventually leaving only a thin ring. Bar = 0.5 mm.

aments on their surface come to lie more and more perpendicular to the fiber axis. Multifilament, helically twisted textile yarns behave in precisely the same manner; consequently, the geometry of the relation between fiber diameter and surface pitch angle has been well known for some time. The progressive change in surface geometry dictated by the rules of packing string-like entities together in a twisted form places a limitation on the size of the structure that can be produced using the strategy found in macrofibers. Thus the building plan cannot be used indefinitely to make ever-larger forms. This limitation has nothing to do with the growth potential of the cells in the fiber. It is a purely physically imposed limitation.

In practice, another physical limitation comes into play beforehand that alters the course of macrofiber morphogenesis. As the number of cell filaments bundled into the fiber shaft increases, rigidity also increases. Fibers eventually reach the point where they are too fat to writhe and bend enough to continue to form loops and ply together. Usually a ball-like form results from which cell filaments are shed. The shed filaments then initiate macrofiber formation anew in the same culture with the decaying ball forms. An example of ball structure and its decay is shown in Figure 12.6.

Primitive Life Cycle in Macrofibers

Macrofibers go through a kind of primitive life cycle during their growth process. Mechanics and physics play a significant role in governing this primitive process. Macrofibers begin life as a double-stranded twisted hairpin form created from a cell filament by a folding or buckling mechanism or by attachment of the cell poles at the ends of a filament to the spore coat from which the initial cell emerged. Once formed, the fibers enter a more of less steady-state period of development that involves growth in length, writhing and bending motions, touching to produce loops, and plying together of the loop to form a twisted bundle that is the shaft of the fiber (Mendelson, 1982). The repetitive folding and plying process may be repeated 10 times or more before the start of ball formation. Each folding cycle increases the number of cell filaments in the shaft of the fiber and consequently its diameter. The forces at play cause individual cell filaments in the fiber to shift and become more perfectly aligned with their neighbors. Fibers pass from this mature, highly ordered state to ball formation. From that point on, the structures become more disorganized. Eventually the ball structure decays, liberating cell filaments and individual cells, which initiate new fiber formation in the same culture that contains the decaying ball structures. The fact that the second-generation fibers arise in the same culture shows that the decay of their parent structure is not caused by the general physiological state of cells in the culture but, rather, something specific to the ball form itself. Physical restraints on growth in the ball coupled to local physiological conditions in the ball structure leading to cell separation are the likely factors.

The transient nature of bacterial macrofibers more closely resembles the plan found in higher organisms that have determinative growth (e.g., animals) than that found in plants. At the cellular level, however, macrofiber cells do not progressively lose their ability to grow as a function of the number of cell cycles they have been through, as do animal cells (Zorn and Smith, 1984) and yeast (Claus et al., 1996). Old macrofibers usually disintegrate, therefore, rather than die. Each cell liberated can start a new fiber. In a given static culture nutrients obviously are eventually exhausted, and metabolic wastes that inhibit growth accumulate; consequently, the number of macrofiber generations supported is limited. If the cells liberated from a macrofiber are transferred to a fresh culture at each generation, they can go on perpetuating descendants indefinitely. Some of our macrofiber strains have been maintained in this way by serial transfer for more than 10 years. Any given fiber has a life-span of only several days, however.

A macrofiber in steady-state growth that is transferred to another culture in which the viscosity has been increased by addition of a polymer such as methylcellulose continues its development but adopts a different mechanism to do so. In the viscous solution it cannot writhe and bend freely, although growth and twisting continue. Fiber morphogenesis proceeds via a buckling/supercoiling process, rather than the usual bending, folding, and plying route (Mendelson and Thwaites, 1990). The outcome of growth with repetitive buckling/supercoiling cycles is again a mature macrofiber and eventually a ball-like form. The growth of individual cells appears not to be affected by the viscosity of the solution, and so it is reasonable to assume that the origin of forces arising by growth of cell filaments is the same regardless of viscosity. The response to forces arising in the growing multicellular entity, in contrast, is strongly influenced by the viscosity. Again the rules of mechanics and physics are the dominant factors governing the multicellular morphogenesis.

Twist and Supercoiling

The behavior of macrofibers in viscous solutions and the fact that fibers always fold or buckle to produce the same helix hand in the new bundle as that present at all earlier stages of the fiber's growth imply that the principles of super-coiling are at play in macrofibers. The mechanics of twist and supercoiling are the heart of the matter. Unlike other systems of twist and supercoiling, however, in macrofibers the forces responsible come from within rather than being imposed from outside. The magnitude of forces involved have been estimated recently using the same physical model (Eq. 1) described earlier for the analysis of snap-opening motions. In this case, however, the forces calculated pertain to twist and supercoiling motions that must drag small wires inserted into macrofiber loops through solutions of increased viscosity. Figure 12.7 shows typical macrofibers carrying short lengths of tungsten wire in their terminal loops. The force of macrofiber twisting can be estimated by determining if it can drag a wire of length A through a solution of given viscosity. Equation (1) applies directly when the angle measured as a function of time is that of the wire being dragged by the twisting loop. In solutions with viscosities of about 2 and 20 poise (normal medium is about 1 centipoise), respectively, values obtained for torque are in the range of 10^{-5} dyne-cm. The amount of force generated per cell cannot yet be estimated because the total number of cells in the fiber examined is not known.

Estimating Forces by Dragging Wires

Macrofibers carrying wires in terminal loops when growing in viscous environments supercoil as do the control fibers that do not carry wires. The super-coil appears as a buckle along the fiber shaft that twists on itself, forming a new double helix out of the fiber shaft that is oriented perpendicular to the original fiber axis. As a result of the consumption of the fiber shaft into the supercoil, the ends of the original fiber are drawn closer to one another. The force of the supercoiling can be estimated therefore by determining if a wire inserted into the loop at one end of the initial fiber can be dragged along with it through a solution of increased viscosity. When dragged perpendicular to the direction of motion (i.e., when dragged so the full length of the wire provides its maximum resistance), the following relation pertains.

$$F = (\gamma)(v)(A) \qquad (2)$$

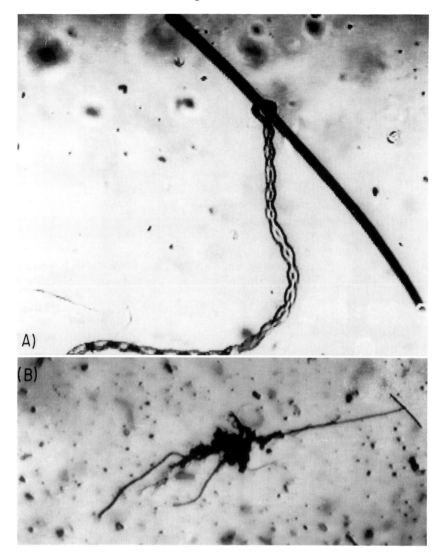

FIGURE 12.7   Macrofibers of strain FJ7 carrying 10 μm diameter wires in their terminal loops. Videographic prints were obtained from time-lapse films of motions in TB medium containing 2% methylcellulose (viscosity 3600 cps). (**A**) Macrofiber with both ends free to rotate. (**B**) Macrofiber with one end tethered in a ball structure that blocks its rotation.

where F is the viscous force that must be overcome, γ is proportional to viscosity, v is the velocity of the wire being dragged, and A is the length of the wire. Wires of 0.024 and 0.042 cm have been observed to move at 3.2×10^{-5} and 2.08×10^{-4} cm/s in solutions of about 20 and 2 poise, respectively. The supercoiling force must therefore be in the range of 10^{-5} dynes. The maximum value of the supercoiling force is being sought by determining the combination of wire length and viscosity that cannot be overcome by it. Under such conditions one end of the fiber remains tethered in place to the wire, and the rest of the fiber is free to supercoil and contract toward the immobilized end.

Hierarchical Nature of Macrofiber Morphogenesis

Macrofibers reveal the extent to which procaryotic multicellularity can mimic developmental phenomena normally found in eucaryotic multicellular organisms. For example, macrofibers undergo specific motions during morphogenesis: They fold or buckle, continuously produce new cells, and reorganize the arrangement of their cells. Macrofiber development follows a program that involves events in which spatial and temporal order are linked. The same holds true for all developing higher organisms, although in each case the details of the developmental program differ. All have in common the fact that development always involves going from simple to complex, from one or few to many cells through a specific progression of transformations to reach a state that can yield the next generation.

Macrofibers show the degree to which cellular organization within the multicellular entity is facilitated by the linking of cells to one another and the geometry of cellular growth. The continuity between molecular, cellular, multicellular, and organismic level processes in macrofibers operates through the production of and response to forces acting on the cellular materials. Similar processes occur in all multicellular organisms but not in single-cell forms that are subject primarily to forces from without rather than to forces generated by themselves. The morphogenesis of higher organisms differs from that found thus far in macrofibers in one important respect: In higher organisms a parallel program of events accompanies the physical processes, giving rise to the production of different cell types within the organism. Differential gene expression comes into play in a precisely regulated spatial and temporal manner. The end result is complexity at many levels, not just at the structural level as in macrofibers. *Bacillus subtilis* is capable of such differential gene expression within populations as shown in the system described below. Similar processes have not yet been found in macrofibers.

Patterns of Gene Expression in Colonies

Presumed Uniformity

Conventional bacterial colonies, because of their presumed purity and uniformity, are used every day in genetics studies, to identify unknown isolates, in biotechnology to initiate industrial processes, and in many other contexts. What could be more homogeneous than the clone descended recently from a single ancestral cell, particularly when all the members of the clone remain together as a single isolated population? Indeed, the inheritance of colony form itself in the progeny derived from individual cells of the same parent colony supports the thesis that the individual cells within a colony, if not genetically identical, must at least be similar to one another. In this regard a bacterial colony is the equivalent of a multicellular eucaryotic organism, itself a clone derived from a single cell zygote. The two differ in the degree of their structural complexity, the range and amount of cell differentiation within their cell populations, and (depending on the eucaryotic organism to which a colony is being compared) whether any cell or only certain cells can initiate the next generation of the organism. These differences may not be as profound as believed. The structure of bacterial colonies is more complicated than expected (Shapiro, 1988). Cell differentiation within colonies has been well documented (Shapiro and Trubatch, 1991). Mutation, gene rearrangements, and mobile genetic elements reduce genetic homogeneity in bacterial clones, and there is now evidence for differential gene expression within colonies (Shapiro, 1984; Salhi and Mendelson, 1993).

Factors Affecting Colony Organization

The structure of a bacterial colony is governed by many inputs. How they work is not understood. Essential factors must include the production of new cells and extracellular materials, whether the cells have translational motility, cell growth geometry, separation or lack thereof of daughter cells after division, and the restraints imposed by the physical environment. Cell surface properties must also play a significant role. For *Bacillus subtilis* the colony form is known to be influenced by the genetic composition of the strain, nutrients used to support growth, growth temperature, physical properties of the substrate on which the colonies grow, cell motility, and in the case of certain derivatives of the common laboratory strain 168 an unusual phase-variation-like instability that results in smooth and rough variants. Various combinations of agar hardness and nutrient concentration levels have been shown to yield a range of colony forms when supporting the growth of motile strains (Ohgiwara et al., 1992; Ben-Jacob et al., 1992, 1994). Fractal patterns were observed in the region of the matrix corresponding to low nutrients and hard agar, suggesting that diffusion-limited aggregation-like processes may be at play during colony formation (Matsushita and Fujikawa, 1990). The underlying causes for the wide variety of colony types observed, collectively called the phase diagram of colony forms, remain to be elucidated. A model has been proposed that ascribes cooperative movements of groups of cells in response to nutrient gradients and chemotactic responses (Ben-Jacob et al., 1994). Appropriate mutants are available to test these ideas, but the experiments have not yet been conducted.

The two-dimensional organization of all bacterial colony forms contains elements of circumferential and sectoral geometry. Circumferential elements stem initially from the isotropic expansion of increased mass and then later from the physiological consequences of the biomass itself. Sectoral elements originate in the phenotypic change of an individual cell that is passed to progeny. The form of the sector depends on the location of the cell from which it originates and how well the new clone competes with the growth of its neighbors within the colony (Shapiro and Trubatch, 1991). Little is known about the three-dimensional organization of colonies beyond the fact that it is much more complex and regulated than would be expected if the cells simply piled on each other during growth. Developing *Escherichia coli* colonies pass through an ordered series of phases in which the three-dimensional contours and cell types within various regions change as a function of colony age (Shapiro, 1987). At least in young colonies the growth axis of cells lies parallel to the agar surface. Where the forces responsible for elevating some cells above others originate and how they work is unknown. Motility cannot be an essential factor, as nonmotile strains also develop three-dimensional colony structure. Motility and the turning of flagella on which it is based are nevertheless sources of force in colonies of motile cells. The influence they may exert on the three-dimensional form of colonies remains to be determined.

Detecting Hidden Variability

Much of the complexity in bacterial colonies is difficult to detect because it involves cellular phenotypes that do not manifest as differences that can be detected in colony form, texture, or gross morphology. Insights have been gained using reporter-gene technology, which facilitates visualization of a gene product within living and growing colonies. Both sectoral and circumferential patterns of gene expression have been observed, the latter indicating that cells respond to cues regarding their location within a colony. Which genes respond and the phenotypes they govern have not yet been identified in these systems as they have in the case of genes that regulate fruiting body formation in *Myxococcus* colonies (Kim and Kaiser, 1992). Nevertheless, studies of expression patterns in conventional colonies clearly show that individual cells sense and respond to their location within the colony and even to the presence of neighboring colonies. One example studied in *B. subtilis* illustrates how the growth of a colony influences the environment surrounding it, which in turn provides a signal for turning on expression of an as yet unidentified gene. Two aspects of this work are described here: (1) that dealing with the establishment and response to nutrient diffusion gradient fields; and (2) the nested patterns of gene expression produced within colonies of complex form that arise within various domains of the colony-form phase diagram (Mendelson and Salhi, 1996).

Studied Strain

Bacillus subtilis strain 5:7 carries the *Escherichia coli lacZ* gene at an unknown location in the host chromosome, having been inserted there by transposition of the Tn917 transposon from a suicide plasmid known as PTV32ts. The reporter gene appears to be under the control of a weak host gene promoter that is expressed late in growth on complex media. When X-gal was used to visualize expression of the reporter-gene colonies grown on complex media, a blue ring developed late in growth near the periphery of the colony. Both the spatial and temporal aspects of the expression pattern were influenced by colony density. When the plating density was high, small colonies arose, and gene expression began early in colony development and appeared throughout the colony rather than just in a peripheral ring. Similar behavior was observed in streaks and colonies grown from toothpick transfers in which the initial inoculum was multicellular. The gene expression pattern of strain 5:7 is therefore not linked to the clonal origin of growth but, rather, reflects the physiological condition of cells with respect to environmental factors including other cells. Multicellularity is the primary factor in the regulation of this particular gene.

Effect of Multicellularity

Multicellularity influences gene expression in 5:7 by influencing the chemical environment in the vicinity of the cell mass and by organizing the spatial locations of cell growth and related physiology, thereby establishing spatial and temporal windows of responsiveness with respect to environmental signals. An example is shown in Figure 12.8. The cross-streak pattern next to which tester colonies were inoculated by toothpick transfer shows the properties of the field surrounding growth with respect to control of gene expression and growth. Tester colonies located close to the streak grew poorly but expressed heavily, in contrast to those located farther away. The distal tester colonies reveal asymmetrical growth from the point of inoculation (the inner circle in each colony), favoring growth away from the streak, whereas expression was restricted to an arc facing the streak. The tester colony patterns reveal that a gradient field

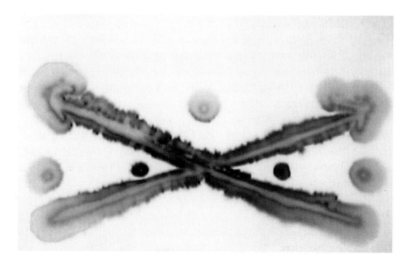

FIGURE 12.8   Cross-streak growth pattern illustrating the influence of neighboring fields on gene expression patterns in strain 5:7. Photographed after 96 hours of incubation at 30°C. Bar = 10 mm. (Reprinted from Salhi and Mendelson, 1993, with permission)

surrounds the streak and that both growth and gene expression were governed by it. The streak itself appears also to have been governed by its own field. Growth of the streak was progressively more robust in regions where the individual arms of the streak were farthest from neighbors. Expression always began first and was more intense in the region where the streaks crossed; it then spread progressively outward with time but remained weakest at the distal ends of the streaks. Global influences of the gradient field are also evident in the higher intensity of expression along the arms of the streak that face closest to one another compared to those on the other side of each arm that face away from neighbors (at the top and bottom of Figure 12.8). The expression pattern within the streak illustrates that only certain cells within the multicellular mass that lie at regular positions relative to the growth of the mass have an active reporter gene. Responsiveness is therefore not only governed by the gradient field but also by growth and position within the cell mass.

Gradient Fields

Three aspects of gradient fields surrounding growth have been examined in the *B. subtilis* 5:7 system: the ability of fields to serve as signals between colonies, the influence of substrate geometry and physics on fields, and whether the fields represent positive or negative gradients with respect to the growth mass. Signaling was studied by examining the influence colonies located at various distances from one another have on each other. The temporal and spatial patterns of expression were characterized in comparison to those appearing in single isolated colonies from the same source grown under identical conditions. Significant interactions were detected between colonies grown in the conventional way on a horizontal agar surface, between colonies grown in opposition to each other on the vertical surfaces at the ends of rectangular agar blocks, between colonies grown in opposition on the top and bottom surfaces of agar slabs, and between colonies, one of which grew on a horizontal surface and the other on a vertical agar surface (Salhi and Mendelson, 1993). Interactions were detected by (1) expression that started earlier in colony development than that observed in single isolated colonies and (2) expression in the form of arcs that always pointed in the direction of the neighboring colony. Gradient fields are therefore an effective means by which colonies can sense the presence of neighboring growth. The results also show that fields surrounding single colonies initiated by toothpick transfer extend at least 16 mm and interact, so colonies 32 mm apart respond to each other's presence.

Gradient fields surrounding growth are contained in the agar gel; consequently, the geometry of the substrate must influence that of the gradient field itself. Indeed using agar blocks and cut-out spaces in conventional agar plates, a strong effect was detected. Colonies surrounding a space where a plug of agar had been removed showed arc patterns of expression pointing toward the missing agar. Others that grew on the wall or the lip edge where the top and vertical surfaces meet had very strong expression. Colonies grown over an edge surface of agar blocks or over a corner point always showed early, heavy expression in cells located above the edge or corner point. It is difficult to see how the physics of an edge or corner surface could exert any influence, given the relative size of the cells with respect to the surface. It is more likely that the strong effects of these surfaces on gene expression reflects the geometry of the diffusion gradient field in the agar gel under the edge and corner point.

The physics of the substrate does come into play when the concentration of agar in the gel is taken into consideration. The 5:7 colonies grown on three times the standard agar concentration (6% rather than 2%) always displayed early, heavy gene expression even though the nutrient concentration in the 6% agar was the same as that in the 2% agar. The pattern of expression also differed in colonies grown on 6% agar compared to those grown on the standard 2%. Figure 12.9 illustrates that the peripheral band of expression is broad in single isolated colonies but double in colonies grown in proximity to one another. In addition, regions of the colony raised high above the 6% agar surface had elevated expression compared to regions at lower elevations, suggesting that diffusion gradient signals may also extend through the cells of the colony itself. If the diffusion gradient fields responsible for triggering gene expression in the 5:7 strain consist of low-molecular-weight components, as seems likely given the distances the fields extend, the pore size differences between 2% and 6% agar gels do not seem to account for the observed differences in expression patterns. The physical or chemical properties of the gel may therefore play some role in the signaling process. Motility is not a factor here, as it is in the case of agar hardness influencing colony form within the phase diagram described earlier, because strain 5:7 is not motile.

Gradient fields surrounding growth could be positive or negative in the sense that they could emanate from the multicellular entity or be the result of

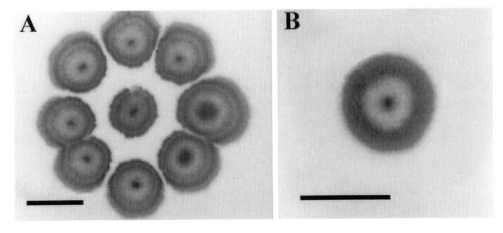

FIGURE 12.9   Colonies of strain 5:7 grown on TBAB medium, 6% agar, containing X-gal. (**A**) Eight colonies surrounding a central colony produced following toothpick transfer by incubation for 72 hours at 30°C. Bar = 10 mm. (**B**) Single colony grown in isolation illustrating the broad region of gene expression within which elevated mounds of cells show the highest expression. Bar = 10 mm.

withdrawal of a diffusible factor from the medium by it. Several lines of evidence suggest that in the 5:7 strain system the latter is the case: Cell growth appears to remove a diffusible component from the medium, creating gradient fields proportional to the amount of growth. When the concentration of this triggering component falls to a critical level, the host gene carrying the *lacZ* reporter becomes derepressed in responsive cells. If the amount of nutrient available for growth was adjusted downward to begin with, expression commenced early and was heavy. The opposite is true if an increased concentration of nutrient was available. At saturation levels of nutrient, expression never appeared, even after the colony had grown to its maximum size and ceased to grow. The negative nature of the triggering stimulus has also been demonstrated in colonies grown on artificial nutrient gradients using agar blocks. Arc patterns of expression arose that always faced the low end of the gradient. An agar block technique was also devised in which the triggering stimulus produced by growth was trapped and assayed by examining its influence on expression in a second tester colony. The trapped signal could be overcome by refreshing the nutrient supply in the block, suggesting that a positive activator is not involved. Multicellularity therefore gives rise to highly regulated gene expression patterns by the consumptive generation of nutrient diffusion gradient fields that act as a gene expression trigger and by generating responsive cells at defined places and times.

Agar Hardness: Nutrient Concentration Influence

Agar hardness and nutrient concentration are factors in the regulation of gene expression in strain 5:7 and the regulation of colony form in other strains of *B. subtilis* that are motile. To examine the relation between the colony forms described in motile strains and gene expression patterns, the 5:7 insertion was transferred to motile recipients by transformation. Three new strains, M8, M18, and C3, and the original 5:7 strain were grown on semisolid media containing nine different amounts of agar (ranging from 0.2% to 6.0%) in combination with five different amounts of nutrients (ranging from 0.1 to 10.0 times that of the standard TB medium). The full matrix of conditions examined was considerably larger and more detailed than that used to produce the phase diagram of colony forms. The results obtained were startling in terms of the complexity of nested patterns. Some examples are shown in Figure 12.10. The inverse relation between nutrient concentration and gene expression revealed in the previous characterization of strain 5:7 also pertains to the motile derivative strains that carry the 5:7 reporter-gene insertion. The variety of colony forms produced by these strains throughout the matrix of conditions examined is similar to that reported in the phase diagram found in other *B. subtilis* strains. Our latest studies reveal chiral forms of colony structure as well as gene expression within colonies. Circumferential and sectoral patterns of expression are present but modified in the context of the colony form. There are clearly some fractal-like elements in the expression patterns observed. The question of why gene expression is not present in certain regions now becomes as sig-

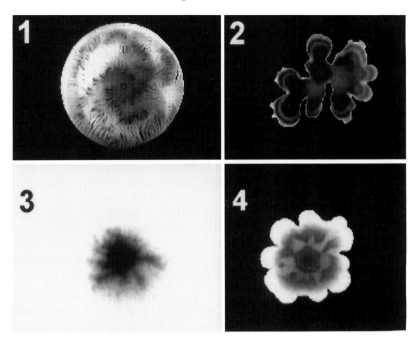

FIGURE 12.10   Single colonies of motile strains carrying the 5:7 insertion grown on various combinations of nutrient and agar concentrations. 1, Strain M8 grown on 1X TB and 0.6% agar; 2, strain M18 grown on 0.5X TB and 1.0% agar; 3, strain C3 grown on 0.5X TB and 2% agar; 4, strain C3 grown on 1X TB and 0.8% agar. 1X TB = 10 g of tryptose, 3 g of beef extract, 5 g of NaCl per liter. Only the tryptose and beef extract components were varied.

nificant as the issue of why the gene is expressed elsewhere. The same could be asked of colony form itself. What is it that prevents growth from occupying the free spaces in deep-fingered forms? The developmental history of the colony and the gene expression pattern within it may provide some clues. Experimental studies of and mathematical models dealing with the collective motility of cell populations on agar surfaces have already given new insight into the processes taking place (Budrene and Berg, 1991; Ben-Jacob et al., 1994). In a collaborative effort with others in applied mathematics, we are currently attempting to build on these foundations by incorporating ideas about nutrient diffusion gradient fields surrounding growth into models that predict the nested patterns of gene expression with complex colony forms. The assumptions used in successful models will be subjected to experimental testing.

Conclusions and Perspective

Multicellularity is thought to have originated independently several times during evolution, suggesting that it may not be difficult to make the transition from life as single autonomous cells to life in an organized population (Margulis, 1981; Kaiser, 1993). Getting nutrients in and distributed and getting rid of wastes are obviously not insurmountable problems. In the two procaryotic multicellular forms described here, macrofibers and conventional colonies, the same diffusion-based mechanism used by their single-celled counterparts provides the inputs and removes the outputs. Thus growth is supported in these multicellular forms and with it comes a spectrum of new phenomena unique to the multicellular situation. Growth in macrofibers is supported throughout the fiber, creating a situation from the mechanical perspective unlike that in any other multicellular form. The inputs and outputs accompanying growth of conventional colonies coupled with their rates of diffusion play a critical role in regulating growth itself and, as shown here, in controlling gene expression within the multicellular entity. In both systems growth of the multicellular form leads to restraints on itself. Such action and reaction provides a means to achieve a higher degree of complexity than can be obtained otherwise from individual cells operating independently and randomly.

The macrofiber self-assembly process illustrates the degree to which forces arising during growth can act on the multicellular biomass from which they originated, helping to organize it into a coherent structure. The magnitude of the force produced is geared to the properties of the materials that respond to it. A sort of complementarity is therefore built into the system. Production and response to force reinforce one another. The self-assembly is not a smooth, continuous process, however. It engenders at least two components, twist and supercoiling. The former gives rise to coherent, aligned, tightly packed multi-filament bundles as a smooth and constant process. When restrained, twist causes supercoiling, a periodic process that creates terminal loops and increases the number of filaments in the shaft of the fiber. Twist has its origin in the geometry of cell growth. That geometry in turn is governed by both genetic and environmental factors. There is clearly a hierarchical relation involving molecular level processes concerned with cell growth, cellular level responses to them, multicellular level ramifications of the cellular level phenomena, and finally multicellular interactions that result in the final three-dimensional macrofiber. At all levels there is an interplay between processes governed by the cells and those governed by the environment or the physical rules of mechanics. The key components are the increase in materials, the properties of the materials, the production of and response to forces, and geometry.

The idea that multicellularity always leads to restraint of one form or another is well illustrated by the periodic and repetitive folding during macrofiber morphogenesis. Folding is the equivalent of negative supercoiling. It appears to be triggered by restraint of the twist that accompanies growth. The balance between twist and supercoiling dictates the course of each fiber's morphogenesis. Neither the mechanics nor the mathematics are well understood, although a good start has been made on the latter (Klapper and Tabor, 1994). It is clear that the degree of twist in macrofibers is positively correlated with the frequency of supercoiling. That is, fibers that have high twist fold at shorter lengths than those with low twist, regardless of their helix hand. The viscosity of the environment in which the fibers grow also strongly influences supercoiling. Viscosities above that of standard growth medium (equivalent to water) increase the frequency of supercoiling, causing fibers to rapidly contract into the ball-like form characteristic of late stages of fiber growth under normal conditions. Individual bacterial cells, because of their small size, are also significantly influenced by viscosity, even by the viscosity of water itself. In the case of multicellular bacteria, a new set of responses to viscosity becomes possible. When the combination of factors—multicellularity, growth, and twist—are all operative, the response is supercoiling. The morphogenesis of other multicellular forms is likely to be driven by various combinations of factors simply because biological materials cannot escape the rules of mechanics. Morphogenesis, the process, must be adapted to the rules of mechanics having evolved under their influence just as the materials of biological organisms are adapted to the physical properties of the environment. The surprising thing learned from macrofibers is the degree to which mechanics can be responsible for bringing order to otherwise chaotic phenomena.

Biological materials cannot escape the rules of chemistry either. Multicellularity therefore also is burdened with chemical restraints. Of particular concern is the local depletion of nutrients in the vicinity of the multicellular mass and the accumulation of metabolic wastes in the same region. The well known starvation and self-poisoning of bacterial colonies illustrates the point. The magnitude and form of a multicellular structure must dictate to some extent the influence it has on the chemical environment surrounding it. Moreover, the feedback of the environment to the multicellular mass in turn is constrained by geometry unless the multicellular form can move away from the location in which it grew or mix the environment in some other way. A sedentary colony has no means to do so. Trapped in place, it can only grow where nutrients are located and metabolic wastes permit. Indeed, even colonies containing motile cells suffer similar restraints. Their internal motions may serve to mix solutions locally but not sufficiently to free them from the chemical restraints of their own growth. Restraints that take the form of diffusion gradient fields reflect the fact that cells grow only in aqueous environments, and the nutrients and metabolic wastes are usually low-molecular-weight compounds that diffuse in aqueous environments even within the interstices of an agar gel.

Given the fact that multicellular growth on a substrate leads to the establishment of diffusion gradient fields surrounding it, and the fact that such fields in turn influence the multicellular mass, the size and geometry of the substrate and the starting concentration of nutrients in it must be significant factors. In addition, the properties of the substrate governing the diffusion of compounds that affect growth and gene expression are also important. Other physical properties of the environment, such as viscosity, are likely to be significant as well. Considering all these factors as initial conditions, one sees that as multicellularity develops it is constrained by the initial conditions, and the final outcome is governed by them. The multicellular form itself is the source of additional conditions, as illustrated in the two *B. subtilis* systems described herein. These conditions in turn further constrain the boundaries of possible outcomes. Complexity in multicellular bacteria arises within the context of such constraints.

It is driven by the cells' reproductive and synthetic capacities interacting with environmental conditions. Reproductive and synthetic capacities themselves are governed ultimately by genes. The picture that emerges, then, is one in which constant adjustments are made by the biological entity in response to physical and chemical environmental conditions, which in turn are altered by the biological entity.

What are the final outcomes of bacterial multicellularity? Both systems in *B. subtilis* described here are ultimately self-limiting. In the case of macrofibers, mature structures either break down into individual cells or die. There does not seem to be a way to maintain a mature fiber indefinitely in a metabolically active state that does not grow. Fibers continuously refreshed with new medium grow to a maximum state and then decay by shedding cells that grow into progeny fibers. The multicellular state can be perpetuated, therefore, but not sustained indefinitely. The same holds true for colony growth and gene expression within it. The temporal relation between the production of colony form and the appearance of gene expression patterns is linked to growth and its cessation. It too follows a program that is self-limiting but can be perpetuated anew if cells escape from the original colony and arrive at a new place of like conditions but with fresh nutrient and no metabolic product inhibitors present. This self-limiting feature of bacterial multicellularity sets the bacterial system apart from both the indeterminate growth mode of higher plants and the determinant growth—the steady-state maintenance strategy—of higher animals. The multicellular bacterial condition appears to reflect the lack of genetic mechanisms in procaryotic cells required to keep cells from growing when confronted with everything needed for growth. Thus the bacterial multicellular forms we have studied are transitory dynamic states in which changing conditions are the rule rather than the exception. One must keep this point in mind when seeking comparisons between bacterial multicellularity and that found elsewhere in the biological world.

Acknowledgments

This work was supported by a grant from the National Institute of General Medical Sciences and the National Center for Research Resources (NIH) and a Shannon Award from the National Center for Research Resources (NIH) to N.H.M. C.L. was supported by the University of Arizona Undergraduate Biology Research Program. Excellent technical assistance was provided by S.D. Whitworth and T. Radabaugh. We thank J.O. Kessler for many valuable suggestions.

References

Ben-Jacob, E., Schochet, O., Tenenbaum, A., Cohen, I., Czirok, A., and Vicesk, T. (1994) Genetic modelling of cooperative growth patterns in bacterial colonies. *Nature* **368**:46–49.

Ben-Jacob, E., Shmueli, H., Schochet, O., and Tenenbaum, A. (1992) Adaptive self-organization during growth of bacterial colonies. *Physica A* **187**:378–424.

Briehl, M.M., and Mendelson, N.H. (1987) Helix hand fidelity in *Bacillus subtilis* macrofibers after spheroplast regeneration. *J. Bacteriol.* **169**:5838–5840.

Budrene, E.O., and Berg, H.C. (1991) Complex patterns formed by motile cells of *Escherichia coli. Nature* **349**:630–633.

Claus, T.J., Kennedy, B., Cole, F., and Guarente, L. (1996) Loss of transcriptional silencing causes sterility in old mother cells of *S. cerevisiae. Cell* **84**:633–642.

Doyle, R.J., and Koch, A.L. (1987) The functions of autolysins in the growth and division of *Bacillus subtilis. C.R.C. Crit. Rev. Microbiol.* **15**:169–222.

Favre, D., Karamata, D., and Mendelson, N.H. (1985) Temperature-pulse-induced "memory" in *Bacillus subtilis* macrofibers and a role for protein(s) in the left-handed-twist state. *J. Bacteriol.* **164**:1141–1145.

Favre, D., Mendelson, N.H., and Thwaites, J.J. (1986) Relaxation motions induced in *Bacillus subtilis* macrofibers by cleavage of peptidoglycan. *J. Gen. Microbiol.* **132**: 2377–3485.

Fein, J.A., and Rogers, H.J. (1976) Autolytic enzyme-deficient mutants of *Bacillus subtilis* 168. *J. Bacteriol.* **127**:1427–1442.

Kaiser, D. (1993) Roland Thaxter's legacy and the origin of multicellular development. *Genetics* **135**:249–254.

Kauffman, S.A. (1993) *The Origins of Order, Self-Organization, and Selection in Evolution.* Oxford University Press, New York.

Kim, S.K., and Kaiser, D. (1992) Control of cell density and pattern by intercellular signalling in *Myxococcus* development. *Annu. Rev. Microbiol.* **46**:117–139.

Klapper, I., and Tabor, M. (1994) A new twist in the kinematics and elastic dynamics of thin filaments and ribbons. *J. Phys. A Math. Gen.* **27**:4919–4924.

Margulis, L. (1981) *Symbiosis in Cell Evolution.* Freeman, San Francisco.

Matsushita, M., and Fujikawa, H. (1990) Diffusion-limited growth in bacterial colony formation. *Physica A* **168**:498–506.

Mendelson, N.H. (1976) Helical growth of *Bacillus subtilis*: a new model of cell growth. *Proc. Natl. Acad. Sci. U.S.A.* **73**:1740–1744.

Mendelson, N.H. (1978) Helical *Bacillus subtilis* macrofibers: morphogenesis of a bacterial multicellular macroorganism. *Proc. Natl. Acad. Sci. U.S.A.* **75**:2478–2480.

Mendelson, N.H. (1982) Dynamics of *Bacillus subtilis* helical macrofiber morphogenesis: writhing, folding, close packing and contraction. *J. Bacteriol.* **151**:438–449.

Mendelson, N.H. (1988) Regulation of *Bacillus subtilis* macrofiber twist development by D-cycloserine. *J. Bacteriol.* **170**:2336–2343.

Mendelson, N.H. (1992) Self-assembly of bacterial macrofibers: a system based upon hierarchies of helices. *Mat. Res. Soc. Symp. Proc.* **255**:43–54.

Mendelson, N.H., and Salhi, B. (1996) Patterns of reporter gene expression in the phase diagram of *Bacillus subtilis* colony forms. *J. Bacteriol.* **178**:1980–1989.

Mendelson, N.H., and Thwaites, J.J. (1989) Do forces and the physical nature of cellular materials govern biological processes? *Comments Theoret. Biol.* **1**:217–236.

Mendelson, N.H., and Thwaites, J.J. (1990) Bending, folding, and buckling processes during bacterial macrofiber morphogenesis. *Mat. Res. Soc. Symp. Proc.* **174**:171–178.

Mendelson, N.H., Favre, D., and Thwaites, J.J. (1984) Twisted states of *Bacillus subtilis* reflect structural states of the cell wall. *Proc. Natl. Acad. Sci. U.S.A.* **81**:3562–3566.

Mendelson, N.H., Thwaites, J.J., Kessler, J.O., and Li, C. (1995) Mechanics of bacterial macrofiber initiation. *J. Bacteriol.* **177**:7060–7069.

Mendelson, N.H., Thwaites, J.J., Favre, D., Surana, U., Briehl, M.M., and Wolfe, A. (1985) Factors contributing to helical shape determination and maintenance in *Bacillus subtilis* macrofibers. *Ann. Inst. Pasteur Microbiol.* **136A**:99–103.

Ohgiwari, M., Matsushita, M., and Matsuyama, T. (1992) Morphological changes in growth phenomena of bacterial colony patterns. *J. Phys. Soc. Jpn.* **61**:816–822.

Paulton, R.J.L. (1970) Analysis of the multiseptate potential of *Bacillus subtilis*. *J. Bacteriol.* **104**:762–767.

Paulton, R.J.L. (1971) Nuclear and cell division in filamentous bacteria. *Nature* **231**: 271–274.

Salhi, B., and Mendelson, N.H. (1993) Patterns of gene expression in *Bacillus subtilis* colonies. *J. Bacteriol.* **175**:5000–5008.

Shapiro, J.A. (1984) The use of Mu*dlac* transposons as tools for vital staining to visualize clonal and non-clonal patterns of organization in bacterial growth on agar surfaces. *J. Gen. Microbiol.* **130**:1169–1181.

Shapiro, J.A. (1987) Organization of developing *Escherichia coli* colonies viewed by scanning electron microscopy. *J. Bacteriol.* **169**:142–156.

Shapiro, J.A. (1988) Bacteria as multicellular organisms. *Sci. Am.* **258**:82–89.

Shapiro, J.A., and Trubatch, D. (1991) Sequential events in bacterial colony morphogenesis. *Physica D* **49**:214–223.

Surana, U., Wolfe, A.J., and Mendelson, N.H. (1988) Regulation of *Bacillus subtilis* macrofiber twist development by D-alanine. *J. Bacteriol.* **170**:2328–2335.

Zorn, G.A., and Smith, B. (1984) Cell clocks and cellular aging. In *Cell Cycle Clocks*, L.N. Edmunds Jr. (ed.), pp. 557–579. Marcel Dekker, New York.

13

Formation of Colony Patterns by a Bacterial Cell Population

MITSUGU MATSUSHITA

Bacteria are said to be typical unicellular organisms. We then tend to think that individual bacteria grow and live independently from each other. Does this simple view express real bacteria life? They certainly seem to move around independently and erratically in water. On the other hand, it is known that in the interfacial environment such as clay or rock surfaces (i.e., their usual living places in nature) bacteria grow, increase in population, and form a group called a colony. For instance, small numbers of typical bacteria (parent cells), such as *Escherichia coli* and *Salmonella typhimurium*, once inoculated on the surface of an appropriate medium such as semisolid agar with enough nutrient and incubated for a while, repeat the growth and cell division many times. Eventually the cell number of the progeny bacteria becomes huge, and they form a visible colony. Colonies differ in size, form, and color, depending on the bacterial species. Moreover, the colony changes its form sensitively with variations in environmental conditions such as temperature (Singleton and Sainsbury, 1981; Singleton, 1992). These facts cannot be understood by assuming that individual bacteria form a colony by simply staying close to each other as the consequence of cell division under the condition of being difficult to move.

Even though bacteria are fundamentally unicellular organisms, they are the outcome of evolution over a few billion years. It is difficult to imagine that they simply stay close together passively under an adverse environment. They may actively live together by collectively interacting with the environmental "field" and in some cases by altering their environment by means of secreting chemicals such as surfactant. In fact, self-chemotaxis, in which they secrete a

366

chemical and react to it themselves, is recognized for some bacteria. When investigation of these so-called multicellular behaviors of bacteria (Shapiro, 1988, 1991) advances, we may be able to understand some part of the mystery of evolution, such as the advent of multicellular organisms. One way to do so must be to look closely at bacterial colonies and elucidate the growth mechanism.

It is also possible to discuss the formation of bacterial colonies from an entirely different viewpoint. Pattern formation in biological systems is usually believed to be much more complicated than that in physical and chemical systems. Biological phenomena take place through complicated intertwinement between inherently complex biological factors and environmental (physico-chemical) conditions. Setting morphogenesis of individual organisms aside, however, the pattern formation of a population of simple biological objects in some cases is dominated by purely physical or chemical factors of environmental conditions. One may then apply to colony formation the research results of statistical physics of growing random patterns such as fractals (Mandelbrot, 1982), an area that has progressed remarkably in the field of physics and chemistry (Feder, 1988; Avnir, 1989; Ben-Jacob and Garik, 1990; Vicsek, 1992). In fact, in some cases the formation of bacterial colonies can be physically ''understood'' much better than the formation of random patterns in physicochemical systems (Matsushita and Fujikawa, 1990). Moreover, in contrast to, for example, dendritic crystal growth, one can easily observe both the macroscopic growth behavior of colony patterns and the microscopic structure and motion of individual cells that constitute the colonies. Hence one may relate microscopic observations of bacterial cells to macroscopic colony formation. It is important to bridge the gap between microscopic and macroscopic observations in order to understand pattern formation in general.

The viewpoints described above may seem contradictory. Here, however, we argue the formation of bacterial colonies from various angles by taking the stand that the bacterial population shows varied and diverse behaviors. Although bacterial colonies have been seen for a long time, we have not looked closely at them and investigated their formation mechanisms. Needless to say, we hope that we can make the latter, the physicochemical standpoint, the starting point of mathematical understanding for the former interesting but difficult biological problem (Murray, 1989).

The study of bacterial colonies seems to lead us into the fantastic area of random patterns. One can say that these colonies are a treasure house for the investigation of growing random patterns. Our present aim is to glimpse a part of the mystery of this richly varied bacterial colony formation (Ben-Jacob et al., 1992, 1994b).

Incubation and Observation of Bacteria

We mainly used the species *Bacillus subtilis*, which is widely distributed in soil, for example. It is rod-shaped (0.5–1.0 μm diameter, 2–5 μm length) with flagella and is motile in water by collectively rotating the flagella. We fixed the strain OG-01 (wild type obtained from food) unless otherwise stated. We also used immotile mutants with no flagella (OG-01b) to study the effect of active cell movement of bacteria. They were obtained from the wild-type strain (OG-01) by means of nitrosoguanidine mutagenesis.

To study the morphological changes due to environmental conditions in the bacterial colony formation, we varied only two parameters in the present experiments: the concentrations of nutrient (peptone) (C_n) and agar (C_a) in a thin agar plate as the incubation medium. Other parameters, such as temperature (35°C), were kept constant.

Experimental procedures are simple. A solution containing 5 g of sodium chloride (NaCl), 5 g of potassium phosphate dibasic (K_2HPO_4), and a specified amount of Bacto-Peptone (Difco Laboratories, Detroit, MI, USA) as nutrient in 1 liter of distilled water was prepared and adjusted to pH 7.1 by adding 6 N hydrochloric acid (HCl). The solution was then mixed with a specified amount of Bacto-Agar (Difco). Values of C_n and C_a as the environmental parameters were determined at this stage. The mixture was autoclaved at 121°C for 15 minutes, and then 20 ml of the solution was poured into each plastic petri dish of 86 mm inner diameter. After solidification at room temperature for 60 minutes, the agar plates were dried at 50°C for 90 minutes. The thickness of the agar plates thus prepared was about 3 mm.

The bacterial strain was point-inoculated at the center of the agar plate surface. The plates were then incubated in a humidified box at 35°C for a designated time. Bacterial colonies grew quasi-two-dimensionally on the agar plate surface.

The colony patterns obtained were recorded on photographs, immediately after taking the petri dishes from an incubator. More often, however, we placed the bacteria in an incubator made of transparent plastic plates and recorded images of the growing colonies on video tapes over a period of time using a CCD camera outside the incubator and a time-lapse video system (Sony, Tokyo). The movement of individual bacterial cells was observed through an inverted microscope (Nikon, Tokyo) and recorded on photographs or video tapes using a CCD camera and the usual video systems. These images were, if necessary (e.g., to calculate fractal dimensions), analyzed by a digital image analyzer Exel (Nippon Avionics, Tokyo) and a personal computer.

Diffusion-Limited Growth of Colonies

Cells of the present bacterium (*B. subtilis*) cannot enter or grow into the agar medium because the average pore size of the agar-gel network is smaller than the cell size. Hence bacterial colonies grow only on an agar surface by consuming nutrient provided from the agar medium that contains it. Colonies develop in the third dimension as well (especially when the incubation medium is hard and the bacteria are well fed). The thickness of colonies in the present case is small compared with the size of the macroscopic colony patterns with which we are concerned. The present colony growth can thus be regarded as quasi-two-dimensional.

The agar concentration (C_a) was first fixed at some value in the range of 10–15 g/L. The resultant agar plates are rather hard, and according to the microscope observations bacterial cells do not move around on the surface of these hard agar plates. In other words, the cells show no active movement. The bacteria grow and perform cell division locally only by feeding on the nutrient peptone, which means that the nutrient diffusion may be important for colony formation under the condition of hard agar medium. (We later discuss the active movement of individual bacterial cells observed when the incubation agar medium is soft.) Incidentally, when a drop of ink is spilled on the surface of these hard agar plates, it takes more than a week for the ink to spread over the petri dish. In other words, the bacteria must be incubated more than a week to observe the effect of nutrient diffusion on colony formation.

First, the colonies had been incubated for 3 weeks after inoculation at various initial nutrient concentrations (C_n) in the range of 0–1 g/L (Matsushita and Fujikawa, 1990). An increase of C_n was found to enhance colony growth. In particular, growth does not occur without nutrient $(C_n = 0)$, which implies that the bacterial growth and cell division at the interface of colonies (local growth processes) are governed primarily by the presence of nutrient where they live. Although this result seems trivial, it is important in that the very existence of nutrient is deeply involved in colony growth.

Because the agar plate is fairly hard as a culture medium, bacterial cells cannot move around actively on the surface of the plate. They repeat cell division only by taking nutrient at their place of ''residence.'' This fact does not mean that the formation of colonies is trivial. The point is that nutrient diffusion may be important for colony growth under the condition of poor nutrient. In order to look at how nutrient diffusion affects the colonies as purely as possible, the diffusion length of the nutrient (l) must be made larger than the size of the colony. The simplest way to do so is to reduce the growth speed (v) of colonies as much as possible by lowering the nutrient concentration (C_n): $l = 2D_n/v$, where D_n is the diffusion coefficient of nutrient in the agar medium.

We fixed the initial nutrient concentration (C_n) at 1 g/L, a concentration that is still much less than normally is used (e.g., 15 g/L). We then observed colony patterns with characteristically branched structure after 3 or 4 weeks of incubation. One example is shown in Figure 13.1A. It has a randomly branched, outwardly open structure. The main branches consist of smaller branches with similar shapes, which also consist of even smaller but similar-shaped branches, and so on.

If any smaller part of a pattern appears similar to the pattern itself when the part is enlarged, the pattern is said to be a self-similar fractal (Mandelbrot, 1982; Feder, 1988; Vicsek, 1992). In other words, a self-similar fractal pattern is one in which similar structures are repeated hierarchically over a wide range of length scales. Such characteristic patterns are common in nature. For instance, looking closely at the surface of a cumulonimbus cloud reveals that it is indistinguishable from that seen from farther away. The bumpy surface of

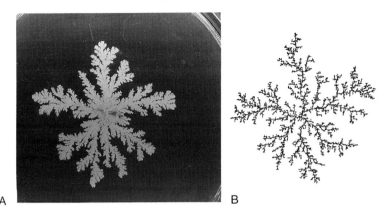

A   B

FIGURE 13.1 (A) Example of DLA-like colonies of *Bacillus subtilis*. Incubated at 35°C for 1 month on the surface of agar plates. C_n = 1 g/L, C_a = 10 g/L. (B) Example of two-dimensional DLA clusters obtained by computer simulations on a square lattice. Particle number $n = 10^4$. Note the similarity between the two.

the cloud seems to be self-similar. A human bronchial tree is another example. The present colony (Fig. 13.1A) also seems to exhibit a characteristically self-similar fractal structure.

On the other hand, Figure 13.1B shows an example of computer simulations in two dimensions of the diffusion-limited aggregation (DLA) model (Witten and Sander, 1981). This model is a simple one, and the growth rule is as follows: Initially, a seed particle of unit size is placed at the origin. A brownian particle of the same size is then released far away from the origin. When the brownian particle reaches the seed particle, it sticks to the seed, forming a two-particle cluster. Another brownian particle is again launched randomly but far away from the origin. When it collides with the cluster, it stops there and becomes a member of a now three-particle cluster. If the brownian particle moves too far from the cluster around the origin, the particle is discarded and another brownian particle is released. This aggregation process of brownian particles, one at a time, is repeated for as long as time and money allow, ultimately creating a large grouping called a DLA cluster. [This model explanation is naive. Much more efficient algorithms are now known for computer simulations of the DLA model (Meakin, 1988).]

The DLA model consists of two processes: the brownian movement of particles and their irreversibly sticking to a cluster. From the viewpoint of general pattern formation, the slowest process governs the pattern growth and determines the global structure. As is understood from the model explanation of DLA, particle sticking is an instantaneous process [the sticking probability of particles (p_s) = 1], and the brownian movement of particles (which is equivalent to the diffusion of particles) is clearly slower. Hence the model is called *diffusion-limited* aggregation and is said to be a prototype for the growth of clusters through diffusion-limited processes. [An opposite situation is also conceivable. Suppose the probability of particle sticking is small ($p_s \to 0$) or the density of brownian particles is high. The sticking process can then be slower than the diffusion. In this case the growth is said to be *reaction-limited*. The Eden model, discussed later, is the simplest reaction-limited growth model.]

When a brownian particle that came from far away approaches a DLA cluster, it may show some tendency to invade the ''fjord'' between branches of the cluster and enter the interior region. The trajectory of a brownian particle is not linear, however, but spread out widely. In fact, the trajectory is a self-similar fractal with the fractal dimension exactly $D = 2$, independent of the dimension of space where a brownian particle moves around. A brownian particle is therefore easily caught by any of the already grown, protruding branches of a cluster and is prevented from entering deep inside the cluster. This phenomenon is called the *screening effect*. The consequence is that the protruding outer branches of the cluster grow more, and the inner branches and interfaces stop growing. Hence a DLA cluster has characteristically ramified structures. The invasion of brownian particles and the screening due to cluster branches do not depend on the branch scales; that is, they take place over every scale of cluster branches. This qualitative assessment explains why DLA clusters are expected to have a characteristic, self-similar fractal structure. It was in fact confirmed by large-scale computer simulations (Meakin, 1988) that the DLA clusters are self-similar and the fractal dimensions are $D \cong 1.71$ and 2.50 in two and three spatial dimensions, respectively.

In this DLA model only one brownian particle is moving around in the space. The growth speed (v) of a DLA cluster is therefore slow; $v \to 0$ and the diffusion length (l) approaches infinity (∞). (Of course one can modify the original DLA model to a version of multibrownian particles in which brownian particles are supplied so as to maintain their constant concentration far away from a growing cluster. The diffusion length then assumes some finite value according to the concentration of brownian particles.) As described above, the brownian motion is equivalent to the diffusion:

$$\frac{\partial c}{\partial t} + D_c \nabla^2 c = 0 \tag{1}$$

where c denotes the concentration of brownian particles and D_c their diffusion coefficient.

In this original DLA case a cluster grows so slowly the time differentiation term in the above diffusion equation can be ignored ($\partial c/\partial t = 0$; the quasistatic approximation). Hence the diffusion equation is reduced to the Laplace equation:

$$\nabla^2 c = 0 \tag{2}$$

Diffusion-limited aggregation is therefore the simplest, or prototype, model for describing the pattern formation in a Laplacian field that satisfies the Laplace equation (Eq. 2). In fact, it is now well known that DLA is an interesting and important model that enables us to explain many seemingly different growing random patterns seen with various phenomena (Meakin, 1988; Avnir, 1989), such as electrodeposition (electrodeposits called metal leaves), dielectric break-down (Lichtenberg's figure, lightening), dendritic crystal growth (natural dendrites), viscous fingering (viscous fingers; e.g., dendritic patterns seen when forcibly separating two glass plates tightly sticking together by water), and chemical dissolution (e.g., of plasters). Mathematically, however, DLA is one of the unsolved problems; that is, no one has succeeded in deriving analytically the fractal dimension formula of DLA clusters despite the simplicity of the model.

It was demonstrated by extensive computer simulations (Meakin, 1986) that when biological growth is governed by DLA processes the growing patterns show characteristic features, such as screening, repulsion, and so on. Conversely, the existence of these effects enables us to confirm the DLA growth in biological systems. It was, in fact, not long before an unambiguous example was discovered in biology that could be understood in terms of the DLA processes (Fujikawa and Matsushita, 1989).

Let us again note the similarity between the two patterns shown in Figure 13.1. The patterns are strikingly similar, except that branches of the colony are thicker than those of the DLA cluster. In fact, averaged over about 25 samples of colony patterns from the same strain, we found the self-similarity to hold well over two orders of length scale and obtained a fractal dimension of $D = 1.72 \pm 0.02$. [See Feder (1988) or Vicsek (1992) about the measurement of fractal dimensions.] This finding is in good agreement with that of DLA clusters grown in two dimensions ($D \cong 1.71$).

In addition, as shown in Figure 13.2, we observed clear evidence of the existence of a screening effect that causes protruding exterior branches of a colony to grow more and prevents interior branches from growing. Note in

A   B

FIGURE 13.2 Growth of a DLA-like colony photographed 12 days (**A**) and 35 days (**B**) after inoculation. C_n = 1 g/L, C_a = 10 g/L. Note that many inner branches, typical examples of which are pointed to by triangles, are seen to stop growing afterward (screening effect).

Figure 13.2 that many interior branches, typical examples of which are indicated by white triangles, are seen to stop growing afterward. As described above, this behavior is exactly what we see during the growth of DLA clusters. We also observed the repulsion behavior of two neighboring colonies inoculated at two points simultaneously, as is clearly seen in Figure 13.3. It is also characteristic for the pattern formation in a Laplacian field (i.e., pattern formation through the DLA mechanism).

Taking these experimental results into account, it is clear that the present bacterial colonies grow through DLA processes. There remains one problem: What creates the Laplacian field or what diffuses? There are two main possibilities: (1) Nutrient diffuses in toward a colony, and (2) some waste material excreted by the bacteria diffuses out from the colony. It is equally possible that

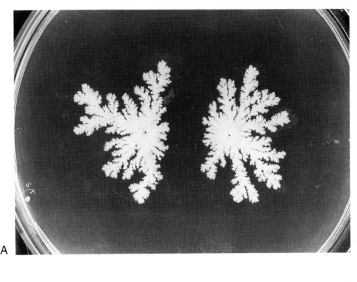

A

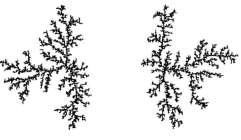

B

FIGURE 13.3   (A) Two neighboring colonies inoculated simultaneously at two points and then incubated for a month. $C_n = 1$ g/L, $C_a = 10$ g/L. (B) Two-dimensional DLA clusters grown from two seed particles. Total number of particles $n = 10^4$. Note the repulsion behavior seen in both (A) and (B).

FIGURE 13.4 Colony showing a clear tendency to grow toward the nutrient. Initially the nutrient peptone was only at the right corner of the dish.

either action can produce the DLA-like pattern, although the effect of nutrient concentration described before strongly suggests the first possibility.

To determine the sequence that probably takes place, we placed the nutrient at a corner of an incubation dish (there was no nutrient elsewhere initially) and inoculated bacteria at the center. It should be remembered that in the present case (i.e., under the condition of hard agar medium with poor nutrient) bacterial cells cannot move around on the surface of the agar plate. Nevertheless, the colony showed a clear tendency to grow toward the area where the nutrient was initially placed, as is clearly seen in Figure 13.4 (Matsushita and Fujikawa, 1990; Fujikawa and Matsushita, 1991).

This result inevitably leads us to accept the first possibility described above: that nutrient diffusion contributes mainly to colony formation under the present condition. We can therefore conclude that for hard agar medium with poor nutrient the present bacterial colonies grow via the DLA mechanism in the nutrient concentration field, which plays the role of the Laplacian field (c) in Eq. (2) (Matsushita and Fujikawa, 1990). In this sense we can assert that the global mechanism of colony formation under this condition is well understood "physically."

We have also confirmed the DLA-like colony morphology for many common rod-shaped bacteria, such as *Escherichia coli* and *Salmonella typhimurium*, under similar conditions (Matsuyama and Matsushita, 1992). It implies that DLA growth is universal in the formation of bacterial colonies.

Morphological Changes During Colony Formation

Let us next discuss the change of colony patterns and the morphological phase diagram under varied environmental conditions (Ohgiwari et al., 1992). Colony patterns were found to change drastically when varying the concentrations of nutrient (C_n) and agar (C_a). Figure 13.5 shows the morphological phase diagram

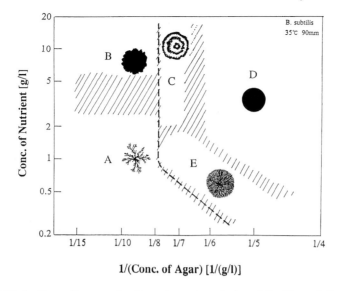

FIGURE 13.5   Phase diagram of pattern change in colonies of *Bacillus subtilis* (wild-type OG-01). Thick broken line indicates the boundary of the active movement of bacterial cells inside colonies.

of colonies observed in the present experiments. The abscissa indicates the inverse of the agar concentration, which implies the softness of the agar incubation medium. Note also that the ordinate is represented by a logarithmic scale. Hence, agar plates become softer as the diagram is followed from left to right, and they become richer in nutrient from bottom to top. Absolute numerical values on the abscissa and ordinate should not be taken too seriously, however. Phase boundaries of colony patterns do not have sharp characteristics, compared with those of ordinary phase diagrams seen in equilibrium thermodynamics, such as the pressure–temperature diagram of water (ice, vapor). We also believe that bacteria seem to become accustomed to (or ''learn about'') harsh environmental conditions such as poor nutrient quality or hard agar medium. We therefore avoid using progeny bacteria over too many generations and from time to time use original ones preserved in a refrigerator.

Here we classify colony patterns into five types, observed in the regions labeled A–E, respectively, in Figure 13.5. Typical patterns observed in the regions are also seen. It can be seen how strongly the variation in environmental conditions influences the colony patterns. In the following sections we discuss the characteristics of colony patterns in each region of Figure 13.5 and their morphological changes between regions. It should be noted in advance, however, that although we describe how colonies grow in each region of the phase diagram or how their morphologies change from one region to another, we still know little about the mechanisms or why. In this sense we believe that there remain many interesting problems to be solved in the future. For the time being, however, we think that the effect of bacterial mutation should not be taken into consideration because the reproducibility of colony patterns shown in the phase diagram was confirmed repeatedly.

Region A: Diffusion-Limited Growth

In region A—at low C_n (poor nutrient medium) and high C_a (hard agar plates)—we obtained DLA-like fractal colony patterns, as described in detail in the previous section.

Region B: Eden-like Growth

Moving from region A to region B—increasing C_n with the C_a more or less fixed (C_a = 10–15 g/L)—colonies showed gradual crossover from highly branched to compact patterns at high C_n (rich nutrient medium). Branches of a colony became gradually thicker as C_n was increased, and eventually they fused to form a compact pattern, an example of which is shown in Figure 13.6A. As seen in this figure, however, an outwardly growing interface still looks rough, and the colonies grew somewhat faster (e.g., they reached a size of about 5 cm within 1 week of incubation). Moreover, two colonies that had been inoculated simultaneously at two points grew and came close together in region B (Fujikawa and Matsushita, 1991), in contrast to region A (Fig. 13.3). This finding implies that the diffusion length (l) of the nutrient is small. This behavior is understandable: If the agar medium contains enough nutrient, its consumption by bacteria does not produce wide nutrient-depletion region around a colony. The width of this depletion region is of the same order as l.

These properties are characteristic of Eden growth. The Eden model (Eden, 1961) is the simplest one that produces clusters whose inner structure is almost completely compact but whose surface is comparatively rough. In fact, this model was proposed to describe the growth of cell colonies (Eden, 1961), tumor (Williams and Bjerknes, 1972), or cancer (Tautu, 1978). The growth rule is simple: Starting from a seed particle (or particles with any alignment such as a line) as an initial cluster, one of the perimeter sites of the cluster is chosen randomly with equal probability and is incorporated into the cluster as its member. This process is simply repeated. The growth process of the Eden model is reaction-limited in the following sense. We can regard an Eden cluster as being surrounded by a sea of diffusing particles. In other words, perimeter sites (or particles) of the cluster surface are diffusing particles that are selected randomly and incorporated into the cluster one by one. The incorporation of adjacent diffusing particles onto the cluster surface is regarded as a chemical reaction. This process is much slower than the diffusion process in this case because diffusing particles are always there and need not diffuse at all. It must again be emphasized that the slower process governs the global properties of cluster formation in general.

As seen from the example in Figure 13.6B, Eden clusters are compact, with the fractal dimension (D) equal to the dimension (d) of the space in which

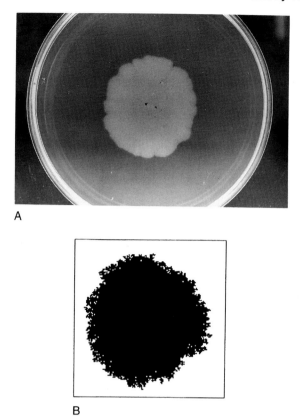

A

B

FIGURE 13.6 (A) Example of Eden-like colonies of *Bacillus subtilis* in region B. $C_n =$ 8 g/L, $C_a = 12$ g/L. This photograph was obtained after 1 week of incubation. (B) Example of Eden clusters obtained by computer simulations. The particle number $n = 5000$.

they are produced ($D = d$). This phenomenon can be explained qualitatively by the steady existence of diffusing particles around an Eden cluster that are ready to be incorporated into it. Eden clusters are therefore not intrinsic self-similar fractals with $D < d$—in contrast to, for example, the DLA clusters shown in Figure 13.1B. However, because the Eden model is the simplest model for reaction-limited growth, Eden clusters, as well as DLA clusters, are one of the universal patterns seen in nature.

Because of the stochastic nature of the particle incorporation described above, the surface of Eden clusters cannot be smooth, as clearly seen in Figure 13.6B. In fact, the surface roughness of Eden clusters has been known to have interesting statistical properties (Family and Vicsek, 1991). Suppose that an Eden cluster (or any other compact cluster) is growing toward a positive y-direction from line seeds on the x-axis in two-dimensional (x, y) space. There is then a curve of growing interface, as shown in Figure 13.7. We can define the height y_i of the interface at the seed site x_i. Let us now take an arbitrary

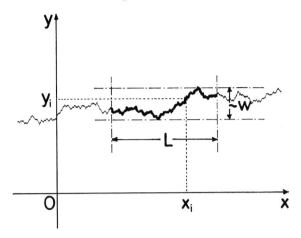

FIGURE 13.7   Growing front of a colony.

interval (L) along the x-direction. The average height (h) is then defined over the interval as

$$h = \frac{1}{L} \sum_{i}^{L} y_i \qquad (3)$$

The sum is taken over L sites belonging to the interval. We can also define the width (w) of a rough interface as the difference between the highest and lowest heights in the interval. However, the standard deviation (w) of the interface height, defined as

$$w^2 = \frac{1}{L} \sum_{i}^{L} (y_i - h)^2 \qquad (4)$$

is statistically a better quantity to characterize the roughness of an interface. An ensemble average should be taken over as many cases as possible even for a fixed value of the interval width L (e.g., by shifting the interval).

It was found in the Eden model (Family and Vicsek, 1985) that for $h \ll L$ (i.e., at the early stages of growth) the width (w) shows the following scaling behavior

$$w \sim h^\beta \qquad (5)$$

whereas for $h \gg L$ (later stages) the width (w) is scaled by L as

$$w \sim L^\alpha \qquad (6)$$

Here the symbol $\sim$ indicates proportionality. The constant α is called the roughness exponent. [If the exponent α in Eq. (6) is not equal to 1.0, it means that the scaling properties of the interface along x direction (parallel to the interface) is different from that along y direction (perpendicular to the interface). This kind of anisotropic scaling behavior is called "self-affinity." On the other hand, when $\alpha = 1$, the scaling behavior is isotropic and independent of directions measured, an example of which is the DLA cluster shown in Figure 13.1B. In this sense the symmetry self-similarity is a special case of the symmetry self-affinity (Feder, 1988; Matsushita and Ouchi, 1989; Vicsek, 1992).] We can unify the relations in Eqs. (5) and (6) as

$$w(L, h) \sim L^\alpha f(h/L^z) \qquad (7)$$

where $f(x)$ is a scaling function satisfying that $f(x) = x^\beta$ for $x \ll 1$ and 1 for $x \gg 1$, and $z = \alpha/\beta$. The interesting point is that the relation in Eq. (7) is now known to hold for various kinds of growing interfaces (Family and Vicsek, 1991). In the case of the Eden model, $\alpha = 0.50$ and $\beta = 0.33$ in two dimensions.

As seen from the comparison of Figures 13.6A and B, the colony pattern shown in Figure 13.6A appears Eden-like. To confirm real Eden growth, however, one must show that the growing interface is self-affine instead of self-similar and obeys characteristic scaling behavior described in Eq. (6) or, if possible, in Eq. (7) with the values of exponents proper to the model. Vicsek et al. (1990) reported that although the self-affinity of the growing interface of colonies of *B. subtilis* and *E. coli* was confirmed, their scaling exponents ($\alpha \cong 0.78$) were not equal to those of the Eden model ($\alpha = 1/2$). Figure 13.8 is an example of the self-affinity analysis for colonies. Here we made use of Eq. (6) and examined the self-affinity of a colony grown from line inoculation. The inoculation line is seen in Figure 13.8A as a blackish line near the bottom. As shown in Figure 13.8B, Eq. (6) seems to hold with $\alpha \cong 0.72$ within some range of length scales (L), supporting the above report. An interesting future problem then is why colonies in region B do not exhibit the scaling characteristic of real Eden growth. In fact, that α is not equal to 0.5 in two dimensions ($d = 2$) implies that there is some long-range correlation in the interface growth; that is, the growth at one place is not independent of the growth at other places far away. Are bacteria communicating with each other in some way or other?

The self-affinity of colonies should definitely be studied further. As a matter of fact, the self-affinity of various growing interfaces has attracted the attention of scientists in diverse fields of science and technology (Family and Vicsek, 1991). Related phenomena include the frontline of wetting of paper and the interface of semiconductors grown by molecular-beam epitaxy (MBE) (Jullien et al., 1992).

Region C: Growth of Concentric Ring Patterns

The agar medium was next softened a little more than in region B by decreasing the agar concentration (C_a) but leaving the nutrient content high (high C_n). Characteristic concentric-ring-like colony patterns were observed after about 1.5 days of incubation (Fujikawa, 1992) (Fig. 13.9). Point-inoculated bacteria repeated cell division and migrated outwardly to create a colony. The colony grew like a disk with a rough surface to some diameter and then stopped growing. After a while bacteria started migrating again from the perimeter of the colony, which grew to some larger diameter and stopped. By repeating

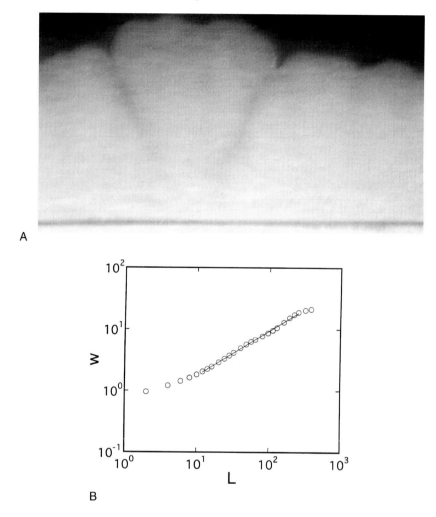

A

B

FIGURE 13.8   Examination of the self-affinity for an Eden-like colony in region B. (**A**) Colony grown from line inoculation. The inoculation line, seen as a black line near the bottom, is about 9 cm long. C_n = 10 g/L, C_a = 10 g/L. Photograph was obtained after 4 days of incubation. (**B**) Log-log plot of width (w) versus interval (L) of the growing interface shown in **A**. The unit of length is 0.23 mm. A linear least-squares analysis (shown as a line) yields the roughness exponent $\alpha \cong 0.72$.

these processes, characteristic colonies such as that in Figure 13.9 were obtained. The outermost part of the colony in Figure 13.9 appears faint and unclear because bacteria at that location are still migrating and have not yet settled.

This kind of concentric ring pattern is similar to Liesegang rings (Henisch, 1988), which emerge when some chemical product (e.g., PbI_2) precipitates as fine crystalline particles due to a chemical reaction in a semisolid medium such

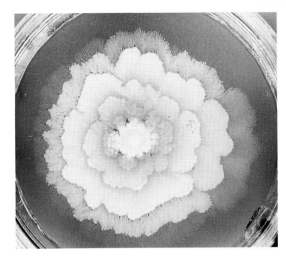

FIGURE 13.9   Concentric-ring-like colonies of *Bacillus subtilis* observed in region C. C_n = 25 g/L, C_a = 7 g/L. Photograph was obtained after incubation for 1.5 days.

as agar (e.g., reaction between $PbNO_3$ and KI in agar medium). The concentric ring-like colonies are commonly seen for various species of *Proteus*, such as *P. mirabilis* (Shapiro, 1988, 1991). The formation mechanism of such colonies should therefore be investigated as an interesting problem inherent to bacteria for the time being. The interested reader should consult Chapter 7 for a detailed discussion.

Region D: Disk-like Growth

In the wide region D, with low C_a (soft agar medium), colonies spread homogeneously with smooth, clear-cut, circular interfaces and no branching, at least macroscopically. They grew rapidly and appeared transparent. The example in Figure 13.10 was obtained after a half-day of incubation. Because in this case one can quantitatively measure the growth rate of the colony interface, the growth mechanism is discussed from a population dynamics viewpoint (see later in the chapter).

Region E: DBM-like Growth

In the narrow region E between A and D, where the nutrient is poor and the medium softness is intermediate, colony morphology took on densely branched patterns clearly reminiscent of the so-called dense-branching morphology (DBM) (Ben-Jacob and Garic, 1990). A typical example (Fig. 13.11) was obtained after about 1 day of incubation. Colonies in this region branch densely, whereas the advancing envelope looks characteristically smooth compared with DLA-like colonies. In this sense the colony patterns seen in region D seem as if dense branches fused to form homogeneously spreading patterns.

The DBM-like patterns were observed not only for the present species (*B. subtilis*) but for other species as well, such as *Serratia marcescens* (Matsuyama

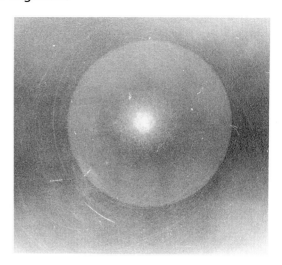

FIGURE 13.10 Simply spreading disk-like colony of *Bacillus subtilis* observed in region D. $C_n = 3$ g/L, $C_a = 5$ g/L. Photograph was obtained after 13 hours of incubation.

and Matsushita, 1993). These patterns are also seen with various inorganic phenomena, such as electrodeposition (Grier et al., 1986; Sawada et al., 1986; Ben-Jacob and Garic, 1990), dendritic crystal growth (Ben-Jacob and Garic, 1990; Yasui and Matsushita, 1992), and viscous fingering (Ben-Jacob and Garic, 1990), described above in the section on DLA growth. In other words, DBM patterns as well as DLA and Eden ones, are regarded as universal patterns seen in nature.

No simple models to describe DBM formation unambiguously have been proposed. [See, however, the modeling attempt by Ben-Jacob et al. (1994a).] We have observed that two DBM-like colonies inoculated at two points grow simultaneously, come close together, and then repel each other with an almost

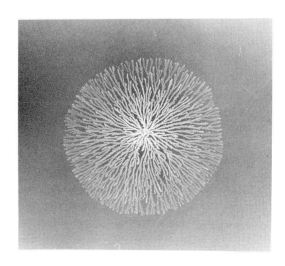

FIGURE 13.11 DBM-like colony of *Bacillus subtilis* observed in region E. $C_n = 0.5$ g/L, $C_a = 5$ g/L. Photograph was obtained after 25 hours of incubation.

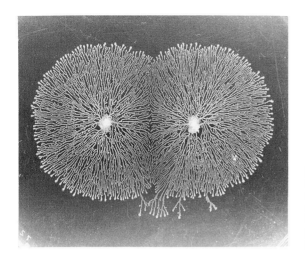

FIGURE 13.12 Two DBM-like colonies in region E were inoculated simultaneously at two points and then incubated for about 1 day. They came close together but eventually repelled each other.

constant gap distance approximately equal to the average distance of neighboring branches (Fig. 13.12). This finding implies that DBM-like patterns seen in region E cannot be described by the diffusion field of nutrient alone. Otherwise the envelope would become rough, similar to the interface of Eden-like colonies in region B (Fig. 13.6A). The diffusion field alone cannot explain the characteristically smooth envelope observed for DBM-like patterns. We must look for an additional mechanism that smooths out the envelope of a growing branched pattern or makes the diffusion length (l) almost constant. The candidate may be a variation of agar surface tension along the envelope due to some surfactant secreted by bacteria. In fact, we have observed a thin, transparent halo some scores of micrometers ahead of the envelope of a growing colony in regions D and E. It is so thin, however, that we have not succeeded in analyzing the material or identifying it chemically. Elucidation of this observation is an interesting problem to be solved in the future.

Effect of Bacterial Cell Movement

Why were such characteristically different colony patterns observed only by varying the softness of the incubation medium and the amount of nutrient contained in it despite the simple shape of bacteria? Because the qualitative reproducibility of patterns themselves was confirmed, it seems unnecessary to consider bacterial mutation. There may be some phenomenological explanation for the pattern change, not to mention biological or essential reasons.

Figure 13.5 showed that the morphological change is conspicuous from region A (DLA-like patterns) to region E (DBM-like), compared with the gradual change from A to B (Eden-like). The growth rate in region E is also much higher than that in region A. It took a colony about 1 month in A and 1 week in B to grow to about 5 cm (half the diameter of a petri dish containing agar medium), whereas it took only 1 day in C and E and a half-day in D. This finding suggests an essential difference in the way colonies grow. In fact, by microscopically observing the growing zones (tips of outwardly growing branches or growing fronts) of colonies, we recognized two distinct growing processes.

The first process exhibits no active moment of individual bacterial cells when growth is taking place on the colony interface at high C_a (hard agar; regions A and B). This type of growth is relatively "static": Only cells comprising the outermost part of a colony feed on nutrient and increase their population mass by cell division. It takes more than a few days to develop a specific pattern. Cells in the inner portion change to spores and enter a rest phase, or hibernation. They never grow afterward unless the nutrient condition is improved. Several layers of bacterial cells in these regions pile up to form colonies.

The other type of growth process exhibits active movement of individual cells inside colonies for intermediate and low C_a (soft agar; regions C, D, E). Colonies obtained in these regions are therefore composed of a monolayer of bacterial cells; that is, the cells usually do not pile up. The bacterium we used (B. subtilis), as well as other bacilli such as E. coli and S. typhimurium, have peritrichous flagella. They move in water by collectively rotating their flagella.

Why can they move around on the surface of extremely sticky agar medium? According to microscopic observations, the colony growth of this type was found to be driven by remarkably "dynamic" active cell movement. For instance, in region E, as seen in Figure 13.13, the outermost growing front of a colony is enveloped by a thin layer (thickness $\simeq$ 5 μm) of bacterial cells whose movement is dull, whereas cells inside the layer move around actively and erratically (which is why the outer layer of the colony in Figure 13.13 looks darker and clear, and bacteria inside look somewhat fuzzy). Sometimes active cells inside the layer of a colony growing front collide with the layer, break through it, and rush out. The rushed-out cells then become immediately dull and part of the layer. (Bacterial cells do not seem able to move around freely on the agar surface where they have never visited before. This behavior is supposedly due to the stickiness of the agar medium, that is, they have to wriggle around to advance toward the virgin sticky area.) As a consequence, the layer or frontline expands somewhat. Although expansion of the frontline is not as speedy as the active movement of individual cells inside, the growth rate of a colony as a whole is still much higher than that of the first type, seen in regions A and B. As imagined from these observations, although bacterial cells invade fresh areas of the surface of agar medium with great difficulty, once they manage to do so they move relatively freely and actively just behind the areas. It looks as if bacteria, when advancing, leave trails soaked by water. Hence they may move around actively wherever they have already visited (i.e., inside a colony). Deep inside the growth front, however, cell movement is again inactive for the colonies grown in regions C and E.

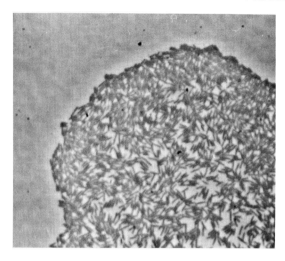

FIGURE 13.13 Photomicrograph of a growing branch tip of DBM-like colonies in region E, such as is shown in Figure 13.11. Width of the photograph is 0.2 mm.

A thick broken line in Figure 13.5 indicates the boundary beyond which the active movement of bacterial cells described above was observed. Note that the line coincides with almost vertical-phase boundaries of morphologies between regions A and E and between B and C. The verticality is understandable because the cell movement is mainly influenced by the softness of the agar plates. The bending from verticality in the lowermost part (boundary between

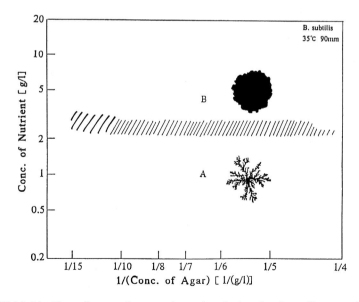

FIGURE 13.14   Phase diagram of pattern change in colonies of an immotile mutant (OG-01b) of *Bacillus subtilis*.

A and E) may be attributed to inactivity of cells due to starvation because of poor nutrient.

Bacterial cells are known to be able to move actively only by using their flagella. Hence if mutants have no flagellum, they cannot exhibit active movement. Because, as described above, the movement seems to induce colony morphology change, we carried out the same experiment already described except we used immotile mutant with no flagellum (OG-01b). The mutant was obtained from the wild-type strain (OG-01) by nitrosoguanidine mutagenesis. We found a surprisingly simple phase diagram consisting of only regions A and B, as seen in Figure 13.14 (Ohgiwari et al., 1992). The observed morphological change is only the gradual crossover from DLA-like to Eden-like colony patterns over the range of agar concentrations examined. We have not observed DBM-like or homogeneously spreading, disk-like patterns. In other words, we found that regions A and B in Figure 13.5 expand laterally to the entire region, and regions C, D, and E disappear. We can therefore conclude that the active movement of bacterial cells seen for the wild type (OG-01) triggers the morphological change.

Population Dynamics Approach to Colony Formation

We noted above in detail that colony formation in region A can be explained by the DLA model. Unfortunately, it is still difficult to understand, even phenomenologically, the formation of concentric-ring-like colonies in region C and DBM-like colonies in region E. If, however, colonies appear macroscopically simple, we may be able to understand the formation mechanism quantitatively. Therefore we discuss the growth mechanism of homogeneously spreading, disk-like patterns seen in region D from another viewpoint (Wakita et al., 1994).

As seen in Figure 13.10, a colony in region D macroscopically appears as a perfect disk with a clear-cut interface. Microscopically, however, it is seen that the growing front is obscure (Fig. 13.15A) compared with that in regions C and E (Fig. 13.13). In other words, the interface cannot be defined clearly in region D. On the other hand, as seen in Figure 13.15B, the distribution of bacterial cells inside a colony looks almost homogeneous, and the bacteria are moving around actively even deep inside the colony. By comparing Figure 13.15B with Figure 13.13, it is seen that the population density of bacterial cells inside a colony in region D is lower even in a monolayer of cells than in region E, which explains why colonies in region D look transparent and are difficult to observe. Inside the colony, active but erratic movement of individual cells is seen almost everywhere. Hence we assume as the 0th approximation that the bacterial cell movement in region D can be described in terms of diffusion in two dimensions. (According to close observations, however, their motion does not look completely brownian but somewhat collective, especially when their population is dense. They tend to make up a group of several cells parallel with each other, which move around together. Of course, one or two cells may join a group and another one or two may leave it from time to time.)

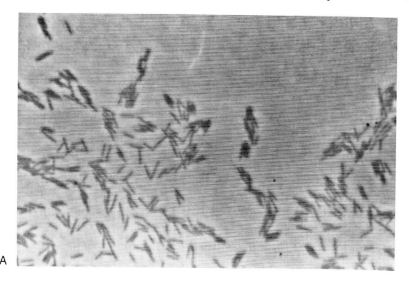

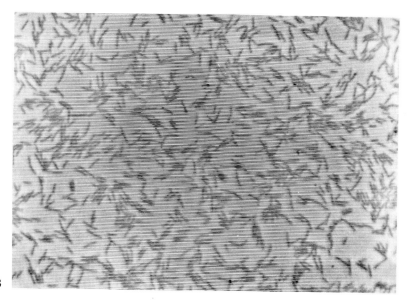

FIGURE 13.15   Photomicrographs of a disk-like colony in region D, shown in Figure 13.10. **(A)** Note the fuzzy interface of the colony. **(B)** Inside the colony. Widths of both figures are 0.2 mm.

Our intention is to regard the colony formation in region D as a combination of the multiplication of cells inside the colony and their outward diffusion.

The spatiotemporal variation of population density of bacterial cells $b(r, t)$ and concentration of nutrient $n(r, t)$ (denoted as C_n for the prepared concentration in the preceding sections) is then represented by reaction-diffusion-type equations.

$$\frac{\partial b}{\partial t} = \nabla \cdot (D_b \nabla b) + f(b, n) \tag{8}$$

$$\frac{\partial n}{\partial t} = D_n \nabla^2 n - vbn \tag{9}$$

where D_b and D_n are the diffusion coefficients of bacterial cells and nutrient, respectively; v is the consumption rate of nutrient by bacteria; and $f(b, n)$ denotes the reaction term due to local bacterial growth. D_b is in general dependent on both b and n. In the following, however, we regard D_b as a constant because we investigate here only in region D. Moreover, we assume that $f(b, n)$ can be described by the following logistic-like form.

$$f(b, n) = [\varepsilon(n) - \mu b]b \tag{10}$$

where the term $\varepsilon(n)$ represents the rate of malthusian growth of individual cells, and μ is the coefficient of competition among cells that describes the suppression of population increase. This form is plausible because too many bacterial cells restrain themselves from increasing their population.

Let us now consider the limiting case in which nutrient is so rich that n can be put spatially constant (i.e., $\varepsilon(n) = \varepsilon$). It should be noted that this does not mean that ε is independent of n. Equations (8) and (9) are then decoupled, and Eq. (8) becomes

$$\frac{\partial b}{\partial t} = D_b \nabla^2 b + (\varepsilon - \mu b)b \tag{11}$$

Known as the Fisher equation (Murray, 1989), this equation and its traveling wave solutions have been widely studied. In particular, the equation asymptotically yields isotropically spreading, homogeneous solutions with stable traveling wavefronts of constant speed in two dimensions. The population density inside and wavefront speed are given, respectively, by $b = \varepsilon/\mu$ and $v = 2(\varepsilon D_b)^{1/2}$.

This Fisher equation may describe the simple, homogeneously spreading colony pattern observed in region D in Figure 13.5. Let us here discuss experimental confirmation for this conjecture. It turned out that it is difficult to estimate the value of ε and especially μ experimentally. Here we assume that μ is independent of the nutrient concentration n. This assumption is plausible because basically μ describes how strongly two bacterial cells repel each other when they encounter. This situation is in clear contrast to the case of ε, the rate of the malthusian population increase of bacteria, which may be strongly dependent on n. We therefore argue the consistency when describing experimental data by the Fisher equation (Eq. 11) under the assumption of constant μ.

We first allowed a pair of colonies that had been inoculated simultaneously to collide with each other. They were observed to fuse, as shown in Figure

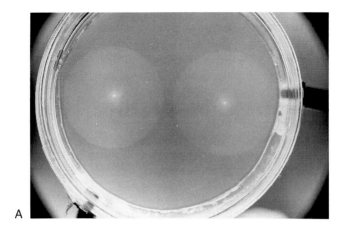

A

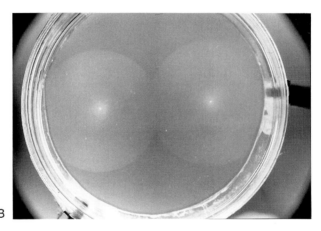

B

FIGURE 13.16   Two disk-like colonies in region D were inoculated simultaneously at two points and then incubated for about 0.5 day. They came close together independently (**A**) and then fused (**B**).

13.16, in striking contrast to the case of DLA-like colonies in region A (Fig. 13.3A) or even DBM-like colonies in region E (Fig. 13.12), which were found to repel each other. Numerical calculations of Eq. (11), started from two points in two dimensions, were found to yield exactly the same behavior. This finding is the first (but qualitative) evidence of the conjecture.

Second, we estimated the population density (b) of bacterial cells. We measured the occupation rate of an area by cells in a homogeneously populated region just inside the colony interface for various nutrient concentrations (n) in region D. We obtained the result that b is approximately proportional to n, which implies that, under the assumption of the Fisher equation with constant μ, ε should be proportional to n, that is, $\varepsilon = \varepsilon_1 n$, where ε_1 is some constant.

This proportionality may not be true in general. However, we are here concerned with the growth rate of bacterial cells only in the specific region D.

Next we measured the diffusion coefficient of bacterial cells (D_b) for various nutrient concentrations (n). As described before, the movement of individual cells does not look like brownian motion over short periods or with high nutrient concentrations because of their elongated shape and high population density, respectively. Nevertheless, we tried to estimate D_b from the ensemble average of squared displacements of individual cells $\langle R^2 \rangle$ during time interval t using the relation $\langle R^2 \rangle = 2D_b t$. We found that D_b does not depend much on the nutrient concentration n, that is, $D_b \sim n^0$.

Finally we measured the interface growth speed (v) for various nutrient concentrations (n). The diameters of growing colonies are not quite proportional to the incubation time. There always seems to be a dormant period of 7–8 hours after incubation during which colonies do not grow. It appears that bacteria need this period to adjust to a new environment and initiate their new life. After this period, however, colonies grow rather fast. Hence we estimated the growth speed (v) when the colony diameter reaches 5 cm. The results clearly showed that v is proportional to $n^{1/2}$, as shown in Figure 13.17. This expression is compatible with the speed $v = 2(\varepsilon D_b)^{1/2} \sim n^{1/2}$ that the Fisher equation gives.

All these experimental results are consistent with the behaviors derived from the Fisher equation. We therefore conclude that the colony formation in region D in Figure 13.5 can essentially be described by the Fisher equation (Wakita et al., 1994). An interesting theoretical problem to be addressed in the near future is if this population dynamics approach can be generalized to other regions, such as C and E, to explain the colony formation observed there.

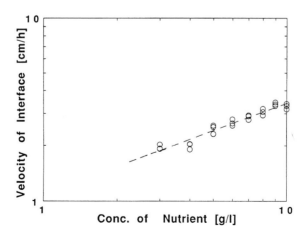

FIGURE 13.17   Growth speed (v) of the interface of colonies in region D, measured as a function of the nutrient concentration ($n = C_n$). Note the double-logarithmic plotting. The slope of a broken line determined by the least-mean-squares fitting is 0.49, clearly implying $v \sim n^{1/2}$.

Summary and Discussion

We have found that colony patterns of *B. subtilis* can be changed on the surface of agar medium by varying the softness of the medium and the concentration of nutrient contained in it. We have also found that there are a few cases where the growth mechanism of colonies can be understood phenomenologically and mathematically, such as DLA-like colonies in region A and disk-like colonies in region D in Figure 13.5. It should be noted that we still know little about the growth mechanism of the interesting colony patterns observed in regions C and E. Their elucidation is important for future exploration. Interestingly and importantly, various characteristic colony patterns reported here, such as the DLA-like, Eden-like, and DBM-like patterns, have been observed for many other species, such as *E. coli* and *S. typhimurium* (Matsuyama and Matsushita, 1992, 1993). This fact certainly indicates that these patterns are common to colonies of bacteria in general.

The observations described so far were mainly concerned with the global or macroscopic aspects of colony formation. There are, of course, many interesting problems with the details. For instance, by investigating pattern changes of colonies from region D to C or E in detail one may be able to understand more about the colony formation mechanism of these regions. Spiral colonies are sometimes obtained in regions C and E. It is possible that the microscopic mode of active motion of individual bacterial cells is reflected in the macroscopic colony patterns, but we have no knowledge about this possibility as yet. We have observed that in region D, but very close to C, the first disk-like colony is followed by the second colony from the original inoculation point. It will be interesting to establish the growth condition for this peculiar behavior and carry out the microscopic observations.

It was mentioned briefly that the active motion of bacterial cells inside a colony in region D looks erratic but is not precisely brownian motion. The cells tend to move around in a group packed parallel with each other. It is an interesting problem to inquire why they do so and to study the quantitative characterization of such bacterial movement.

What we have described so far are, in all cases, multicellular and nontrivially cooperative behaviors of bacteria usually called unicellular organisms. Bacteria such as *B. subtilis* and *S. marcescens* secrete surfactants when forming their colonies (Matsuyama et al., 1989, 1992, 1993) that drastically change colony morphology. In fact, mutants defective in producing surfactants were found to exhibit colony patterns different from those of the wild type. It was also shown that surfactants isolated from other species of bacteria have the same effect; that is, mutants defective in producing surfactants form a colony of almost the same pattern as the wild type when the surface of the agar medium is coated by surfactant (Matsuyama et al., 1993). Thus by producing surface-active materials bacteria manage to relax the surface tension of water, which strongly influences them because of their small size and enhances their freedom of active movement. The consequence may be strikingly characteristic colony patterns, such as that shown in Figure 13.11.

At some stage of their evolution bacteria acquired the ability to physically improve or chemically relax difficult environmental conditions by secreting surfactants or amino acids. Collective motions of bacteria were a natural prerequisite for this development. Such multicellular, cooperative behaviors of bacteria might have been the origin of the multicellular organisms.

Acknowledgments

The author thanks T. Matsuyama for the many useful conversations and joint research. He is also grateful to H. Fujikawa, M. Ohgiwari, J. Wakita, A. Nakahara, K. Komatsu, Y. Shimada, H. Itoh, N. Shigesada, K. Kawasaki, and many others for their kind contributions to the present work.

References

Avnir, D. (ed) (1989) *The Fractal Approach to Heterogeneous Chemistry.* Wiley, Chichester.

Ben-Jacob, E., and Garik, P. (1990) The formation of patterns in non-equilibrium growth. *Nature* **343**:523–530.

Ben-Jacob, E., Shmueli, H., Shochet, O., and Tenenbaum, A. (1992) Adaptive self-organization during growth of bacterial colonies. *Physica A* **187**:378–424.

Ben-Jacob, E., Shochet, O., Tenenbaum, A., Cohen, I., Czirok, A., and Vicsek, T. (1994a) Generic modelling of cooperative growth patterns in bacterial colonies. *Nature* **368**:46–49.

Ben-Jacob, E., Tenenbaum, A., Shochet, O., and Avidan, O. (1994b) Holotransformations of bacterial colonies and genome cybernetics. *Physica A* **202**:1–47.

Eden, M. (1961) A two-dimensional growth process. In *Proceedings of the Fourth Berkeley Symposium on Mathematical Statistics,* Vol. IV: *Biology and Problems of Health,* J. Neyman (ed.), pp. 223–239, University of California Press, Berkeley.

Family, F., and Vicsek, T. (1985). Scaling of the active zone in the Eden process on percolation networks and the ballistic deposition model. *J. Phys. A Math. Gen.* **18**: L75–L81.

Family, F., and Vicsek, T. (eds) (1991) *Dynamics of Fractal Surfaces.* World Scientific, Singapore.

Feder, J. (1988) *Fractals.* Plenum Press, New York.

Fujikawa, H. (1992) Periodic growth of *Bacillus subtilis* colonies on agar plates. *Physica A* **189**:15–21.

Fujikawa, H., and Matsushita, M. (1989) Fractal growth of *Bacillus subtilis* on agar plates. *J. Phys. Soc. Jpn.* **58**:3875–3878.

Fujikawa, H., and Matsushita, M. (1991) Bacterial fractal growth in the concentration field of nutrient. *J. Phys. Soc. Jpn.* **60**:88–94.

Grier, D., Ben-Jacob, E., Clarke, R., and Sander, L.M. (1986) Morphology and microstructure in electrochemical deposition of zinc. *Phys. Rev. Lett.* **56**:1264–1267.

Henisch, H.K. (1988) *Crystals in Gels and Liesegang Rings.* Cambridge University Press, Cambridge.

Jullien, R., Kertész, J., Meakin, P., and Wolf, D.E. (eds) (1992) *Surface Disordering: Growth, Roughening and Phase Transitions.* Nova Science, Commack, New York.

Mandelbrot, B.B. (1982) *The Fractal Geometry of Nature.* Freeman, San Francisco.

Matsushita, M., and Fujikawa, H. (1990) Diffusion-limited growth in bacterial colony formation. *Physica A* **168**:498–506.

Matsushita, M., and Ouchi, S. (1989) On the self-affinity of various curves. *Physica D* **38**:246–251.

Matsuyama, T., and Matsushita, M. (1992) Self-similar colony morphogenesis by gram-negative rods as the experimental model of fractal growth by a cell population. *Appl. Environ. Microbiol.* **58**:1227–1232.

Matsuyama, T., and Matsushita, M. (1993) Fractal morphogenesis by a bacterial cell population. *C.R.C. Crit. Rev. Microbiol.* **19**:117–135.

Matsuyama, T., Harshey, R.M., and Matsushita, M. (1993) Self-similar colony morphogenesis by bacteria as the experimental model of fractal growth by a cell population. *Fractals* **1**:302–311.

Matsuyama, T., Kaneda, K., Nakagawa, Y., Isa, K., Hara-Hotta, H., and Yano, I. (1992) A novel extracellular cyclic lipopeptide which promotes flagellum-dependent and -independent spreading growth of *Serratia marcescens*. *J. Bacteriol.* **174**:1769–1776.

Matsuyama, T., Sogawa, M., and Nakagawa, Y. (1989) Fractal spreading growth of *Serratia marcescens* which produces surface active exolipids. *FEMS Microbiol. Lett.* **61**:243–246.

Meakin, P. (1986) A new model for biological pattern formation. *J. Theor. Biol.* **118**:101–113.

Meakin, P. (1988) The growth of fractal aggregates and their fractal measures. In *Phase Transitions and Critical Phenomena*, Vol. 12, C. Domb and J.L. Lebowitz (eds.), pp. 335–489. Academic Press, Orlando, FL.

Murray, J.D. (1989) *Mathematical Biology*. Springer-Verlag, Berlin.

Ohgiwari, M., Matsushita, M., and Matsuyama, T. (1992) Morphological changes in growth phenomena of bacterial colony patterns. *J. Phys. Soc. Jpn.* **61**:816–822.

Sawada, Y., Dougherty, A., and Gollub, J.P. (1986) Dendritic and fractal patterns in electrolytic metal deposits. *Phys. Rev. Lett.* **56**:1260–1263.

Shapiro, J.A. (1988) Bacteria as multicellular organisms. *Sci. Am.* **256**:82–89.

Shapiro, J.A. (1991) Multicellular behavior of bacteria. *ASM News* **57**:247–253.

Singleton, P. (1992) *Introduction to Bacteria*, 2nd ed. Wiley, New York.

Singleton, P., and Sainsbury, D. (1981) *Introduction to Bacteria*. Wiley, New York.

Tautu, P. (1978) Mathematical models in oncology: a bird's-eye view. *Z. Krebsforsch.* **91**:223–235.

Vicsek, T. (1992) *Fractal Growth Phenomena*, 2nd ed. World Scientific, Singapore.

Vicsek, T., Cserzö, M., and Horváth, V.K. (1990) Self-affine growth of bacterial colonies *Physica A* **167**:315–321.

Wakita, J., Komatsu, K., Nakahara, A., Matsuyama, T., and Matsushita, M. (1994) Experimental investigation on the validity of population dynamics approach to bacterial colony formation. *J. Phys. Soc. Jpn.* **63**:1205–1211.

Williams, T., and Bjerknes, R. (1972) Stochastic model for abnormal clone spread through epithelial basal layer. *Nature* **236**:19–21.

Witten, T.A., and Sander, L.M. (1981) Diffusion-limited aggregation, a kinetic critical phenomenon. *Phys. Rev. Lett.* **47**:1400–1403.

Yasui, M., and Matsushita, M. (1992) Morphological changes in dendritic crystal growth of ammonium chloride on agar plates. *J. Phys. Soc. Jpn.* **61**:2327–2332.

14

Cooperative Formation of Bacterial Patterns

ESHEL BEN-JACOB & INON COHEN

Many natural phenomena, in living and nonliving systems alike, display the spontaneous emergence of patterns; growth of snowflakes, aggregation of soot particles, solidification of metals, formation of corals, growth of bacterial colonies, and cell differentiation during embryonic development are just a few. The question that comes to mind is whether each phenomenon is unique, its pattern resulting from special causes and effects, or if there is a unifying picture in which they share similar underlying principles. The exciting developments in the understanding of diffusive patterning in nonliving systems (Kessler et al., 1988; Langer, 1989; Ben-Jacob and Garik, 1990; Ben-Jacob, 1993) contain a promise for a unified theoretical framework (for nonliving systems) that might also pave the road toward a new understanding of processes in living systems.

Motivated by the above, we set out to study cooperative microbial behavior under stress. Traditionally, bacterial colonies are grown on substrates of high nutrient level and intermediate agar concentration. Under such conditions, the colonies develop simple compact patterns with a smooth envelope. This behavior fits well the view of bacterial colonies as a collection of independent unicellular organisms. However, in nature bacterial colonies must regularly cope with hostile environmental conditions (Stainer et al., 1957; Shapiro, 1988). We created hostile conditions in a petri dish by using a low level of nutrients, a hard surface (high concentration of agar), or both, and observed the patterns exhibited by the stressed bacteria (Fig. 14.1). Drawing on the analogy of diffusive patterning in nonliving systems, we expected complex patterns. The bacterial reproduction rate, which determines the growth rate of the colony, is

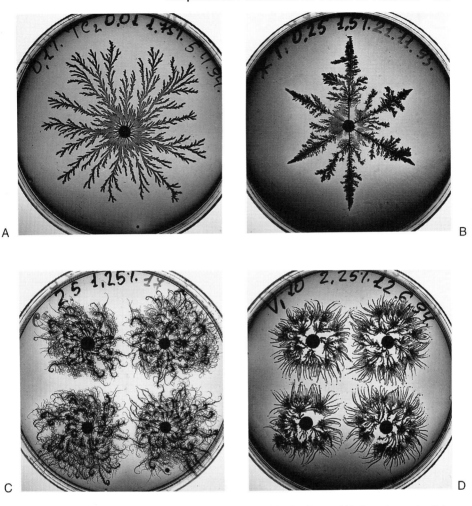

FIGURE 14.1   Patterns observed during growth on thin, hard agar. **(A)** Fractal growth of the **T** morphotype at a peptone level of 0.001 g/L and 1.75% agar concentration. **(B)** "Bacterial snowflake"; growth of **T** morphotype in the presence of imposed sixfold anisotropy (Ben-Jacob et al., 1996) at peptone 0.25 g/L and 1.5% agar. **(C)** Example of the chiral morphotype for peptone 2.5 g/L and 1.25% agar. Note that the branches have a twist of the same handedness. **(D)** Example of the vortex morphotype for peptone 10 g/L and 2.25% agar. Note that each branch has a leading droplet. The latter is consistent of many bacteria moving coherently around the center of the droplet.

limited by the nutrient concentration, and the latter is limited by the diffusion of nutrients toward the colony. Hence the growth of colonies appears to resemble diffusion-limited growth in nonliving systems, such as solidification from a supersaturated solution or electrochemical deposition (Ben-Jacob and Garik, 1990; Ben-Jacob, 1993). Indeed, bacterial colonies do develop patterns reminiscent of those observed during growth in nonliving systems (Fujikawa and

Matsushita, 1989; Matsushita and Fujikawa, 1990; Ben-Jacob et al., 1992, 1994c; Matsuyama et al., 1993; Matsuyama and Matsushita, 1993). Still, one should not conclude that complex patterning of bacterial colonies is yet another example (albeit more involved) of a spontaneous emergence of patterns that may be explained according to the theory of patterning in nonliving systems.

Bacterial colonies exhibit a far richer behavior than patterning of nonliving systems, reflecting the additional levels of complexity involved. The building blocks of the colonies are themselves living systems, each with its own autonomous self-interest and internal degrees of freedom. At the same time, efficient adaptation of the colony to the imposed growth conditions requires adaptive self-organization, which can be achieved only via cooperative behavior of the individual bacteria.

To achieve this state, the bacteria have developed various communication channels (Adler, 1973; Lackiie, 1986; Devreotes, 1989): from direct (by contact) bacterium–bacterium physical and chemical interaction, through indirect physical and chemical interactions via marks left on the agar surface and chemical (chemotactic) signaling, to genetic communication via exchange of genetic material (Shapiro, 1995). Looking at the colonies, it becomes evident that they should be viewed as adaptive cybernetic systems or multicellular organisms possessing fantastic capabilities to cope with hostile environmental conditions and survive them.

The organization of the colony during complex patterning might directly affect the genetic changes of the individual bacterium and vice versa (Ben-Jacob et al., 1994c, 1995a,b). We present a picture of genome cybernetics based on new conceptual elements, the cybernators. These elements can cause genetic changes in the genome of an individual bacterium but are regulated by the state of the colony as a whole. Thus they provide the colony with means to effect genetic changes in the genome of the individual bacteria, changes that in turn allow the colony more efficient self-organization in its attempt to cope with environmental conditions.

Growth Patterns of T Morphotype

Isolation of the T Morphotype

Ben-Jacob et al. (1992, 1994c, 1995a) performed numerous experiments in which colonies of *Bacillus subtilis* were grown under conditions of low nutrient concentration (peptone level about 1 g/L) and mild agar (1.5% agar concentration). Occasionally, bursts of a new mode of growth exhibiting branching patterns were observed. This new mode of tip-splitting growth was found to be inheritable and the geometrical property transferable by a single bacterium. We named this mode **T** morphotype. [The term morphotype refers to the inheritable

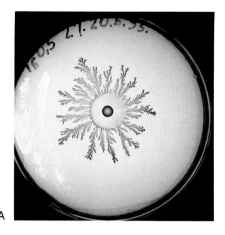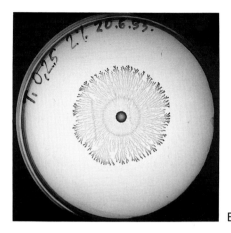

A   B

FIGURE 14.2   Change from a fractal pattern (**A**) to a dense, more organized structure (**B**) with 2% agar and peptone 0.5 g/L (**A**) and peptone 0.25 g/L (**B**).

geometrical (morphological) character of the colony, which can be transferred by a single bacterium. This terminology was suggested by D. Gutnick (personal communication).]

The **T** morphotype can assume different shapes as the growth conditions are varied. [For more detailed classification of the "essential" observed patterns see Ben-Jacob et al. (1994b,c, 1995a).] In general, the patterns are compact at high peptone levels and become more ramified (fractal-like) at low levels. A closer look reveals qualitative changes in the patterns. For example, at 2% agar concentration and a peptone level below 0.25 g/L the colonies adopt a denser, more organized structure with a well defined circular envelope (Fig. 14.2).

Patterns Formed by Density Variations

Additional features are observed during the growth of the **T** morphotype. A closer look at an individual branch (Fig. 14.3A,B) reveals density variations within the branch. These variations are accumulations of layers of bacteria (and so are referred to as three-dimensional structures).

Colonies of **T** morphotype grown on hard substrate exhibit branching patterns with a global twist of the same handedness, as shown in Figure 14.3C. Similar observations during growth of other bacterial strains were reported by Matsuyama and Matsushita (1993) and Matsuyama et al. (1993). For reasons to be clarified below, we refer to such growth patterns as having weak chirality. At relatively high agar concentrations (>1.75%) and at high peptone levels (>10 g/L) a pronounced structure of concentric rings appears (Fig. 14.3D).

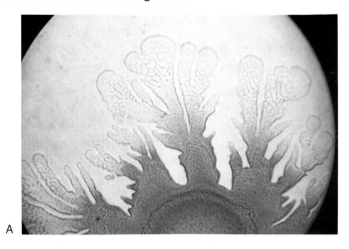

A

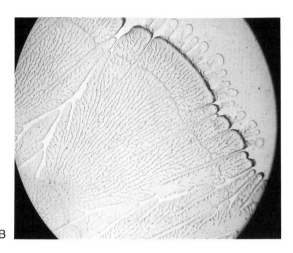

B

FIGURE 14.3   Additional features exhibited by the **T** morphotype. (**A, B**) Density variations within a single branch. Figure continues.

Microscopic Observations

Microscopic studies reveal that the bacteria perform a random walk-like movement in a fluid (referred to as a wetting fluid) excreted by the bacteria or drawn by the bacteria from the agar. Isolated bacteria spotted on agar do not move. The bacterial movement is confined to the fluid whose boundaries define an envelope for the colony. We proposed that ''The envelope propagates slowly as if by the action of effective internal pressure produced by the collective movement of the bacteria'' (Ben-Jacob et al., 1994b). The observations also reveal that the bacteria within the colony are more active at the leading tips of the branches, whereas further down the cells slow down and eventually sporulate.

C

D

FIGURE 14.3 *Continued* (**C**) Weak chirality (global twist of the colony) on hard agar. (**D**) Structure of concentric rings for peptone 10 g/L and 1.75% agar.

Modeling the Growth

We promote the generic modeling approach (Kessler, 1985; Kessler and Levine, 1993; Ben-Jacob et al., 1994b, 1995c), wherein we try to elicit from the bacteria the generic features to be included in the model. For example, in the model for patterning of the **T** morphotype (below) we include the following features: (1) diffusion of nutrients; (2) movement of bacteria; (3) nutrient consumption, reproduction, and sporulation; (4) cooperative movement.

Communicating Walkers Model

The communicating walkers model (Ben-Jacob et al., 1994b) is a hybridization of the ''continuous'' and ''atomistic'' approaches used to study nonliving systems (Shochet et al., 1992). The diffusion of the chemicals is handled by solving a continuous diffusion equation (including sources and sinks) on a triagonal lattice, while bacteria are represented by walkers, allowing a more detailed description. In a typical experiment there are 10^9-10^{10} bacteria in a petri dish at the end of growth. Hence it is impractical to incorporate into the model each and every bacterium; instead, each of the walkers represents about 10^4-10^5 bacteria, so we work with 10^4-10^6 walkers during one run.

The walkers perform random walks on a plane within an envelope representing the boundary of the wetting fluid. This envelope is defined on the same triagonal lattice where the diffusion equations are solved. Based on the microscopic observations, we identified the movement as swimming, although at this point we did not show by direct experiments the existence of flagella. To incorporate the swimming of the bacteria into the model, at each time step each of the active walkers (motile and metabolizing, as described below) moves a step of size d at a random angle Θ. Starting from location $\vec{r}_i$, it moves to a new location $\vec{r}_i'$ given by:

$$\vec{r}_i' = \vec{r}_i + d(\cos\Theta; \sin\Theta) \tag{1}$$

If $\vec{r}_i'$ is outside the envelope, the walker does not move. A counter on the segment of the envelope that would have been crossed by the movement $\vec{r}_i \rightarrow \vec{r}_i'$ is increased by one. When the segment counter reaches a specified number of hits (N_c), the envelope propagates one lattice step, and an additional lattice cell is added to the colony. This requirement of N_c hits represents the colony propagation through wetting of unoccupied areas by the bacteria. Note that N_c is related to the agar concentration, as more wetting fluid must be produced (more "collisions" are needed) to push the envelope on a harder substrate.

Motivated by the presence of a maximal growth rate of the bacteria even for optimal conditions, each walker in the model consumes food at a constant rate (Ω_c) if sufficient food is available. An estimate of Ω_c (Ben-Jacob et al., 1995a) is about 0.2 ng/s.

We represent the metabolic state of the ith walker by an "internal energy" (E_i). The rate of change of the internal energy is given by

$$\frac{dE_i}{dt} = \kappa C_{\text{consumed}} - \frac{E_m}{\tau_R} \tag{2}$$

where κ is a conversion factor from food to internal energy ($\kappa \cong 5 \times 10^3$ cal/g), and E_m represents the total energy loss for all processes over the reproduction time τ_R, excluding energy loss for cell division. C_{consumed} is

$$C_{\text{consumed}} \equiv \min(\Omega_C, \Omega_C') \tag{3}$$

where Ω_C' is the maximal rate of food consumption as limited by the locally available food (Ben-Jacob et al., 1995a).

When sufficient food is available, E_i increases until it reaches a threshold energy. Upon reaching this threshold, the walker divides into two. When food is deficient for an interval of time, causing E_i to drop to zero, the walker "freezes."

Although the food source in our experiments is peptone (and not a single carbon source), we represent the diffusion of nutrients by solving the diffusion equation for a single agent whose concentration is denoted by $C(\vec{r}, t)$:

$$\frac{\partial C}{\partial t} = D_C \nabla^2 C - \sigma C_{\text{consumed}} \tag{4}$$

where the last term includes the consumption of food by the walkers (σ is their density). The equation is solved on the triagonal lattice. We start the simulation with a uniform distribution C_0 (denoted in the figures by P for peptone, P = 10 corresponds approximately to 1 g/L). The diffusion constant D_C is typically (depending on agar dryness) $10^{-4}-10^{-6}$ cm²/s.

Results of numerical simulations of the model are shown in Figure 14.4. One time step in the simulations corresponds to about 1 second. A typical run is up to about 10^4-10^5 time steps, which translates to about 2 days of bacterial growth. Such simulations on a RISK machine take about 2.5 hours, which is shorter by a factor of 20 relative to "real life." As in the case of real bacterial colonies, the patterns are compact at high peptone levels and become fractal with decreasing food level. For a given peptone level, the patterns are more ramified as the agar concentration increases. The results shown in Figure 14.4 capture some features of the experimentally observed patterns. However, some critical features, such as the ability of the bacteria to develop organized patterns at low peptone levels and the three-dimensional structures, are not accounted for by the model at this stage. We believe that chemotactic responses must be included in the model to produce these features.

Generally, chemotaxis means changes in the movement of the bacteria in response to a gradient of certain chemical fields (Adler, 1969; Berg and Purcell, 1977; Lackiie, 1986). The movement is biased along the gradient in either the gradient direction or the opposite direction. Usually, a chemotactic response means a response to an externally produced field, as in the case of food che-

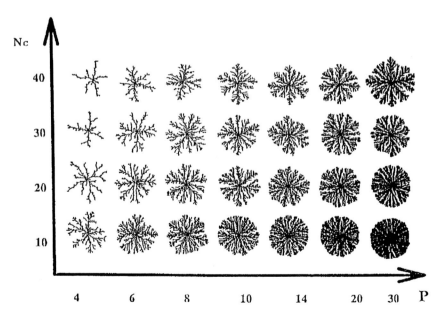

FIGURE 14.4 Results of numerical simulations of the communicating walkers model. P denotes the initial level of nutrient in the simulations; P = 10 corresponds to a peptone level of 1 g/L in the experiments.

motaxis. However, the chemotactic response may also be to a field produced directly or indirectly by the bacteria. We refer to such response as chemotactic signaling. As we show below, it provides an effective means for self-organization of the colony.

We proposed (Ben-Jacob et al., 1994b) that the stressed bacteria inside the colony emit a chemorepellent. That is, under starvation bacteria emit a material (either purposely or as a by-product of an adaptation process) that causes other bacteria to move away.

The equation describing the concentration field $R(\vec{r}, t)$ of the chemorepellent is:

$$\frac{\partial R}{\partial t} = D_R \nabla^2 R + \sigma^* P_R - R_{\text{consumed}} \tag{5}$$

where σ^* denotes the density of stressed walkers, P_R is the production rate, and R_{consumed} is given by $\min(\Omega_R \sigma, R)$, expressing that the decomposition of R is limited by a maximal rate (Ω_R).

The movement of the active bacteria changes from pure random walk (equal probability to move along any direction) to a random walk with a bias along the gradient of the communication field (higher probability to move in the direction out of the signaling material).

The dramatic effect of repulsive chemotactic signaling is demonstrated in Figure 14.5. Note the similarity to the observations shown in Figure 14.2. The pattern becomes much denser with a smooth circular envelope, whereas the branches are thinner and radially oriented. This structure enables the colony to spread over the same distance with fewer walkers. Thus the emission of chemorepellent by the stressed walkers serves the interest of the colony as a whole under low-nutrient conditions.

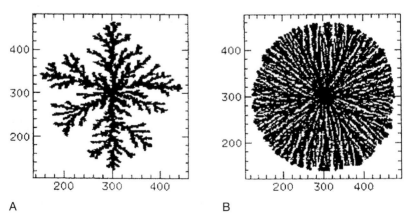

A   B

FIGURE 14.5 Effect of chemorepulsive signaling. (**A**) Pattern produced by the model in the absence of chemosignaling for $P = 10$ and $N_c = 40$. (**B**) For the same parameters as in **A** but with chemorepulsive signaling included. Note that the pattern becomes densely radial. These results are similar to the observations in Figure 14.2.

It is proposed that the wealth of observed patterns is attained by varying the relative strength of three kinds of chemotactic responses (Ben-Jacob et al., 1995a,b,c): long range, chemorepellent, and nutrient chemotaxis, and short-range autocatalytic chemoattraction. The interplay between the three kinds of chemosignaling provides the colony with a means for self-organization, both short and long range. The balance is tuned by the bacteria in response to the environmental conditions.

To model this strategy, we include an additional field that describes chemoattractant signaling. The production of the chemoattractant is autocatalytic; that is, a walker is stimulated to produce the chemical when its concentration in the surrounding of the walker exceeds a minimum value. The chemoattractant is initially activated when the concentration of a triggering field (could be waste products or other hazardous materials) exceeds a threshold value. We also propose that seemingly unrelated patterns result from employment of the same generic strategies of repellent–attractant interplay (Ben-Jacob et al., 1995c).

Formation of Chiral Patterns

Bursts of C Morphotype

During growth of the **T** morphotype on a soft substrate (about 1% agar concentration), bursts of new patterns that overgrow the original branching pattern are observed (see figures in Ben-Jacob et al., 1994c). The new patterns consist of thinner branches, all with the same handedness of strong twisting. We refer to these new patterns as having strong chirality (to distinguish from the weak chirality of **T** morphotype). The new morphological character is inheritable, transferable by a single bacterium, and is stable at a range of growth conditions; therefore we view it as a distinct morphotype (**C** morphotype). We have also observed bursts of the **C** morphotype directly from *Bacillus subtilis* colonies grown of soft agar (Ben-Jacob et al., 1995a).

Closer Look at the Colonies

The **C** morphotype exhibits a wealth of patterns, depending on the growth conditions (Fig. 14.6). As with the **T** morphotype, the patterns are generally compact at high peptone levels and ramified (fractal) at low peptone levels.

Optical microscope observations indicate that during growth of strong chirality, as in the case of **T** morphotype, the bacteria move within a wetting fluid that sets the colony's boundaries. The bacteria are long (sometimes exceeding 10 μm), much longer than the **T** morphotype bacteria (1–2 μm). The layer of fluid is usually thinner than the bacterial length, so the motion is quasi-two-dimensional (i.e., there is limited upward and downward motion). There is also only limited twisting in the plane of motion: The movement of neighboring

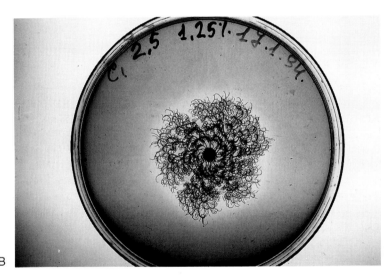

FIGURE 14.6   Examples of patterns exhibited by the **C** morphotype grown on 1.25% agar. (**A, B, C**) For peptone concentrations of 1.0, 2.5, and 5.0 g/L, respectively. Figure continues.

bacteria appears to be correlated in orientation, and the bacteria mainly move forward and backward, parallel to the average orientation of the surrounding bacteria. At the leading tips of the branches, there is more pronounced rotation in the motion of the bacteria. Most of the time the twists have a specific handedness of rotations. Electron microscope observations do not reveal a chiral structure on the bacterial membrane, which might explain the specific handedness (see figures in Ben-Jacob et al., 1995d).

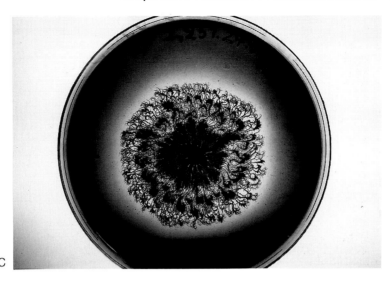

C

FIGURE 14.6   *Continued*

Observed Transitions Between the **T** *and* **C** *Morphotypes*

It was proposed that the growth velocity (the rate of spread of the colony) is
the selective pressure leading to the **T** → **C** transitions (Ben-Jacob et al.,
1994a,c). Indeed, the transitions always occur when the **C** morphotype is faster
than the **T** morphotype. Moreover, the reverse **C** → **T** transitions are observed
at growth conditions for which the **T** morphotype is faster. The bursts of **C**
morphotype during growth on soft agar are frequent: About 60% of **T** morphotype colonies grown on suitable substrate (i.e., conditions in which **C** morphotype is much faster than **T** morphotype) show such bursts. As transitions
are frequent and in both directions, it is suggested that they are associated with
activation and deactivation of a simple biological property. Yet, a property that
can lead to a dramatic change in the growth patterns.

Proposed Mechanism Based on Flagellar Handedness

It is known that flagella have specific chirality (Eisenbach, 1990; Stock et al.,
1990; Shaw, 1991). Ben-Jacob et al. (1995d) proposed that the latter is the
origin of the observed macroscopic chirality. Ordinarily, as the bundles of flagella unfold, the bacteria tumble and end up at a new random angle (relative
to the original one). The situation is different for quasi-two-dimensional motion
(motion in the thin layer of fluid). We assume that in this case (rotation on the
agar surface) the turning has a well defined handedness of rotation. In addition
to flagellar chirality, the bacteria must be able to distinguish between up and
down. We tested experimentally whether the bacteria rely on gravity to distinguish up from down and found that growth in an upside-down petri dish shows

the same chirality, whereas growth of bacteria on the other side of the substrate leads to opposite chirality. Therefore we concluded that the determination of up versus down is accomplished via the vertical diffusion of nutrients or signaling materials inside the substrate or via attachment of the bacteria to the agar surface.

To cause the observed chirality, the rotation must also be, on average, smaller than 90 degrees relative to a specific direction and with a small stochastic part around the average value. We assume that in the case of strong chirality (long bacteria) a bacterium–bacterium co-alignment (orientational interaction) limits the average rotation and its stochastic part. We further assume that the rotation is relative to the local mean orientation of the surrounding bacteria.

Modeling the New Mechanism

To test the above assumptions we included the additional assumed features in the communicating walkers model (Ben-Jacob et al., 1994a). As before, the bacteria are represented by walkers, each of which should be viewed as a mesoscopic unit; but in this case each walker represents only 10–1000 bacteria. Again, the metabolic state of the ith walker (located at $\vec{r_i}$) is represented by an "internal energy" E_i. The time evolution of E_i, food consumption, and food diffusion are the same as described for the **T** morphotype.

To represent the bacterial orientation we assign an angle θ_i to each walker. Every time step each of the active walkers ($E_i > 0$) rotates to a new orientation θ_i', which is derived from the walker's previous orientation (θ_i) by

$$\theta_i' = P(\theta_i, \Phi(\vec{r_i})) + Ch + \xi \tag{6}$$

Ch and ξ represent the new features of rotation due to tumbling; Ch is a fixed rotation, and ξ is the stochastic part of the rotation, chosen uniformly from the interval $[-\eta, \eta]$; $\Phi(\vec{r_i})$ is the local mean orientation in the neighborhood of $\vec{r_i}$; P is a projection function that represents the orientational interaction that acts on each walker to orient θ_i along the direction $\Phi(\vec{r_i})$. P is defined by

$$P(\alpha, \beta) = \alpha + (\beta - \alpha) \bmod \pi \tag{7}$$

Once oriented, the walker advances a step d in either the direction θ_i' (forward) or the direction $\theta_i' + \pi$ (backward). Hence the new location $\vec{r_i'}$ is given by

$$\vec{r_i'} = \vec{r_i} + \begin{cases} d(\cos\theta_i', \sin\theta_i') & \text{with probability } 0.5 \\ d(-\cos\theta_i', -\sin\theta_i') & \text{with probability } 0.5 \end{cases}$$

As for the **T** morphotype, the movement is confined within an envelope defined on a triagonal lattice. The step is not performed if $\vec{r_i'}$ is outside the envelope. Whenever this is the case, a counter on the appropriate segment of the envelope is increased by one. When a segment counter reaches N_c, the envelope advances one lattice step, and a new lattice cell is occupied.

Next we specify the mean orientation field Φ. To do so, we assume that each lattice cell (hexagonal unit area) is assigned one value of $\Phi(\vec{r})$, representing the average orientation of the bacteria in the local neighborhood. The value of Φ is set when a new lattice cell is first occupied. In a more detailed model, the value of Φ changes with slow dynamics. Here we simply set the value of Φ to be equal to the average over the orientations of the N_c attempted steps that led to occupation of the new lattice cell.

Results of the numerical simulations of the model are shown in Figure 14.7A. These results capture some important features of the observed patterns: The microscopic twist Ch leads to a chiral morphology on the macroscopic level. The growth is via stable tips, all of which twist with the same handedness and emit side branches. The dynamics of the side branches' emission during the time evolution of the model is similar to the observed dynamics. The patterns are dense at high peptone levels and become more ramified with decreasing nutrient level. For a given peptone level, the patterns become more ramified with increasing agar concentration. These findings are in agreement with experimental observations.

In Figure 14.7B we show that for large η, corresponding to weak orientation interaction in the case of shorter bacteria, the chiral nature of the pattern gives way to a branching pattern. This provides a plausible explanation for the branching patterns produced by **C** morphotype grown on high peptone levels (Ben-Jacob et al., 1995d), as the bacteria are shorter when grown on a rich substrate.

Plausible Explanation of Weak Chirality

Colonies of **T** morphotype grown on hard substrate (about 2.5% agar concentration) exhibit branching patterns with a global twist with the same handedness, as shown in Figure 14.3C. Similar observations during growth of other bacterial strains have been reported by Matsuyama and Matsushita (1993) and Matsuyama et al. (1993). We refer to such growth patterns as having weak chirality, in contrast to the strong chirality exhibited by the **C** morphotype.

It has been proposed that, in the case of **T** morphotype, it is the high viscosity of the wetting fluid during growth on hard agar that limits the rotation of tumbling, instead of the bacterium–bacterium co-alignment of the **C** morphotype. We further assume that the rotation is relative to the direction of the gradient of a chemotaxis signaling field, which serves as a specific direction instead of the local mean orientation field in the case of **C** morphotype. Here we specifically assumed the chemorepellent to be the ''directing'' field. In Figure 14.7C we show that inclusion of the above features indeed leads to the observed weak chirality. It also provides a plausible explanation for the observations of weak chirality by Matsuyama and Matsushita (1993) in strains defective in the production of wetting fluid.

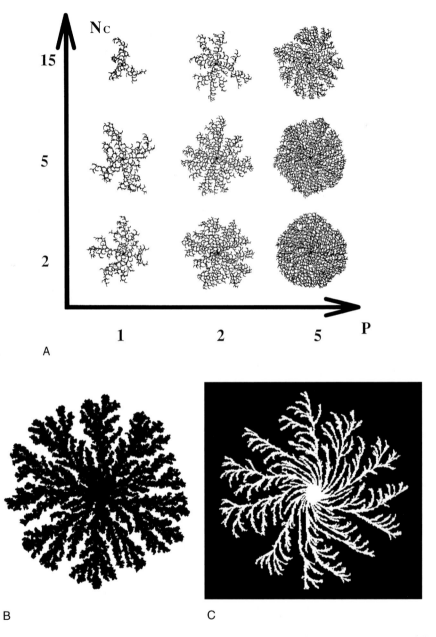

FIGURE 14.7 Results of numerical simulations of the communicating walkers model. (**A**) Chiral patterns produced when orientation interaction is included. (**B**) Effect of increasing the noise level (η) up to $\eta = \pi/2$. The pattern loses its chiral nature. (**C**) Weak chirality produced by the model when the features mentioned in the text are included. Note the similarity to the observations shown in Figure 14.3C.

Vortex Morphotype

Observations of vortices formed by bacterial colonies of *Bacillus circulans* were reported more than a half-century ago. Vortices are only one of several phenomena produced by *B. circulans*; these bacteria also exhibit collective migration, "turbulent-like" collective flow, complicated vortex dynamics (e.g., merging and splitting of vortices, attraction and repulsion), rotating "bagels," and more. A fascinating movie of *B. circulans* (Wolf, 1968) shows some of those phenomena and ends with a remark that the observed phenomena might be too complicated to ever be explained. Indeed, even the simplest phenomenon of collective migration of schools of organisms posed a challenge for many years; only recently have satisfying models been devised (e.g., Reynolds, 1987; Hammingson, 1995; Toner and Tu, 1995; Vicsek et al., 1995).

One of the morphotypes we have isolated from *B. subtilis* exhibits the same behavior as *B. circulans*. That morphotype was given the name vortex (**V** morphotype). It was isolated from *B. subtilis* in a manner similar to that of the **T** morphotype but at different growth conditions—most noticeably at lower temperatures (Ben-Jacob et al., 1995a). Again we performed numerous experiments in which colonies of *B. subtilis* were grown at 30°C at a peptone level of 5 g/L and 1.5–2.0% agar concentration. Occasionally, bursts of the new mode of growth, exhibiting collective migration and vortex formation, were observed (Ben-Jacob et al., 1994a,c). The new mode was found to be inheritable, and the morphological property was transferable by a single bacterium; hence it deserves acknowledgment as a distinctive morphotype.

All sorts of fascinating patterns are exhibited by the **V** morphotype as growth conditions are varied (Figs. 14.1, 14.8). Typically, the **V** morphotype exhibits branching patterns. Each branch is produced by a leading droplet (a vortex) and has side branches, each with its own leading droplet. In many cases the branches have a well pronounced global twist (weak chirality), always with the same handedness.

Each droplet consists of many bacteria that spin around a common center (hence the name vortex) at a typical velocity of 10 μm/s, which varies according to the growth conditions and the location in the colony. Depending on growth conditions, the number of bacteria in a vortex may vary from hundreds to millions. In some cases we observe a single long bacterium that glides in a circle. Both clockwise and anticlockwise vortices are observed within a given colony. The vortices also vary from single- to many-layered structures. Sometimes we observe vortices with an empty core, that is "bagel"-shaped. The latter usually spin in one direction, but in some cases the inner bacteria and outer bacteria spin in opposite directions. During some stages of growth we observed gliding droplets, which are reminiscent of the "worm" motion of slime mold or schools of multicellular organisms. Typically, a collective movement is observed in the trail behind the leading droplet.

Ben-Jacob et al. (1995a) proposed that a new kind of collective chemoattractive response is needed for the vortex to form: The individual bacterium weakly varies its velocity according to the local concentration of the chemo-

FIGURE 14.8   Some patterns produced by the **V** morphotype. Figure continues.

tactic material, which imposes a torque, or local vorticity, on the collective motion. When a glider (which replaces the walker, as the motion in this case is gliding) moves perpendicular to the chemical gradient, it is subject to a force acting to twist the motion toward the high concentrations of the chemoattractant. Thus in the case of positive "chemotaxis," $d\vec{v}/dt$ may be written in the

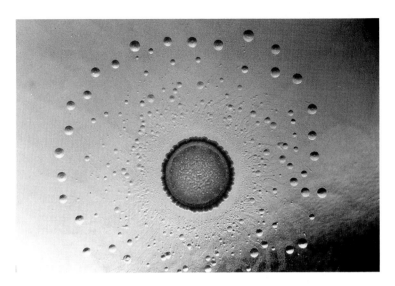

FIGURE 14.8 *Continued*

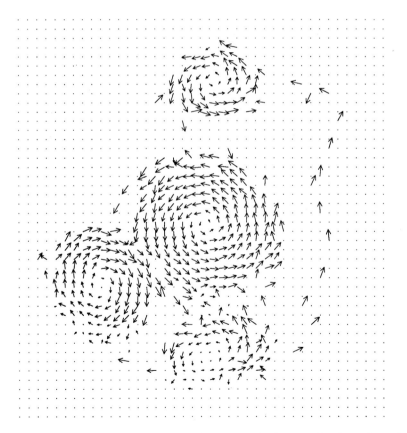

FIGURE 14.9 Numerical simulations of vortices. Arrows indicate the direction and magnitude of the gliders' velocity.

form of:

$$\frac{d\vec{v}}{dt} \sim -\frac{1}{v}\vec{v} \times (\vec{v} \times \nabla c_A) \tag{8}$$

To indicate the special nature of this chemotactic response, we call it rotor chemotaxis.

It has been shown that such rotor chemotaxis can indeed lead to the formation of stationary vortices (Fig. 14.9) (fixed in size and location), rotating "bagels," moving "worms," and other elements on a length-scale smaller than an individual branch. During growth of the colony, these elements organize to form the observed global pattern. Clearly, additional features are employed by the colony to provide the required control and regulatory mechanisms for efficient self-organization.

Concluding Remarks

Many results have been presented, yet we are far from the end of the story. Many studies are waiting to be performed with outcomes likely to be even more exciting than what was already found. We describe here only one of all the fascinating examples. In Figure 14.10 we show growth of the **C** morphotype with 0.1% of **T** bacteria added. The growth conditions are soft agar, so the **C** morphotype is the preferred one. Naively, we expected under such conditions that the addition of 0.1% **T** would have no effect at all. Examining Figure 14.10, the result is clearly in contrast to the expectations. It demonstrates that fascinating observations and intellectual challenges are indeed waiting to be resolved.

FIGURE 14.10 Result of a reconstruction experiment where **C** morphotype bacteria were mixed before inoculation with 0.1% of **T** morphotype bacteria. The growth is initially chiral. It transforms to a tip-splitting growth, from which bursts of chiral growth are observed.

If proved to be real transitions and not contaminations, the observed morphotype bursts from *B. subtilis* to **T**, **C**, and **V** morphotypes are the most fascinating and the most important of all the phenomena described in this chapter. Many wild types of *B. subtilis* are motile. *B. subtilis* 168 is an almost "artificial creature" that has been spoiled in the laboratory for years and has lost its ability to move on agar surfaces. When a colony of *B. subtilis* 168 is grown on a poor substrate, it is of great advantage to the colony to revive the ability to move. Hence an adaptive mutation of such a stressed bacterium should include nonmotility ↔ motility transitions. The *B. subtilis* 168 cells are motile in a fluid environment, so we think the motility of the new morphotype results from the cells' ability to extract a wetting fluid. Cooperative processes involving an exchange of genetic information might play a role in morphotype adaptation or adaptive morphogenesis (Ben-Jacob et al., 1994c, 1995a,b). The above theory is supported by experiments demonstrating that genetic communication is required for adaptive mutations (Galitski and Roth, 1995; Rasicella et al., 1995).

Precisely here lies the point where the interlevel information transfer cannot be analyzed from the "inside" perspective of the genome alone, and a more general cybernetic framework is needed. If the genome is a "smart system" (Shapiro, 1992), what is needed are hypotheses about the tasks it must perform to adapt to changing adverse environments, taking into account the range of simultaneous readjustments and autocatalytic emergences.

Based on our work (Ben-Jacob et al., 1992, 1994c), we propose to view the autonomous elements in the genome as cybernetic units (cybernators) in order to describe their functional role. An element may be, for example, a specific single macromolecule, a combination of molecules, or even a collective excitation of the genome performing a specific function. Generally, it should be viewed as a conceptual unit regardless of its actual nature. The cybernators are cybernetic elements whose function is regulated by colony parameters such as growth kinetics, bacterial density, density variations, level of stress, and so on. The crucial point is that because the cybernator's activity is regulated by the state of the colony, it can produce in the genome of the individual bacterium changes that are beneficial to the colony as a whole. Through cybernators, the colony organization can directly affect the genetic metamorphosis of the individuals in various forms, ranging from synchronized or autocatalytic genetic changes to cooperative ones. In this new picture, the genome is more than an information storage unit; it has the capability of functioning as a data-processing and problem-solving unit with its own direct channels for external information, including information about the state of the colony. In addition, it is capable of restructuring itself according to the outcome of the data processing. For these reasons, the genome might be viewed as an adaptive cybernetic unit (Ben-Jacob et al., 1992, 1994c).

We have presented an hypothesis describing a cybernetic capacity of the bacteria, a capacity that serves to regulate three levels of interactions: those of the cybernator, the bacterium, and the colony. The "interest" of the cybernator serves the "purpose" of the colony by genetically readjusting the genome of the single bacterium. The cybernator provides a singular feedback mechanism as the colony uses it to induce changes in the single bacterium, thus leading to consistent adaptive self-organization of the colony.

This concept differs from the present mechanical-statistics description of nonliving systems at equilibrium, which is based on two levels: the atomic (micro) level and the system (macro) level. Perhaps instead of applying concepts of statistical mechanics to describe processes in living systems (e.g., bacterial patterning and adaptive mutagenesis), we should try to use the lesson learned from bacterial colonies as a hint for the development of new understanding of evolving living and nonliving systems.

Acknowledgments

This chapter describes the results of research ventures with many collaborators. O. Shochet has been a key player in the collaboration until he finished his Ph.D. thesis during the summer of 1995. Modeling the complex patterning of **T**, **C**, and **V** morphotypes was done in collaboration with T. Vicsek and A. Czirók. The morphotype transition experiments were performed with D. Gutnick, R. Rudner, and E. Freidkin. The idea of genome cybernetics has been developed with A. Tenenbaum. We are most thankful to I. Brains for her technical assistance. We are also grateful to D. Gutnick, R. Rudner, J. Shapiro, and E. Ron for many discussions and their help in guiding us to the microbiological world. The research endeavor described here was supported in part by a grant from the Israel USA Binational Foundation (BSF 92-00051), a grant from the German Israel Foundation (GIF 090102), and a grant from the Israeli Academy of Sciences.

References

Adler, J. (1969) Chemoreceptors in bacteria. *Science* **166**:1588–1597.

Adler, J. (1973) A method for measuring chemotaxis and use of the method to determine optimum conditions for chemotaxis by *Escherichia coli. J. Gen. Microbiol.* **74**: 77–91.

Ben-Jacob, E. (1993) From snowflake formation to the growth of bacterial colonies. Part I. Diffusive patterning in non-living systems. *Contemp. Phys.* **34**:247–273.

Ben-Jacob, E., and Garik, P. (1990) The formation of patterns in non-equilibrium growth. *Nature* **343**:523–530.

Ben-Jacob, E., Cohen, I., and Czirók, A. (1995a) Smart bacterial colonies: from complex patterns to cooperative evolution. *Fractals* (in press).

Ben-Jacob, E., Cohen, I., and Czirók, A. (1995b) Smart bacterial colonies: In *Physics of Biological Systems: From Molecules to Species. Lecture Notes in Physics.* Springer-Verlag, Berlin, Heidelberg (in press).

Ben-Jacob, E., Cohen, I., Shochet, O., Aranson, I., Levine, H., and Tsimiring, L. (1995c) Complex bacterial patterns. *Nature* **373**:566–567.

Ben-Jacob, E., Cohen, I., Shochet, O., Czirók, A., and Vicsek, T. (1995d) Cooperative formation of chiral patterns during growth of bacterial colonies. *Phys. Rev. Lett.* **75**:2899–2302.

Ben-Jacob, E., Shmueli, H., Shochet, O., and Tenenbaum, A. (1992) Adaptive self-organization during growth of bacterial colonies. *Physica A* **187**:378–424.

Ben-Jacob, E., Shochet, O., Tenenbaum, A., Cohen, I., Czirók, A., and Vicsek, T. (1994a) Communication, regulation and control during complex patterning of bacterial colonies. *Fractals* **2**:15–44.

Ben-Jacob, E., Shochet, O., Tenenbaum, A., Cohen, I., Czirók, A., and Vicsek, T. (1994b) Generic modelling of cooperative growth patterns in bacterial colonies. *Nature* **368**:46–49.

Ben-Jacob, E., Tenenbaum, A., Shochet, O., and Avidan, O. (1994c) Holotransformations of bacterial colonies and genome cybernetics. *Physica A* **202**:1–47.

Ben-Jacob, E., Shochet, O., Tenenbaum, A., Cohen, I., Czirók, A., and Vicsek, T. (1996) Response of bacterial colonies to imposed anisotropy. *Phys. Rev. E.* **53**:1835–1843.

Berg, H.C., and Purcell, E.M. (1977) Physics of chemoreception. *Biophys. J.* **20**:193–219.

Devreotes, P. (1989) *Dicytostelium discoideum*: a model system for cell–cell interactions in development. *Science* **245**:1054–1058.

Eisenbach, M. (1990) Functions of the flagellar modes of rotation in bacterial motility and chemotaxis. *Mol. Microbiol.* **4**:161–167.

Fujikawa H., and Matsushita, M. (1989) Fractal growth of *Bacillus subtilis* on agar plates. *J. Phys. Soc. Jpn.* **58**:3875–3878.

Galitski, T., and Roth, J.R. (1995) Evidence that plasmid transfer replication underlies apparent adaptive mutation. *Science* **268**:421–423.

Hammingson, J. (1995) Modellization of self-propelling particales with a coupled map lattice model. *J. Physica A* **28**:4245–4250.

Kessler, J.O. (1985) Co-operative and concentrative phenomena of swimming microorganisms. *Contemp. Phys.* **26**:147–166.

Kessler, D.A., and Levine, H. (1993) Pattern formation in *Dictyostelium* via the dynamics of cooperative biological entities. *Phys. Rev. E* **48**:4801–4804.

Kessler, D.A., Koplik, J., and Levine, H. (1988) Pattern selection in fingered growth phenomena. *Adv. Phys.* **37**:255.

Lackiie, J.M. (ed.) (1986) *Biology of the Chemotatic Response*. Cambridge University Press, Cambridge.

Langer, J.S. (1989) Dendrites, viscous fingering, and the theory of pattern formation. *Science* **243**:1150–1154.

Matsushita, M., and Fujikawa, H. (1990) Diffusion-limited growth in bacterial colony formation. *Physica A* **168**:498–506.

Matsuyama, T., and Matsushita, M. (1993) Fractal morphogenesis by a bacterial cell population. *Crit. Rev. Microbiol.* **19**:117–135.

Matsuyama, T., Harshey, R.M., and Matsushita, M. (1993) Self-similar colony morphogenesis by bacteria as the experimental model of fractal growth by a cell population. *Fractals* **1**:302–311.

Rasicella, J.P., Park, P.U., and Fox, M.S. (1995) Adaptive mutation in *Escherichia coli*: a role for conjugation. *Science* **268**:418–420.

Reynolds, C.W. (1987) Flocks, herds, and schools: a distributed behavioral model. *Comput. Graphics* **21**(4):25–34.

Shapiro, J.A. (1988) Bacteria as multicellular organisms. *Sci. Am.* **258**(6):62–69.

Shapiro, J.A. (1992) Natural genetic engineering in evolution. *Genetica* **86**:99–111.

Shapiro, J.A. (1995) Adaptive mutation: who's really in the garden. *Science* **268**:373–374.

Shaw, C.H. (1991) Swimming against the tide: chemotaxis in *Agrobacterium*. *Bioessays* **13**(1):25–29.

Shochet, O., Kassner, K., Ben-Jacob, E., Lipson, S.G., and Müller-Krumbhaar, H. (1992) Morphology transition during non-equilibrium growth. I. Study of equilibrium shapes and properties. *Physica A* **181**:136–155.

Stainer, R.Y., Doudoroff, M., and Adelberg, E.A. (1957) *The Microbial World*. Prentice-Hall, Englewood Cliffs, NJ.

Stock, J.B., Stock, A.M., and Mottonen, M. (1990) Signal transduction in bacteria. *Nature* **344**:395–400.

Toner, J., and Tu, Y. (1995) Long-range order in a two-dimensional dynamical XY model: How birds fly together. *Phys. Rev. Lett.* **75**:4326–4329.

Vicsek, T., Czirók, A., Ben-Jacob, E., Cohen, I., Shochet, O., and Tenenbaum, A. (1995) Novel type of phase transition in a system of self-driven particles. *Phys. Rev. Lett.* **75**:1226–1229.

Wolf, G. (ed.) (1968) *Encyclopaedia Cinematographica*. Institute for Wissenschaftlichen Film, Göttingen.

15

Collective Behavior and Dynamics of Swimming Bacteria

JOHN O. KESSLER &

MARTIN F. WOJCIECHOWSKI

Concentrated assemblies of swimming microorganisms exhibit collective behavior. Biological and physical factors conspire to generate dynamic patterns of concentration and convection. Most easily observed are the autonomously generated time-dependent spatial variations in concentration of the organisms. These variations are caused and maintained by swimming and by flows that are collectively shaped and energized by swimming. The motion of the water within which the organisms live can be inferred from trajectories of inert suspended particles and of the microbes themselves. Spatial and temporal patterns of dissolved molecular concentrations that are important for maintenance of the organisms' life processes have been inferred by combining mathematical methods with observations of motile behavior. These "bioconvection patterns" depend on the force of gravity. They occur when the concentration of microorganisms is sufficient and when the average direction of their swimming is upward. Bacterial and various algal populations generate such patterns. Examples of bacterial bioconvection are shown in Figure 15.1. Kessler (1985, 1986, 1989) and Pedley and Kessler (1992a,b) have provided further illustrations and references and have summarized progress on the theory relating to algae. Protozoan bioconvection patterns have been shown and discussed by Childress et al. (1975) and Levandowsky et al. (1975). Pfennig (1962) examined bioconvection of the anaerobic bacterium *Chromatium okenii*. An early reference is Wager (1911).

The reason for providing this short summary of previous work is to emphasize that the phenomenon of bioconvective self-organization is ubiquitous.

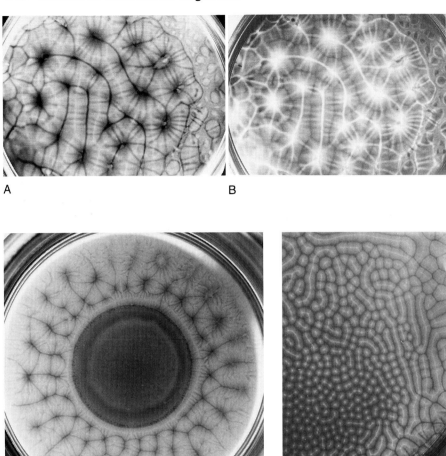

FIGURE 15.1 Bioconvection patterns of the swimming bacteria *Bacillus subtilis*. The concentration of cells is approximately 5×10^8 cells/cm^3; the petri dish is 5.5 cm in diameter, and the depth of the fluid is $\simeq 3$ mm. (**A**) Video image acquired using darkfield illumination. (**B**) Identical image but with reverse contrast, to show that the regions of highest cell concentration (dark in (**A**, bright in **B**) are of similar width as low concentration regions (bright in **A**, dark in **B**). (**C**) A centrally located coverslip, 22 mm diameter, is held up by surface tension. Note the convection roll at the boundary. There is no pattern underneath the glass because there is no oxygen gradient. (**D**) Depth of fluid is $\simeq 2.5$ mm.

Bioconvection provides benefits that are far from trivial to populations of aerobic bacteria. More generally, this type of pattern formation serves as a model for primitive biological self-organization operating through indirect interaction among cells in a fluid environment. Physical conservation principles govern the global system dynamics that create persistent groupings of swimming cells. The coherence lengths of these dynamic relationships extend over distances of $10^{4\pm1}$ cell diameters.

Direct chemical communication among organisms is generally unimportant. The basic coupling among organisms involves large segments of the population all at once. It originates from initial environmental asymmetry that may in part be compounded or altered as part of the process of pattern formation and maintenance.

The collectively energized, regularly patterned destruction of uniformity arises from the suspended swimmers' average motile behavior within the locally prevalent constraints and symmetries (Kessler, 1989). Because the emergent patterns change some of the symmetries of the local ecosystem they also change behavior and so on, regeneratively. The mathematical equations that model these multiply interactive nonlinear systems are presented in the next section. These coupled nonlinear partial differential equations connect the conservation of molecules and swimming cells to force and motion. For bacteria, they feature consumption, diffusion, and advection of oxygen, taxis, and advection of organisms. They show that the source of fluid motion is the force of gravity acting differentially on relatively heavy (i.e., high concentration of cells) and light (few cells) subvolumes of fluid.

The fluid mechanical equations for bioconvection are almost the same as the analogous ones that govern thermal and solutal convection (Tritton, 1988). Dynamic variation in local mass density is the common thread. However, there are distinctly different mechanisms that, through associated equations and boundary conditions, govern those mass variations. For example, thermally driven convection passively transports heat that enters and leaves through bounding surfaces. Bioconvection is isothermal. For the bacterial case, the flow is energized from within, and the transported entity is an assembly of swimmers. Each cell attempts to swim with varying accuracy and speed, up the local gradient of molecular concentration.

The descending plumes of bacteria that develop owing to upswimming are reminiscent of those situations in thermal convection where there is a distributed heat source within the fluid. For bioconvection there is of course no heat source; rather, the density inversion is caused by upswimming of the organism population. Straughan (1993) has provided instructive insights into externally energized but otherwise somewhat similar sorts of convection. Bioconvection diverges substantively from the thermal variety once the convective nonlinear regime is under way: (1) The fluid's boundaries neither maintain nor extract a density-upsetting flux; and (2) after convective plumes have begun to descend the (denser than water) organisms whose accumulation causes and reinforces convection swim transversely to gravity, into the oxygen gradient transported by the plumes, thus reinforcing and sharpening the rather complex convection (see Fig. 15.4).

At any particular locality each cell is also subject to passive, physical co-orientation by the spatial variation of the local velocity field, acting through differential viscous drag. This combination of actively and passively oriented swimming, or gyrotaxis, was presented by Kessler (1985, 1986) and treated more comprehensively by Pedley and Kessler (1987), in the context of algal swimming oriented by combinations of gravitational torque acting on aniso-

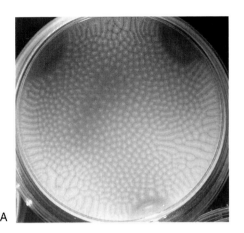

A

B

C

D

tropic cells, phototaxis, and rotational drag due to vorticity. Gravitational torques are probably not important in bacteria, but the gyrotaxis that encompasses aerotaxis and drag due to shear must occur. Swimming of the bacteria and orienting influences that have tensor character fundamentally distinguish bioconvection from thermal or solutal convections.

The bacterial bioconvection phenomena described are based on the behavior of bacterial cells exposed to a temporally and spatially varying concentration of oxygen molecules. Extensive measurements have been performed of aerotaxis in still water, in an environment of changing dissolved oxygen concentration that develops as a result of bacterial respiration and unilateral diffusion from an air interface. The measured distributions of motile behavior, presented below, are the basic biological ingredients to be combined with the equations that model the dynamics of bacterial populations.

Suspensions of the swimming aerobic bacterium *Bacillus subtilis* were used in the experimental investigations. Under conditions of good nutrient supply and aeration, they typically contain $10^7 - 10^9$ organisms/cm^3; the spacing between them is $\geq 10^{-3}$ cm^3, about 10 cell diameters. The volume fraction of bacteria is then in the range $10^{-4} - 10^{-2}$. The hydrodynamic properties of these suspensions, the average density and the viscosity, are virtually the same as those of water. The mass density of the individual cells is about 10% more than water; and they consume oxygen at a substantial rate, approximately 10^6 molecules per second per cell (Berg, 1983).

When the organisms consume dissolved oxygen within the bulk of the fluid, it is replenished by diffusion from the air interface. In this way, consumption and supply generate an oxygen concentration gradient within the fluid. The primary ''global'' communication among cells is this collectively generated gradient. The gradient is initially directed toward the interface, usually upward. After convection sets in, the gradient still points generally upward. In the appropriate locations, it points toward the fluid that descends from the vicinity of the air interface, as it contains more dissolved oxygen than the bulk.

During the time when the oxygen concentration first decreases owing to consumption, the mean direction of the cells' velocity distribution points in the direction of the oxygen concentration gradient (see Fig. 15.7B, below). As a result of this behavior, and because the swimming cells cannot penetrate the

---

←───

FIGURE 15.2 Other symmetries. The patterns are seen in plain view. *Bacillus subtilis* ~ 5 × 10^8 cells/cm^3. (**A, B**) In 3.5 cm diameter petri dishes. The top surface is curved at the edges owing to surface tension; the central regions are therefore shallower than the periphery. (**A**) Dot pattern is usual for cultures of depth $(d) \leq 2$ mm. The finger-like rolls extending inward from the edge are characteristic of deeper cultures. Their orientation perpendicular to the boundary is standard. The threefold symmetry of the peripheral cell sweep-out regions is usual, but twofold and fourfold cases are also common. (**B**) This culture was deeper $(d \simeq 3$ mm). It exhibits one form of dendrite-like convection that in this case sweeps cells from the periphery into central regions. (**C, D**) Temporal development of spatial structure of a *B. subtilis* culture in a shallow watch glass. Overall diameter is about 4 cm. Cells are transported along the dendrites from the periphery to the center. Cell concentration in the central region is probably $\geq 10^{10}$ /cm^3.

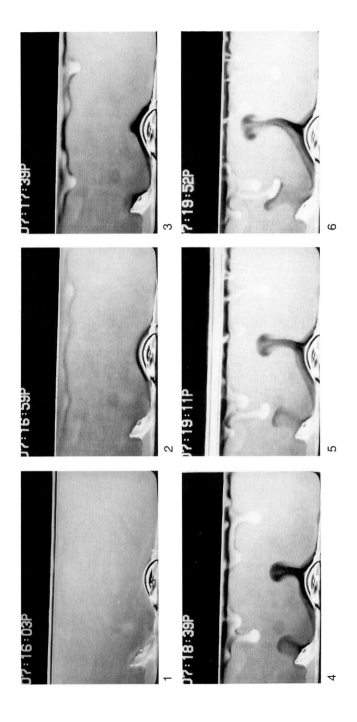

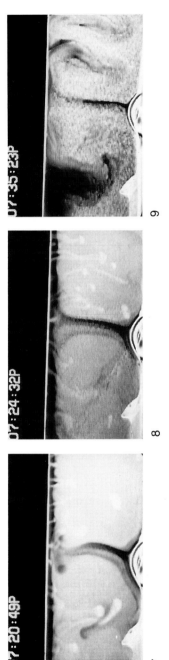

FIGURE 15.3 Development of light and heavy plumes. *Bacillus subtilis* culture is contained in a cuvette made from microscope slides separated by 1 mm. The illumination is darkfield: Bright regions contain many cells, dark regions few. The height of the fluid is ~ 8 mm. Dark line near the top marks the approximate location of the air–water interface. An air bubble is trapped at the bottom. The sequence of events is indicated by the clock: (1) Approximately uniform initial cell concentration. (2) Cells have swum toward the bubble, leaving a depletion zone surrounding it. Cells that have swum toward the air interface have generated an unstable layer that is beginning to form sedimentation waves. (3) Depleted fluid near the bubble rises; it is less dense than the surrounding cell culture. Unstable layer at the top is forming plumes. Top depletion layer is also clearly visible. (4–7) Continuation. (8) Rotary convection begins. (9) Continuation: mixing flows. Granularity is characteristic of self-chemotactic accumulation of slightly anoxic cells.

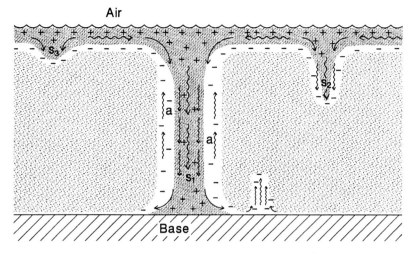

FIGURE 15.4 Buoyant convection. Schematic cross section through a fluid that contains regions of low density (white, minus signs), medium density (light gray), and high density (dark gray, plus signs). The region near the air interface has become denser than the lower regions, possibly because it is cold or because organisms have swum into it. Plumes s_2 and s_3 are beginning to form; s_1, formed earlier, has already reached the bottom and is beginning to spread along it. To the right of s_1, a small low density region is beginning to rise, similar to the black plume in Figure 15.3. Below the top higher density stratum, there are low density layers that could have been generated by organisms swimming upward. These low density strata are analogous to the black, more or less horizontal subsurface strata in all the side-view illustrations. The white low density sheaths that surround the dense plumes are moving upward (squiggly arrows) because they are lighter than the average (light gray) fluid. These sheaths are caused by entrainment of substrata in the flow of the descending plumes. For the case of aerotactic bacteria, they are also partly caused by cells swimming into the denser (+, dark gray) water, which transports oxygen from the layer adjacent to air. The upflow (a), adjacent to s_1, illustrates a typical situation, as seen for example in Figures 15.5G and 15.5H. Because of conservation of volume (Eq. 1), the locations surrounding descending plumes must support flow toward the plumes. This flow is indicated by squiggly arrows in the top stratum, parallel to the air interface. The flows away from the foot of the plume s_1 into the ascending plume (and from the nether regions generally to the upper regions generally) is not shown. The circulating flow of Figure 15.3, step 9, illustrates these connections.

air interface, they accumulate in its vicinity. Where the cells have swum up-gradient, the cell concentration diminishes (see Figs. 15.3, 15.5, and 15.7B, below). The increased cell concentration below the air interface, the decreased concentration to less than average in a depletion layer just underneath, and maintenance of the original average concentration beyond reach of the oxygen gradient completes the first phase of pattern generation (Hillesdon et al., 1995).

The density of bacteria is greater than that of water. As discussed in the Appendix, accumulation of cells in some region of fluid therefore increases the mean mass density of that region of fluid. Because the effect of gravity is to cause regions that are denser than average to sink and lighter ones to rise, the local accumulation and depletion of bacteria due to their directional swimming generates gravitational instability. Slowly varying yet chaotic convection-concentration patterns are the eventual result. They may consist of plumes, columns, or complex rolls, depending on the specifics of the boundary geometry and cell concentration (Figs. 15.1–15.6).

Governing Equations

Bioconvection can be mathematically modeled by continuum conservation equations that link the fluid mechanics, the cells' swimming behavior, and the distribution, consumption, and supply of some quantity (Q) that elicits directional swimming. For algal populations, Q is illumination. For bacteria, Q could be a nutrient, an exudate from the cells, or various attractants and repellents. The oxygen concentration (c) is the only Q-factor explicitly considered here.

Volume is approximately conserved when

$$\nabla \cdot \boldsymbol{u} = 0 \tag{1}$$

where $\boldsymbol{u}$ is the fluids's velocity. The expression becomes exact as $(V/u)vn \rightarrow 0$. The ratio of cell swimming speed to fluid speed is (V/u); the volume fraction of cells is a single cell's volume (v) $\times$ the cell concentration (n) (see Appendix at the end of the chapter). The field variables $\boldsymbol{u}$, V, c, n, and p are all functions of $\boldsymbol{x}$ and t.

The fluid momentum conservation is given by

$$\bar{\rho}\,\frac{\partial \boldsymbol{u}}{\partial t} + \boldsymbol{u} \cdot \nabla \boldsymbol{u} = \mu \nabla^2 \boldsymbol{u} - \nabla p + gnv\Delta\rho, \tag{2}$$

the driving body force being supplied by the last term, which is the force of gravity acting on the buoyancy-corrected mass of the swimming cells. The pressure is p; the viscosity is μ; the mean mass density of the suspension is $\bar{\rho}$; and $\Delta\rho$ is the difference between the mass densities of cells and water.

The conservation, consumption, and transport of the oxygen concentration c (molecules cm^{-3}) is given by

$$\frac{\partial c}{\partial t} = D\nabla^2 c - \nabla \cdot (\boldsymbol{u}c) - \gamma n. \tag{3}$$

The bacterial consumption of oxygen is γ (molecules $cell^{-1}\ s^{-1}$); and the diffusion coefficient is $D(cm^2 s^{-1})$. If only the vertical direction is of interest, $\nabla \rightarrow \partial/\partial z$. The fluid, velocity $\boldsymbol{u}$, transports oxygen; hence the flux term $\boldsymbol{u}c$ (molecules $cm^{-2}s^{-1}$). The quantity γ may be a function of c and the time (t). The diffusion coefficient may be enhanced by cell activity and collisions, then becoming a function of V and n. The conservation equation for cells is

$$\frac{\partial n}{\partial t} = -\nabla \cdot [(\boldsymbol{u} + V)n]. \tag{4}$$

The swimming velocity $V(c)$ is a function of c, its derivatives, and possibly the history of c in the cells' reference frame. In this version of cell conservation, both deterministic and stochastic motile behaviors are included in $V(c)$. When $c < c$ (threshold), $V(c) = 0$.

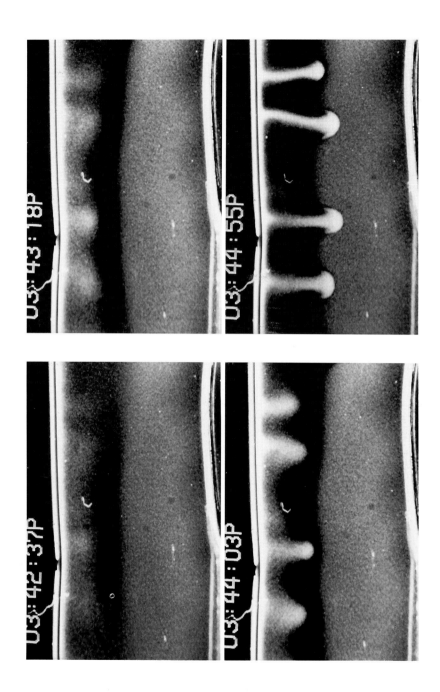

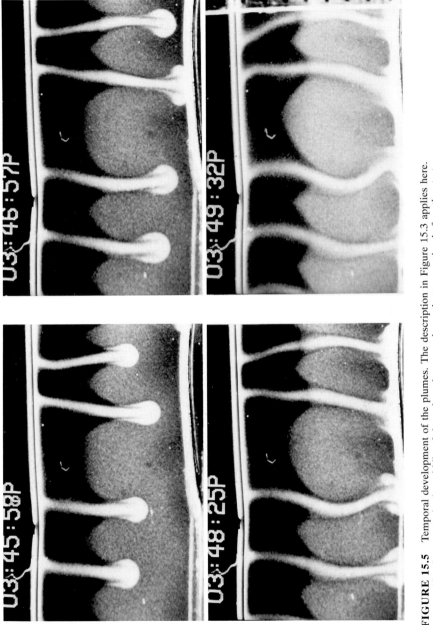

FIGURE 15.5 Temporal development of the plumes. The description in Figure 15.3 applies here. The stratification occurs in the middle of the culture because, due to better aeration before the start, the initial oxygen concentration was higher than in the case of Figure 15.3. A millimeter scale is located in the final image.

427

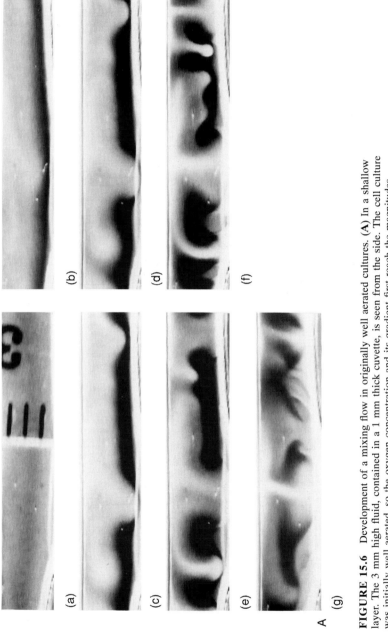

FIGURE 15.6 Development of a mixing flow in originally well aerated cultures. (**A**) In a shallow layer. The 3 mm high fluid, contained in a 1 mm thick cuvette, is seen from the side. The cell culture was initially well aerated, so the oxygen concentration and its gradient first reach the magnitudes where oxygen taxis begins at the *bottom* of the fluid. The dark layer near the bottom of image (b) is due to decline in cell concentration owing to upswimming. The sequence (a–g) provides some insight relating to the development of large area shallow patterns and their eventual mixing flows. The times of the images in minutes:seconds, were: a, 0:0; b, 1:26; c, 2:07; d, 2:50; f, 3:44; g, 6:16. Figure continues.

428

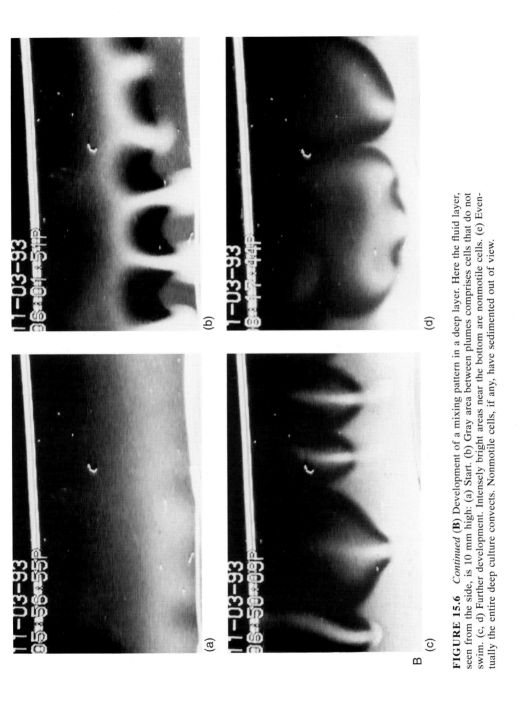

FIGURE 15.6 *Continued* (**B**) Development of a mixing pattern in a deep layer. Here the fluid layer, seen from the side, is 10 mm high: (a) Start. (b) Gray area between plumes comprises cells that do not swim. (c, d) Further development. Intensely bright areas near the bottom are nonmotile cells. (e) Eventually the entire deep culture convects. Nonmotile cells, if any, have sedimented out of view.

429

Equation (4) requires averaging, using a probability density $f(V)$ for the distribution of V. A Fokker-Planck equation may be used (Pedley and Kessler, 1990) to supply $f(V)$, given suitable experimentally or theoretically derived ingredients. The statistical data presented below are the ingredients for a first direct experimental estimate of $f(V)$.

The conservation equation may also be written so cell swimming is separated into deterministic and stochastic parts (Keller and Segel, 1971; Childress et al., 1975; Kessler, 1985, 1986), the latter modeled by a diffusion coefficient D_c for cell swimming. Then

$$\frac{\partial n}{\partial t} = -\nabla \cdot [(\boldsymbol{u} + \boldsymbol{V})n - D_c\nabla n]. \tag{5}$$

This formulation is useful for estimating the degree of instability via the Rayleigh number (see Appendix). In Eq. (5) the magnitude and direction of $\boldsymbol{V}$ depend explicitly on c, $\boldsymbol{u}$, and their derivatives. The coefficient D_c also depends on $\boldsymbol{u}$ and c. More generally, the diffusion term may have tensor character (Childress et al., 1975; Pedley and Kessler, 1990, 1992b). Unlike algae, bacterial cells are probably not individually oriented by gravitationally supplied torque (Kessler, 1992).

Calculations of $c(\boldsymbol{x}, t)$ and $n(\boldsymbol{x}, t)$, based on Eqs. (3) and (5), with $\boldsymbol{u} = 0$, are in qualitative accord with the experimentally observed development of stratified upward density inversion (Hillesdon et al., 1995). The object of the calculations presented in that paper was to test mathematically simple heuristic rules for aerotaxis and diffusion and to model the density stratification during the time interval preceding the onset of convection. Similar calculations were done by Boon and Herpigny (1984).

The diffusion flux $(D_c\nabla n)$ that appears in Eq. (5) is in standard form. A useful, important discussion of other forms was undertaken by Schnitzer et al. (1990). When the cell concentration becomes large enough, collisions among cells can become an important factor in changing the orientations of trajectories. The diffusion coefficient due solely to collisions (D_b) has the form $A[f(V)]V\lambda$, where A is a number that depends on the local general motion $f(V)$, and λ is a mean free path, proportional to $(nR^2)^{-1}$, where n is the cell concentration and R is an effective collision radius. When cells pass each other at a distance less than $2R$ apart, they disorient each other's trajectories.

The diameters of bacterial bodies are approximately 10^{-4} cm, but flagella are much longer (e.g., 5×10^{-4} to 10×10^{-4} cm). When bacteria tumble they are "larger" than when they swim smoothly. During smooth swimming their flagella neatly stream behind in a rotating bundle. Furthermore, when most of the cells swim in the same direction there are few collisions but a good possibility of vortical interactions—hence the dependence of A on $f(V)$. Under conditions when the cell concentration is high, $D = D_cD_b(D_c + D_b)$ is the combined diffusion coefficient (Kessler, 1986).

Materials and Methods

Experiments were concerned with the swimming behavior of *Bacillus subtilis*, a gram-positive, endospore-forming, aerobic eubacterium commonly found in soils. The strain used in this work is YB886 (*trpC2*, *metB5*, *amyE*, *sigB*, *xin-1*, SPβ⁻), a derivative of the naturally transformable *B. subtilis* strain 168 (Yasbin et al., 1980). Liquid cultures of *B. subtilis* YB886 were grown in GM1/GM2, Spizizen's minimal glucose salts medium (Spizizen, 1958) supplemented with 0.1% yeast extract, 0.02% casein hydrolysate, 5 mM $MgCl_2$ (GM2), 0.5 mM $CaCl_2$ (GM2), and the appropriate amino acids at a final concentration of 50 μg/ml. Cultures were grown in GM1 at 37°C with vigorous shaking (250 rpm) in a waterbath until 90 minutes after the end of exponential growth (designated T_0), with growth monitored by measuring turbidity using a Klett-Summerson colorimeter (no. 66 filter). At T_0 + 90 minutes the cells were diluted 10-fold into warm GM2 medium and incubated at 37°C with aeration for a minimum of 60 minutes before measurements of swimming activity were begun. Cell densities at this time are typically 8×10^7/ml to 1×10^8/ml.

All experiments described here were performed at 20° ± 1°C. During experimental runs, only bioconvection stirred the cultures being investigated. To conduct successive experiments stock cultures were maintained on a shaker. The cell concentration increased during these experiments, usually within the range 10^8–10^9 cells/cm^3. The "steady-state" experiments often lasted several days. No nutrients were added during that time. Patterns were observed in plan view using darkfield illumination from a high frequency fluorescent ring light (LOA). Bacterial cultures were placed in a covered plastic petri dish (Falcon). A waterbath below the petri dish absorbed heat radiation from the illumination. As a control for thermal convection, we showed that dead cells did not produce patterns. Slight thermal gradients could have somewhat affected pattern symmetry.

The petri dishes were covered so as to inhibit evaporation. The slight amount of cooling associated with evaporation strongly affects pattern symmetry and dynamics. It adds thermal convection to bioconvection. The effect may be observed by removing the lid from a closed petri dish containing well established bioconvection. The original pattern usually sharpens, then disappears. A new pattern that combines thermal with bioconvection then reappears.

After a bacterial culture was pipetted into a petri dish, the first indications of plume formation occurred after 20–300 seconds depending on the degree of aeration and cell concentration of the culture. Patterns began as plumes, more or less regularly spaced. These plumes eventually generated a widespread series of rolls traversed by narrow regions that contained few cells and were seen as black areas on the darkfield images (Fig. 15.1). In shallow cultures the plumes, seen in plain view as "dots," were the final pattern. In cultures shallower than about 1 mm, patterns were not observed.

The rate of convective motion was estimated by observing the periodic motions and drifts across convection rolls of small particles mixed into the fluid. The small particles preferred for this purpose were 6 μm diameter latex spheres and clumps thereof. Because their sedimentation rate is low they could be observed throughout many cycles. This technique will eventually be used to provide quantitative information on the efficiency of mixing processes.

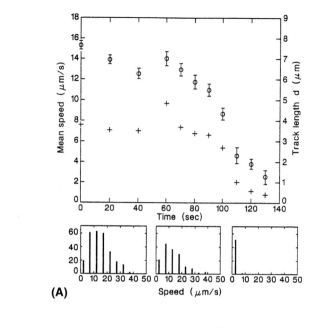

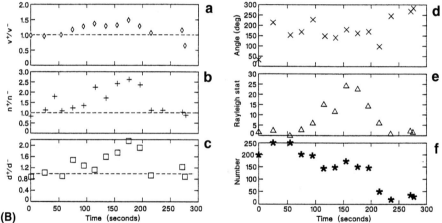

FIGURE 15.7 Statistical data on individual cell dynamics in collectively created environments. (**A**) Choking, the collapse of cell motility as consumption decreases the oxygen concentration. These data were acquired at a location 2985 μm from the air–fluid interface, beyond the range of diffusion of significant amounts of oxygen, during the time available. No directionality was observed in the cell trajectories. The diamonds indicate speed, and error bars are the standard error of the mean. The crosses are the mean lengths (*d*) of the observed trajectories. Each point is the mean of a distribution. Histograms for the distribution of swimming speeds are presented in the insets. The first corresponds to *t* = 0, the second to 70 seconds, the third to 130 seconds. Although the distributions shift to low values of speed as the oxygen runs out, a few fast cells seem to remain. (**B**) Six measures of motional asymmetry as a function of the time during which consumption causes an oxygen gradient. The air–fluid interface is located at 180 degrees, 1300 μm to the left of the location where these data were obtained. The times on the abscissa start at the beginning of measurements (see **D**). (a) Ratio of speeds of cells swimming up-gradient to speeds down-gradient. The plotted points correspond to the sequence in **C**. (b) Ratio of numbers of cells swimming up/down-gradient. Three corresponding angular distributions of velocity are shown in **E**. Figure continues.

432

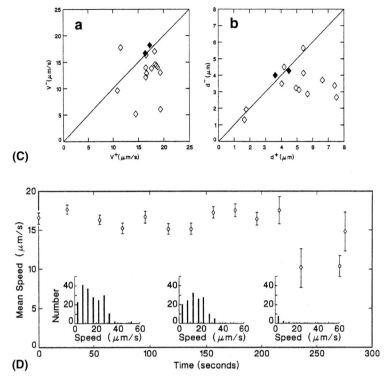

(C)

(D)

FIGURE 15.7 *Continued* (c) Ratio of mean observed track lengths. The track length (*d*) is related, in a rather complex way, to the length of a cell's trajectory; it is the projection of the portion of the trajectory within the depth of field onto the plane of observation. (d) Mean direction of cell trajectories. Cells seem to be heading mostly toward 180 degrees, up-gradient, at all times. However, the statistical confidence for *t* < 100 seconds and *t* > 200 seconds is low [see (e) and **E**]. (e) Rayleigh statistic ($r^2 n$), a measure of the statistical sharpness of directionality. This measure implies that the mean angle [see (d)] is most meaningful for 100 < *t* < 200. (f) Total sample (i.e., the number, *n*, of cell trajectories on which the data are based). The decrease is due to the tactic swimming up-gradient, demonstrated by the data in (a) to (e). (**C**) Asymmetrical response to the time-varying gradient of oxygen concentration. Experimental measurements, at a fixed location, of the average magnitudes of speeds *V* (a) and track length *d* (b) up (+) and down (−) the gradient of oxygen concentration. The air–fluid interface is located 1300 μm to the left (i.e., at 180 degrees). Then *V*(+) is the average speed of all cell tracks whose direction of travel Θ lies in the interval 90–270 degrees; *V*(−) is the average speed of tracks within −90 degrees < Θ < 90 degrees. The two filled points are "early" in the development of the oxygen gradient; they correspond to the first two points, where *t* < 50 seconds in **B** and **D**. These diagrams indicate that there is a strong bias, in track length and swimming, for travel up the oxygen gradient. This bias complements the directional bias in the mean direction of travel Θ, plotted in **B**. (**D**) Swimming speed, averaged over all directions of travel, as a function of the time during which the oxygen concentration decreases and the gradient develops. These data are taken at a fixed location 1300 μm from the air–fluid interface. Averages are over all cells in the field of view. The error bars are the standard error of the mean i.e., standard deviation divided by number of cells in the sample. That number decreases with time because of the tactic migration up-gradient (see **B**). This decrease is responsible for the growth of the error bars at the longer times. The time specified on the abscissa is the time from the beginning of measurements, approximately 2 minutes after the bacterial culture is placed in the microslide. Oxygen consumption and diffusion have therefore taken place well before *t* = 0. The initial concentration of dissolved oxygen is unknown. Average swimming speeds are derived from the distributions of speeds. Three representative histograms of these distributions are shown, located near the associated mean speed data point. The first histogram corresponds to the first point, at *t* = 0, the second to *t* = 130 seconds, the third to *t* = 230 seconds. Figure continues.

433

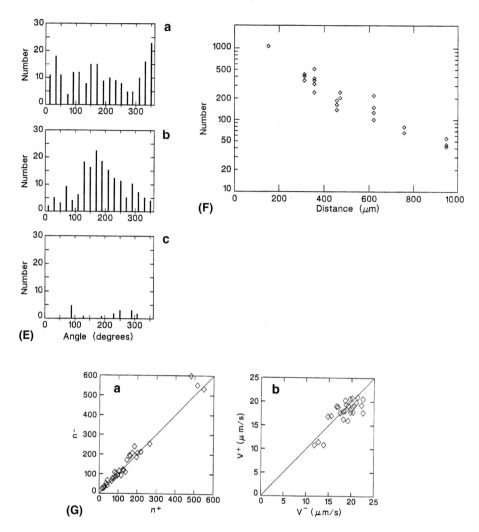

FIGURE 15.7 *Continued* (**E**) Angular distribution of trajectory directions. These histograms are examples of the distributions that yield the average directions and the sum-unit vector (*r*) that is included in the Rayleigh statistic. Histogram (a) corresponds to *t* = 0 in Figure 15.4: (b) to *t* = 156 seconds and (c) to *t* = 236 seconds. The height of each of the bars on this histogram represents the number of tracks within its 20 degree range. Each bar is centered in its range. These diagrams demonstrate that directional motility develops during formation of the oxygen concentration gradient. They are the fundamental ingredient of the angular part of the velocity distribution function. (**F**) Spatial distribution of bacterial cells under time-independent conditions. The number of swimming bacterial cells is recorded within an area of fixed size during a mea-surement interval of 5 seconds. The abscissa is the distance from the air–fluid interface. These steady-state data were obtained about 20 hours after the cuvette was first filled with the bacterial culture; the observed cell numbers include the effects of tactic accumulation and the change in total cell numbers due to cell division and cell decay. Collectively driven roiling and mixing of the fluid occurs for locations $x < 100$ μm. (**G**) Lack of asymmetry within the time-independent oxygen gradient of Figure 15.6A(a). The number of cells that swim away from the oxygen source are shown as $n(-)$; $n(+)$ represents those that swim toward it, as detailed in **C**. The points correspond to eight distances from the air–fluid interface. (b) $V(+)$ and $V(-)$ are the correspond-ing speeds. The distribution of swimming directions (not shown) is also flat. Figure continues.

434

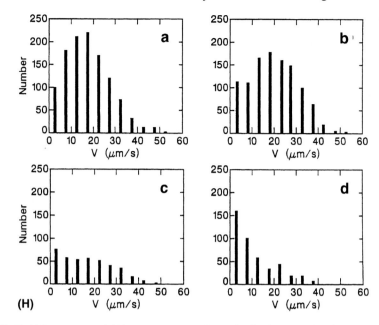

(H)

FIGURE 15.7 *Continued* **(H)** Four samples of the distribution of swimming speeds under approximately time-independent conditions. The distributions were measured at various distances from the air–fluid interface. The number of bacteria per trial at the measuring sites are shown in F. Each of the histograms is the sum of three trials. Three single points not visible in (a) are located at $V = 58$, 63, and 78 μm^{-1}. The distances (x) from the interface, the mean speeds ($\bar{V}$) and the standard deviations (SD), derived from averages of several trials comprising N tracks, at a particular x are:

Trial	x (μm)	$\bar{V}$ (μms^{-1})	$\pm$ SD (μms^{-1})	N
a	156	17.9	0.7	1143
b	364	18.7	1.2	1079
c	626	18.1	1.7	409
d	947	11.3	0.4	461

The standard deviations of the *distributions* of V were independent of x despite the changes in shape. They ranged from 10 to 12 μms^{-1}.

To observe plumes from the side, standing cuvettes were prepared from microscope slides and 1 mm spacers, using Vaseline or silicone grease as sealants. Plumes were also usually illuminated with skew light beams; that is, the darkfield system was used.

For measurements of the velocity spectra of many individual cells, the bacterial culture was filled half-way into a "flat microslide" (Vitro Dynamics). The air–fluid meniscus served as a source of oxygen. The ends of the microslide were sealed with silicone vacuum grease (Dow Corning). This preparation was then placed flat on a standard microscope stage. A 40× LWD Olympus objective directly imaged the cells onto a camera chip. The cell trajectories were recorded on a VCR for later measurement. The lumen of the microslide was 100 μm. Cell tracks were usually recorded near the top of the lumen to provide approximately constant contrast and to restrict upward swimming of the cells. Thermal convection was absent, judged by the fact that small inert particles did not move. The statistical analysis of the cell trajectories was carried out by a computer system and software supplied by the Motion Analysis Corporation (Santa Rosa, CA), modified and augmented.

The stored frame-by-frame sequence of images of projected cell trajectories were processed to generate speeds and angles for inclusion in the statistical analysis of velocity distributions. It was decided to adopt a fixed head-to-tail strategy for the entire data set presented here. The distance between the appearance and disappearance of a swimmer was divided by the time interval, and that was taken as the measured speed. The orientation was inferred from the vector joining tail to head. A frame-by-frame analysis averaged over an entire swimming path usually provided considerably higher speeds than head-to-tail but similar directionality. The effect is explained by the tortuosity of the trajectories. It will be investigated further as a method for assisting in the derivation of experimental values of the bacterial diffusion coefficient (D_c).

The difference in speed between the head-to-tail methods and the frame-by-frame, or multiple frame, methods might be thought to be due to the generally accepted run/tumble strategy of bacterial taxis. However, run/tumble behavior was rarely observed in these cultures. Oxygen taxis and pattern formation are currently being investigated in *B. subtilis* cultures grown in GM-1 medium only. When such populations age, run/tumble behavior becomes common. In these populations the cells are shorter (approximately 2 μm) than the ≥ 4 μm average length of the cells whose velocity statistics are presented here.

The speeds and directions of cells are measured in two-dimensional projection. The statistical graphs present that data without any attempt at correction. When the direction is strongly peaked within the plane of observation, angular corrections for projection are small. For an isotropic distribution, as in Figure 15.7A, the tracks measured to be nominally the fastest probably lie in the projection plane. Analyses that categorize the whole sample of trajectories into up-gradient and down-gradient are independent of projection.

It is well known that some aerobic bacterial cells swim up an oxygen gradient (Shioi et al., 1987), but how does one know that these cells do so? When a cell culture is placed into an inert gas atmosphere (argon) no convection patterns developed. Furthermore, when one floats a microscope coverslip on a culture (Fig. 15.1), whatever convection patterns may have existed before soon die out. That the convection is not driven by surface tension variations is demonstrated by its persistence near the edges of the coverslip.

The Data

Convection Patterns: Collective Behavior

Figures 15.1–15.6 show convection patterns and their development. Although minor thermal perturbations cannot be ruled out, the time development and general symmetry are reproducible. It appears that regions that contain relatively few cells compress regions with more cells (Fig. 15.2). The mean intensity of scattered light and the pattern morphology are evidence for that inference. A considerable amount of motion occurs in the form of traveling "black" bands (few cells) and the general changes of shape. Narrow black bands that cross rolls are characteristic of bacterial bioconvection; they are probably related to horizontal oxygen taxis.

The data presented in this article, on patterns and distributions of swimming speeds, was acquired with bacterial cultures toward the end of rapid growth, or in stationary phase. Some net increase of cell concentration may occur during the recording of data. Eventually, accumulation of extracellular products and depletion of nutrients causes senescence, characterized by generation of inert precipitating detritus. The convection rolls driven by the remaining swimming cells sweep that detritus into regularly spaced windrows. Convection patterns may also be observed during exponential growth phase, as soon as the concentration of cells exceeds the threshold of hydrodynamic instability. Such "early" patterns are most easily observed by innoculating fresh growth medium with a few bacteria and waiting, typically a few hours.

Figures 15.3–15.6 show the depletion of cells near the air interface (or below it) due to upswimming. The initial phase is followed by a descent of arrays of bacteria-heavy plumes. Eventually these arrays become more irregular, generating complex mixing-flow systems. In shallow layers such as that shown in Figure 15.6A, the mixing flows are often in the form of standing waves. Mixing has been traced using dissolved dyes and suspended small particles.

Oxygen Taxis and Choking: Statistics of Individual Behavior

The basis of pattern formation is the bacterial response to the spatiotemporally varying oxygen concentration created by consumption and supply. The measurements of behavior reported here were made in the setting of cell concentrations for which the pattern formation occurs. The results of these measurements are the ingredients required for a quantitative theory based on Eqs. (1) through (5). They were not yet available for the analysis presented by Hillesdon et al. (1995).

Consumption and supply create a time-varying dissolved oxygen concentration $c(x, t)$. Bacteria that are an appropriate distance from the interface experience the gradient of c and swim in that direction. When the cells are so far from the interface that they deplete the local dissolved oxygen before sufficient supply arrives by diffusion, they lose motility. Figure 15.3 shows the mean speed, the speed spectra, and the mean track length for that case.

The correlation of c with consumption, supply, and diffusion are discussed by Kessler et al. (1994). Direct measurements of c are still in the planning stage. Figures 15.7B–E show various ways of quantifying oxygen taxis. The length of directional runs, the directional speed, and the orientation of tracks all cause accumulation up the gradient. The asymmetry of directional cell trajectories $n(+)/n(-)$ is a projection-independent measure.

Each of the plots in Figure 15.7B emphasizes a single distinct aspect of the bacterial cell population's behavior. The time axis, which is common to all, is correlated with the decline of the local concentration of oxygen owing to consumption and the appearance of an oxygen concentration gradient as oxygen diffuses in from the air–fluid interface, approximately 1.3 mm away. The spatial and temporal dependence of the oxygen concentration is a function of the consumption rate of the cells and their shifting location.

Figures 15.7B and C shows that cells swim *more quickly* up-gradient than down-gradient. That there are *more cells* that swim up-gradient than down-gradient is shown in Figure 15.7B. The mean length d for which tracks remain visible is also skewed toward the oxygen sources. Tracks remain visible if they have no sudden turns and if their orientation is flat (i.e., within the depth of field). The sum of unit vectors of a sample of N directional paths specifies Θ. The unit vectors point from the beginning to the end of each path. The length of that direction vector, divided by N, is called r (Zar, 1984). For completely randomly oriented directions, $r = 0$; when all unit vectors coincide, $r = 1$. The Rayleigh statistic r^2n is a measure of directionality that takes into account both the distribution of directions and the sample size. Both Θ and the Rayleigh statistic are plotted in Figure 15.7B. The latter shows that the data for $100 \leq t \leq 200$ seconds are much more strongly unimodal than the rest. Because the air–fluid meniscus is curved and the flat microslide cuvette is much wider than the area within which tracks are observed, exact orientation toward 180 degrees could not be expected, even for completely deterministic swimmers.

Figure 15.7C demonstrates the asymmetry of the cells' response to the oxygen gradient, when it arrives. Figure 15.7D shows that the mean swimming speed ($\langle V \rangle$) does not vary significantly during the time $100 < t < 200$ seconds when Figure 15.7B shows that the most assymetrical response occurs. This observation is surprising; one might think that $\langle V \rangle$ ought to increase with time because the up-gradient flux of swimmers is greater than the down-gradient one, that is, $n(+)V(+) > n(-)V(-)$. (The symbols are defined in the caption of Figure 15.7B.) To resolve this problem, one may define $n(\pm) = n/2 \pm \Delta n$ and $V(\pm) = V^* \pm \Delta V$. Then

$$n\langle V \rangle = (n/2 + \Delta n)(V^* + \Delta V) + (n/2 - \Delta n)(V^* - \Delta V), \qquad (6)$$

or

$$\langle V \rangle = V^* + 2(\Delta n/n)\Delta V. \qquad (7)$$

The independence of $\langle V \rangle$ on time and oxygen concentration (whereas Δn and ΔV are zero at the beginning, before the oxygen gradient develops, and then both positive) implies that V^* decreases with time. Figure 15.7A shows motility far from the meniscus, beyond the diffusion range of oxygen. The speeds plotted in Figure 15.7A can probably be identified with V^*.

The response to the appearance of the oxygen gradient is shown in Figure 15.7E. These data yield the angular part of the velocity probability density

$f(V)$. The rose-track diagrams can be used to derive estimates of D_c (N. A. Hill, 1994, personal communication).

Long-Term Behavior: Steady-State and Locally Generated Turbulence

Asymmetry of directional swimming nearly vanishes after the initial migration up the oxygen gradient and loss of motility of the more remote population. Eventually, both n and dn/dx become large (Fig. 15.7F).

A new collective behavior pattern then appears in the microslide preparation. It does not depend on gravity. Close to the meniscus, where oxygen taxis has concentrated the population, the volume fraction of cells approaches unity (Fig. 15.8). Collisions, flagellar interactions, and possibly some sort of chemical signaling generate violent turbulent vortical motion that can be seen as large, often rather evenly spaced, rotating clouds of cells and intermittent jets that appear near the meniscus (Fig. 15.9B). The reach, away from the meniscus, of these collective dynamic interactions is 50–100 μm. A range of smaller scales of collective movements reaches down to a few cell diameters, especially near the triple line, where water, glass, and air meet. Figure 15.8 shows the region where these phenomena occur.

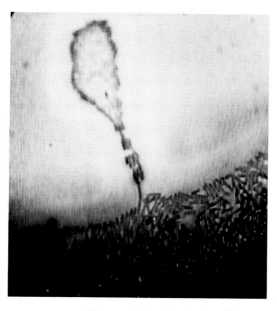

FIGURE 15.8 Accumulation of cells at the meniscus. *Bacillus subtilis* that have swum toward the air interface accumulate in the air-glass-water-defined wedge of the meniscus. The closely packed cells are seen where they form a monolayer. The dark regions contain many cells. The droplet of cells beyond the meniscus is an ejected wetting layer. The illumination is brightfield, using a 40× Olympus objective to image the scene at the video camera chip.

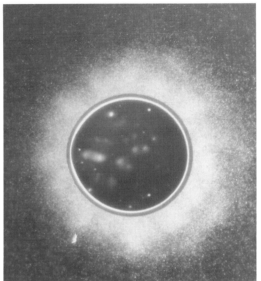

B

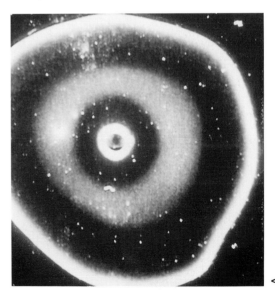

A

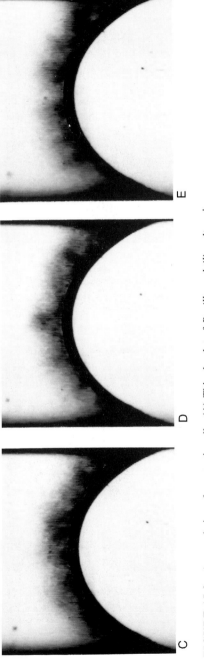

C   D   E

FIGURE 15.9   Accumulation of aerotactic cells. (**A**) Thin droplet of *Bacillus subtilis* culture is confined between glass microscope slides. Initially, the cells were uniformly dispersed. After oxygen consumption and supply established radial gradients, inward toward the air trapped in a central bubble (1 mm diameter) and outward to the fluid–atmosphere interface, the cells swim up the gradients. The eventual accumulations are seen with darkfield illumination. (**B**) Coronal turbulent rotation in a cell culture that has concentrated itself around an air bubble (diameter ~ 1 mm) below a floating coverslip (darkfield). (**C–E**) Higher magnification of analogous preparation in a flat microslide (bright field). Intervals between C, D, and E were about 30 seconds, to illustrate the time dependence. Air is below the curved interface, the bacterial culture is above, but most cells have swum toward the oxygen source.

The direct contact interactions among the cells appears sometimes to be augmented by surface tension gradient-driven motions of the interface, presumably caused by intermittent cell exudates. Circumstantial evidence for this source of motion is provided by the occasional rapid shuttling back and forth of large groups of cells adjacent to the meniscus.

The roiling and mixing action often continues for days. Then, at the distal end of the dynamic cell clouds, one can observe individuals that have inadvertently swum away from the densely populated region suddenly turn and swim back. This behavior is probably due to a chemotactic attraction among the cells, augmenting the effect of the oxygen gradient. Oxygen, chemical signals, and the cell population itself are all transported and mixed by these collective dynamics.

The statistics that govern the "steady state" in the calm regions > 50 μm away from the meniscus suggest a balance between taxis and diffusion, accounting for the approximately exponential spatial variation of n. For the model steady-state situation, where diffusion D_c due to random swimming balances directional swimming with average velocity (V), the solution of Eq. (5) is

$$n(x) = n(0)\exp[-(V/D_c)x]. \tag{8}$$

The more complicated situation, where large n causes collisions, is discussed by Kessler (1986). The observed length D_c/V is $200-300$ μm. The joint solution of Eq. (8) and the time-independent version of Eq. (3) shows that, in the steady state, the cell concentration (n) varies linearly with the oxygen concentration (see Appendix).

Figure 15.7G shows that for the steady state there is little or no directionality in the individuals' swimming dynamics. Within measurement accuracy the mean speed V was not a function of distance from the meniscus. Nevertheless, the distribution of speeds varied markedly (Fig. 15.7H). The steady-state accumulation of cells near the air−water interfaces is dramatically shown in Figure 15.9A. Grishanin et al. (1991) have also demonstrated such accumulation patterns.

Discussion

Bacterial populations can generate dynamic, collective, "multicellular" action through interplay of biological and physical factors that operate on a macroscopic level. In a bounded, still environment, the individual organisms' metabolism (i.e., consumption and rejection of molecules) generates the first macro level of organization, the concentration gradient. If confined to some general area within the gradient, members of the population can react to the locally prevailing molecular concentrations by adjusting their life processes and possibly emitting some new molecules. Of course they thereby affect further development of the gradient and its components. Standard theories of biological self-organization involve such considerations (Frankel, 1989). The diffusive transport of molecules to and from boundaries and interfaces and within the cells' field of action is governed by physical law.

When a population of swimming bacteria is suspended in a bounded fluid, metabolism and diffusion still operate. Physical laws now enter in a new mode. The force of gravity and the rules of fluid mechanics catalyze collective multicellularity, generating apparently chaotic global dynamic patterns that mix and qualitatively and quantitatively improve both long-range and short-range transport.

For the case of *B. subtilis*, the principal consumed molecule is oxygen. Bioconvection occurs when $n \geq 10^8/cm^3$ and the depth of the fluid culture is sufficient (e.g., > 1 mm). When these conditions are not met, the effective Rayleigh number (see Appendix) is too low and the oxygen gradient may not be steep enough. The fluid must be bounded, by the air–water interface(s) and by various solid surfaces. The boundaries and overall size determine the formation and direction of the molecular concentration gradient that causes the accumulation of cells whose mass and spatial distribution generates the variations in mean fluid density required for driving convection. The force and direction of gravity are other ingredients of convection, and they cause upswimming toward the major air–water interface.

A lung seems a suitable metaphor for the bioconvective system, except that this organ and its associated systems comprise many types of cell, whereas in the bacterial case there is just one. In the respiratory system, muscle action causes large conduits to transport air through ever more intricate branches to the alveoli. Initially the flow is all convective; eventually, toward the smallest divisions it is all diffusive. At the alveolar level, diffusive exchange of gas takes place, into and out of the smallest blood vessels. The oxygenated blood is then convected to the muscles, that, among other things, power the airflow.

In the case of multicellular bacterial patterns, convection and large-scale mixing (Fig. 15.6) are the qualitative, endogenously generated improvements of transport, analogous to the macro airflow of breathing. Mixing and diffusion at the smallest scale of the patterns to regions that could not be reached by diffusion from the outside are analogous to molecular transport at the smallest bronchial branches and the alveoli.

The local turbulence collectively generated by cells that have accumulated toward an air–fluid interface (Figs. 15.8, 15.9) constitutes another sort of multicellular transport-enhancing action. In that case gravity plays no discernible role. The cells are so densely packed that interactions through contacts of the cell bodies or the flagella are common. One may observe local whorls and mass movements, either at the meniscus or many cell diameters distant from it. These direct cellular interactions cause a general roiling motion that clearly augments transport toward or away from the meniscus.

The investigations reported here have been conducted so as to minimize or eliminate thermally driven or wind-driven components of convection. Such additions to the system would complicate and possibly mask the basic phenomena. However, in an outdoor setting or in rich puddles that might be associated with air-conditioning systems, for example, one can expect facilitation or merely joint operation of an entire range of convective phenomena.

Pfennig (1962) showed that the anaerobic photosynthetic bacteria *Chromatium okenii* form bioconvection patterns that are similar to those of *B. subtilis*. In that case, the cells swam upward toward a light source. They accumulated near a boundary of oxygen that had diffused in from the top surface of the fluid. Upswimming appeared to be purely phototactic. It is of interest to note that these cells did not appear to run/tumble. When reaching the oxygen interface they simply turned and swam away from it. *C. okenii*, like most of the *B. subtilis* used in this investigation, are rather large cells. It may be that when cells are large enough gradient-sensing mechanisms operate, even for bacteria. Populations of the aerobic bacteria *Serratia liquefaciens* and *Pseudomonas putida* KT2442 produce convection patterns that are similar to the ones produced by *B. subtilis*. They differ in details of geometry and time dependence and may involve signalling via exudates.

Managed microbial ecosystems have been studied in the past and some of that work has been summarized by Wimpenny (1988). There has also been much illuminating work on bacterial taxes where the attractant or repellent, and the organisms, are well controlled for maximum insight into some particular aspect of the problem (Armitage, 1992). The research presented here differs from this previous work by its method, which allows the microorganisms to create their own fluid environment that varies in time and across the space available to the organisms. Physical and biochemical factors come jointly into play. Whatever dependence there may be on the history of the cells' surroundings is automatically accounted for. The experimental results concerning time-dependent and directional motility and the associated population frequency spectra are therefore ''natural.'' They are also the appropriate ingredients for an eventual quantitative description of bioconvection patterns.

Acknowledgments

This work was aided by NASA grant NAG 9442 and by the generous support of Ralph and Alice Sheets through the University of Arizona Foundation. We also thank Mitzi de Martino for her help with this manuscript. Robert Strittmatter, Amy Doud, Diahn Swartz, and David Wiseley contributed greatly to the analysis of the data. The many discussions with Neil Mendelson have been most useful and instructive.

Appendix

Convection

Solid objects sink or rise when their mass density (ρ) is greater or smaller than (ρ_w) the density of the fluid in which they are suspended. Parcels of heavy or light fluid behave similarly. Such blobs may deform during their motion, but the basic phenomena of buoyancy, shown in Figure 15.4, are the same. Regions of fluid might be "light" because they are hot or contain fewer solutes than their surrounds or "heavy" when they are cold or contain more solutes. Freshwater regions embedded in a salty water surround, or vice versa, are an example. Buoyancy effects are not altered even when the contents of a heavy blob of fluid move about within it. The brownian motion of molecules or the swimming of microorganisms, regardless of whether directionally skewed, have no effect on buoyancy. Generally, sinking is more rapid than the swimming rate, so the population in a typical sinking region of a bioconvection pattern remains approximately constant.

At the end of its travel, a sinking blob, shown as a column in Figure 15.4, spreads along the bottom, at the boundary between the fluid and the container. A rising blob of empty fluid spreads out at the air interface.

Because fluid volume is conserved, a parcel of fluid that sinks in one location must displace another one upward. Fluid in the upper regions must exactly and concurrently replace the sinking fluid. In this way a localized descending plume causes a widespread rotational convective flow. A spatially distributed assembly of plumes not only transports fluid and suspended matter both down and up, it mixes the fluid by stretching, folding, and overturning it. When the plumes are due to variations in microorganism concentration, these flows act to transport and mix both cells and molecules. Swimming cells that escape particular blobs eventually start new ones.

One might think that, as usual, these mixing processes would tend to produce uniformity, which is the trend for the passively advected components of the fluid such as oxygen molecules or tracer dyes. However, the living suspended particles continually unmix themselves. Through oriented motility they create a succession of fluid parcels whose density is greater or less than the average. These blobs and surface layers become the plumes and rolls that maintain the chaotic flow structure throughout a cell culture over durations that are vastly longer than the time required for one swimming organism to traverse the depth of the fluid. The term "chaotic" is used to indicate that the observed bioconvective flow patterns are quasiperiodic, but that details such as the location and shape of the narrow cell-free bands appear to vary with time and in space in an apparently random or unpredictable manner. This qualitative experimental result is not surprising. Local fluctuations in density due to slight variations in temperature seem unavoidable. Moreover, cell motility and taxis are intrinsically stochastic in character, even when there is an unambiguous mean direction of swimming. During typical observations there is always time for amplification of small perturbations by the nonlinear global dynamics.

For standard thermal convection, chaotic flows are observed when the gravitational inversion (i.e., hot light bottom and cold dense top) considerably exceeds the threshold of instability (see Tritton, 1988, pp. 171–176 and 362–377 for definitions and an excellent discussion). Such flows can be described by $Gr \equiv g(\Delta\rho/\rho)L^3 v^{-2}$ (the Grashof number) and $Pr \equiv v/D_c$ (the Prandtl number). The fractional density difference driving convection is $\Delta\rho/\rho$, $L \approx$ the depth of the fluid layer, and $v = \mu/\rho$, the kinematic viscosity. D_c is the diffusivity of the entity that causes the density difference—heat in the case of thermal convection, swimmers for bioconvection. For bacterial convection D_c $\leq 10^{-6}$ cm^2/s. Then when $n \simeq 10^9$ cells/cm^3, Gr ≥ 10 and Pr $\geq 10^4$. The product Gr $\times$ Pr $\equiv$ Ra is the Rayleigh number. When Ra $\geq 10^5$, thermal convection tends to become chaotic, a tendency that presumably holds also for bioconvection. The stochastic behavior of the bacteria probably shifts the onset of chaotic motion to lower the effective Ra.

The data of Figure 15.7B were used to provide a rough estimate of D_c, as speed $\times$ path length for the swimmers. Early estimates based on swimming path roughness statistics give the same result.

Sinking or Rising Blobs

The static buoyancy-corrected downward gravitational force on an object is

$$F = vg(\rho_c - \rho) \tag{1A}$$

where v is its volume, g the acceleration of gravity, ρ_c the density of the object, and ρ the mean density of the fluid surrounding the object. The vertical motion of the object is resisted by the viscous drag F_d of the fluid. The vertical speed (V) is constant when $F = F_d$. At small Reynolds number ($Va\rho/\mu$), the drag force on a sphere of radius a in fluid of viscosity μ is

$$F_d = 6\pi a\mu V \tag{2A}$$

Then from $F = F_d$, as $v = (4/3)\pi a^3$ one obtains

$$V = \frac{2ga^2(\rho_c - \rho)}{9\mu} \tag{3A}$$

Tritton (1988, p. 174) arrived generally at the same dependence on parameters for a fluid with density variations and a balance between viscous and buoyancy forces. The factor 2/9 applies only to solid spheres. Without the 2/9, $V \cong 2.5 \times 10^{-2}$ cm/s when $a = 0.05$ cm, $(\rho_c - \rho) \cong 10^{-4}$ g/cm^3, and $\mu = 10^{-2}$ g/cm.

Density Relations

In a suspension containing n particles/cm^3, each having a volume of v cm^3, a cubic centimeter of suspension contains nv cm^3 of particles, called the volume fraction, and $(1 - nv)$ cm^3 of empty fluid. Then if the particles' density is ρ_c and that of the fluid is ρ_w, the density of the suspension is

$$\rho_s = \rho_x nv + \rho_w(1 - nv) \tag{4A}$$

The difference between the densities of the suspension and of pure fluid is then

$$\Delta\rho_{ws} = (\rho_c - \rho_w)nv \equiv nv\Delta\rho \tag{5A}$$

When a blob contains n bacteria/cm^3, each having density ρ_c and volume v, and the concentration in the suspension is n_0 cm^{-3}, the buoyancy-corrected gravitational force is

$$f = g(\rho_c - \rho_w)(n - n_0)v. \tag{6A}$$

Figure 15.3 shows a suspension of bacteria in which a plume of fluid that is depleted of bacteria rises, and plumes that contain more than the average number descend. In the depleted plume, $n < n_0$ and in the others $n > n_0$. Note that the acceleration of gravity g is positive downward. In the previous section the density offset $\rho_c - \rho$ corresponds to $\Delta\rho_{ws}$. The value 10^{-4} g/cm^3 was derived from $\Delta\rho \equiv \rho_c - \rho_w = 10^{-1}$ g/cm^3, $n = 5 \times 10^8$ cells/cm^3, and $v = 2 \times 10^{-12}$ cm^3/cell.

Mixing and Transport

An inert particle suspended in moving fluid travels with it. When n particles/cm^3 flow with speed u cm/s, the particle flux is nu particles/cm^2/s. Thermal agitation causes brownian movement (i.e., diffusion of particles relative to the fluid). The diffusion flux is $D\nabla n$ particles/cm^2/s, where D is in cm^2/s, and ∇ is the gradient operator. One may define L(cm) as a scale length for change of the concentration n; then the time t_d of dispersal by diffusion scales as $t_d \sim L^2/D$. The time for convective transport is $t_c = L/u$; then the ratio $t_d/t_c = Lu/D = $ Pe (the Péclet number) provides an estimate of the dominant transport mode. When Pe $\gg 1$, $t_d \gg t_c$ (i.e., diffusion takes much longer than flow). Evidently, large L favors convective dispersal.

Diffusion alone takes too long when fluid components in a large container are to be mixed. Conventional mixing uses stirring and sloshing processes to reduce the distances (L) between the concentrated and dilute regions of fluid. One may think of these processes as accomplishing a series of stretchings and foldings until adjacent thinned layers are close enough for diffusion to operate expeditiously (Ottino, 1989). The final, most localized step in molecular mixing must always be diffusion. The irregular foldings and stretchings that accompany chaotic flows generally improve the efficiency of mixing.

Bacterial bioconvection transports and mixes oxygen and motile and non-motile cells. In a 2- to 3-mm deep layer, fluid velocities of approximately 10^{-2} cm/s have been observed. Then, for this L and for oxygen ($D \cong 2 \times 10^{-5}$ cm^2/s), Pe $\simeq 0.2 \times 10^{-2}/2 \times 10^{-5} = 10^2$; that is, transport from the air interface to the bottom of a typical fluid layer is much more rapid than it would be by diffusion alone. The stretching and folding that accompanies transport during bacterial bioconvection generate close contact between oxygen and bacteria.

Cells that have become nonmotile for lack of oxygen are also transported by flow energized by the cells that swim. The diffusion coefficient for the passive cells is negligible. Translocation and mixing of swimmers are complex, as the convection velocity of the fluid is similar in magnitude to the swimming speed of the bacteria.

Steady-State Accumulation of Cells

The experimentally observed spatial variation of cell concentration (Fig. 15.7F) is

$$n(x) = n(0)e^{-x/q}. \tag{7A}$$

Inserting this expression into Eq. (3), setting u and $\partial c / \partial t$ equal to zero, and assuming that at the air–water interface $x = 0$, the oxygen concentration $c(x)$ becomes fixed at $c(0)$. One then finds that

$$c(x) = c(0) + \frac{\gamma n(0) q^2}{D} (e^{-x/q} - 1). \tag{8A}$$

Combining Eqs. (7A) and (8A)

$$n(x) = n(0) + \frac{D}{\gamma q^2} [c(x) - c(0)]. \tag{9A}$$

It seems appropriate, and it was not so planned, that using experimentally observed values of q and $n(0)$ the numerical magnitude of $\gamma n(0) q^2 / D$ is about 10^{17}, that is, about the same as $c(0)$. This fact implies consistency of experiment and theory.

Oxygen Uptake

The cells' rate of uptake of molecules is governed by the diffusion flux from remote regions to the edge of a cell. Assuming the steady consumption of molecules at the rate γ by spherical cells of radius R_c, and ignoring their locomotion, one obtains for a single cell the flux differential equation

$$-4\pi r^2 D \frac{dc}{dr} = -\gamma \tag{10A}$$

Its solution is

$$c(r) = \bar{c} - \frac{\gamma}{4\pi D r}, \tag{11A}$$

where $\bar{c}$ is the mean concentration far from the cell. If γ is known, one may infer a scale for the radial dependence of the concentration. It is $R_D = \gamma/(4\pi D \bar{c})$. The concentration at the cell surface (R_c) is

$$c(R_c) = \bar{c}(1 - R_D/R_c) \tag{12A}$$

When $\gamma \simeq 10^6 (O_2/\text{cell/s})$, $\bar{c} \simeq 10^{17}(O_2/\text{cm}^3)$, and $D \simeq (2 \times 10^{-5} \text{ cm}^2/\text{s})$, $R_D \simeq 0.6 \times 10^{-7}$ cm. This result says that when the concentration is near the maximum, determined by solubility, and when the uptake rate is limiting, the concentration at the cell surface is a fraction $R_D/R_c \approx 0.6 \times 10^{-3}$ down from $\bar{c}$, using $R_c = 10^{-4}$ cm. Oxygen taxis is then rather unlikely. When $R_D = R_c$, $c(R_c) = 0$. This condition was considered by Berg (1983). The uptake rate is then limited by supply. It is given by $\gamma = 4\pi D \bar{c}_t R_c$. Then $\bar{c}_t \simeq 0.6 \times 10^{14} O_2$ molecules/cm^3, or approximately $10^{-3} \times \bar{c}(\text{saturated})$. Taxis ought to begin after consumption has caused c to decrease to approximately that level. This effect is observed and clearly demonstrated in Figure 15.6A.

How much time is required for respiration to reduce $\bar{c}(\text{saturated})$ to $\bar{c}_t$? Because the oxygen loss rate far from any air–water interface is simply $n\gamma$, the time for 5×10^8 cells/cm^3 to exhaust the oxygen is about 200 seconds. This finding is experimentally borne out as a typical time period for the start of pattern formation in an initially well agitated culture. In cultures that contain a lower concentration of oxygen, the oxygen taxis begins almost immediately. The plumes then descend from near the top of the preparation, as in Figure 15.5.

References

Armitage, J.P. (1992) Behavioral response in bacteria. *Annu. Rev. Physiol.* **43**:683–714.

Berg, H.C. (1983) *Random Walks in Biology*. Cambridge University Press, Cambridge.

Boon, J.P., and Herpigny, B. (1984) Formation and evolution of spatial structures. *Springer Lect. Notes Biomath.* **55**:13–29.

Childress, S., Levandowsky, M., and Spiegel, E.A. (1975) Pattern formation in a suspension of swimming micro-organisms. *J. Fluid Mech.* **69**:595–613.

Frankel, J. (1989) *Pattern Formation*. Oxford University Press, Oxford.

Grishanin, R.N., Chalmina, I.I., and Zhulin, I.B. (1991) Behavior of *Azospirillum brasiliense* in a spatial gradient of oxygen and in a 'redox' gradient of an artificial electron acceptor. *J. Gen. Microbiol.* **137**:2781–2785.

Hillesdon, A.J., Pedley, T.J., and Kessler, J.O. (1995) The development of concentration gradients in a suspension of chemotactic bacteria. *Bull. Math. Biol.* **57**:299–344.

Keller, E.G., and Segel, L.A. (1971) Travelling bands of chemotactic bacteria: a theoretical analysis. *J. Theor. Biol.* **30**:235–248.

Kessler, J.O. (1985) Cooperative and concentrative phenomena in swimming microorganisms. *Contemp. Phys.* **26**:147–166.

Kessler, J.O. (1986) The external dynamics of swimming micro-organisms. *Prog. Phycol. Res.* **4**:257–307.

Kessler, J.O. (1989) Path and pattern: the mutual dynamics of swimming cells and their environment. *Comments Theor. Biol.* **1**:85–108.

Kessler, J.O. (1992) Theory and experimental results on gravitational effects on monocellular algae. *Adv. Space Res.* **12**:33–42.

Kessler, J.O., Hoelzer, M.A., Pedley, T.J., and Hill, N.A. (1994) Functional patterns of swimming bacteria. In *Mechanics and Physiology of Animal Swimming*, L. Maddock, Q. Bone, and J.M.V. Rayner (eds.), pp. 3–12. Cambridge University Press, Cambridge.

Levandowsky, M., Childress, S., Spiegel, E.A., and Hutner, S.H. (1975) A mathematical model of pattern formation by swimming micro-organisms. *J. Protozool.* **22**:296–306.

Ottino, J.M. (1989) *The Kinematics of Mixing: Stretching, Chaos and Transport*. Cambridge University Press, Cambridge.

Pedley, T.J., and Kessler, J.O. (1987) The orientation of spheroidal microorganisms swimming in a flow field. *Proc. R. Soc. Lond.* **B231**:47–70.

Pedley, T.J., and Kessler, J.O. (1990) A new continuum model for suspensions of gyrotactic micro-organisms. *J. Fluid. Mech.* **212**:155–182.

Pedley, T.J., and Kessler, J.O. (1992a) Bioconvection. *Sci. Prog.* **76**:105–123.

Pedley, T.J., and Kessler, J.O. (1992b) Hydrodynamic phenomena in suspensions of swimming micro-organisms. *Annu. Rev. Fluid Mech.* **24**:313–358.

Pfennig, N. (1962) Beobachtungen über das Schwärmen von *Chromatium okenii. Arch. Mikrobiol.* **42**:90–95.

Schnitzer, M.J., Block, S.M., Berg, H.C., and Purcell, E.M. (1990) Strategies for chemotaxis. In *Biology of the Chemotactic Response*, J.P. Armitage and J.M. Lackie
(eds.), pp. 15–33, Cambridge University Press, Cambridge.

Shioi, J., Dang, V., and Taylor, B.L. (1987). Oxygen as attractant and repellent in
bacterial chemotaxis. *J. Bacteriol.* **169**:3118–3123.

Spizizen, J. (1958). Transformation of biochemically deficient strains of *Bacillus subtilis*
by deoxyribonucleate. *Proc. Natl. Acad. Sci. U.S.A.* **44**:1072–1078.

Straughan, B. (1993) *Mathematical Aspects of Penetrative Convection*. Longman Scientific, Harlow.

Tritton, D.J. (1988) *Physical Fluid Dynamics*, 2nd ed. Clarendon Press, Oxford.

Wager, H. (1911) On the effects of gravity upon movements and aggregation of *Euglena
viridis* and other micro-organisms. *Philos. Trans. R. Soc. Lond.* **B210**:333–390.

Wimpenny, J.W.T. ed. (1988) *CRC Handbook of Laboratory Model Systems for Microbial Ecosystems*, vols. 1 and 2. CRC Press, Boca Raton, FL.

Yasbin, R.E., Fields, P.I., and Andersen, B.J. (1980) Properties of *Bacillus subtilis* 168
derivatives freed of their natural prophages. *Gene* **12**:155–159.

Zar, J.H. (1984) *Biostatistical Analysis*, chs. 24 and 25. Prentice-Hall, Englewood Cliffs,
NJ.

Index

451